# Ancient Astronomy

# Ancient Astronomy:

## An Encyclopedia of Cosmologies and Myth

Clive Ruggles

ABC CLIO

Santa Barbara, California  Denver, Colorado  Oxford, England

Copyright 2005 by Clive Ruggles

All rights reserved. No part of this publication may be reproduced, stored in a retrieval system, or transmitted, in any form or by any means, electronic, mechanical, photocopying, recording, or otherwise, except for the inclusion of brief quotations in a review, without prior permission in writing from the publishers.

Library of Congress Cataloging-in-Publication Data

Ruggles, C. L. N. (Clive L. N.)
  Ancient astronomy : an encyclopedia of cosmologies and myth / Clive Ruggles.
    p. cm.
Includes bibliographical references and index.
ISBN 1-85109-477-6 (acid-free paper) — ISBN 1-85109-616-7 (eBook)
1. Astronomy, Ancient—Encyclopedias. 2. Cosmology—Encyclopedias.
3. Astrology and mythology—Encyclopedias. I. Title.

QB16.R84 2005
520'.93'03—dc22                                     2005018209

08 07 06 05   10 9 8 7 6 5 4 3 2 1

This book is also available on the World Wide Web as an eBook.
Visit abc-clio.com for details.

ABC-CLIO, Inc.
130 Cremona Drive, P.O. Box 1911
Santa Barbara, California 93116-1911

This book is printed on acid-free paper.

Manufactured in the United States of America

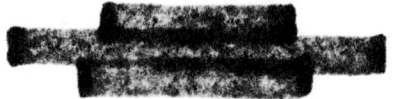

# Contents

*Introduction, ix*

## Ancient Astronomy

Aboriginal Astronomy, 1
Abri Blanchard Bone, 5
Acronical Rise, 7
Acronical Set, 7
Alignment Studies, 8
Altitude, 8
Ancient Egyptian Calendars, 9
Andean Mountain Shrines, 13
Angkor, 14
Antas, 17
Antizenith Passage of the Sun, 19
Archaeoastronomy, 19
Archaeotopography, 23
Astro-Archaeology, 24
Astrology, 24
Astronomical Dating, 27
Avebury, 29
Axial Stone Circles, 30
Azimuth, 33
Aztec Sacred Geography, 33

Babylonian Astronomy and
  Astrology, 37
Ballochroy, 41

Barasana "Caterpillar Jaguar"
  Constellation, 43
Beltany, 44
Borana Calendar, 45
Boyne Valley Tombs, 46
Brainport Bay, 48
Brodgar, Ring of, 50
"Brown" Archaeoastronomy, 52
Bush Barrow Gold Lozenge, 52

Cacaxtla, 55
Cahokia, 57
Calendars, 59
Callanish, 61
Caracol at Chichen Itza, 64
Carahunge, 65
Cardinal Directions, 67
Casa Rinconada, 69
Catastrophic Events, 72
Celestial Sphere, 74
Celtic Calendar, 75
*Ceque* System, 77
Chaco Canyon, 80
Chaco Meridian, 83

Chaco Supernova Pictograph, 86
Chinese Astronomy, 90
Christianization of "Pagan" Festivals, 95
Church Orientations, 96
Circles of Earth, Timber, and Stone, 99
Circumpolar Stars, 102
Clava Cairns, 103
Cobo, Bernabé (1582–1657), 105
Coffin Lids, 106
Cognitive Archaeology, 108
Comets, Novae, and Meteors, 110
Compass and Clinometer Surveys, 112
Constellation Maps on the Ground, 113
Cosmology, 115
Crucifixion of Christ, 117
Crucuno, 119
Cumbrian Stone Circles, 122
Cursus Monuments, 123
Cusco Sun Pillars, 126

Declination, 129
Delphic Oracle, 130
Diurnal Motion, 132
Dresden Codex, 132
Drombeg, 134

Easter Island, 137
Eclipse Records and the Earth's Rotation, 141
Ecliptic, 142
Egyptian Temples and Tombs, 143
Emu in the Sky, 147
Equinoxes, 148
Ethnoastronomy, 152
Ethnocentrism, 152
Extinction, 153

Fajada Butte Sun Dagger, 155

Field Survey, 158
Fiskerton, 160

Governor's Palace at Uxmal, 163
GPS Surveys, 165
Grand Menhir Brisé, 166
"Green" Archaeoastronomy, 169
Gregorian Calendar, 169
Group E Structures, 170

Ha'amonga-a-Maui, 175
Hawaiian Calendar, 176
Heliacal Rise, 180
Hesiod (Eighth Century B.C.E.), 181
Hopewell Mounds, 183
Hopi Calendar and Worldview, 186
Horizon Calendars of Central Mexico, 188
How the Sky Has Changed over the Centuries, 190

Inferior Planets, Motions of, 191
Inuit Cosmology, 193
Iron-Age Roundhouses, 195
Is Paras, 197
Islamic Astronomy, 199
Island of the Sun, 203

Javanese Calendar, 207
Julian Calendar, 208

Kintraw, 211
Kukulcan, 213
Kumukahi, 215
Kumulipo, 217

Lakota Sacred Geography, 219
Land of the Rising Sun, 221
Landscape, 223
Ley Lines, 224
Lockyer, Sir Norman (1836–1920), 227

Lunar and Luni-Solar Calendars, 228
Lunar Eclipses, 230
Lunar Parallax, 234
Lunar Phase Cycle, 235

Maes Howe, 237
Magellanic Clouds, 240
Mangareva, 241
Maya Long Count, 243
Medicine Wheels, 246
"Megalithic Astronomy," 248
"Megalithic" Calendar, 248
Megalithic Monuments of Britain and Ireland, 250
Megalithic "Observatories," 252
Meridian, 253
Mesoamerican Calendar Round, 255
Mesoamerican Cross-Circle Designs, 259
Methodology, 261
Mid-Quarter Days, 265
Minoan Temples and Tombs, 266
Misminay, 267
Mithraism, 269
Monuments and Cosmology, 271
Moon, Motions of, 272
Mursi Calendar, 274

Nā Pali Chant, 279
Nabta Playa, 282
Namoratung'a, 284
Nasca Lines and Figures, 286
Nationalism, 292
Navajo Cosmology, 293
Navajo Hogan, 295
Navajo Star Ceilings, 296
Navigation, 298
Navigation in Ancient Oceania, 300
Nebra Disc, 304
Necker Island, 307
Newgrange, 309

Nissen, Heinrich (1839–1912), 312
Nuraghi, 313

Obliquity of the Ecliptic, 317
Orientation, 319
Orion, 321

Palaeoscience, 325
Pantheon, 328
Pawnee Cosmology, 329
Pawnee Earth Lodge, 331
Pawnee Star Chart, 332
Pilgrimage, 333
Polynesian and Micronesian Astronomy, 336
Polynesian Temple Platforms and Enclosures, 340
Power, 343
Precession, 345
Precision and Accuracy, 347
Prehistoric Tombs and Temples in Europe, 348
Presa de la Mula, 351
Pyramids of Giza, 353

Quipu, 357

Recumbent Stone Circles, 361
Refraction, 364
Roman Astronomy and Astrology, 365
Rujm el-Hiri, 366

Sacred Geographies, 369
Sarmizegetusa Regia, 370
Saroeak, 372
Science or Symbolism?, 375
Short Stone Rows, 376
Sky Bears, 378
Solar Eclipses, 380
Solstices, 384
Solstitial Alignments, 385

Solstitial Directions, 388
Somerville, Boyle (1864–1936), 389
Son Mas, 390
Space and Time, Ancient Perceptions of, 390
Star and Crescent Symbol, 392
Star Compasses of the Pacific, 394
Star Names, 396
Star of Bethlehem, 396
Star Rising and Setting Positions, 398
Statistical Analysis, 399
Stone Circles, 401
Stonehenge, 405
Sun, Motions of, 409
Superior Planets, Motions of, 410
Swedish Rock Art, 412
Symbols, 414

Taulas, 417
Temple Alignments in Ancient Greece, 419
Teotihuacan Street Grid, 421
Theodolite Surveys, 423
Thom, Alexander (1894–1985), 425
Thornborough, 427
Tri-Radial Cairns, 429

Venus in Mesoamerica, 433

Wedge Tombs, 435

Years B.C.E. and Years before 0, 439
Yekuana Roundhouses, 439

Zenith Passage of the Sun, 443
Zenith Stars in Polynesia, 444
Zenith Tubes, 445

*Bibliography, 449*
*Glossary, 483*
*Topic Index, 487*
*Geographical Index, 491*
*Chronological Index, 495*
*Cultural Index, 499*
*Index, 503*

# Introduction

## The Wonder of the Skies

The sight of a truly dark night sky is simply breathtaking. In the modern world it is denied to all but those who still live, or might occasionally venture, well beyond the glare of city lights and the effects of atmospheric pollution. For people in the past, though, the panoply of thousands of stars was a familiar sight. Some twinkled particularly brightly; others hovered on the borderline of visibility. A few were tinted orange or blue. They formed distinct patterns that moved around but never changed. The Milky Way was a familiar sight to all, snaking across the heavens with its light and dark patches forming recognizable shapes. The sky also contained numerous isolated faint and fuzzy objects (nebulae), fixed among the stars, as well as bright wandering stars (planets), and of course the moon, dominating the night sky, lighting the way by night, and rendering many of its fainter companions temporarily invisible. Added to all this was the occasional appearance of something rare or totally unexpected: a shower of meteors (shooting stars), a lunar eclipse, an aurora (at high latitudes), or—more seldom still—a comet (often perceived as a tailed or plumed star) or a new star (nova).

The exact appearance of the heavenly vault differed from place to place and time to time. Whether it was visible on a particular night was dependent, of course, upon local weather conditions. However, the visual impact of the sky on a clear night was always stunning, and even in the cloudiest and wettest of places there were a good many clear periods. As a result, human communities the world over, and for many thousands of years, have recognized familiar patterns and cycles of change in the skies, as well as unexpected sights and events, and struggled to make sense of them. The inhabitants of the night sky—whether perceived as animals, fantastical creatures, legendary characters, ancestors, or other entities—were ever present. The cycles of their appearance and disappearance provided the basis for, and reinforced, countless creation stories and cultural myths.

The lack of opportunity to experience the true night sky is, for a great many people in the modern Western world, something beyond their control. The same cannot be said, though, of the daytime sky, and yet few people pay much attention to it beyond worrying about the weather. Only a meager proportion—and this does not just apply to city-dwellers—could, for example, orient themselves from the time of day and the position of the sun, or could accurately describe the way in which the sun's daily arc through the sky changes with the seasons.

As a result, it can be difficult for us to imagine just how great an impact the sky, whether by night or day, had upon human cultures in the past. Yet we need look no further than the many indigenous peoples today for whom the sky continues to be of immense importance. In addition to this, certain world religions such as Islam continue to tie daily and annual observances very explicitly to the appearance of the sun and moon (see ISLAMIC ASTRONOMY). And one only has to talk to some of the older people living in rural areas, even in the world's most densely populated and developed nations, to discover that an immense amount of practical, everyday sky knowledge was possessed even within living memory. Who, sky watchers of the past might wonder, could fail to notice the regular cycle of lunar phases keeping time with the human menstrual cycle? Who could fail to recognize the changing path taken by the sun through the sky, higher in summer and lower in winter? Who, even in an unfamiliar place, could fail to be able to tell directions and hence find their way about by night or day using the sun or stars?

## Perceptions of the Skies in the Past

ARCHAEOASTRONOMY can be defined as the study of beliefs and practices concerning the sky in the past, especially in prehistory, and the uses to which people's knowledge of the skies were put. It has been recognized in recent years as a legitimate and worthwhile academic pursuit, but it also strikes a popular chord, bringing together two of the most attractive sciences, archaeology and astronomy. It combines the excitement of the cosmos with the romance of the past. Books making spectacular claims about ancient astronomical knowledge often find themselves on bestseller lists. But why should studying people's perceptions of the skies in history, let alone in prehistory, be of particular interest or importance to scholars, not to mention anyone else, in the twenty-first century?

One reason is that archaeoastronomy gives us important insights into how people in the past made sense of the world within which they dwelt. People's lives were governed by observations of objects and events in the world around them. On one level, it is obvious that human communities needed to keep track of various seasonal markers in order to control and maintain their food supply through the year. This would be true whether they subsisted by harvesting plants, fruits, and tubers, by fishing, by hunting, by herding animals or by growing crops, or (most commonly) by a combination of these. CALENDARS, however rudimentary, were necessary for survival in all types of human society: for the smallest bands of hunter-gatherers; for farming villages controlled by local chiefs; and for complex urban societies seeking to support the higher social echelons of elite specialists, craftsmen, priests, and kings.

But humans share a deeper need to understand the world that they inhabit. This does not necessarily mean seeing humankind (or the human mind) as existing within the "objective reality" of the natural environment, as is the modern scientific view, but more generally perceiving the cosmos as a conceptual whole with oneself and one's own community embedded within it. In such a view of the world, the human body, or the dwelling house, or the sacred temple, is often seen as a microcosm reflecting and reinforcing the nature and structure of the cosmos as a whole. Such a worldview makes people strive to be at one with the cosmos of which they are a part and achieve a harmony of existence that enables individuals and whole communities to survive and flourish. This harmony is sustained by continually observing what is happening in the world and keeping one's own actions in tune.

The sky was an integral and prominent part of the perceived cosmos. Its familiar sights and unceasing cycles were part of the fabric of life. Its rhythms were correlated with the time of day and with the seasons. In fact, although we know these cycles to be unfailingly regular and reliable, to past peoples they may not have appeared any more so than a host of other seasonal indicators in the world around them. The important point for us is that, unlike the rest of their perceived world, the sky is a part that we can visualize directly. Landscapes change: patterns of settlement and land use alter, people move around, plants constantly grow and die. We only have indirect knowledge of the appearance of past landscapes. But thanks to modern astronomy, we can reconstruct mathematically the actual appearance of the sky (or, at least, the regular aspects of it—that is, the appearance of the sun, moon, planets, and stars) at any place on earth at any time in the past and visualize it on a computer display or in a planetarium. This gives us a "direct view" of a prominent part of the world that was perceived by peoples in the past.

By studying knowledge passed down orally through myth and story, as well as a variety of other types of evidence available to the historian and the archaeologist, we can begin to appreciate the many different ways in which human societies came to understand what they saw in the skies, and how they came to "use" that knowledge for cultural ends. The sky could, for instance, affirm the POWER of cultural elites, as in the case of the Inca rulers, whose right to rule stemmed from their avowed relationship to the sun god himself (see CEQUE SYSTEM; CUSCO SUN PILLARS). Then again, it could reinforce the status of a shaman or priest who, by performing the appropriate rites at the appropriate times, could be seen as able to affect what happened in the sky—for example by reversing the direction of the sun at the winter solstice and preventing the days from getting ever shorter and colder. Likewise, people tend to lay out buildings in tune with the cosmos, in other words, reflecting perceived links with the wider world (see, for example, NAVAJO HOGAN; PAWNEE EARTH LODGE). Where these links relate to the sky—for instance, to the rising or setting position of the sun, moon, or stars—then we have particular hope of recognizing them, since we know where these celestial bodies appeared in the past (see HOW THE SKY HAS CHANGED OVER THE CENTURIES). By studying architectural alignments upon sky phenomena, we can obtain valuable insights into the worldviews that engendered them.

There is no reason investigations of people's perceptions of the skies should be limited to past societies, and a field known as ETHNOASTRONOMY has emerged in recent years alongside archaeoastronomy, concerned with studies of beliefs and practices relating to the sky among modern indigenous peoples. Since there is no hard line between past and present, many scholars prefer to merge the two fields and to speak instead of *cultural astronomy.* Regardless of how the subject is divided up, the study of people's perceptions of the skies has a deep resonance and helps us appreciate the richness and diversity of human cognition and belief (see COGNITIVE ARCHAEOLOGY).

This said, our knowledge of particular practices is always indirect, whether the evidence we are working with is ethnographic, historical, or archaeological, and questions of METHODOLOGY are highly important. For example, when studying an astronomical alignment (such as a building or temple oriented upon solstitial sunrise) it is necessary to have reasonable confidence that the alignment was actually intentional. This is not self-evident, since everything must, after all, point somewhere. In the 1970s, an abyss developed between two groups of academics regarding the interpretation of British megalithic monuments (see MEGALITHIC MONUMENTS OF BRITAIN AND IRELAND). On the one side were those who paid great attention to STATISTICAL ANALYSIS but little to anthropological theory and tended to argue that "MEGALITHIC ASTRONOMY" was highly mathematically sophisticated. On the other were those who did the reverse and reached the diametrically opposite conclusion: that ancient peoples had no interest in the sky at all.

To get past this impasse, it is important to distinguish between "our" science, which provides tried and tested methods for fairly selecting and assessing evidence, and "their" science, the worldview (ancient, historical, or indigenous) we are interested in (see PALAEOSCIENCE; SCIENCE OR SYMBOLISM?). Those who are mathematically adept but anthropologically naive tend to try to show, in a proprietary way, that people in the past were our intellectual equals by demonstrating that they were capable of sophisticated mathematics and astronomy. The flaw, and also the irony, in this approach lies in the desire to measure the achievements of a past society against the yardstick of our own. This tendency, well known to anthropologists as ETHNOCENTRISM, amounts to putting our own culture on a pedestal. It is necessary to recognize that the worldview we are studying might well be logical enough in its own terms even though it does not conform in every respect to our rationality, since it is built upon different assumptions and principles. A few thousand years is nothing in human evolutionary terms, and thus people living a few thousand years ago were undoubtedly our intellectual equals: however, their way of thinking was different. We pay respect to that difference by trying to understand their way of thinking in its own terms rather than trying to make it conform to ours. It is for this reason that many prefer to avoid using the singular term *astronomy* to describe ancient peoples' interest in the skies, speaking instead, if they use the term at all, of ancient *astronomies*.

All people develop a personal view of the world. It is influenced by their own memories and experiences in which any things that may be perceived—objects, places, events, people and their actions, other living creatures, and plants—can acquire particular meanings, as can the relationships perceived to exist between them. What is more, among the countless things that make up an individual's experience, some come to acquire more significance than others, in a selection process that can seem, to an outsider, highly arbitrary. A "discovery" that seems important to one person may be entirely overlooked or simply of no interest to another. Thus one of the Mursi, a modern group of cultivators and herders living in southwestern Ethiopia, determined the number of days that had elapsed between the planting and first harvesting of his sorghum crop by wearing a cord round his ankle and tying a new knot in it each day. However, those to whom he proudly announced the result were mildly surprised that anyone should have taken so much trouble to deduce something so irrelevant to their daily lives—something that added nothing to their total stock of knowledge about the world.

Groups of people living together do, however, tend to develop common *mindsets*—sets of common perceptions and understandings. These worldviews are reinforced by shared experiences and social convention, and by regular communication, which in small groups is likely to take place by means of a common language. In the framework of modern Western thought, we organize what we perceive in nature according to the principles of Linnaean classification (deriving from the work of the eighteenth-century Swedish botanist Carolus Linnaeus)—that is, into a hierarchical structure of groups (taxa) based upon their observable characteristics. We also tend to believe that the universe is an empirical reality existing independently of the human mind that can progressively be understood through rational argument and scientific experimentation. Other worldviews, however, do not generally classify objects and phenomena into taxa familiar from a Western viewpoint: categories and relationships are not determined by "objective" criteria but simply by whether enough people agree on them. As a result there is generally no clear distinction between the sacred and the mundane, the animate and the inanimate, the empirically real and the fantastic; nor indeed between the terrestrial and the celestial. In traditional cultures, creatures such as fish-man or bird-woman may be seen as fellow beings, every bit as much a part of the perceived universe as entities a Westerner would identify as physically real. As an example, Old Star, chief protector of the inhabitants of the sacred *He* world of the Barasana of the Colombian Amazon, is at once a short trumpet, a constellation corresponding to ORION, the fierce thunder jaguar, and a human warrior.

The sky is a crucial component in practically every indigenous worldview. For earlier peoples and for many indigenous groups around the world today, what we in the modern Western world separate out and categorize as astronomical and meteorological phenomena were an integral part of the environment as a whole. This total environment, including things both real and imagined, *was* the cosmos; the words COSMOLOGY (as widely used in anthropological literature) and *cosmovisión* are broadly synonymous with worldview. In non-Western worldviews, direct associations between the terrestrial and the celestial are commonplace. A good example of this is the BARASANA "CATERPILLAR JAGUAR" CONSTELLATION, whose behavior is believed to relate directly to that of earthly caterpillars, which fall out of trees and provide an important food source at a certain time of the year.

## Myths and Monuments

Where we have first-hand informants or can actually still witness long-standing cultural practices, we are in the strongest position to recognize aspects of other worldviews. Examples such as the modern Yucatec Maya village of Yalcobá in southern Mexico, where the structure of the cosmos is reflected in a whole variety of aspects of social behavior, show just how rich and complex these practices can be. Yet ethnographers can be misled by informants, especially if they ask the wrong questions—all too easy if they have very little initial understanding of the nature of the worldview they are studying. Added to this, sacred information is often withheld, or the anthropologist may not be at liberty to pass it on. There is also the additional danger that the ethnographer may succeed, unwittingly, in influencing the very worldview he or she is trying to investigate, so that a subsequent investigator is misled into thinking that certain modern knowledge was in fact indigenous. Finally, most of us are limited to approaching the cultural information indirectly, at best at second hand, which imposes a selectivity that is not of our choosing and a filter—that of the ethnographer's interpretations—through which we are forced to view everything.

Historical accounts, whether by indigenous people themselves or by past ethnographers, are subject to all the problems just mentioned together with some additional ones. For one thing, an author who is no longer alive cannot be questioned, so there is no possibility of clarification or elaboration. For another, in interpreting a historical account it may be critical to appreciate the context in which it was produced.

Ancient written records directly relating to astronomy exist not only among the civilizations of the Middle and Far East (see for example CHINESE ASTRONOMY). Perhaps the most extraordinary example produced in the American continent is the Maya DRESDEN CODEX, a pre-Columbian astronomical (or, more accurately, astrological) almanac. Both its complexity and level of detail are exceptional. In some cultures, other types of recording device encapsulated sacred information, including astronomical knowledge or calendrical data. One intriguing example is the QUIPU, bundles of knotted strings used in the Inca empire. A much more widespread practice, which did not produce any form of written record, was to embed sky knowledge and cosmological beliefs within myths that were transmitted from generation to generation orally. Storytelling may have been entertaining, but it could also have the deeper purpose of passing on wisdom. Creation myths often served to confirm a community's rightful place in space and time, or to establish the genealogical credentials, and hence the social standing, of a king or leader. (Genealogies were not limited to human forbears. The KUMULIPO, a 2,000-line long Hawaiian creation chant composed in the eighteenth century, recounted in detail how chief Ka-ʻĪ-i-mamao was related, ultimately, to everything in the world.) Some sacred stories were carefully learned and recounted, often in the context of formal ceremonies. Other tales might change in the telling, bringing many variations down to us but leaving intact the essential underlying substance and meaning.

The further we delve into the past, the more we find ourselves limited to the archaeological record—the material remains of past human activity. Silent alignments of stone temples and tombs, interplays of sunlight and shade that light up dark spaces only on rare occasions, symbols with unfathomable meanings but which resemble the sun or moon or familiar groups of stars—these form many of the most famous manifestations of ancient astronomy. But they also present the serious scholar with serious methodological problems. Every oriented structure must point towards *some* point on the horizon, and in all likelihood to one or more identifiable astronomical targets. Similarly, the majority of entrances and openings will let in a shaft of sunlight at *some* time of the year and day. The mere existence of (say) a solar alignment is no guarantee that this meant anything to the builders of a house, temple, or tomb. Since astronomical alignments can—and frequently will—occur completely by chance, we must do more than simply "butterfly collect" them if we are interested in what they actually meant to people in the past. There are two ways of proceeding: either to seek statistical confirmation—for example, by identifying a group of several monuments in which a certain type of astronomical alignment occurs repeatedly—or by finding corroborating archaeological evidence of different forms. An example of the former is the RECUMBENT STONE CIRCLES of northeast Scotland, which have a consistent orientation relating to the moon. Two different examples of the latter, also from Neolithic and Bronze Age Britain, are the THORNBOROUGH henges (large round embanked enclosures) in Yorkshire, England, and the cairns at Balnuaran of Clava in Inverness, Scotland (see CLAVA CAIRNS).

In some cases, such as pre-Columbian Mesoamerica, we have both archaeological and other forms of evidence, including ethnohistory (accounts recorded during the early years of European contact), iconography, and written records (see "BROWN" ARCHAEOASTRONOMY). Integrating these diverse forms of evidence can be a considerable challenge. This problem is illustrated by the prolonged controversy that surrounded the putative Venus alignment at the so-called Governor's

Palace at the Maya city of Uxmal (see GOVERNOR'S PALACE AT UXMAL). Inscriptions on the building attest to a strong interest in the planet Venus, but the apparent orientation of the building toward an extreme rising point of the planet has generated much debate. In ancient Mesoamerica in general, the historical evidence relating to astronomy and calendrics is strong, and the archaeological evidence—for example in the form of alignments—tends to strengthen and corroborate this. If it were not for the accounts of the early Spanish chroniclers, the calendrical inscriptions, and the vital bark books (codices), we would simply have no idea of the sophistication and complexity of Mesoamerican astronomy.

In other places there can be a finer balance. The origins of a piece of oral history are typically much more obscure than the historical context of written records or inscriptions. It can be extremely difficult to separate genuine traditions and stories that stretch back over many generations from more modern inventions or infiltrations. In Hawai'i, a very rich oral heritage existed until the early nineteenth century, of which only fragments now survive (see HAWAIIAN CALENDAR; POLYNESIAN AND MICRONESIAN ASTRONOMY). Yet in assessing the significance of temple alignments in these islands, some of those surviving fragments may contain vital gems of information (see POLYNESIAN TEMPLE PLATFORMS AND ENCLOSURES). They are of uncertain provenance, so must be used with extreme caution, but one would be unwise to simply ignore them and revert to statistics alone.

Where the only available evidence is archaeological, it is possible to use analogies from different cultures where other forms of evidence exist to suggest interpretations. Thus in the 1970s, the Scottish archaeologist Euan MacKie attempted to use the analogy of Maya society to argue (in support of theories, popular at the time, that many of the British megaliths were high-precision observatories) for the existence of highly skilled astronomer-priests in Neolithic Britain (see "MEGALITHIC ASTRONOMY"; "MEGALITHIC" CALENDAR). However, this direct use of ethnographic or historical analogy is generally fraught with problems. Most archaeologists concluded, for a variety of reasons, that MacKie's efforts were badly misguided. On the other hand, analogies can be very useful in challenging assumptions that might have been made too unquestioningly, and hence in suggesting new interpretative possibilities. It is generally assumed, for example, that any people who observed the changing rising and setting position of the sun on the horizon over the year must have understood that this regulated seasonal events. Yet while it is true that some of the Mursi, already mentioned, track the changing rising position of the sun on the horizon, the MURSI CALENDAR is based upon the moon. The sun is regarded as no more reliable a seasonal indicator than the behavior of various birds, animals, or plants. The Mursi example stops us leaping to conclusions about practices in the past by showing us that the range of possibilities is wider than we might have imagined.

In summary, where we have only archaeological evidence to go on, it may be possible to gain insights into prehistoric worldviews by recognizing symbolic associations in the material record. Analogy may be useful in the process of interpretation, but our conclusions will never be categorical. We can only ever have a certain degree of belief that a specific association had meaning to a particular group of people, although our belief may be strengthened or weakened by further evidence.

## From Tally Marks to Calendars

How did people begin to make links between different cycles of activity they observed around them and hence begin to understand and control the periodic changes in the natural world as they perceived it? At what stage did they start consciously to plan their own actions in accordance with those perceptions?

Groups of people from the earliest times would certainly have varied their subsistence activities in accordance with the seasons, whether these involved searching for edible plants and animals, fishing, or hunting. But in this alone they were little different from many others in the animal kingdom: many birds, after all, undertake seasonal migrations. Two characteristics that have distinguished humankind for several tens of thousands of years are people's capacity for abstract thought—making connections between things in order to satisfy the desire to make sense of them—and the use of symbolic representation to express ideas. There is ample evidence of the latter in the Upper Palaeolithic, in the form of systematic markings on many small portable objects such as pieces of bone. Their meaning is much debated, but it has been suggested that some at least represented rudimentary lunar calendars—tallies of days or some other sort of symbolic notation relating to the lunar phase cycle. The most famous example is a fragment of eagle's wing found in a cave at Abri Blanchard in southern France, dating to around 30,000 B.C.E. (see ABRI BLANCHARD BONE). Some much more recent, and less controversial, examples of possible lunar tallies among hunter-gatherers are found in rock art from northeastern Mexico. The site of PRESA DE LA MULA, near Monterrey, is one of two petroglyphs that seem to record counts of days stretching over seven synodic (lunar-phase-cycle) months, divided up according to the appearance of the moon. However, these designs were not calendars: they were clearly not intended for regulating future activities. Rather, they were first-hand observational records, with all the vagaries and imperfections this implies.

Counting days, as such, is unlikely to have been an overriding issue for most people in the past. They had to be aware of many different cycles of activity in their lives. One of the most effective ways of keeping track of things is through myth and appropriately timed ritual performance. These were invariably tied in, along with many aspects of living and being, with the unchanging entities in the sky and the ceaseless cycles in which they were seen to move. Stories help describe the world and explain why things are as they are. Ceremonies serve in a very active way to affirm the natural order. Stories and ceremonies also involve an element of control, in that their successful performance, with adherence to strict protocols, is often seen as necessary—even vital—in order to ensure continuity, renewal, and growth. The appropriate rituals may ensure the arrival of the sun on a given morning, or bring about the reversal of the sun's winterward movement so that it will return and give greater warmth again each spring. Whether performed by a lone shaman, by whole communities working in unison, or by a powerful priest with the active participation of a controlling elite, rituals, ceremonies, and other performances can ensure that the order of the cosmos, and of the lives of the people within it, is duly maintained. The Blessingway and other sacred rites of the Navajo people are just one example of performances of this kind recorded in modern times and, in some cases, continuing to the present day, both among native Americans and among indigenous peoples around the world. The Blessingway ceremony serves to ensure good health, prosperity, and so on by preventing misfortune, and involves detailed renditions of the Navajo creation story (see NAVAJO COSMOLOGY).

What may be an extremely old tradition of myths and performances relates to the identification of two bears circling around the northern celestial pole. The evidence to support this assertion, paradoxically, is found in modern oral traditions across Europe. In the Basque country, the bears' motions around the celestial pole are linked to an annual cycle of storytelling and song and dance performances involving whole communities. Other variants are found throughout Europe. Constellation myths within the framework of Greek and Near Eastern thought form an integral part of the modern Western heritage, but the prevalence of indigenous bear myths suggests that two circumpolar bears are star figures emanating from much older cosmological traditions (see SKY BEARS).

In the 1969 book *Hamlet's Mill*, Giorgio De Santillana and Hertha Von Dechend suggested that specific astronomical knowledge—namely the gradual shifting of the entire mantle of stars over the centuries due to a phenomenon known as PRECESSION—had been systematically "encoded" in mythological narratives all over the world over a period of many millennia. There are various problems with the evidence presented in support of this idea, but there is also a fundamental difficulty with the idea itself. It presupposes an almost universal concern in ancient times with a particular concept (precession) familiar to ourselves. This is very different from arguing that some communities in certain places and times might have had stable enough cosmologies—sky knowledge passed down accurately enough over several generations—to have noticed the gradually changing appearance of the sky and attempted to come to terms with it in one way or another.

The point about the sky bears, on the other hand, is that we may be glimpsing an ancient framework of understanding very different from the Western one. According to the American linguist Roslyn Frank, who has investigated the relevant European folk traditions in detail, it was a framework in which earthly bears were venerated. She argues that a human individual playing the role of a bear shaman acted to maintain the cosmic order, organizing the cycle of earthly ceremonial in tune with the annual cycles of the stars. If Frank is right, then we should not look back to some sort of universal, all-prevailing myth or cosmology, diluted with time, but rather to one that always coexisted with many others. One would have to argue that it achieved particular prevalence at high northern latitudes because it brought together what was seen on earth and in the sky in a particularly effective and understandable way; thus it tended to propagate, encountering and mixing with other traditions, and eventually leaving traces visible in folk traditions to this very day.

Another way in which a people can affirm the perceived cosmic order is by carefully timed movement through the landscape. The Lakota people of South Dakota, for example, traditionally keep their subsistence activities in tune with other seasonal cycles by moving from one place to another in an annual pattern that, as they see it, reflects the sun's movement through the constellations (see LAKOTA SACRED GEOGRAPHY). Their annual round combines subsistence activities focused upon the terrestrial movements of the buffalo, with sacred rites focused upon the celestial movements of the sun. In doing so, it ties together earth, sky, and people into a comprehensible whole, thereby keeping everything in harmony. The Mursi of Ethiopia perceive a direct connection between the successive disappearance of four bright stars in the southern sky and events on the ground related to successive floods of the river Omo. This enables them to time their annual migration to the banks of the Omo precisely enough to carry out the vital planting of their crop of cowpeas within a few days of the river's final flood.

Can we hope to find, in the material record, evidence of astronomically timed movement through the landscape in the past? If we could determine the times of year when certain sites or locations were occupied, and if we could spot symbolic links between these places and the sun or stars, then it is certainly possible that this could give us some clues. Symbolism relating to a solstice, at a site that is occupied around the time of solstice, is one possibility. It has been suggested that CHACO CANYON in New Mexico was a focus for sacred pilgrimage for Anasazi (ancestral Pueblo) communities in the surrounding area around the eleventh century C.E. The archaeological evidence for this is supported by archaeoastronomical evidence that people in the outlying communities used astronomical observations to synchronize their convergence upon Chaco. Another, much older, example of a similar phenomenon occurs at THORNBOROUGH, a group of Neolithic henges in Yorkshire, England. Here the archaeoastronomical evidence consists of a series of alignments upon the rising position of the three stars of ORION's belt, suggesting that it might have been the first predawn appearance (HELIACAL RISE) of this asterism that triggered an annual PILGRIMAGE for an autumn festival at the henges.

In modern indigenous societies, we see ample evidence of the various ways in which people strive to harmonize what they do with cycles of events in the natural world. Studies of tally counts, astronomically related myth and ritual performance, and patterns of ritual movement have begun to reveal some of the ways in which this may have been done in the past. CALENDARS—self-consistent systems of marking time—developed from these more rudimentary perceptions of correlations between different cycles of activity in nature, and the desire to keep human activities—ones that we might identify as sacred, mundane, or having aspects of both—in tune with natural events. The cycles of change of the celestial bodies—regular, immutable, and reliable—are clearly of particular importance in this context.

It is commonly assumed that there is a natural progression of calendrical development relating to astronomical observation. The first step is to develop a simple month-by-month calendar based on the phase cycle of the moon, the most obvious cycle in the night sky. However, because there are between twelve and thirteen synodic (phase-cycle) months in a solar year, it follows that in order to keep in phase with the seasons, one needs to have twelve months in some years and thirteen in others. The second step, then, is to take note of phenomena that are related to the seasons, including astronomical ones, using them to keep lunar calendars in pace with the seasonal year by inserting or omitting a month from time to time as required (see LUNAR AND LUNI-SOLAR CALENDARS). The ancient Egyptians, for example, used the first predawn appearance (heliacal rise) of Sirius, the brightest star in the sky, to keep the start of the new year in step with the annual flood of the Nile. Sirius would have been first seen approximately eleven days later in the twelfth month each year. Whenever it did not appear until the last few days of the twelfth month, then an extra or *intercalary* month was added so that it would continue to rise in the final month of the following year.

A seasonally related phenomenon of particular importance is the changing rising or setting position of the sun along the horizon (see SOLSTICES, SOLSTITIAL DIRECTIONS). Once a society recognizes this, they can use it to regulate a solar calendar that is entirely independent of the moon. If the horizon is sufficiently distant and contains a good many distinctive points, then such a calendar can quite easily be kept accurate to within two or three days of the "true" solar year. One of the classic modern examples of a solar horizon calendar is that of the Hopi people of Arizona. As first recorded by the ethnographer Alexander Stephen at the end of the nineteenth century, solar horizon observations were used by the Hopi both to regulate crop-planting activities and to pinpoint events within an elaborate ceremonial calendar (see HOPI CALENDAR AND WORLDVIEW).

The final step in the development of luni-solar calendars is to replace the ad hoc insertion or omission of intercalary months by a systematic procedure. By about the fifth century B.C.E., the Babylonians had developed a fixed system whereby a thirteenth month was inserted into seven different years, in a fixed pattern, within every period of nineteen years (see BABYLONIAN ASTRONOMY AND AS-

TROLOGY). This *Metonic cycle,* named after the Greek astronomer Meton, keeps the lunar calendar in step with the solar year to within an error of just one day in every 200 years.

The Babylonian calendar was an impressive achievement made possible by systematic astronomical observations recorded over many generations. However, ancient calendars do not inevitably follow the progression just described (and even in Babylonia itself, calendrical developments were more complicated). The Mursi yet again provide a good example. They have what, to an outsider, looks like a thoroughly haphazard calendar in which no one ever seems to know for certain what month it is, although everyone believes there are "experts" around. In practice, different opinions always exist, and the calendar is effectively adjusted "on the fly" according to various seasonal markers—though no one is aware that these adjustments are being made. The calendar is completely self-consistent in its own terms, and there is no need for intercalary months. The nearby Borana have a completely different and utterly distinctive luni-stellar calendar that reckons the time of the month and year by observing the moon in relation to the stars, completely ignoring the sun (see BORANA CALENDAR). The *Works* part of HESIOD's *Works and Days* describes farmers' rules of thumb in eighth-century B.C.E. Greece that related exclusively to seasonal astronomical phenomena such as the heliacal rising of stars; the lunar phases are only mentioned in the separate *Days* part. The Roman civic calendar, upon which the modern (Western) calendar is based, only emerged from chaos when it ignored the moon completely (see ROMAN ASTRONOMY AND ASTROLOGY). On the other hand, uncorrected lunar calendars remain of considerable importance to this day, one of the most obvious examples being the Islamic calendar (see ISLAMIC ASTRONOMY). Finally, the ancient Mesoamerican calendar, arguably the most sophisticated and complex of all the world's calendars, operated by combining cycles as diverse as the 365-day year, a 260-day cycle (whose astronomical derivation, if it is astronomically derived at all, remains unclear), and the 584-day synodic cycle of the planet Venus (see MESOAMERICAN CALENDAR ROUND).

In short, there is no inevitable path in the development of calendars. Instead, they advance in diverse ways according to local conditions and needs. This means, for one thing, that they cannot be used as yardsticks of cultural achievement. It also means that they cannot be considered as abstractions, divorced from the social context in which they developed and the social needs that they fulfilled. The Hopi calendar, for example, had (as we would see it) both a pragmatic and a sacred function, but from the Hopi perspective it functioned holistically to ensure the well-being of the community in all respects. Different calendars can have different purposes and even run alongside one another, as—it appears—did the religious and administrative calendars in ancient Egypt (see ANCIENT EGYPTIAN CALENDARS). To the historian or archaeologist, understanding the technical aspects of an ancient calendar is often of limited interest. It is much more intriguing (and often much more challenging) to understand how a calendar operated in its social context, what it meant to people, and what its social implications were.

Modern folk calendars and associated traditions that still exist in many parts of the world, particularly in rural communities, often preserve an inherently holistic worldview, integrating earth and sky in a mixture of what the modern scientist would be inclined to view as rationality and superstition. In the Baltic States of Latvia and Lithuania, for instance, a mixture of prognostications survives that relate people's character, health, and happiness to the phase of the moon at the time of their birth and at important moments in their lives. This Baltic tradition leads to a related issue: the role of ASTROLOGY. The history of astronomy is clearly intimately bound up with the history of astrology; the distinction between the two is only an issue in the context of the modern scientific way of understanding the world. Taking an anthropological viewpoint, we must realize that systems of thought including beliefs that we might describe as "astrological" may themselves constitute ways of understanding the world that are perfectly coherent and logical in their own terms. Thus by considering the biblical skies from the point of view of the astrologers of the time, rather than simply as modern astronomers, a very plausible solution has recently been suggested to the riddle of the identity of the STAR OF BETHLEHEM, which has perplexed astronomers for years.

Ancient peoples generally did not share the modern conception of time as an abstract entity, as a line along which we move through our lives (see SPACE AND TIME, ANCIENT PERCEPTIONS OF). The earliest calendars almost certainly did not come into being as a result of simply conceived astronomical observations undertaken in order to mark the passage of abstract time. They emerged in the context of complex views of the universe in which many aspects of natural and human activity were seen as tied together in fundamental ways. This concept is of vital importance when we try to interpret archaeological monuments that appear to have incorporated alignments marking sunrise or sunset on particular calendrical dates. A fundamental question is: Which regular astronomical events might have been significant to people in the past? Alongside the SOLSTICES, the EQUINOXES are widely assumed to have had inherent significance, yet many investigators have been surprised to discover the absence of equinoctial markers in the culture they are studying. From the point of view of the observers themselves, however, there is a clear difference between the two. The solstices are tangible: at these times, the sun reaches the physical extremes of its motions along the horizon. The SOLSTITIAL DIRECTIONS mark the boundary between those parts of the horizon where the sun can rise or set, only passes over, or is never seen. These define a natural division of the world, as seen from any central point of observation, into four parts. The equinoxes, on the other hand, would in practice only have been observable as halfway points between the solstices. They would generally have been no more likely than any other date in the year to be correlated with any of the other seasonal events that would actually have meant something to people. The midpoint, whether in time or space, seems significant only if one views time and space in the abstract. This is not to deny that particular groups of people in the past may have chosen to divide the year into four roughly equal divisions, but we can not simply assume that the equinox was a significant calendrical date for all.

## Interpreting the Archaeological Record

As we have seen, astronomy helps us learn how the cosmos was perceived and understood in the past based on evidence we can retrieve from the material record. The "raw resource" is directly accessible to us: within certain boundaries of error we can reconstruct the positions of the sun, moon, stars, and planets in the night sky at any place and any time in the past. This means that we can readily identify ways in which ancient architecture could have made reference to celestial objects and events. But we must retain a sense of proportion. The astronomical ORIENTATION of a monument, to take an example, might only have conveyed meaning to a few people around the time of its construction, whereas the monument would also have been significant to people in many other ways, communicating meaning by way of its form, the materials used, and other aspects of its location.

In recent years archaeologists have applied a variety of interpretative approaches to the study of how people in the past conceptualized the LANDSCAPE. Yet *landscape* itself is a limiting if not excluding term. Its use reinforces a Western conceptualization of space that divides the world into three distinct parts—land, sea, and sky—and then, more often than not, leads us to ignore two out of three of these. Contrast this concept of landscape with that of the indigenous person who would, without hesitation, identify numerous connections between objects and events in the sky and many other aspects of his or her experience. In dealing with past as well as modern indigenous societies, the term *worldview* should mean exactly that: a people's understanding of the totality of the perceived environment.

The evidence available to the archaeologist consists of a present-day landscape containing scattered traces of past human activity, the end result of centuries—even millennia—of processes of attrition that have served to modify and to destroy. Archaeologists can look at the remains of permanent structures such as houses, tombs, and temples, and examine their location as well as architectural features that seem to express associations with features in the visible landscape and objects in the sky. Other human activities leave traces that are fixed in space, such as rock art (see PRESA DE LA MULA; SWEDISH ROCK ART). Beyond this are numerous types of material evidence, such as small artifacts, which are not fixed in space but may still tell us a great deal in various ways.

A well-tried approach is to try to identify alignments that were deliberately incorporated in architecture or in the location of manmade structures in relation to other prominent features in the landscape. The latter approach has been successful in central Mexico, where the day of sunrise or sunset behind prominent mountains as viewed from certain pyramids and palaces has been found to correlate with calendrical ceremonies performed in sanctuaries built on those mountains. The hilltop palace of CACAXTLA, built in what is now the Mexican state of Tlaxcala in the seventh century C.E., was located directly on a preexisting alignment connecting an older ceremonial site with La Malinche, a prominent volcanic peak still important today as a source of rain and symbol of fertility. The times in the year when the sun rose behind this mountain from Cacaxtla and other temples in the alignment marked two important calendrical festivals and times of ritual pilgrimage. The observances survive, in the form of Christian festivals, to this day.

In this instance ethnohistoric and ethnographic evidence confirms the meaning of symbolic alignments. Where such evidence does not exist, identifying meaningful alignments is much more problematic. Clearly, it is possible to identify alignments between manmade structures and natural features that are in fact totally fortuitous and had no meaning whatsoever to anyone in the past. One possibility is to look for (what seem to us to have been) the most prominent manmade and natural features in the landscape in order to examine possible alignments of significance between them. But even this may be misleading. Mursi sun watchers stand by a favorite boulder or tree to make observations rather than erecting a more permanent marker that would be evident in the archaeological record of the future. Many important dates in the Hopi solar horizon calendar are marked by tiny, inconspicuous skyline notches, while many more conspicuous horizon features are unused. Furthermore, there is generally no obvious natural feature (such as a prominent notch or peak) in the solstitial directions, important as these actually were and are to the Hopi. The reason is that traditional Hopi villages from which observations were made were not located with regard to astronomy and calendrics. Walpi, for example, is perched on a narrow cliff-top mesa: its inhabitants had a given horizon and could not move their position of observation except within very narrow margins in and around the village. They did make precise horizon observations, but this fact would not be recoverable from the archaeological record alone. Any future archaeologist proposing that these particular alignments were significant would be open to the accusation of being highly selective with the data.

In other cases, however, evidence of horizon observations might be more readily identifiable in the archaeological record of the future. The Zuni, for example, had sun-watching stations used for horizon observations prior to the summer solstice. And the inhabitants of the Polynesian island of MANGAREVA established observation places for noting the summer and winter solstices against suitable landmarks, such as adjacent islets or mountain ridges, often erecting stones upon them as foresights. This example bears some similarity to the interpretation of many of the British megaliths, erected in the third and second millennia B.C.E., put forward in the mid-twentieth century by the Scottish engineer, Alexander Thom (see THOM, ALEXANDER). Basing his conclusions upon surveys of many hundreds of STONE CIRCLES, SHORT STONE ROWS, and single standing stones, Thom concluded that "megalithic man" had undertaken high-precision observations of the sun and moon using distant mountainous horizons as the observing instrument. The megaliths, according to Thom, marked where to stand and, in many cases, pointed out the horizon foresight (such as a conspicuous notch between two mountains) that was to be viewed.

Thom's results did not stand the test of time. One reason is that his data set included a wide variety of sites spanning a wide geographical area and a period of some two millennia (see MEGALITHIC MONUMENTS OF BRITAIN AND IRELAND). Subsequent analyses that have been restricted to certain areas during particular periods, and have taken account of a wider range of archaeological evidence, have been more successful. The Scottish RECUMBENT STONE CIRCLES (RSCs) are particularly enlightening in this respect. These are a group of several dozen stone circles confined to an area within about fifty kilometers (thirty miles) of Aberdeen in northeastern Scotland. Their distinguishing feature is a single recumbent stone flanked by two tall uprights, which are without exception oriented between west-southwest and south-southeast, that is, within a quarter of the available horizon. Furthermore, they are consistently aligned upon the midsummer full moon, suggesting that ceremonials were carried out there when the midsummer full moon was passing low over the recumbent stone. This is a conclusion backed up at one excavated site, Berrybrae, where scatters of quartz and burned flint—white stones whose color is similar to the light of the moon—were found in the vicinity of the recumbent stone. However, more recent excavations at other sites have confounded the issue.

The RSCs were modest monuments, apparently serving relatively small farming communities around 2000 B.C.E. Aligning them upon the moon at an important time, and also if possible upon a conspicuous feature in the landscape such as a prominent hill (as is the case at many but not all of the circles), tied them into nature in two different ways. This almost certainly served to reinforce the sacred status of the monuments for the people they served. It seems likely that society in this area at the time never became centrally organized and controlled, and this has bequeathed to us a set of small, similar monuments among which we can easily spot repeated patterns. This enables us to catch a glimpse of some aspects of ritual tradition and worldview—a glimpse that points strongly to the moon as a principal focus of attention. Similar conclusions have been reached in investigations of SHORT STONE ROWS in southwest Ireland.

A step on from identifying and interpreting monumental alignments is to ask why the monuments were placed where they were and not elsewhere. Answering this question involves a detailed investigation of whole landscapes to identify potential alternative locations. Such an investigation was carried out in the late 1980s in the northern part of the Isle of Mull, off the west coast of Scotland. It showed that a set of five stone rows found there, apart from being consistently aligned upon the moon, were all placed so that Ben More—the most conspicuous mountain on the horizon in the area—was on the very margin of visibility, clearly in sight to one side of each row but completely hidden by intervening ground from the other. One suggestion is that the stone rows' main significance was as some sort of symbolic boundary marker between areas from which this (sacred?) mountain was, and was not, visible.

One supposition that emerges from these studies of monuments incorporating astronomical alignments is that many of them became "special" when the astronomical body in question appeared at the appointed place. At these times, their sacred power was surely reinforced. Another way in which a similar effect could be achieved—creating a very considerable visual impact at certain special times—was through the interplay of sunlight and shadow. Widespread evidence indicates an interest in creating carefully orchestrated interplays of shadow and light at sacred places, sometimes producing special effects visible on only very rare occasions. A famous example occurs at the passage tomb of NEWGRANGE in Ireland. Here, for a few precious minutes after sunrise on a few days around winter solstice, the dark interior of the tomb becomes lit up by sunlight shining directly down the passage. Even in the present day, such *hierophanies* can capture the imagination and become the focus of great public spectacles, whether or not they were actually intended by the builders. A case in point involves the pyramid of KUKULCAN (El Castillo) at Chichen Itza. It contains a staircase on each of its four sides, and at the base of the northern staircase is the carved head of a serpent. On days close to the equinoxes, the light of the late afternoon sun falling across the stepped corner of the pyramid creates the effect of a serpent's body, which only "appears" at these times. This spectacle now attracts tens of thousands of visitors. In contrast to Kukulcan, no serious doubts exist that the Newgrange hierophany was actually intentional; though, by its nature, it could never have formed a great public spectacle, since the space inside the tomb was confined. This dramatic effect was intended for the ancestors, or for ancestral spirits.

What was the purpose and meaning of such hierophanies? There is no simple answer, but further clues can be found by looking at more modest examples, often to be found in rock art. By carefully placing rock art designs, sunlight could be made to play across them at certain times, with impressive effect. A number of interesting examples are to be found in California, which was densely populated by hunter-gatherer groups prior to the European conquest. The Luiseño, for example, had an intense ceremonialism, a rich sky lore, and a calendar regulated by various astronomical observations. Although their seasonal calendar was lunar-based, they observed and celebrated the solstices, attaching particular importance to the winter solstice, which they regarded as a time of cosmological crisis. Solar imagery is evident at various Luiseño rock art sites, and light-and-shadow effects have been discovered at three or four of them, more than one involving daggers of sunlight that bisect painted discs.

One of the most famous shadow-and-light phenomena concerns a petroglyph situated toward the summit of Fajada Butte, in Chaco Canyon (see FAJADA BUTTE SUN DAGGER). A carved spiral hidden away behind three slabs of rock leaning against a vertical rock face is occasionally lit up by the light of the sun shining through cracks in the rocks in front of it. Around noon on the summer solstice, however, a thoroughly distinctive dagger of light suddenly appears and bisects the carving. How should this be interpreted? Assuming that the slabs were carefully placed rather than falling into their positions naturally (and there has been some debate on this issue), it appears that a good deal of care was taken in positioning the spirals so that the summer solstice, and possibly other times of the year also, were clearly marked. The location is difficult to access but good for astronomical observation. Comparisons with the practices of historic Pueblos suggest that it might have been a sun shrine—a sacred place where a sun priest came to deposit offerings at certain important times in the ceremonial year. In this sense it would have been far more than a simple calendrical device. It was more likely something that, through its powerful symbolism, helped to affirm the sacredness of this inaccessible place, and by so doing helped to reinforce the power of the person or people who understood its meaning and had the privilege of using it.

As with astronomical alignments, we again encounter a methodological problem: just because we observe a shadow-and-light phenomenon does not mean that it was intended by the builders or had any special significance before our "discovery" of it. Given any pattern carved on a rock, what are the chances that the sun will play across it at some time, quite fortuitously, creating an unintentional effect? In the case of the Chaco petroglyph, more secure interpretations can only stem from wider evidence about Anasazi culture, perhaps with valuable clues from historic and modern Pueblo groups.

A different form of shadow-and-light phenomenon has been noted in Mesoamerica: the so-called zenith tube. ZENITH TUBES are vertical tubes incorporated in specialized ceremonial structures that marked the biannual passage of the sun through the zenith at noon, at which time sunlight could pass directly down the tube and light a chamber below. One of the most impressive examples is found at the ruins of Xochicalco in Mexico, where a tube more than five meters (sixteen feet) long opens into the roof of a natural cave artificially extended to form a rectangular gallery with three central pillars. The use of such tubes for the purposes of astronomical observation cannot be established with certainty but seems likely given the known significance of the ZENITH PASSAGE OF THE SUN in Mesoamerica, both from studies of the ancient Mesoamerican calendar and from modern ethnography.

Displays of light and shadow, especially if they occur only infrequently, can be highly impressive. It is easy to see how they could have conveyed symbolic power, reinforcing sacred associations that formed part of the fabric of the prevailing worldview. Their presence in the archaeological record has the potential to reveal aspects of that worldview to us. The power to impress is conveyed to the modern investigator through direct experience—"being there" and experiencing the event for oneself has proved popular and is useful in generating tentative interpretations. But it also brings dangers. There is considerable potential for us to experience shadow-and-light phenomena that had no significance whatsoever to people in the past.

Great monuments also have the power to impress, none more so than STONEHENGE in England. Over the years it has attracted more than its fair share of theories, from the plausible to the plainly ridiculous. However, the general perception prevails that there was some sort of connection between Stonehenge and the skies, which accounts for its frequent appearance on the front covers of books on ancient astronomy. In the meantime, archaeological excavations have put some of these theories in a more secure context. The bluestone and sarsen circles, constructed around the middle of the third millennium B.C.E., replaced earlier timber constructions built within a circular ditch and bank that had been completed some 500 years earlier. The later stone monument was certainly a place of great power, and the solstitial alignment of the main axis, which all commentators agree was deliberate, represented a shift of several degrees from the earlier orientation. The transformation of the monument from timber into stone, the use of exotic stones from faraway places, and the sheer size and scale of the edifice strongly suggest a process of enriching the ritual symbolism of the site around 2500 B.C.E. to legitimize its place at the center of the cosmos and hence its ultimate power. This transformation probably served in turn to reinforce the earthly power and influence of a social elite. The change in the axis to incorporate the solar alignment was almost certainly part of the same process.

Modern Western science is only one of many frameworks of thought within which people all over the world, from Palaeolithic times to the present day, have gained an understanding of natural phenomena, including what could be seen in the sky. Modern science only seems the pinnacle of intellectual achievement from a Western perspective, and seeing it as such creates a tendency, when trying to comprehend non-Western modes of understanding, to single out particular "advances," seeing them as steps along the road. From an anthropological point of view, it seems stiflingly restrictive to proceed in this way when we can seek a much broader appreciation of different developments in human thought. This can only be achieved by studying the great many ways in which people have striven to comprehend the world that they inhabit and the many different contexts—physical and social—within which particular developments in thought have occurred. This applies just as much to sky knowledge as to all other knowledge. Above all, we need to recognize the magnificent diversity of human worldviews that has existed through time, and to respect them for what they were and are, even though we will inevitably continue to describe them, and try to make sense of them, in the framework of our own.

# Using This Book

The entries in this encyclopedia fall into three categories. First, there are those that elaborate upon some of the key *themes and issues* relating to ancient astronomy, many of which have been introduced above. They include, for example, LUNAR AND LUNI-SOLAR CALENDARS, PALAEOSCIENCE, and ASTRONOMICAL DATING. Other entries in this category relate to particular objects or phenomena in the sky and their cultural significance (for example, ORION and LUNAR ECLIPSES). Also included are broad definitions and explanations of key concepts (such as ARCHAEOASTRONOMY, COSMOLOGY, and ALIGNMENT STUDIES), brief descriptions of procedures and techniques (for example, FIELD SURVEY and STATISTICAL ANALYSIS), and a few people who have been important in the overall development of the subject, such as SIR NORMAN LOCKYER and ALEXANDER THOM. The second category comprises *case studies* of a variety of types of human society, from hunter-gatherer groups to urban city-states, and spanning a chronological and geographical range that stretches from Upper Palaeolithic Europe through Polynesia and other remote parts of the world prior to European contact. Though the case studies are far from exhaustive, they have been carefully selected to illustrate particular issues and themes, with an emphasis upon cultures and places that are less well known outside the specialist literature of archaeoastronomy. The final category of entries consists of short, nontechnical explanations of *basic concepts* like SOLSTICES, HELIACAL RISE, and DECLINATION. At the end of each entry is a list of related entries divided into three parts: themes, case studies, and basic concepts. Straightforward definitions of terms are given in the glossary.

We have avoided drawing sharp divisions between *ancient* and *modern*—divisions that only serve to mask the rich variety of human ways of understanding and appreciating the sky. For this reason, we include topics such as CHURCH ORIENTATIONS and ASTROLOGY, which extend into relatively recent and even modern times, together with examples of traditions that live on still among modern indigenous peoples, such as BARASANA "CATERPILLAR JAGUAR" CONSTELLATION, MURSI CALENDAR, and PAWNEE COSMOLOGY.

# Acknowledgements

The author would like to record his grateful thanks to the following colleagues and friends who have been kind enough to read and check certain entries, and to suggest corrections: Juan Antonio Belmonte, Von Del Chamberlain, David Dearborn, Michael Hoskin, Stephen McCluskey, Michael Parker Pearson, Bradley Schaefer, Keith Snedegar, Ivan Šprajc, John Steele, Richard Stephenson, and Gary Urton. Needless to say, the responsibility for all remaining shortcomings rests squarely with the author.

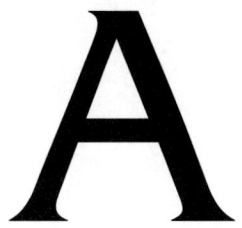

## Aboriginal Astronomy

When Captain James Cook first sailed up the east coast of New Holland (what is now Australia) in 1770, he encountered native people who had little interest in their strange visitors and none whatsoever in trading—people who surprised him by being generally unclothed, constructing little in the way of shelter around their camp fires, and even sleeping in the open. Yet their lifestyle, while short of so many of the conveniences viewed by Europeans as essential, also seemed free from many of the cares that accompanied them. As Cook saw it, the earth and sea satisfied all their needs.

As we now know, the Aboriginal way of life had gradually developed over many thousands, and indeed tens of thousands, of years. Australia's first inhabitants traveled to the continent more than 50,000 years ago from Asia, via the islands of what is now Indonesia. Despite lower sea levels at that time, the journey still involved crossing straits as wide as eighty kilometers (fifty miles) or more. During the long history of the Aboriginal peoples, factors such as climate change almost certainly caused significant upheavals, but by the time the Europeans arrived, Aboriginal groups were to be found in all corners of this vast and largely arid continent. Their lifestyle had become well adapted to an environment that, from a European perspective, was immensely inhospitable. Never settling for long in one place, they lived by gathering food, hunting, and (for people in the coastal fringes) fishing. This hunter-gatherer style of subsistence had never been superseded by agriculture, let alone affected by technological innovation such as metallurgy.

Since Aboriginal culture and tradition had developed independently of the rest of the world for millennia, it is tragic that much of it was utterly destroyed in such a short time after the arrival of the Europeans. It is estimated that at the time of contact there were more than 750 different Aboriginal languages, few of which now survive. Though our knowledge of the ways in which different native Australians viewed the world in pre-contact times is extremely fragmentary, there was clearly considerable diversity between different groups. Yet it is evident that certain fundamental concepts and

Aboriginal engraving at America Track, Kuringai Chase National Park, New South Wales, showing the head and neck of the culture hero Daramulan. (Courtesy of Clive Ruggles)

principles underlay most Aboriginal systems of thought. The most important of these is the *Dreamtime* or *Dreaming*. The Dreamtime is sometimes portrayed as a sort of parallel, more fundamental reality in which ancestor spirits created the world and continue to exist within it. Certain places and paths through the landscape have deep significance within the Dreamtime. The actions of ancestors are re-enacted in many aspects of life, and oftentimes these re-enactments involve being at certain places and following certain paths.

Aboriginal peoples identified strongly with the landscape. Some of the places charged with sacred power were natural, such as distinctive rocks, termite mounds, or water-holes. Others were created, such as "ringed" eucalyptus trees (branches growing apart were trained together again until they rejoined, forming a ring in the trunk). In some locations, "stone configurations"—constructed arrangements of rocks and boulders—also appear to mark sacred places in the landscape, especially in regions devoid of other prominent natural landmarks.

However, to see the Dreaming as something in the Aboriginal mind that is superimposed upon the physical world is a quintessentially Western view of things. From an Aboriginal perspective, the motions and actions of the ancestors are an integral part of the world and can be perceived in places, in the objects that are found there, in the pathways between them, and in events that are constantly happening ("signs," which must be perpetually watched for). Living people also leave their mark by their passage and conduct, not necessarily in a way that a Westerner would recognize as "physical," but by ascribing new meanings to places and pathways in the world. Naturally that world includes the sky. The actions of the ancestors can also be perceived in the appearance and motions of the sun, moon, and stars: sky events affect people, and people affect sky events.

Celestial mythology features extensively in the surviving Aboriginal cultural heritage. A wealth of stories relate to sky objects, including Orion, the Pleiades, the Southern Cross and Pointers, and the Magellanic Clouds. A recent collaboration between ethnographer Hugh Cairns and Yidumduma (Bill Harney), a senior elder of the Wardaman community in Northern Territory, has shown how the night sky acts as a repository of ancient traditions and learning in this community. This learning was taught using *song lines* that cross through the constellations just as the so-called *dreaming paths* pass through the earthly landscape below.

Extrapolating back in time is much more difficult. Among the few tangible remains from earlier epochs, the most ubiquitous are the rock art sites found in several parts of Australia: over 100,000 such sites have been recorded. Their dates are generally indeterminable within wide bounds, although some may be several thousand or even tens of thousands of years old. Some include what are clearly depictions of ancestor spirits, people, animals, and activities such as hunting and dancing, while others are more abstract. To what extent modern practices can inform us about the meaning and use of art in the past is a contentious issue. Some contemporary bark paintings certainly include depictions of the sun, moon, Venus, various constellations, the Milky Way, or the Magellanic Clouds, although in many cases it would be impossible to recognize them for what they are if we did not have informants to tell us. In some cases we know something of the stories to which they relate and the context in which they were used to tell those stories. Rock art, on the other hand, is tied to a particular place. For Yidumduma, rock art sites are places (but by no means the only places) where ancestor spirits reside, and the significance of each is revealed in the context of the whole Wardaman creation story. The sky is an intimate part of this story:

> Dreaming for the cosmos, the land and sky together for people that travel. The Creation routes connected to the people on the wall of the rock, and the connection up on the top, on the sky, down the bottom, and how everything is changing in this country. All the animal[s] and . . . the ceremony of the people in the rock . . . and all these other ones divided from the bottom and up to the top in the sky, with all the stars all part of human life from the sky. (Cairns and Yidumduma 2003, p. 31)

Taken out of context, as this example shows, the complexity of the connections between the content of rock art, its location, and the sky would be quite unfathomable to an archaeologist without an informant to interpret them.

Can more direct links between rock art and the sky sometimes be evident, even where we lack "the big picture"? A number of rock art sites contain inexplicable groupings of round holes or cup marks, which have not, it seems, arisen naturally; and some of these, it has been suggested, could represent constellations. However, it is disconcertingly easy to fit random collections of cup marks with patterns of stars in the sky if one is prepared to be sufficiently flexible in one's selection criteria, so the case remains unproven. There is little other evidence of direct relationships between rock art motifs and the appearance of celestial objects, although it has been suggested that the stylized forms of emu on some panels reflect the shape of the emu in the sky, a conglomeration of dark clouds in the Milky Way recognized as such in a number of places across Australia. John Morieson, an Australian archaeoastronomer, has suggested that a number of stone configurations in northwestern Victoria related directly to the rising or setting sun and incorporated deliberate alignments toward sunrise or sunset in particular seasons in the year.

There is certainly evidence that a number of different Aboriginal groups took note of the annual changes in the position of certain stars or constellations as indicators of seasonal events that affected their food supply. Arcturus is the fourth brightest star in the sky. For the Boorong people of northern Victoria, this star was Marpeankurrk, the ancestor who discovered *bittur*, or termite larvae, a delicious and valuable source of protein available at a time of year (August and September) when other food sources were scarce. Every year Marpeankurrk taught the living people when to find the *bittur* by appearing in the north during the evening. But night by night she appeared for a shorter and shorter time until she disappeared completely—and so did the *bittur*.

Although this is one of many similar examples, there is no evidence of any Aboriginal group having reckoned time using a calendar based, for example, on the phase cycle of the moon. Another curious feature of Aboriginal astronomy—at least, as far as we can tell from the available evidence—is that, despite living in wide open landscapes and having an intimate knowledge of the skies, Aboriginal peoples do not seem to have used the stars to navigate at night. Their astronomical knowledge, and their knowledge of the world in general, was qualitative rather than quantitative. It consisted of a rich network of associations confirmed in myth and tied to places and pathways both in the land and in the sky. Their worldview integrated land, sky, and social structure and kept human action intimately in tune with the cosmos.

Despite our fragmentary knowledge of Aboriginal astronomy, it is of great importance in showing how sky knowledge among hunter-gatherers can be rich and complex, despite leaving little or no trace in the material record. This fact has implications for how we interpret what extremely fragmentary archaeological evidence we have pertaining to astronomical knowledge in Upper Palaeolithic and Mesolithic times.

**See also:**
Constellation Maps on the Ground; Emu in the Sky; Landscape; Magellanic Clouds; Navigation.

**References and further reading**
Cairns, Hugh, and Bill Yidumduma Harney. *Dark Sparklers: Yidumduma's Wardaman Aboriginal Astronomy*. Merimbula, NSW: H.C. Cairns, 2003.
Chamberlain, Von Del, John Carlson, and Jane Young, eds. *Songs from the Sky: Indigenous Astronomical and Cosmological Traditions of the World*, 358–379. Bognor Regis, UK: Ocarina Books and College Park, MD: Center for Archaeoastronomy, 2005.
Flood, Josephine. *Rock Art of the Dreamtime*. Sydney: HarperCollins, 1997.
Ingold, Tim. *The Perception of the Environment*, 52–58. New York and London: Routledge, 2000.
Johnson, Dianne. *Night Skies of Aboriginal Australia: A Noctuary*. Sydney: Oceania Publications/University of Sydney, 1998.
Morieson, John. *The Night Sky of the Boorong*. Melbourne: Unpublished MA thesis, University of Melbourne, 1996.
Mountford, Charles P. *Nomads of the Australian Desert*, 449–483. Adelaide: Rigby, 1976.
Ruggles, Clive, ed. *Archaeoastronomy in the 1990s*, 136–152. Loughborough, UK: Group D Publications, 1993.
Selin, Helaine, ed. *Astronomy across Cultures*, 53–90. Dordrecht, Neth.: Kluwer, 2000.

# Abri Blanchard Bone

The earliest indications of the use of a symbolic notation to represent or visualize an astronomical cycle come from the Upper Palaeolithic period. At this time, in addition to striking cave art, engravings were made on small portable objects such as stones and pieces of bone and antler. Thousands of examples are known. A number of these are in the form of series of marks, and several were subjected to meticulous microscopic analysis by the American researcher Alexander Marshack. He concluded that the marks were not a simple decoration but carefully accumulated, often using different tools and techniques, over a period of time.

One famous example is part of an eagle's wing discovered in a cave at Abri Blanchard in the Dordogne valley, France. Dated to around 30,000 B.C.E., it contains a series of notched marks in a serpentine pattern. Marshack proposed that these represent a tally of days. The assumption is that the earliest marks are those in the center of the pattern, and that marks were accumulated around existing ones. By following the line outwards and back and forth, we discover that there are about fifteen marks in each sweep before the direction changes.

The most prominent astronomical cycle is the phase cycle of the moon. In addition to being readily observable, it coincides with the female menstrual cycle. The moon's phase cycle is certainly recognized among modern hunter-gatherer groups, although not necessarily universally (an apparent exception being Australian Aboriginals). The period of the lunar phase cycle (synodic month) is between twenty-nine and thirty days, so one interpretation of the Abri Blanchard bone is that it represents a tally in which the days of the waxing moon are marked off in one direction and those of the waning moon in the other; in other words it forms a rudimentary lunar calendar, maintained for about two months. Marshack suggested a similar lunar-calendar interpretation for patterns on a number of other Upper Palaeolithic portable artifacts.

Several criticisms can be made of the interpretation of the Abri Blanchard bone as a lunar calendar. Two assumptions upon which it rests are that the number of marks between each "turn" in the line was significant and, second, that this represented the period between successive new or full moons. How easy might it be to fit other explanations? How can we judge a particular interpretation against the alternatives? At least one of the turns is not sharp, which gives greater flexibility in interpretation. Two of the lines could easily be interpreted as separate straight lines rather than part of the serpentine pattern. And there is the question of what exactly we mean by "new moon": there is a one- or two-day period each month when the moon is not visible at all (astronomical new moon occurs in the middle of this), but it is the first reappearance of the crescent moon in the evening sky (the popular concept of "new moon") that is the most significant event in visual terms, widely recognized even in the today's world, from small indigenous groups to major religious calendars. Finally, although some of the marks appear round and others crescent-shaped, there is no apparent correlation between the shape of the marks themselves and the lunar phases.

All of these points introduce doubts in the interpretation of the Abri Blanchard bone as a lunar calendar. It is possible to address them by increasing the complexity of the explanation: for example, the left-to-right sequences contain more marks and can be taken to represent the days from full moon to new moon in the popular sense. But the potential for speculative argument seems endless. Indeed, the underlying microscopic evidence that the engravings represent tally marks or some other form of notation in the first place has been vigorously questioned.

And yet the underlying idea—that certain people in the Upper Palaeolithic would have recognized the phase cycle of the moon and may have attempted to record it—seems plausible enough. One approach would be to try to ascertain, hypothetically, how easy we would find it to "recognize" patterns in sets of markings that were in fact unintentional (for instance, caused by people sharpening tools) or had other meanings entirely. The conclusions could then be used as the basis for a formal statistical test. Yet even if lunar tallies or calendars were quite commonplace in the Upper Palaeolithic, they may not have been recorded at all consistently, in which case any attempt to identify sets of calendrical tallies recorded in a systematic way would be doomed to failure.

Studies of some of the many other Upper Palaeolithic engraved artifacts may clarify the issue. One example, another bone fragment known as the Täi plaque, has been interpreted by Marshack as a more sophisticated lunar calendar. And yet more complex designs give us greater flexibility in interpretation. This is not to say that such explanations are necessarily misguided, but rather that assessing them is no trivial matter. Identifying a methodology that will satisfy both scientists and social scientists is a challenge that has yet to be met by archaeologists.

**See also:**
Methodology; Palaeoscience.
Aboriginal Astronomy; Presa de la Mula.
Lunar Phase Cycle.

**References and further reading**
d'Errico, Francesco. "Palaeolithic Lunar Calendars: A Case of Wishful Thinking?" *Current Anthropology* 30 (1989), 117–118, 494–500.
Knight, Chris. *Blood Relations*. New Haven: Yale University Press, 1991.
Marshack, Alexander. *The Roots of Civilization*. New York: Weidenfeld and Nicolson, 1972.
———. "The Täi Plaque and Calendrical Notation in the Upper Palaeolithic." *Cambridge Archaeological Journal* 1 (1991), 25–61.

# Accuracy
*See* Precision and Accuracy.

# Acronical Rise
Alternatively spelled "acronychal," for example, in British usage. *See* Heliacal Rise.

# Acronical Set
Alternatively spelled "acronychal," for example, in British usage. *See* Heliacal Rise.

## Alignment Studies

Archaeoastronomy owes its emergence in the 1970s largely to the furor caused by the controversial and often spectacular claims made by Alexander Thom and others about astronomical alignments at British megalithic monuments. Although archaeoastronomy itself soon grew to encompass a much wider range of evidence, "alignment studies" remain at the heart of a great many archaeoastronomical investigations, particularly those concerning prehistoric Europe.

Where we are searching for evidence of astronomical concerns in prehistory, alignments of monumental architecture remain at the forefront of most investigations. Yet alignments can arise fortuitously, since every oriented structure must point somewhere. Hence the importance of repeated trends, which can be identified and/or verified statistically; good exemplars are the short stone rows in western Scotland and the recumbent stone circles in eastern Scotland. (The stone circle at Drombeg in Ireland provides a cautionary case study.) On the other hand, where other types of evidence are available to us (such as written documents or ethnohistory), studies of the significance of particular alignments may be carried out in a broader context with little or no need for statistical verification. A good example of this is the alignment of the so-called Governor's Palace at the Maya site of Uxmal.

The term *alignment studies* is not limited to the architecture of large monuments and public buildings but also includes studies of the layout of cities, as in pre-Columbian Mesoamerica (Teotihuacan, for example) or the ancient classical civilizations, and their possible relationships to celestial objects.

**See also:**
Archaeoastronomy; "Brown" Archaeoastronomy; "Green" Archaeoastronomy; Methodology; Thom, Alexander (1894–1985).
Drombeg; Governor's Palace at Uxmal; Recumbent Stone Circles; Short Stone Rows; Teotihuacan Street Grid.

**References and further reading**
Aveni, Anthony F. *Skywatchers*. Austin: University of Texas Press, 2001.
Ruggles, Clive. *Astronomy in Prehistoric Britain and Ireland*. New Haven: Yale University Press, 1999.

## Altitude

Altitude is the vertical angle between a given direction—such as the direction toward a particular point on the horizon from a given place—and the horizontal plane through the observer. A positive altitude indicates that the point being observed is above the observer; if it is below, then the altitude will be negative. Thus the altitude of a horizon point level with the observer is 0°. That of the summit of a high or nearby hill might be as much as 5° or 10°, but that of a sea horizon viewed from a high place might be −0.5° or −1°.

One can also speak of the altitude of a star in the sky, but this will not generally be the same as the angle of the star above the horizon, since the horizon altitude will not normally be 0°.

There is considerable confusion between the terms *altitude* and *elevation*. Elevation is normally taken to mean the height of a given place above sea level, but the two terms are quite often transposed, for example when pilots speak of the altitude of an airplane to mean its height above sea level and use an altimeter to measure it, while some engineers and astronomers use "elevation" to mean "altitude" as we have defined it here.

**See also:**
Compass and Clinometer Surveys; Field Survey; Theodolite Surveys.
Azimuth.

**References and further reading**
Ridpath, Ian, ed. *Norton's Star Atlas and Reference Handbook* (20th ed.), 4. New York: Pi Press, 2004.
Ruggles, Clive. *Astronomy in Prehistoric Britain and Ireland*, ix, 22. New Haven: Yale University Press, 1999.

# Ancient Egyptian Calendars

The kingdom of Ancient Egypt existed for over three millennia and for much of this time was remarkable in having two different calendars in simultaneous operation. Each arose in response to different social needs and developed a distinct function. At least, this is the standard interpretation of the evidence.

The oldest Egyptian calendar was lunar. It arose in Predynastic times (prior to c. 3000 B.C.E.) from the simple need to keep agriculture in track with the seasons. Twelve lunar phase-cycle (synodic) months only amount to 354 days, so a mechanism is needed for adding an additional (intercalary) month every two to three years in order to keep the calendar in track with the seasons. It is generally supposed that in Upper Egypt (the Nile valley) the calendar was regulated by Sothis, or Sirius, whose first appearance before dawn each year (heliacal rise) coincided with the regular annual flood of the Nile, the most critical event in the agricultural year. Whenever Sothis was not seen until late in the twelfth month, an additional (intercalary) month was added.

Some scholars have argued that a lunar calendar had also developed independently in Lower Egypt (the Nile Delta), but that there it was regulated by observations of the sun. A myth of central importance for the Egyptians was the daily rebirth of the sun god Ra by the sky goddess Nut, who stretched from one side of the sky to the other. This legend was played out

*Inscribed gray granite fragment from the "Naos of the Decades," an Egyptian shrine dating to the fourth century B.C.E., part of a text giving astrological prognostications associated with different decades (ten-day periods). (H. Lewandowski/Art Resource)*

in the sky. If the Milky Way was seen as Nut herself, as has been suggested, the legend would doubtless have accounted for its shifting position with respect to the rising and setting positions of the sun, according to the time of year. This would have provided the conceptual basis for what we might too easily see as the purely pragmatic process of keeping track of the seasons by tracking the movement of the sunrise and sunset along the horizon. When the two parts of Egypt became unified into a single kingdom, these luni-stellar and luni-solar calendars would have merged.

The Early Dynastic period (c. 3000–2600 B.C.E.) was the time of the first pharaohs and the development of written records. The lunar calendar was fine for regulating agricultural activity on a local scale but became unworkable for satisfying the needs of a complex economy and state bureaucracy. Recording and regulating the movement of perishable commodities, for example, demanded absolute agreement about the date. Yet the beginning of each month, and the insertion of intercalary months, was determined by observation, and observations could differ; as a result, it was often difficult to be certain either about the month or the day in the month. The result was the development of a quite independent civil calendar used for administration purposes. This comprised twelve "months" of exactly thirty days each, divided into ten-day periods known (rather confusingly, given that the word is commonly taken to mean "ten years") as *decades,* followed by a five-day period known as the additional or *epagomenal* days, making a total of 365. Once the calendar was set up, it was defined indefinitely; there was no need to regulate it by observations of nature. It was so fit for its purpose that it survived well beyond the end of ancient Egypt itself and was still being used by astronomers in medieval times.

It may seem surprising to suggest that the lunar calendar did not die out when the civil calendar was introduced. Yet this is precisely what has generally been believed for many years, following the work of the great Egyptian scholar R. A. Parker in the 1940s. According to this view, the purpose of the lunar calendar became the regulation of religious observances. Responsibility for determining the correct month and day within this calendar, and with it the timing of religious festivals, became the responsibility of certain priests. At some stage, and possibly as early as Old Kingdom times (c. 2600–2100 B.C.E.), Egyptians started to use the heliacal rise of other stars to help regulate the calendar and to develop the system of *decans*. These were thirty-six stars or star groups whose heliacal risings occurred in succession, roughly ten days apart—in other words, marking successive decades (ten-day periods) in the civil calendar. This meant that the two calendars could easily be synchronized.

Only recently has a different view emerged, proposed by the Spanish archaeoastronomer Juan Antonio Belmonte. This is that different Egyptian calendars did not in fact coexist. In Belmonte's view, sun observations were used initially to establish the 365-day calendar, but thereafter the dates of lunar-related religious festivals were simply established within this calendar, rather as Christians determine the annual date of Easter to this day.

Whatever the outcome of this debate, it is clear that the decans also heralded a hugely important innovation: they could be used as "clocks" to mark the passing of time during the night. Instead of focusing just on the heliacal rise, that is, on the decan rising immediately before dawn, one simply had to view the risings of successive decans through the night. On this basis, the Egyptians were the first to divide the night into approximately equal time intervals. Theoretically, at the equinoxes, when sunrise and sunset are exactly twelve hours apart, eighteen ideally placed decans should rise between sunset and sunrise at intervals of 40 minutes. In practice, however, the length of the night varies through the year, and twilight renders the first and last risings invisible. Furthermore, it could not have been possible to find stars to use as decans that were precisely evenly spaced across the sky. Finally, the Egyptian kingdom stretched sufficiently far from north to south for latitude to make a difference. The most decans that could actually be seen in one night, during the longest nights of the year around the winter solstice, was twelve. This fact, it is widely believed, led to the concept of the night being divided into twelve "hours," whose length varied through the year, but which were nonetheless a precursor to the modern concept of dividing the day-night cycle into twenty-four equal hours. There is a scholarly consensus on the identification of several of the decans, but the identification of others remains extremely speculative.

The Egyptian civil calendar was simple, useful, and all-pervasive. It worked unfailingly for almost three millennia, and although it gradually slipped with respect to the seasons (historical sources confirm that the civil new year once more coincided with the heliacal rising of Sirius in C.E. 139), this never seems to have created a problem during any particular epoch. Not until the third century B.C.E. was any attempt made to add leap years.

See also:
Lunar and Luni-Solar Calendars.
Coffin Lids; Egyptian Temples and Tombs.
Heliacal Rise; Lunar Phase Cycle.

**References and further reading**
Belmonte, Juan Antonio. "Some Open Questions on the Egyptian Calendar: an Astronomer's View." *Trabajos de Egiptología [Papers on Ancient Egypt]* 2 (2003), 7–56.
Clagett, Marshall. *Ancient Egyptian Science: A Source Book, Vol. 2: Calendars, Clocks, and Astronomy.* Philadelphia: American Philosophical Society, 1995.
Depuydt, Leo. *Civil Calendar and Lunar Calendar in Ancient Egypt (Orientalia Lovaniensia Analecta).* Leuven, Belgium: Departement Oosterse Studies, 1977.
Hoskin, Michael, ed. *The Cambridge Illustrated History of Astronomy,* 24–25. New York: Cambridge University Press, 1997.
McCready, Stuart, ed. *The Discovery of Time,* 82–83, 122–123, 158. Naperville, IL: Sourcebooks, 2001.
Neugebauer, Otto, and Richard A. Parker. *Egyptian Astronomical Texts, I: The Early Decans.* Providence, RI: Brown University Press, 1960.
———. *Egyptian Astronomical Texts, III: Decans, Planets, Constellations and Zodiacs.* Providence, RI: Brown University Press, 1969.
Parker, R. A. *The Calendars of Ancient Egypt.* Chicago: University of Chicago Press, 1950.
Selin, Helaine, ed. *Astronomy across Cultures,* 480–484. Dordrecht, Neth.: Kluwer, 2000.
Spalinger, Anthony, ed. *Revolutions in Time: Studies in Egyptian Calendrics (Varia Aegyptiaca).* San Antonio, TX: Van Siclen Press, 1994.
Walker, Christopher, ed. *Astronomy before the Telescope,* 33–35. London: British Museum Press, 1996.

# Andean Mountain Shrines

Traditional belief systems that still persist in a number of remote Andean villages link mountains, ancestor worship, ritual pilgrimage, and the sea as the ultimate source of water and fertility. The landscape is "animated" in the sense that unusual or prominent features are perceived to have a supernatural aspect. Local communities often regard themselves as the descendents of mountain deities; consequently mountain peaks—and especially volcanoes—occupy a prominent place in their cosmic beliefs and communal rituals. Mountains and mountain gods are seen as the controllers of rain, and their summits sometimes remain an important focus of ceremonial activity today. These metaphysical convictions do have a foundation in the physical world, in that prominent mountains have a strong effect on local meteorological phenomena.

These modern mythologies represent the remnants of belief systems stretching back at least as far as Inca times. Numerous shrines, both Incaic and more modern, have been discovered on hills and mountain peaks in Peru, Bolivia, Chile, and Argentina. Sea shells and river stones were commonly offered to deities for water, but more macabre offerings have also been uncovered: human (including child) sacrifices. We know from the accounts of Spanish chroniclers that elaborate pilgrimages were involved in reaching these sacred places. But the intensity of conviction that motivated the expeditions still defies the imagination. Offerings have been found on mountain peaks as high as 6,000 meters (20,000 feet).

There is every reason to believe that some of the mountain pilgrimages were tied to specific calendrical rituals and that they were astronomically timed. Broadly similar sets of beliefs and practices in central Mexico provide more concrete evidence of calendrical and astronomical associations. However, the specific associations that regulated the Andean rituals remain largely unknown.

See also:
Pilgrimage.
Cacaxtla.

### References and further reading
Constanza Ceruti, María. *Cumbres Sagradas del Noroeste Argentino.* Buenos Aires: Editorial Universitaria de Buenos Aires, 1999. [In Spanish.]
D'Altroy, Terence. *The Incas,* 169–171. Oxford: Blackwell, 2002.
Reinhard, Johan. *The Nazca Lines: A New Perspective on their Origin and Meaning* (4th ed.). Lima: Editorial Los Pinos, 1988.
———. "Sacred Mountains, Human Sacrifices, and Pilgrimages among the Inca." In John B. Carlson, ed., *Pilgrimage and the Ritual Landscape in Pre-Columbian America.* Washington, DC: Dumbarton Oaks, in press.
Reinhard, Johan, and María Constanza Ceruti. *Investigaciones Arqueológicas en el Volcán Llullaillaco.* Salta, Argentina: Ediciones Universidad Catolica de Salta, 2000. [In Spanish.]

Saunders, Nicholas, ed. *Ancient America: Contributions to New World Archaeology*, 145–172. Oxford: Oxbow Books, 1992.

## Angkor

The ancient Khmer civilization flowered in southeastern Asia for over five centuries. A number of interacting and competing kingdoms had existed in the area since the early centuries C.E., and both the Buddhist and Hindu religions had been widely adopted. But in C.E. 802, several of these kingdoms joined together to form a powerful state. The capital, Angkor, situated in modern Cambodia, became the center of an empire that, at its height, stretched over 1,000 kilometers (600 miles) from what is now Burma in the west across Thailand and Laos to central Vietnam in the east, and a similar distance north to south, from the Khorat plateau in northern Thailand both to the tip of southern Vietnam and down the long isthmus toward the Malay peninsula. After the early fourteenth century, the Khmer empire went into decline and Angkor collapsed. However, a new capital emerged downstream. Khmer culture endured and in many respects thrived until the arrival of the Europeans and the establishment of the French Protectorate in the mid-nineteenth century.

The word *Angkor* means "Holy City" in Sanskrit. The arrangement of magnificent buildings and reservoirs (*barays*) that survives to this day resulted from construction projects initiated by a succession of rulers, and includes both Buddhist and Hindu temples, according to the religious persuasion of the king in question. At its heart is Angkor Thom, a huge square compound—some 3.5 kilometers (2.2 miles) on a side—containing a royal palace surrounded by numerous public buildings, platforms, and courts, and including two temple-pyramids, the Baphuon and the Bayon. Angkor Thom is flanked both to the east and west by enormous rectangular reservoirs, each over 7 kilometers (4.3 miles) long and 1.5 kilometers (1 mile) wide. To the south is a small hill, Phnom Bakheng (Mighty Mount Ancestor), with a temple on its summit. Beyond that to the south is the most magnificent temple complex of all: Angkor Wat, arguably the largest and most impressive religious structure ever built. Although it has been a Buddhist temple (*wat*) for many centuries, it was originally dedicated to the Hindu god Vishnu. Scattered in its vicinity, but mostly found close to the western and southern sides of the East Baray, are a number of additional temple enclosures, including three that functioned as Buddhist monasteries.

Although it is Angkor Wat's artistic and architectural splendor that so impresses the modern visitor, many of the temples built there also expressed vital principles of cosmic harmony. Some of these expressions are obvious: the large numbers of towers pointing at heaven; the cardinal orientation of

Angkor Wat, Cambodia. (PhotoDisc, Inc.)

the temple precincts and *barays;* and the westward orientation of Angkor Wat, generally supposed to express a link between death and the setting sun, but possibly symbolizing the direction associated with Vishnu. The central tower of the Bayon—at the center of Angkor Thom, which is itself at the center of Angkor—represented Mount Meru, the dwelling place of the gods, placed at the center of a symbolic model of the cosmos. Other principles and relationships may have been expressed more subtly, encoded numerically and geometrically in a variety of ways. Thus the Bakheng temple, with seven levels to represent the layers of heaven, takes the form of a stepped pyramid supporting 108 towers, symmetrically arranged. The visitor mounting the central staircase on any side, however, can only see thirty-three of the towers—the number of principal gods in the Hindu pantheon. The design of Angkor Wat incorporated alignments upon solstitial and equinoctial sunrise, and it has even been suggested that the dimensions of the central structure encoded, in Khmer units of measurement, the number of days in the year. Although many of the specific suggestions are unproven and controversial, they remain plausible, given the ways in which cosmological principles are known to have been encapsulated in Hindu temples throughout history and still are today.

    The author Graham Hancock has gone further, arguing that the entire layout of temples at Angkor had another purpose. It formed a gigantic model on the ground of the constellation Draco, just as (he and others have supposed) the Pyramids of Giza in Egypt modeled Orion's Belt. Although this may sound like a simple extension of the idea of human constructions modeling the cosmos, it is ill conceived at a number of levels. For one thing, the idea of producing literal "maps" of the stars in the sky, as opposed to symbolic models of the gods in heaven, has no known place in Hindu beliefs and practice. In other words, this is an idea totally divorced from the social context in which it is supposed to have operated. Second, the construction of successive temples would have needed to have conformed to a very specific grand plan conceived before any of the building started and strictly adhered to by one ruler after another. Third, in order to make the orientation fit the appearance of the constellation in the sky, one has to assume that the plan was conceived around the year 10,500 B.C.E. Though it is not beyond the bounds of possibility that sacred sites might have been recognized for one reason and then retained their sacredness, it is sheer fantasy to suggest that a fixed plan could have been perpetuated for more than 11,000 years. Likewise, the idea of a "lost golden age" at such an early date flies in the face of well-established archaeological evidence from all over the world.

But how, then, does one explain the apparent fit of the temple positions with the pattern of stars in Draco? The answer is that apparently impressive matches like this can easily arise fortuitously, especially if one is prepared to be selective with the data. There are many temples at Angkor, many bright stars in the sky to choose from, and one can project the curved surface of the sky onto the flat surface of the ground in various ways. Certain temples, such as the Baphuon and Bayon, do not fit stars, and certain bright stars in the relevant part of Draco, such as Eltanin ($\gamma$ Dra) and Altais ($\delta$ Dra), do not fit temples. Furthermore, the fit in the remainder of cases is not always impressive. It is possible to argue that there were inevitable errors in the process of identifying the correct location, multiplicities of purpose, and other difficulties in practice. But these possibilities (for which there is no direct evidence) provide no justification for selecting data arbitrarily in order to obtain more impressive fits. This is a flawed game that one plays at one's peril.

The example of Angkor serves to show the severe dangers of seeking astronomical correlates of spatial patterns in archaeological data too keenly and taking interpretations far too far without heeding the constraint of cultural evidence. We do not need to go to such lengths to acknowledge that astronomical and cosmological symbolism was deeply engrained in this extraordinary group of monuments.

**See also:**
Cardinal Directions; Constellation Maps on the Ground; Methodology.

Pawnee Star Chart; Pyramids of Giza.
Star Names.

### References and further reading

Coe, Michael D. *Angkor and the Khmer Civilization.* London: Thames and Hudson, 2003.
Hancock, Graham, and Santha Faiia. *Heaven's Mirror,* 115–198. London: Penguin, 1998.
Higham, Charles. *The Archaeology of Mainland Southeast Asia: from 10,000 BC to the Fall of Angkor,* 321–355. Cambridge: Cambridge University Press, 1989.
Krupp, Edwin C. *Skywatchers, Shamans and Kings,* 303–310. New York: Wiley, 1997.
Malville, J. McKim. "Angkor Wat: Time, Space and Kingship." *Archaeoastronomy: The Journal of Astronomy in Culture* 17 (2003), 108–110.
Stencel, Robert, Fred Gifford, and Eleanor Morón. "Astronomy and Cosmology at Angkor Wat." *Science* 193 (1976), 281–287.
Wheatley, Paul. *The Pivot of the Four Quarters: A Preliminary Enquiry into the Origins and Character of the Ancient Chinese City,* 257–267, 436–451. Edinburgh: Edinburgh University Press, 1971.

# Antas

Antas are a distinctive type of dolmen concentrated in the Alentejo region of central Portugal and the bordering provinces of western Spain. *Anta* in Portuguese simply means "dolmen," but the word has been appropriated to signify a particular method of construction. A large, flattish stone was set in the ground on its side to act as the backstone. Two further slabs were then placed upright leaning against it, one on each side, then a further pair was added with each one leaning against one of the earlier pair, and so on, to form a chamber with a single entrance. Most commonly, seven stones were used in total. A flat capstone was finally placed horizontally on top. Antas vary considerably in size, the smallest chambers being under 2 meters (6 feet) wide and no more than 1.5 meters (5 feet) high, while the largest measure over 5 meters (16 feet) wide and 4 meters (13 feet) high. A magnificent example of the latter is the Anta Grande do Zambujeiro near Évora.

A remarkable feature of the antas is their pattern of orientation. Michael Hoskin, a British historian of astronomy, surveyed 177 of them in the 1990s, using a compass and clinometer to determine their axial orientation (the direction along the axis of symmetry toward the entrance) and their astronomical potential in terms of the declination of the indicated horizon point. According to Hoskin, not a single orientation falls outside that part of the horizon on which the sun rises at some date in the year. Many different groups of later prehistoric monuments within western Europe show distinctive *orientation signatures,* but the implications of this group's signature seem particularly clear: the antas faced sunrise on some day in the year, per-

*The seven-stone anta at Mellizo, one of a number found near Valencia de Alcántara in Extremadura, western Spain, close to the border with Portugal. (Courtesy of Clive Ruggles)*

haps the day on which construction commenced.

If so, then we must add a qualification. The spread of orientations does not correspond to what one would expect if these dolmens were constructed at random times in the year. If that were the case, we would expect more orientations to fall toward the two ends of the solar range (where its rising position changes less from day to day, and so it spends more days) than in the middle. The actual anta orientations, on the other hand, fall mainly toward the middle of the range. This could be explained in one of two ways. The first possibility is that they were merely oriented roughly eastwards. In this case, the fact that the most extreme orientations happen to fall close to the edges of the solar range, that is, in the solstitial directions, would be seen as purely fortuitous. The other possibility is that tombs were constructed preferentially at certain times of the year, namely spring and/or autumn. Hoskin suggests the latter, arguing that in the spring and summer the needs of agriculture or animal herding would take precedence, but if work on dolmen construction commenced in the fall, it could reliably be completed during the ensuing winter.

**See also:**
Solstitial Directions.
Prehistoric Tombs and Temples in Europe.
Declination; Solstices.

**References and further reading**
Belmonte, Juan Antonio, and Michael Hoskin. *Reflejo del Cosmos*, 35–40. Madrid: Equipo Sirius, 2002. [In Spanish.]
Chapman, Robert. *Emerging Complexity: The Later Prehistory of South-East Spain, Iberia and the West Mediterranean.* Cambridge: Cambridge University Press, 1990.
Hoskin, Michael. *Tombs, Temples and their Orientations*, 85–100. Bognor Regis, UK: Ocarina Books, 2001.

## Antizenith Passage of the Sun

Just as the sun passes through the zenith, that is, directly overhead, twice a year as viewed from any location within the tropics, so it also passes twice a year through the point directly beneath the observer. This point is known as the antizenith or nadir. The dates of solar antizenith passage occur six months away from those of zenith passage so that, for example, the antizenith passage dates at a location a little way south of the Tropic of Cancer will be shortly before and after the December solstice.

The antizenith passage of the sun may seem a totally esoteric event since, unlike zenith passage, it is not directly observable. Yet some human cultures do seem to have had an interest in identifying the dates of solar antizenith passages and marking them with appropriate observances. It has been suggested, for example, that some of the pillars erected on the western horizon at the Inca capital city of Cusco were aligned upon sunset on the day of antizenith passage.

**See also:**
Cusco Sun Pillars.
Zenith Passage of the Sun.

**References and further reading**
Aveni, Anthony F. *Between the Lines: The Mystery of the Giant Ground Drawings of Ancient Nasca, Peru*, 131. Austin: University of Texas Press, 2000. (Published in the UK as *Nasca: The Eighth Wonder of the World*. London: British Museum Press, 2000.)
———. *Stairways to the Stars: Skywatching in Three Great Ancient Cultures*, 25–27, 172–176. New York: Wiley, 1997.

## Archaeoastronomy

Archaeoastronomy is best defined as the study of beliefs and practices concerning the sky in the past, and especially in prehistory, and the uses to which people's knowledge of the skies were put. It can be misleading to think of archaeoastronomy as the study of ancient astronomy, since people in the past might have related to the sky in very different ways from people in the modern Western world. For this reason many people prefer to avoid the word *astronomy* altogether. Some speak of *astronomies* to emphasize this point, as is evident in the title of books such as *Astronomies and Cultures* (Ruggles and Saunders 1993).

There have been scientific investigations of the possible astronomical significance of spectacular ancient monuments ever since the later nineteenth century—in the work of Sir Norman Lockyer and Alexander Thom, for example. However, the term *archaeoastronomy* has only been in existence since 1969 and soon thereafter came to take on a much broader meaning. Some popular authors such as Gerald Hawkins, whose book *Stonehenge Decoded* brought Stonehenge into the limelight in the mid-1960s, preferred the term *astro-archaeology,* and this for some became synonymous with *alignment studies,* causing a good deal of confusion that has still not entirely abated. Nonetheless, it was archaeoastronomy that emerged as a recognized academic "interdiscipline" a decade later, marked most significantly by the appearance of two academic journals: *Archaeoastronomy: The Journal of the Center for Archaeoastronomy* in the US and, in the UK, the *Archaeoastronomy* supplement to the *Journal for the History of Astronomy.* It was also signaled by a landmark international conference in archaeoastronomy at Oxford in 1981, the first of a series that has continued ever since.

By that time, archaeoastronomy had already expanded beyond the mere study of monumental alignments. This development was led from the Americas in the 1970s, particularly by studies of astronomy in pre-Columbian Mesoamerica. These studies leaned heavily upon ethnohistorical evidence—the writings of chroniclers in the early years after European contact and conquest—and also, at least in the case of the Maya, upon written evidence in the form of native "books," most notably the Dresden Codex. Alignment studies formed an important part of the new integrated approach to studying Mesoamerican archaeoastronomy but did not drive it. On the other hand, European and especially British archaeoastronomers in the 1970s were largely obsessed with archaeological and statistical reappraisals of Alexander Thom's work and were busy developing formal fieldwork procedures and quantitative techniques for resolving the huge controversy surrounding Thom's conclusions. The American archaeoastronomer Anthony Aveni would later come to characterize these two contrasting approaches within archaeoastronomy as "brown" and "green," after the colors of the covers of the two volumes (*Archaeoastronomy in the New World* and *Archaeoastronomy in the Old World,* respectively) that emerged from the 1981 Oxford conference. Finding the right balance between these two methodological approaches remains one of the major challenges for archaeoastronomy to this day.

Archaeoastronomers, then, are prepared to consider a range of types of evidence—not simply archaeological—and they certainly do not restrict themselves to alignments of monuments. In this sense the word *archaeoas-*

*tronomy* is misleading, even though it is the term that has stuck. Furthermore, there is no clear boundary between the past and present. Almost from the outset, "brown" archaeoastronomers who studied astronomical traditions within North America realized that myths and traditions surviving in modern indigenous Native American groups could be highly relevant to our understanding of astronomical traditions of the past. When we study reports of ethnographers who worked, say, fifty years ago on a tradition that has since died out, it is clearly pointless to worry about whether to classify this as ethnohistory or ethnography. We may even be able to identify longer threads of continuity stretching from the past into the present, so that, for example, the study of a modern Andean village such as Misminay in Peru may give us some insights into worldviews prevailing in Incaic or even pre-Incaic times. In short, what was once (and sometimes still is) identified as *ethnoastronomy* blends seamlessly into *archaeoastronomy*. Many would prefer simply to combine the two fields of study, referring to them together by a term such as *cultural astronomy*.

Why was archaeoastronomy so controversial in its formative days in the 1960s and 1970s? Partly because Thom's own interpretations, though based upon high-quality surveys of many hundreds of British megalithic sites and supported by rigorous statistical analysis, struck most archaeologists as clear examples of ethnocentrism. In seeing prehistoric Britain as populated by "megalithic man" whom he described as "a competent engineer with an extensive knowledge of practical geometry," Thom was merely seeing his own reflection in the past, the skeptics claimed. In like manner, they pointed out, several prominent astronomers were seeing their own reflections when they described Stonehenge as a sophisticated observatory and eclipse predictor. But there was also a deeper reason. In the 1970s, most archaeologists in the forefront of theoretical developments were trying to describe the actions of human communities in prehistory and the processes of social change in terms of human responses to ecological or environmental constraints. These "processual archaeologists" saw no reason why prehistoric people would have bothered incorporating astronomical alignments into communal monuments.

In the decades that followed, both archaeoastronomy and theoretical archaeology moved forward. Archaeoastronomy ceased to be dominated by astronomers and began to incorporate more archaeologists and anthropologists. This would, at the very least, help to avoid naive pitfalls such as ethnocentrism. As a result, archaeoastronomy began to lose its cavalier attitude toward the body of theory that had been developed (and was continuing to develop) within archaeology and anthropology for interpreting the actions of any past community from what we can find in their material remains. At the same time, the so-called "post-processual" revolution was taking place in theoretical archaeology. A new generation of archaeologists was beginning to confront issues such as cognition. What was going on in people's minds, claimed these interpretative archaeologists, was just as vital in determining people's actions in the past as any environmental constraints. Put another way, how a group of people viewed the world—their cosmology—was just as important in determining their actions as what a modern archaeologist might consider "rational" considerations. This conclusion was supported by numerous case studies from modern indigenous communities such as the Hopi and Pawnee in North America. To speak of "rational" behavior is itself ethnocentric, since what we consider rational is a product of our own worldview.

Why are archaeoastronomy (and ethnoastronomy) worth doing? If we want to understand more about why certain human communities did what they did in the past, then we need to try to understand aspects of their cosmologies—the ways in which they perceived the world. Astronomy is an essential part of nearly every cosmology, because all human communities have a sky, and the sky and the objects in it form an integral part of the perceived world. The objects in the night sky are immutable, regular, reliable, predictable; for communities who did not live inside brightly lit buildings and whose skies were not polluted by modern city lights, they were there for people to watch and contemplate night after night, season after season. They become imbued with meaning.

But what different groups of people perceive as important in the sky, and what significance they ascribe to it, is highly culture-dependent. Two good examples that illustrate this in different ways are the calendar of the Borana of Ethiopia, and the emu "dark cloud constellation" recognized by certain Aboriginal groups in Australia. Nevertheless, people generally try to keep their actions in harmony with the cosmos as they perceive it, which may be the reason many prehistoric monuments were deliberately aligned with objects in the wider landscape, including objects in the sky. Recognizing associations that certain human communities considered important can help us understand something of the worldviews that gave rise to them.

The advantage that the sky holds for us in trying to spot such associations is that it forms a part of the ancient environment that is directly accessible to us, unlike the terrestrial landscape, which is often transitory. We can use modern astronomy to reconstruct, with considerable accuracy, the appearance of the night sky at any place and time, including the motions of the sun, moon, stars, and planets. And this capability offers a considerable advantage, whether we are concerned with myths and beliefs relating to the sky, monumental alignments upon celestial objects, or other kinds of evidence.

See also:

Astro-Archaeology; "Brown" Archaeoastronomy; Cosmology; Ethnoastronomy; Ethnocentrism; "Green" Archaeoastronomy; Lockyer, Sir Norman (1836–1920); Methodology; Thom, Alexander (1894–1985).

Borana Calendar; Dresden Codex; Emu in the Sky; Hopi Calendar and Worldview; Misminay; Pawnee Cosmology; Stonehenge.

**References and further reading**

*Archaeoastronomy*, the supplement to the *Journal for the History of Astronomy*. Nos. 1 (1979) to 27 (2002). Published by Science History Publications, Cambridge, England.

*Archaeoastronomy*, the bulletin, subsequently journal, of the Center for Archaeoastronomy, College Park, MD. Vols. 1 (1977) to 11 (1993). Published by the Center for Archaeoastronomy, College Park, MD. Superseded by the University of Texas Press journal, see below.

*Archaeoastronomy, The Journal of Astronomy in Culture*. Vols. 14 (1999) to date. Published by the University of Texas Press.

Aveni, Anthony F., ed. *Archaeoastronomy in the New World*. Cambridge: Cambridge University Press, 1982.

———, ed. *World Archaeoastronomy*. Cambridge: Cambridge University Press, 1989.

Heggie, Douglas C., ed. *Archaeoastronomy in the Old World*. Cambridge: Cambridge University Press, 1982.

Renfrew, Colin, and Paul Bahn, eds. *Archaeology: The Key Concepts*, 11–16. Abingdon, UK: Routledge, 2005.

Ruggles, Clive, ed. *Archaeoastronomy in the 1990s*. Loughborough, UK: Group D Publications, 1993.

Ruggles, Clive. *Astronomy in Prehistoric Britain and Ireland*, 1–11. New Haven: Yale University Press, 1999.

Ruggles, Clive, and Nicholas Saunders, eds. *Astronomies and Cultures*. Niwot, CO: University Press of Colorado, 1993.

# Archaeotopography

One of the criticisms of archaeoastronomy is that archaeoastronomers who investigate the reasons why ancient buildings and monuments were situated and oriented as they are often seem concerned only with the possibility that the main influencing factors were astronomical. But many different considerations, some quite unrelated to astronomy, can determine the orientation of a monument. One possibility is alignment based on prominent topographic features in the surrounding landscape. Accordingly, the term *archaeotopography* was coined by Michael Hoskin in 1997 to describe the collection of orientation data, as opposed to its (exclusively astronomical) interpretation. However, since the term would seem, similarly, to imply a necessarily topographical interpretation, and since topographical and astronomical motivations are only two among numerous possible reasons for orienting a structure in a particular direction, the term is rather misleading and has not been adopted widely.

See also:
Archaeoastronomy; Cosmology; Orientation.

**References and further reading**
Belmonte, Juan, and Michael Hoskin. *Reflejo del Cosmos*, 21–24. Madrid: Equipo Sirius, 2002. [In Spanish.]
Hoskin, Michael. *Tombs, Temples and their Orientations*, 13–15. Bognor Regis, UK: Ocarina Books, 2001.

## Astro-Archaeology

This confusing term has sometimes been used as an alternative to *archaeoastronomy* and sometimes to mean the study of astronomical alignments at ancient monuments—in other words, to represent only a segment of the wider endeavor of archaeoastronomy. It has now largely fallen into disuse, at least among academics working in this area.

See also:
Archaeoastronomy.

**References and further reading**
Aveni, Anthony F. *Skywatchers*, 2. Austin: University of Texas Press, 2001.
Ruggles, Clive. *Astronomy in Prehistoric Britain and Ireland*, 226. New Haven: Yale University Press, 1999.

## Astrology

To a modern astronomer, astrology is anathema. The idea that there can be any direct connection between the configuration or appearance of distant heavenly bodies in the sky and current or future events in the terrestrial world flies in the face of laws of physics that have been established beyond question over many centuries. This is not to deny that in a few cases an astrological relationship might actually have a physical basis: some claim, for example, that a correlation can exist between the growth of plants and the phase of the moon, because the level of moisture in the soil is related to the lunar phase through a tidal effect.

From the perspective of the archaeologist or anthropologist, whose ultimate interest is in human behavior rather than the laws of the universe, whether such an argument is scientifically verifiable or not is not the point. What interests these scientists is the fact that people throughout the ages have drawn direct connections between the appearance of the sky and events on earth, and that this forms an integral part of their understanding of how the world works. Even in the modern Western world, popular astrology represents a widespread perception of how celestial events influence terrestrial ones and challenges the "institutional" view represented by scientific astronomy. Modern astronomers may dismiss astrology as nonsense, but the direct associations it presupposes between celestial and terrestrial events may well be far closer to the ways people throughout history have managed to make sense of the world than the explanations provided by modern science.

Outside the Western scientific tradition there is no meaningful distinction between astronomy and astrology; indeed, archaeoastronomy might equally well be named archaeoastrology. Modern indigenous worldviews (cosmologies) commonly feature the idea that good fortune on earth depends upon keeping human action in harmony with what is happening in the skies, and there is every reason to assume that this has been true since early prehistory.

The term *astrology* can be applied to three rather different but not entirely separable ways in which people have perceived connections between the configuration of the heavens and events on earth: the belief that particular celestial configurations can portend future events; the belief that they can determine or influence the characteristics and lives of people, most commonly at their moment of birth; and the belief that they are directly connected to (that is, influence and/or reflect) current terrestrial events. Each type of perceived connection could provoke a variety of actions in response to certain observed celestial events.

The belief that what is seen in the sky may foreshadow the future—typically stemming, as for the ancient Greeks, from a belief that it can indicate the intentions of the gods—underlies celestial divination: the use of observations of sky phenomena to predict future earthly events. Such a belief is very widespread in human history. Unique, unexpected, and imposing events such as solar and lunar eclipses were widely seen as portents of disaster. But more regular celestial events also indicated auspicious or inauspicious times for planting crops, having children, going to war, and so on. In city-state or empire, astronomers and astrologers were employed to identify good and bad omens for the benefit of their nations and their rulers. This happened in ancient Babylonia, China, Greece, and Rome, as well as in pre-Columbian Mesoamerica. Observation was followed by prognosis (interpretation) and then by action (prescription). A good example of celestial divination is found in the early Chinese artifacts known as *oracle bones*. Oracle bone inscriptions (a subset of which relate to astronomical observations) followed a prescribed format: a preface describing the action taken by the diviner was followed by the resulting prognostication and then by a verification describing what actually came to pass.

Predictability did not necessarily detract from the divinatory power of a celestial event. The motions of the five visible planets are regular and predictable, though quite complex. Planetary astrology, which can assign divinatory significance to the cycles of appearance of particular planets as well as to their positions with respect to the background stars and constellations and even to one another, extends back to Babylonian times. The Maya went to extraordinary lengths to reduce the workings of the cosmos to a series of interacting regular cycles. The Dresden Codex—a surviving almanac containing tabulations of the cycles of appearance of the moon, Venus and possibly Mars, and even of lunar eclipses—seems to have been an attempt to make the various motions of the heavenly bodies predictable. In doing so, the Maya reached a remarkable level of mathematical sophistication, yet their ultimate motive seems to have remained divinatory. For the ancient Chinese, on the other hand, predicting celestial events through systematic observations and recording fell within the domain of calendrics. This discipline acted almost independently of, and in some senses in competition with, astrology. Here, once lunar eclipses began to become predictable around the year 0, they lost their divinatory significance.

The idea that people's lives can be permanently influenced by the celestial—and particularly the planetary—configurations at the moment of their birth is one of the defining characteristics of modern astrology, and also one that modern scientists find particularly indigestible. (It should be said that planetary birth charts go far beyond, and are considerably more complex than, the popular perception of horoscopes based solely upon the sun's position within the zodiac at birth ["birth sign"].) Planetary birth charts have their origins in ancient Babylonia and reached an apex in Greece and Rome. When astrologers began to generate horoscopes, this fundamentally altered their role. Instead of observing the skies and waiting for calamitous astronomical events, they were now required to work out planetary configurations at specified moments in the past, something that demanded considerable technical skill. Given the particularity of the idea of the planetary birth chart as against the myriad ways one might envision the influence of the celestial bodies on human lives, it is surprising that the practice has proved so persistent.

Insofar as it maintains that a person's destiny is determined or influenced by the configuration of the heavens at the time of their birth, horoscopic astrology is actually a form of divination. It also introduces awkward issues about free will: if one's fate is already sealed, there seems little point in trying to alter it. However, variants are evident in modern folklore that largely overcome this problem. Thus in the Baltic states of Lithuania and Latvia, well into the twentieth century, it was commonly believed among rural communities that the phase of the moon at birth influenced various aspects of a person's character—their propensity to strength or timidity, cleverness, long life, a joyful or gloomy disposition, and so on. But such characteristics could be modified throughout life by choosing, for example, the correct phase of the moon for weaning, baptism, marriage, or building a house.

The most fundamental connection between objects and events in the sky and those on earth that we might term *astrological* relates to the here and now. Belief in the direct interconnectedness of things is evident among modern indigenous communities and surely extended far back into prehistory. Modern examples include the Barasana of the Colombian Amazon, who understand that the celestial caterpillar causes the proliferation of earthly caterpillars; the Mursi of Ethiopia, for whom the flooding of the river they call *waar* can be determined, without going down to the banks, by the behavior of the star of the same name; and those native Hawaiians who still carry on the ancient practice of planting taro and other crops according to the day of the month in the traditional calendar (i.e., the phase of the moon). The extent to which such mental connections might be considered astrological is arguable, but if our interest is in the practices themselves, and what was going on in the minds of the people who practiced them, then the question is largely irrelevant—as irrelevant as the question of which practices might have a rational basis in modern scientific terms. What one might choose to term *science* and what one might choose to term *astrology* are both rather subjective in the context of an alternative rationality, and the distinction between them is certainly meaningless.

See also:
Archaeoastronomy; Cosmology; Lunar Eclipses; Science or Symbolism?; Solar Eclipses.
Babylonian Astronomy and Astrology; Barasana "Caterpillar Jaguar" Constellation; Chinese Astronomy; Dresden Codex; Hawaiian Calendar; Mesoamerican Calendar Round; Mursi Calendar; Roman Astronomy and Astrology; Star of Bethlehem.
Inferior Planets, Motions of; Superior Planets, Motions of.

**References and further reading**
Aveni, Anthony F. *Conversing with the Planets*, 128–177. New York: Times Books, 1992.
Campion, Nicholas. *The Great Year: Astrology, Millenarianism and History in the Western Tradition*. London: Arkana/Penguin, 1994.
Kasak, Enn. "Ancient Astrology as a Common Root for Science and Pseudo-Science." *Folklore* 15 (2000), available electronically at http://www.folklore.ee/folklore/vol15/ancient.htm.
Ruggles, Clive, Frank Prendergast, and Tom Ray, eds. *Astronomy, Cosmology and Landscape*, 158–166. Bognor Regis, UK: Ocarina Books, 2001.
Ruggles, Clive, and Nicholas Saunders, eds. *Astronomies and Cultures*, 1–31. Niwot, CO: University Press of Colorado, 1993.
Selin, Helaine, ed. *Astronomy across Cultures*, 443–452, 509–553. Dordrecht, Neth.: Kluwer, 2000.
Swerdlow, Noel M., ed. *Ancient Astronomy and Celestial Divination*. Cambridge, MA: MIT Press, 1999.

# Astronomical Dating

Is it possible to date an archaeological site by astronomical means? In theory, the answer is yes, for if we can identify an alignment and the intended astronomical target, measure where the alignment points, and then use modern astronomy to calculate where that event occurred at different times in the past, then we can fit the best date to the alignment. Stellar alignments would seem to be most promising, because the rising and setting positions of most stars change significantly over the centuries owing to precession. The rising and setting position of the sun at the solstices and the moon at the standstill limits also alter with time, although by much less, owing to the slow change in the obliquity of the ecliptic.

In practice, however, astronomical dating is rife with pitfalls. The main one is that we rarely have reliable (e.g., historical) evidence to tell us either that a particular alignment was actually deliberate, or what is was aligned upon in the first place. Where we do have historical evidence, we are likely to know the date fairly accurately already. (A case where we do have some historical evidence, but not as much as we might like, and the alignment does tell us something, is the Venus alignment of the Governor's Palace at the Maya city of Uxmal.)

More often, we are dealing with prehistoric structures where we have no evidence other than the alignments themselves. If we find an alignment that we suspect to be stellar, we can try different stars and different dates to see if any combination fits particularly well. The problem here is that if, say, we are willing to consider the fifteen brightest stars and a five-hundred-year date range, then there is approximately a one-in-three chance that we will be able to find a star and a date to fit *any* alignment. It is frighteningly easy, then, to fit a star and a date fortuitously and made all the easier when we consider that only rarely does one single alignment at a particular monument stand out as the obvious astronomical candidate. In order to be fair with the data—one of the most fundamental methodological principles—we should consider all possible alignments. Added to this is the problem of extinction (the dimming of a star at low altitude due to the earth's atmosphere), which may mean that most if not all of our fifteen stars wouldn't have been visible all the way down to the horizon in the first place. If we are willing to postulate (as some people have done) that structures were aligned upon the point of appearance or disappearance of a star rather than its actual rising and setting point, it increases the chances of our being able to fit a date and star fortuitously to any given alignment (especially at high latitudes, where the astronomical bodies rise and set at a fairly shallow angle). Overall, the potential for circular argument is obvious.

Yet despite these problems, there have been cases where postulated stellar alignments can help in the process of dating a site. One relates to the Pyramids of Giza in Egypt and another to a perplexing later prehistoric sanctuary in Mallorca, Son Mas.

Solar and lunar alignments don't suffer from these problems but have others of their own, mainly that the change over time is very small. The change in the declination of the solstitial sun, for example, amounts to only about one arc minute per century. This means that only if we have good reason to suspect that a solstitial alignment had very high precision, such as that envisaged by Alexander Thom at British megalithic sites, can such an alignment be used to indicate a date accurate to within a few centuries. Even then, variations in refraction may provide an insurmountable obstacle.

See also:
Methodology; Thom, Alexander (1894–1985).
Ballochroy; Governor's Palace at Uxmal; Pyramids of Giza; Son Mas.
Declination; Extinction; Moon, Motions of; Obliquity of the Ecliptic; Precession; Refraction; Solstices; Star Rising and Setting Positions.

References and further reading
Aveni, Anthony F. *Skywatchers,* 102. Austin: University of Texas Press, 2001.
Belmonte, Juan, and Michael Hoskin. *Reflejo del Cosmos,* 155–157. Madrid: Equipo Sirius, 2002. [In Spanish.]
Hoskin, Michael. *Tombs, Temples and their Orientations,* 49–51. Bognor Regis, UK: Ocarina Books, 2001.
Ruggles, Clive. *Astronomy in Prehistoric Britain and Ireland,* 227, 230. New Haven: Yale University Press, 1999.

# Avebury

Situated in Wiltshire, England, a few kilometers south of the modern town of Swindon, Avebury is one of the largest and most impressive Later Neolithic henge monuments in Britain. Built around the middle of the third millennium B.C.E., it measures some 350 meters (1,150 feet) across and surrounds an entire modern village. The outer ditch and bank, when first constructed, were some 9 meters (30 feet) deep and 8 meters (25 feet) high, respectively, and it is estimated that almost a hundred stones encircled the interior. There were also two inner circles, each around 100 meters (300 feet) in diameter. At the center of the northernmost of these was a configuration of three massive stones arranged like three of the four sides of a huge rectangular box, known as the Cove. Running away from two of the four entrances of the henge, to the west and south, were two long stone avenues. Much of the first 800 meters (half mile) of the latter (the Kennet Avenue) survives, but it originally ran for over 2.3 kilometers (1.5 miles), connecting Avebury to a site whose construction had commenced many centuries earlier: a set of concentric timber circles (possibly a roofed building) later replaced by two circles of small stones, known as the Sanctuary.

Surprisingly, given the obvious importance of the monumental landscape in and around Avebury in Later Neolithic times, few astronomical

*Part of the great henge and stone circles at Avebury, Wiltshire, England. (Courtesy of Clive Ruggles)*

alignments have ever been claimed here. Alexander Thom's only interest in the site was geometrical. The British archaeologist Aubrey Burl has suggested that the Cove was aligned upon the most northerly rising position of the moon, but the few other known examples of coves show no astronomical consistency.

**See also:**
Thom, Alexander (1894–1985).
Circles of Earth, Timber, and Stone; Hopewell Mounds.

**References and further reading**
Burl, Aubrey. *Prehistoric Avebury*. New Haven: Yale University Press, 1979.
Gillings, Mark, and Joshua Pollard. *Avebury*. London: Duckworth, 2004.
Malone, Caroline. *Avebury*. London: Batsford/English Heritage, 1994.
Pollard, Joshua, and Andrew Reynolds. *Avebury: Biography of a Landscape*. Mount Pleasant, SC: Tempus, 2002.
Ruggles, Clive. *Astronomy in Prehistoric Britain and Ireland*, 133. New Haven: Yale University Press, 1999.

## Axial Stone Circles

Axial stone circles (ASCs), or more correctly axial-stone circles, are a distinctive type of stone circle found only in the southwestern corner of Ireland, in counties Cork and Kerry. Only one other regional group of stone circles bears any resemblance to the axial stone circles: the recumbent stone circles (RSCs) found several hundred kilometers away on the opposite side of the British Isles, in northeastern Scotland. The Irish axial stone circles (ASCs), like their Scottish counterparts, consist of a single stone placed on its side, with all the other stones being set upright. Without exception, this recumbent stone is found in the southwestern or western part of the circle. But unlike the typical Scottish RSC, the recumbent in Irish ASCs is modest in size. The remaining stones in the circle tend to increase (not decrease, as with the RSCs) in height around to the opposite side, where we find a pair of uprights, generally the tallest, known as *portals,* because it is easy to imagine them forming an entrance into the circle. All of the remaining stones are placed symmetrically in pairs about the axis passing between the portals, across the circle and through the middle of the recumbent stone. This distinctive symmetry means, for one thing, that there is always an odd number of stones in these circles (from five to nineteen). Perhaps more importantly, it also emphasizes the axial orientation. This raises the intriguing question of why it seems to have been so important to orient this axis in a westerly or southwesterly direction (taken from the portals toward the recumbent stone).

The nature of the link between the two groups of monuments is unknown. The Scottish RSCs appear to be older, dating perhaps toward the end of the third millennium B.C.E., while the ASCs appear to extend well into the second millennium. However, secure archaeological dating evidence is still scarce. Postulating a direct link, some authors refer to the axial stone circles as recumbent stone circles; others warn against prejudging the nature of the interrelationship between the two groups and argue in favor of a distinctive term for a distinctive group of monuments.

Despite some systematic differences, the strong resemblance—both in form and orientation—between these two geographically separate groups of stone circles is undeniable. It is perhaps surprising, then, that when we look more closely at the individual orientations of the ASCs we find that—quite unlike the RSCs—they were not guided by particular topographic or astronomical considerations in any consistent way. Unlike their Scottish counterparts, where there is a clear preference for a reasonably distant horizon behind the recumbent, the ASCs are found in a variety of topographic situations, some with high ground rising close behind the recumbent and blocking the distant view completely. And unlike the RSCs, where there is a consistent pattern of orientation in relation to the moon, the Irish sites show no consistent pattern in relation to any astronomical body.

The lesson to be learned here is that a variety of considerations may have influenced the location and orientation of any prehistoric ritual monument, including topographic factors (relationships to prominent natural features such as visible mountains, sacred places, water sources, and so on) as

The axial stone circle at Reenascreena South, Co. Cork, Ireland, viewed along its axis in the direction from the portals towards the recumbent stone. (Courtesy of Clive Ruggles)

well as the heavenly bodies. In different places and different times, different considerations might have held sway according to the dominant tradition. While some elements of prevailing traditions may be carried forward from place to place and through time, others may be modified or be abandoned. The sky forms only one part of a much bigger picture, even where we do have clear and consistent astronomical alignments. Whatever the links between these two very distinctive stone circle traditions at the opposite sides of Scotland and Ireland, it is clear that the strong astronomical (and specifically, lunar) associations of the Scottish RSCs were not transmitted across to their Irish counterparts.

The lack of consistent astronomy among the axial stone circles does not necessarily mean that one or two individual examples could not have been deliberately astronomically oriented, for reasons specific to a particular place and time. The most quoted examples are Drombeg, near Ros Ó gCairbre (Ross Carbery) on the south coast of County Cork, a well-preserved circle whose axis points toward midwinter sunset; and nearby Bohonagh, which faces more or less due west and has been claimed as aligned upon sunset at the equinox. However, the fact that Drombeg is the only one of several dozen large axial stone circles with a clear solstitial alignment must raise the possibility that this could have arisen by chance, and the equinoctial interpretation of Bohonagh is subject to certain difficulties that pertain to all putative equinoctial alignments.

> See also:
> Equinoxes; Solstitial Alignments.
> Drombeg; Recumbent Stone Circles; Stone Circles.
>
> References and further reading
> Burl, Aubrey. *The Stone Circles of Britain, Ireland and Brittany*, 262–273. New Haven: Yale University Press, 2000.
> Ruggles, Clive. *Astronomy in Prehistoric Britain and Ireland*, 99–101. New Haven: Yale University Press, 1999.

# Azimuth

Azimuth means the bearing of a direction—such as that toward a particular point on the horizon from a given place—measured clockwise around from due north. Thus the azimuth of due north is 0°, that of due east is 90°, that of due south 180°, and that of due west 270°. The azimuth of a point one degree to the west of north is 359°. Azimuth is measured in the horizontal plane through the observer. To fully specify the direction of an observed point, one must also specify the vertical angle or *altitude* of the point.

It is possible to specify the position of a star or other object in the sky by its azimuth and altitude, but this will change with time owing to the diurnal motion of the celestial sphere. Even at a given moment, the azimuth and altitude values of a particular star will be different for observers at different places on the earth.

> See also:
> Compass and Clinometer Surveys; Field Survey; Theodolite Surveys.
> Altitude; Celestial Sphere; Diurnal Motion.
>
> References and further reading
> Aveni, Anthony F. *Skywatchers*, 50. Austin: University of Texas Press, 2001.
> Ridpath, Ian, ed. *Norton's Star Atlas and Reference Handbook (20th ed.)*, 4. New York: Pi Press, 2004.
> Ruggles, Clive. *Astronomy in Prehistoric Britain and Ireland*, 22. New Haven: Yale University Press, 1999.

# Aztec Sacred Geography

The Aztec (also known as the Nahua, or Mexica) empire—the last of the great Mesoamerican civilizations—dominated the highlands of central Mexico at the time of the arrival of Hernando Cortés in 1519. It had risen to power following a series of military conquests just a century or two earlier and maintained economic control by extracting tributes in the form of foodstuffs and raw materials (as well as personal services) from conquered populations. The Aztec capital of Tenochtitlan, situated at the center of present-day Mexico City on an island approached along three long causeways, had an estimated population of 250,000. The mass sacrifices to their war and sun god Huitzilopochtli, which took place at the great Templo Mayor in the center of the city, are legendary.

The landscape around the Aztec capital is characterized by strings of mountains and towering volcanoes that surround and dominate flat valleys, creating obvious associations between mountains, clouds, rains, fogs, thunderstorms, springs, and rivers. Under the valley floors and mountain slopes are numerous caves created by ancient lava flows. And before time took its toll and the suburbs of Mexico City spread through the landscape, the terrain was also peppered with magnificent human constructions, both the temples of the Aztecs themselves and the conspicuous remains of earlier temples dating back to the Preclassic period (at least as far as the mid-first millennium B.C.E.). This combination created, in the Aztec mind, a vibrant perceived world strewn with the abodes of powerful spirits: mountains were sources of water and rain; caves were entrances to the underworld; and the huge ceremonial center of Teotihuacan with its enormous pyramids, a thousand years old by this time, was itself seen as a magnificent creation of the gods.

Tributes to the gods had to be made in the appropriate place but also at the correct time. One of the most critical actions was to appease Tlaloc, the god of rain and fertility, and persuade him to send water for the year's maize crop. Petitions to Tlaloc were timed in relation to calendar festivals and often involved the sacrifice of children. Thus on the first day of the month Atlcahualo (in the 365-day calendrical cycle or *xihuitl*), or around the middle of February in the Gregorian calendar at the time of the Conquest, the bodies of sacrificed children were thrown into caves close to mountain sanctuaries, since the water was thought to remain there, inside the mountain, until released by the gods as rain. On the summit of Mount Tlaloc itself, at an elevation of more than 4,000 meters (13,000 feet), was a shrine containing an idol of the rain god where the nobles from Tenochtitlan and adjacent cities converged during a great festival in the month Hueytozoztli, at the end of April. Here, a young boy was sacrificed while, in a complementary ceremony taking place at a nearby lake, a similar fate awaited a young girl, who was appropriately dressed in blue.

Many different calendrically timed rituals such as these, taking place all over the Aztec empire, generated a network of relationships in people's minds between sacred places in the landscape (particularly mountains), the activities that took place there, and the timing of those activities. Evidence suggests that those perceptions were reinforced both by the positioning of temples in the landscape and by solar alignments deliberately built into those temples. It has been proposed that the Templo Mayor was (at least approximately) aligned upon Cerro Tlaloc on the horizon to the east, that the sun would have risen more or less behind that mountain on the equinox, and that two prominent mountains on the eastern horizon from Cerro Tlaloc itself, across the next valley, aligned with sunrise on two important days when mountain ceremonies were taking place on that peak. More recent work suggests that the later phases of the temple were in fact oriented upon sunset at the feast of Tlacaxipehualiztli, which coincided with the Julian vernal equinox in 1519 and was duly recorded by the chronicler Motolinia. Whatever the details in this case, the combination of ethnohistorical accounts relating to the nature and timing of ceremonies, archaeological evidence of votive offerings at sites such as mountain shrines, and archaeoastronomical data on orientations and alignments makes a convincing case that many relationships such as these were real enough in the Aztec mind.

**See also:**
Sacred Geographies.
Cacaxtla; Horizon Calendars of Central Mexico; Mesoamerican Calendar Round.

**References and further reading**
Aveni, Anthony F. *Skywatchers,* 235–244. Austin: University of Texas Press, 2001.
Boone, Elizabeth H. *The Aztec Templo Mayor,* 211–256. Washington, DC: Dumbarton Oaks, 1987.
Carmichael, David, Jane Hubert, Brian Reeves, and Audhild Schanche, eds. *Sacred Sites, Sacred Places,* 172–183. London: Routledge, 1994.
Carrasco, Davíd, ed. *To Change Place: Aztec Ceremonial Landscapes.* Niwot, CO: University Press of Colorado, 1991.
Iwaniszewski, Stanisław, "Archaeology and Archaeoastronomy of Mount Tlaloc, Mexico: A Reconsideration." *Latin American Antiquity* 2 (1994), 158–176.
Ruggles, Clive, and Nicholas Saunders, eds. *Astronomies and Cultures,* 253–295. Niwot, CO: University Press of Colorado, 1993.
Šprajc, Ivan. "Astronomical Alignments at the Templo Mayor of Tenochtitlan, Mexico." *Archaeoastronomy* 25 (supplement to *Journal for the History of Astronomy* 31) (2000), S11–S40.
———. *Orientaciones Astronómicas en la Arquitectura Prehispánica del Centro de México,* 383–405. Mexico City: Instituto Nacional de Antropología e Historia (Colección Científica 427), 2001. [In Spanish.]

# Babylonian Astronomy and Astrology

Ancient Babylonia occupies a pivotal place in the history of modern scientific astronomy. In great part this is due to the conscientious nature of the astronomical observations that were made there and the meticulous way in which they were recorded for generation after generation. In time, the existence of a huge, cumulative database of past observations made possible the development of mathematically based rules for predicting future events. The Babylonian legacy of careful observation and recording combined with mathematical modeling went on to influence developments in ancient Greece and beyond. The other reason ancient Babylonia is so important to modern historians of astronomy is the fortunate choice of medium on which many of the ancient astronomical observations (along with many other documents) were recorded. The method used was to press wedge-shaped marks into smooth, damp tablets of clay using a stylus. Subsequently, the tablets were dried in the sun or fired in kilns for permanence. Clay tablets do not tend to disintegrate with time like (say) parchments or papyri and are unaffected by subsequent fire, so they frequently survived the looting or destruction of buildings and other cataclysmic events of history. The Babylonian cuneiform script was deciphered in the nineteenth century. In short, many high-quality records have survived, and they can be read.

The ancient city-state of Babylon lay some 90 kilometers (55 miles) south of modern Baghdad. Its power and influence came to cover all of lower (southern) Mesopotamia—the region of modern Iraq between the Tigris and Euphrates rivers down to the Persian Gulf—in the eighteenth century B.C.E., after which it followed a turbulent history under a succession of dynasties until its conquest by the Persians in 539 B.C.E. Subsequently, Babylon became part of greater empires: the Persian until 331 B.C.E., when it was conquered by Alexander the Great; then (after Alexander's death) the Seleucid Empire; it ultimately fell to the Romans in 63 B.C.E. The latest known cuneiform tablet dates to C.E. 75.

Most of the written evidence that comes down to us is in the form of clay tablets from the Seleucid period from 311 B.C.E. onwards. Those including astronomical data are of various types: astronomical diaries containing nightly observations; records of sightings of astronomical events such as planetary conjunctions and eclipses; and (increasingly with time) almanacs containing predictions of the length of the month, the positions of the planets among the fixed stars, and many other things. These documents demonstrate beautifully how the systematic accumulation of carefully recorded passive observations led in time to the ability to predict using mathematical models. One thing that made this development possible was the Babylonian system for representing numbers: like ours it used a fixed base, but instead of ten, the base was sixty. In other words, each "digit"—itself a set of strokes representing tens and ones—represented a value from zero to fifty-nine, with subsequent digits representing "units," multiples of sixty, multiples of sixty times sixty, and so on. (The Maya, in contrast, used a base of twenty.)

In the Babylonian calendar, the new day began at sunset and the new month began when the thin crescent moon was first sighted in the evening sky after sunset. Back in the third millennium B.C.E., two calendars seem to have existed in parallel: an "ideal" calendar that was theoretical, with each month containing thirty days, and a common calendar based on actual lunar observations. The first of these calendars is used in early clay tablets that are essentially business documents: this is hardly surprising, since people had to agree on the day something had been signed or a commitment had been made, and this could not be dependent upon the vagaries of whether or not the new crescent moon had been seen (which could be a matter of some dispute, especially if it had been cloudy in certain places at critical times). This civil calendar needed to be the same all over Babylonia and not subject to disputes between different local officials.

Nonetheless, it seems that the actual astronomical calendar (rather than the abstract "ideal" calendar) was used for civil purposes from the second millennium B.C.E. onwards, despite the attendant problems. Gradually, the analysis of accumulated observations of first sightings of the new crescent moon enabled Babylonian astronomers to develop mathematical "rules of thumb" that permitted month lengths to be accurately and reliably predicted. Furthermore, by about the fifth century B.C.E., the nineteen-month (Metonic) intercalation cycle had been discovered and established. It provided a rule whereby the additional (intercalary) months needed to keep the lunar calendar in step with the seasonal (solar) year could be added in a mechanical, deterministic, yet reliable way. Astronomers no longer had to depend upon independent astronomical observations, such as the heliacal rising of Sirius.

One of the most remarkable consequences of the Babylonian astronomers' attention to detail and the sheer volume of records that they accumulated over many generations was the recognition of the so-called Saros cycle: it described the fact that if a lunar eclipse occurs, others will tend to follow at regular intervals of eighteen years and eleven days for many centuries thereafter. This discovery was no mean feat, since approximately forty different Saros cycles run simultaneously, and the conditions for visibility of any particular eclipse vary considerably—many are completely invisible from any given location on the earth (if they occur during daytime, for example). They simply cannot be revealed by casual observations of lunar eclipses over a few years or even decades, even if clear skies were permanently assured.

In view of its undoubted influence on the development of modern astronomy, it is tempting to view the Babylonian tradition as the birthplace of scientific investigation of the heavens. But this would be highly misleading. The main motivation for the Babylonians' intense interest in the skies was astrological. Different days in the common calendar were associated with different prognostications, and by the seventh century B.C.E., scholars were advising the Assyrian king of the calendrical omens. A particular concern was the issue of whether the forthcoming month would have twenty-nine or thirty days. Other concerns included the length of day and night, the heliacal events of stars, the positions of the planets, and of course the occurrence of solar and lunar eclipses. One of the most important series of celestially related clay tablets has become known as the *Enuma Anu Enlil*, meaning "When [the great gods] Anu and Enlil . . ." (the first line of the text). This document, which dates to around the end of the second or the beginning of the first millennium B.C.E., runs to seventy tablets, and multiple copies exist. It contains around seven thousand omens accumulated from past experience and provides advice as to whether certain celestial configurations and events—signals from the gods—would indicate their pleasure or displeasure.

This was not astrology in the sense that celestial configurations were perceived as the direct cause of terrestrial events (although this did become a widespread philosophy from the fourth century B.C.E. onwards in Hellenistic Greece) but rather that they provided portents of events that could then, if necessary, be averted by taking appropriate action. In this, astronomical predictions were used along with a variety of other forms of divination.

Ancient Babylonia was also the birthplace of modern horoscopic astrology, or at least the earliest known example of the belief in the predictive capabilities of charts recording planetary positions at the moment of a person's birth. (Actually, the horoscopes of modern popular astrology represent a revival of this belief rather than any sort of continuity of tradition.) An important prerequisite was the division of the zodiac (through which the planets move) into twelve regions of equal size. Birth charts began to appear in the second half of the first millennium B.C.E. and represented a move away from the astrologers having to watch the skies passively, waiting for omens to appear, to the more active pursuit (performed on demand) of calculating where the planets would have been among the stars at a particular time. It also represented a shift away from observational astronomy toward intensive mathematics. It is ironic, in view of the way in which modern astrology is seen as the very antithesis of modern science—irrational and unscientific—that this astrological innovation in Babylonian times necessitated making full use of the most up-to-date scientific knowledge and methods that had been developed by this time.

From the late third century B.C.E. onward, two fundamentally different schools of thought emerged for generating predictions from the extensive records of existing observations. These seem to have coexisted throughout the final few centuries of ancient Babylon (until the late first century C.E.), to judge by two types of works—mathematical ephemerides and almanacs known as Goal Year Texts—that were evidently produced in parallel. The ephemerides represent the height of Babylonian scientific achievement, using sophisticated mathematical models to predict phenomena of the moon and planets with remarkable accuracy. The Goal Year Texts, on the other hand—each one a sort of astrological handbook for a given year—seem to represent an independent tradition of prediction based upon repeating cycles that had been discovered by studying the existing diaries of observations (the Saros cycle was one of these).

There is a great deal still to be learned about the nature of astronomical and astrological knowledge in ancient Babylonia and the social context in which it operated. Though about three thousand fragments of clay tablets containing astronomical information are currently known to exist, a huge amount of basic data simply remains unexplored. There are tens of thousands of fragments of clay tablets in the British Museum alone, many tens of thousands more in other museums around the world, and untold quantities still buried under the ground in modern Iraq. Since a sizeable proportion of the clay tablets that have been studied contain astronomical information, there is every reason to expect the same to be true in the future. And while many of the museum specimens are of unknown provenance, only eventually having found their way into the public domain after progressing along tortuous routes, some of those still waiting to be discovered may be excavated in a context that will yield valuable archaeological information about their broader function and significance.

See also:
Astrology; Eclipse Records and the Earth's Rotation; Lunar and Luni-Solar Calendars; Lunar Eclipses; Mithraism; Solar Eclipses.

Fiskerton; Maya Long Count.
Heliacal Rise; Zodiacs.

**References and further reading**

Aaboe, Asger. *Episodes from the Early History of Astronomy,* 24–65. New York: Springer, 2001.

Hunger, Hermann, and David Pingree. *Astral Sciences in Mesopotamia.* Boston and Leiden: Brill, 1999.

Neugebauer, Otto. *The Exact Sciences in Antiquity,* 92–138. Princeton, NJ: Princeton University Press, 1951. (Second edition published 1957 by Brown University Press, Providence, RI, 97–144; further corrected edition published 1969 by Dover, New York, 97–144.)

Rochberg, Francesca. *The Heavenly Writing: Divination, Horoscopy, and Astronomy in Mesopotamian Culture.* Cambridge: Cambridge University Press, 2004.

Steele, John. *Observations and Predictions of Eclipse Times by Early Astronomers.* Dordrecht, Neth.: Kluwer, 2000.

Thurston, Hugh. *Early Astronomy,* 64–81. Berlin: Springer-Verlag, 1994.

Walker, Christopher, ed. *Astronomy before the Telescope,* 42–67. London: British Museum Press, 1996.

# Ballochroy

Ballochroy is one of many hundreds of small megalithic monuments found in western Britain. The casual visitor is unlikely to be greatly impressed at the sight of it: a row of three standing stones, one broken off, occupying an unassuming location in a field behind a barn at Ballochroy farmhouse on the west coast of the Kintyre peninsula, Argyll, Scotland. There is, however, a good view over the coast to the west, and this view is key to understanding its significance, for this modest monument is one of the earliest and most famous examples of a megalithic "observatory" put forward by Alexander Thom, during the 1950s. It assumed a central place in the controversies that raged for more than two decades over Thom's theories.

Ballochroy encapsulates Thom's idea that prehistoric Britons used features on distant horizons as astronomical foresights in order to observe and record the motions of the sun and moon to remarkable precision. The central stone at Ballochroy has a broad, flat face oriented across the alignment that points northwest, directly at the slopes of Corra Bheinn, a mountain on the Island of Jura some 31 kilometers (19 miles) away. On the summer solstice, the tip of the sun's disc twinkled down the indicated slope; a couple days before or after, when the sun's path was just slightly lower, it would not have been visible. The row of three stones itself points southwestward toward a small island called Cara Island about 12 kilometers (7 miles) away. Close to the winter solstice, the tip of the sun's disc gleamed to the right of the island as it set; on the solstice itself this would not have been the case.

The best evidence supporting the theory that Ballochroy was a "solar observatory" is that there are not one but two foresights at the same site, marking the setting sun at both of the solstices. Surely such a coincidence could not have arisen by chance? And yet many critics raised doubts. One of the other standing stones in the row also has a broad, flat face pointing northwestwards, but this one points at a different mountain. And the alignment along the row is very broad, encompassing not only the right-hand end of Cara Island but also its left-hand end and central peak as well. If we are fair with the data, then we should admit the existence of at least a few other candidates for foresights that are equally plausible but have no ready astronomical explanation.

On the other hand, the fact that the claimed alignments are so precise means that we can use the slow change in the setting path of the solstitial sun century by century (due to the gradual change in the obliquity of the ecliptic) to calculate the best-fit dates for the two foresights and see if they coincide. The result is stunning: the best-fit date is pretty much the same for both foresights, around 1600 B.C.E., and this is a date that is certainly plausible archaeologically.

Yet archaeological and historical evidence has all but destroyed the idea that Ballochroy is a precise solstitial observatory. There is, and was, more to this monument than three stones in a row. A drawing by the antiquarian Edward Lhuyd in 1699 clearly shows three cairns in line with the three stones, one of them so large that it would have blocked the view to Cara Island, together with a fourth stone beyond. The burial cist (a box-shaped tomb with four side slabs) that was originally covered by this mound is still visible at the site, although the mound and the other features sketched by Lhuyd have been destroyed. The remains of this cairn were excavated in the 1960s, and the archaeological evidence indicates that it is very unlikely to have been constructed as late as the mid-second millennium B.C.E. If the stones were erected at the date indicated by the two alignments, then one of them was always blocked by the cairn. Though a few have argued that people might have stood atop the cairn to make the observations, this is special pleading. Being realistic, we are forced to conclude that the solstitial alignment of the stone row at Ballochroy, if intentional at all, was only of a low precision: it was not an observing instrument.

The example of Ballochroy remains important to archaeoastronomers because it demonstrates the dangers of enthusiastically endorsing alignments that seem to fit an astronomical theory while ignoring other possibilities because they don't. This isn't being fair with the evidence. It also shows the importance of the broader context of archaeological and, where we have it, historical evidence.

See also:
Astronomical Dating; Megalithic "Observatories"; Methodology; Thom, Alexander (1894–1985).
Short Stone Rows.
Obliquity of the Ecliptic; Solstices.

**References and further reading**
Burl, Aubrey. *Prehistoric Astronomy and Ritual*, 7–11. Princes Risborough, UK: Shire, 1983.
Ruggles, Clive. *Astronomy in Prehistoric Britain and Ireland*, 19–25. New Haven: Yale University Press, 1999.

# Barasana "Caterpillar Jaguar" Constellation

The Barasana are a group of forest-dwellers in the Colombian Amazon. They survive by a mixture of fishing, hunting, and gathering, supplemented by slash-and-burn agriculture. June, July, and August (in our calendar) are difficult months for them, since their regular food sources are scarce. But at this time of year pupating caterpillars fall down from the trees and provide a much-needed source of nutrition. The date this happens coincides with the time when the Caterpillar Jaguar, a constellation regarded by the Barasana as (among other things) the Father of Caterpillars, rises higher and higher in the sky at dusk. Since the Caterpillar Jaguar is formed by stars in our constellations of Scorpius and Cetus, it is easy to explain, from a Western perspective, the association that the Barasana observe: the time of year following the acronical rising of Scorpius and Cetus happens to be the time of year when several species of caterpillar pupate.

The Barasana can explain the phenomenon too. They say that the Father of Caterpillars, by rising higher and higher in the sky at dusk, is directly responsible for the increasing numbers of earthly caterpillars. In the Barasana worldview, there exists a direct correspondence between two entities that from a Western perspective are quite distinct: the position of a constellation in the sky and the behavior of terrestrial creatures.

This is a good example of how a non-Western worldview can make sense of the cosmos by drawing direct connections between things that we would regard as quite unconnected. It is also a good example of a way of understanding the world that (at least potentially) has predictive capability: when the constellation rises, then the caterpillars will start to appear. This characteristic of Barasana cosmology, some would argue, means that we could regard it as a rudimentary form of science. However, it might also be described by others as astrology in the sense that events of earth are perceived to be determined by, or at least directly linked with, the configuration of objects in the sky.

**See also:**
Astrology; Cosmology; Science or Symbolism?
Heliacal Rise.

**References and further reading**
Aveni, Anthony F., and Gary Urton, eds. *Ethnoastronomy and Archaeoastronomy in the American Tropics*, 183–201. New York: New York Academy of Sciences, 1982.
Hugh-Jones, Stephen. *The Palm and the Pleiades: Initiation and Cosmology in Northwest Amazonia*. Cambridge: Cambridge University Press, 1979.

## Basque Stone Octagons
*See* Saroeak.

## Beltany

The very name of this large stone circle in County Donegal, northwest Ireland, is of astronomical interest because it suggests an association with the Celtic calendrical festival of Beltaine, or Bealltaine, which is associated with the mid-quarter day of May 6. A conspicuous alignment at the site backs up the idea of such an association. If one stands by the tallest stone in the circle, which is on the southwestern side, then one can see on the far side a distinctive triangular stone covered with several cup marks. This stone is aligned with a point on the horizon where the sun rises on May 6. As a consequence, the cup-marked stone is often dubbed the Beltane Stone.

This combination of a 5,000-year-old astronomical alignment and a more recent mythical tradition that may have given rise to the name, and their possible linkage through Celtic calendrical festivals, is an attractive possibility. However, we cannot be sure. Several other astronomical alignments have been postulated at the circle, showing that the Beltane alignment may not be the most convincing. For example, there is a small hill summit at the relevant point on the horizon, but a much more conspicuous hill, Binnion Hill, appears just to the left, where the sun rises on about May 21. The idea of a precise Celtic calendar and its linkage back to an earlier "megalithic" calendar is also problematic in a number of ways. The Beltane alignment at Beltany may, after all, just be a coincidence.

Nonetheless, cup-marked stones do seem to be associated with astronomical alignments elsewhere—for example, at the Scottish recumbent stone circles—so the idea may not be so far-fetched that this particular springtime sunrise alignment was significant, for some reason, back in the Neolithic.

**See also:**
Celtic Calendar; "Megalithic" Calendar; Mid-Quarter Days.
Recumbent Stone Circles; Stone Circles.

**References and further reading**
Burl, Aubrey. *The Stone Circles of Britain, Ireland and Brittany*, 83–86. New Haven: Yale University Press, 2000.

# Borana Calendar

The Borana are a group of nomadic cattle herders inhabiting an area that lies partly in southern Ethiopia and partly in northern Kenya. Their calendar, which regulates both subsistence and ceremonial activities, is of vital importance to them and is regulated by experts on sky observation known as *ayantu*. The Borana calendar is conventional in some respects. It is based upon the phase cycle of the moon, with each new month signaled by the appearance of the new, crescent moon in the evening sky. Both months and days are named, and the *ayantu* generally know the current month and day from memory. But if there is any uncertainty they make observations of the moon and stars. Seven months of the year are identified by a star or group of stars that rises side by side with the crescent moon when it first appears in the evening sky. The seven stars and asterisms used are Triangulum, the Pleiades, Aldebaran, Bellatrix, Orion's belt and sword, Saiph, and Sirius. During the other half of the year, at least some of the *ayantu* watch to see at what phase the moon is side by side with the first star group in the list, Triangulum (although others may use different asterisms for this purpose). "Side by side" is a concept that only makes sense because the Borana live very close to the equator. This means that all celestial objects rise vertically somewhere in the eastern sector of the horizon and set vertically in the western sector. On any particular night, all the stars and asterisms that rise together continue to climb up into the sky "side by side."

The Borana's obsession with Triangulum is interesting in itself. It is often argued that the brightest stars and asterisms are likely to have been the most important to prehistoric peoples. Thus, people who seek stellar alignments at prehistoric monuments tend to examine the brightest stars. The Borana show that this is not necessarily the case. Triangulum is a relatively faint group of three stars, yet it is of prime importance to them.

What renders the Borana calendar highly distinctive, and confused anthropologists and archaeoastronomers for over a decade, is that day names do not start afresh with each month but progress in an endless cycle of twenty-seven names. Since the phase-cycle (synodic) month is between twenty-nine and thirty days long, two or three days appear twice in each month, once at the beginning and again at the end. The reason for this seemingly incomprehensible practice becomes clear when we consider the monthly passage of the moon through the stars. Night by night, the moon moves slowly relative to the stars, completing a circuit through the stars in 27.3 days. This is known as the sidereal month. At the equator, this means that if the moon rises on one night side by side with a certain set of stars, by the next it will have progressed and rise level with a different set. After 27 nights, it will be rising side by side once again (more or less) with the original set of stars. The Borana day is determined by the stars that the moon is level with on a given night, regardless of its phase. Twenty-seven day names suffice, with one being repeated in roughly every third cycle as observations dictate.

The fact that the *ayantu* ignore the sun in determining the time of year seems surprising until we realize the implications of their location close to the equator. Here, the sun's behavior changes little through the year: the length of the day is effectively the same all year round, and the annual swing in the sun's rising or setting position between the solstices is relatively small. On the other hand, the vertical motion of the celestial bodies at night makes it natural to notice which stars are level with the moon, especially just after the moon has risen or before it sets.

See also:
Lunar and Luni-Solar Calendars.
Mursi Calendar; Namoratung'a.
Lunar Phase Cycle; Solstices.

**References and further reading**
Bassi, Marco. "On the Borana Calendrical System." *Current Anthropology* 29 (1988), 619–624.
Legesse, Asmerom. *Gada: Three Approaches to the Study of African Society.* New York: Macmillan, 1973.
Ruggles, Clive. "The Borana Calendar: Some Observations." *Archaeoastronomy* 11 (1987), S35–53.
———, ed. *Archaeoastronomy in the 1990s,* 117–122. Loughborough, UK: Group D Publications, 1993.
Tablino, Paul. "The Reckoning of Time by the Borana Hayyantu." *Rassagna di Studi Ethiopici* 38 (1996), 191–205.

## Boyne Valley Tombs

The northern banks of the river Boyne in County Meath, Ireland, at a spot called the Bend of the Boyne, are the site of a remarkable concentration of Neolithic tombs dating to the late fourth millennium B.C.E. These include three large passage tombs: Newgrange, Knowth, and Dowth.

At Newgrange a single 19-meter (60-foot) passage leads in from an entrance on the southeast side of a huge mound (some 80 meters [260 feet] across), famously oriented so that the sun shines directly along it, lighting up the central chamber, just after dawn on days around the winter solstice. Each of the other major tombs has two passages. Those at Knowth run deep into the interior from entrances opposite each other, facing more or less due east and west respectively. Dowth has passages of different lengths, both in its southwestern quarter. The longer one runs in from an entrance in the west-southwest, the shorter one from the southwest.

Apart from Newgrange, only the shorter passage at Dowth could conceivably be interpreted as aligned upon the solstitial sun, and the target here is winter solstice sunset rather than sunrise. However, it has also been pointed out that the three monuments, taken together, might have a broader calendrical significance. Each of their five passages is oriented close to sunrise or sunset on a solstice, equinox, or mid-quarter day: that is, upon one of the dates obtained by dividing the year into eight exactly equal parts starting at either of the solstices. This division is reflected in the traditional Celtic calendrical festivals, which fall close to these dates. It is also reflected in the precise alignments upon sunrise or sunset on these dates at many British megalithic sites that were claimed by Alexander Thom in the mid-twentieth century and seemed to imply that a calendar dividing the year into eight equal parts had been in extensive use throughout Neolithic Britain.

However, as later reassessments showed, the archaeological and statistical evidence simply does not support Thom's "megalithic" calendar, and the idea of any all-pervasive Celtic calendar in later, Iron Age times has proven highly questionable. Astronomical and calendrical practices throughout later prehistory were much more variable and localized. Furthermore, the supposedly calendrical alignments of the Boyne valley tomb passages (Newgrange aside) are not exact. All this suggests that the Boyne valley tombs may have done no more than to fit within a broad general pattern of orientation practice that prevailed locally and perhaps extended to Irish passage tombs further afield. This much would accord with what one finds among local groups of later prehistoric tombs and temples throughout western Europe. Indeed, we need look no further than the sixteen smaller passage tombs that surrounded Knowth in order to see a broad range of passage orientations varying from northeast around through east and south to southwest, but avoiding the northwest and north.

Yet it remains possible that more particular practices relating to the seasons and the skies may show up archaeologically in other ways, as "one-off" phenomena, such as the solstitial hierophany at Newgrange itself. Thus, one of the decorated kerbstones at Knowth resembles a sundial, while another contains a cyclic arrangement of twenty-nine circles and crescents that could be a representation of the phase cycle of the moon.

See also:
Celtic Calendar; Equinoxes; "Megalithic" Calendar; Mid-Quarter Days; Solstitial Directions; Thom, Alexander (1894–1985).
Newgrange; Prehistoric Tombs and Temples in Europe.
Obliquity of the Ecliptic; Solstices.

**References and further reading**
Eogan, George. *Knowth and the Passage-Tombs of Ireland*. London: Thames and Hudson, 1986.

O'Kelly, Michael. *Newgrange: Archaeology, Art and Legend*. London: Thames and Hudson, 1982.
Prendergast, Frank, and Tom Ray. "Ancient Astronomical Alignments: Fact or Fiction?" *Archaeology Ireland* 16 [2], 60 (2002), 32–35.
Ruggles, Clive. *Astronomy in Prehistoric Britain and Ireland*, 129. New Haven: Yale University Press, 1999.
Stout, Geraldine. *Newgrange and the Bend of the Boyne*, 40–57. Cork: Cork University Press, 2002.
Waddell, John. *The Prehistoric Archaeology of Ireland*, 59–65. Galway: Galway University Press, 1998.
Whittle, Alasdair. *Europe in the Neolithic*, 244–248. Cambridge: Cambridge University Press, 1996.

## Brainport Bay

The landscape of much of the western highlands of Scotland is one of mountains dissected by long narrow lakes, similar to Norwegian fjords, extending inland from the coast. Loch Fyne in Argyll is one of the longest of these, stretching for more than 50 kilometers (30 miles). On its shores, about halfway along, is one of the most convincing candidates for a prehistoric solar observation platform.

The site known as Brainport Bay or Minard occupies a spectacular setting on a low eminence jutting out into the lake. It achieved notoriety in the late 1970s when, following excavations by the local archaeological society, a curious alignment of artificial structures came to light, dating to the Bronze Age in the mid second millennium B.C.E. It includes a *back platform,* a flat area built around a natural rock outcrop; two large boulders standing on end, known as the *observation boulders,* with a flat cobbled area between; and a *main platform,* again artificially enhanced. The main platform contains what appears to have been an extraordinary sighting device. Two slender standing stones a little over 1 meter (3 feet) tall stood upright in clefts between rocks (they had fallen and been moved but were re-erected after the excavations had located the stone-holes). As viewed from the observation boulders, these stones would have lined up, with the precision of a rifle barrel, upon a notch between two mountains on the only distant horizon visible from the site, some 45 kilometers (30 miles) along the lake to the northeast.

In the Bronze Age, the midsummer sunrise would have occurred just a little to the left of this notch. As the sun moved steadily upwards and to the right, it would have crossed the exact alignment, passing just above the notch, a few minutes later. Furthermore, the appearance of the sun in the alignment would not have been confined to the solstice itself, but would have been equally impressive for a period of several days, like the famous midwinter alignment at Newgrange. This means that unreliable weather is unlikely to have prevented at least one successful sighting of the event in most years.

As far as the evidence goes, this is one of the most convincing cases of a monumental solstitial alignment constructed in the Bronze Age in the whole of Britain and Ireland. Compared with the possible observing platform at Kintraw, for example, the structures at Brainport Bay show clear evidence of human activity and enhancement of the natural features, but not of habitation or burial. It is a prime candidate, in other words, for a ceremonial or ritual site. We are still a long way from knowing what the solstice meant to the people who, we assume, came and observed or worshipped here at dawn on or around the longest day of the year. It is tempting, though, to imagine a priest or two, or perhaps a couple of other people of special social standing, observing the event from the special vantage of the observation boulders while a larger audience had to content themselves with standing on the back platform up the hill behind. On the basis of current evidence, though, this remains speculative.

Despite this relatively straightforward and impressive alignment, the site nevertheless became controversial. This is because of additional suggestions by Euan MacKie, the archaeologist who had originally excavated at Kintraw and was also responsible for bringing Brainport Bay to the attention of a wider audience. One of MacKie's ideas was that the Brainport Bay alignment was not imprecise and not just used for ceremonial purposes: it was a precision observing instrument for determining the exact date of the solstice. It is difficult to determine the exact date of the solstice, because the change in the rising position of the sun on days near to the solstice is only minuscule. The horizon notch that the "rifle barrel" stones align upon actually marked the first gleam of sunrise about fifteen days before and after the solstice. The prehistoric observers, according to MacKie, would have determined the solstice by counting the number of days between the two times when the sun rose in the notch and halving the difference.

The problem with this argument is that it cuts both ways. Counting in its favor is the fact that if prehistoric people *were* trying to determine the solstice, then using a horizon notch displaced from the actual solstitial rising position of the sun and halving the difference would have been a very sensible way to do it, because they were focusing on a place on the horizon where the day-to-day change in the position of sunrise was easily perceptible. Counting against the idea, though, is the fact that any notch within a short distance of the solstice is susceptible to a similar argument, since we do not have direct knowledge about the interval of time around the solstice that would have to be halved. In mountainous country there are lots of notches, and the chances of a fortuitous alignment are very considerable. The upshot is that on the basis of the evidence available to us we can only return one verdict—and fortunately it is one that is allowed in the Scottish legal system—namely, "not proven."

MacKie went further still. There are a variety of other signs of human activity in the immediate area around the Brainport Bay alignment, including a 3.4 meters (11 feet) long fallen standing stone and cup markings on rock outcroppings. The whole area, MacKie suggested, formed a "calendrical complex," with several precise alignments upon sunrise or sunset on epoch dates in the "megalithic" calendar. But critics have pointed out a number of difficulties, particularly the fact that the alleged calendrical alignments at Brainport Bay are marked in very different ways: across a platform to a pyramid-shaped stone; from a standing stone to a cup mark on a rock; and along the line of a cup-and-groove mark (but not along a similar cup-and-groove mark on another stone). A great many potential alignments of various types exist here, which implies that the calendrical ones could easily have arisen fortuitously.

**See also:**
"Megalithic" Calendar; Methodology; Solstitial Alignments.
Kintraw; Newgrange.
Obliquity of the Ecliptic; Solstices.

**References and further reading**
Ruggles, Clive. *Astronomy in Prehistoric Britain and Ireland*, 29–34. New Haven: Yale University Press, 1999.
———, ed. *Records in Stone: Papers in Memory of Alexander Thom*, 213–224. Cambridge: Cambridge University Press, 2002.

## Brodgar, Ring of

The Ring of Brodgar is one of the most impressive, and one of the two most northerly, examples in Britain of a henge monument containing a stone circle. The stone circle is over 100 meters (330 feet) across, standing within a ditch over 120 meters (400 feet) in diameter (the outer bank has now almost completely disappeared). Several of the stones rise to a height of over 3 meters (10 feet). Located on the largest of the Orkney Islands (Mainland Orkney) off the northern coast of Scotland, it occupies a magnificent situation toward the end of a long, narrow isthmus projecting into a large freshwater lake (the lochs of Harray and Stenness). Just 1.5 kilometers (1 mile) away are the even taller Stones of Stenness, another henge and stone circle in a matching situation: Stenness occupies a shorter isthmus on the other side of a 200 meters (700 feet) gap that one can now cross by a road bridge. The Ring of Brodgar was clearly of paramount importance in the prehistoric landscape in around 3000 B.C.E., when it was probably constructed. Standing sentinel over what was almost certainly an important route across the island, and clearly visible from all around, it occupied a position of great power and influence. That it remained a conspicuous ancestral monument revered for many generations seems evident from the numerous

*The Ring of Brodgar, Orkney Islands, Scotland. (Corel Corp.)*

round cairns that were built around it in the Bronze Age, more than a millennium later.

The site achieved astronomical notoriety in the 1970s when Alexander Thom and his son Archie first published the theory that it was a sophisticated lunar observatory. According to the Thoms, various alignments at the site, mostly between the outlying cairns, were deliberately aligned upon horizon foresights that marked limiting rising and setting positions in the moon's complex cycles. So precise were the alleged alignments—to less than two arc minutes—that the Thoms were forced to conclude that programs of observation lasting many generations must have been needed.

The Ring of Brodgar, more than any other single monument, demonstrates the dangers of over-enthusiastic interpretation in archaeoastronomy. Time has dealt harshly with the "lunar observatory" hypothesis. The problem lies in the selection of data: the selection of alignments, the selection of foresights, and the selection of lunar "targets" deemed significant are all highly questionable. The four foresights, for example, are of varying degrees of conspicuity, varying from the crest of the impressive distant cliffs of Hellia 13 kilometers (8 miles) away on the island of Hoy, down to an almost imperceptible blip on the horizon at a place called Ravie Hill. There is no reason (other than their lunar potential) to pick out the four chosen from the many dozens, if not hundreds, of other possibilities.

Above all, the example of Brodgar shows the paramount importance of a field methodology that ensures the fair selection of data and its critical assessment. It is all too easy, even for a professional academic working outside his or her own discipline, to end up—unwittingly—merely picking out the evidence that seems to support a favored theory while ignoring the rest.

**See also:**
Megalithic "Observatories"; Methodology; Thom, Alexander (1894–1985). Circles of Earth, Timber, and Stone.
Moon, Motions of.

**References and further reading**
Burl, Aubrey. *The Stone Circles of Britain, Ireland and Brittany*, 210–214. New Haven: Yale University Press, 2000.
Renfrew, Colin, ed. *The Prehistory of Orkney*, 118–130. Edinburgh: Edinburgh University Press, 1993.
Ruggles, Clive. *Astronomy in Prehistoric Britain and Ireland*, 63–67. New Haven: Yale University Press, 1999.
Thom, Alexander, and Archibald S. Thom. *Megalithic Remains in Britain and Brittany*, 122–137. Oxford: Oxford University Press, 1978.

## "Brown" Archaeoastronomy

This term denotes an approach in archaeoastronomy that is not primarily focused upon alignment studies but is concerned with a much broader range of types of evidence, such as written documents or ethnohistorical accounts. This approach emerged in North America during the 1970s, particularly in the context of studies of astronomy in native North America and pre-Columbian Mesoamerica. It involved attempts to integrate approaches from a range of humanities and social science disciplines such as history, cultural anthropology, art history, ethnography, folklore studies, history of religions, and many more. This broad, multidisciplinary approach contrasted starkly with the pursuit of statistical rigor that absorbed most Old World archaeoastronomers at the time.

**See also:**
Alignment Studies; Archaeoastronomy; "Green" Archaeoastronomy.

**References and further reading**
Aveni, Anthony F., ed. *World Archaeoastronomy*, 3–12. Cambridge: Cambridge University Press, 1989.

## Bush Barrow Gold Lozenge

The Bush Barrow is one of a number of spectacular burial mounds in the vicinity of Stonehenge in southern England, many of them clustering along the low ridges that surround the famous site itself. These round barrows, the burial places of prominent Bronze Age chieftains in the Wessex region, were built around 2000 B.C.E., several centuries after the main construction activity had ceased at Stonehenge. Few doubt that their position, at the boundary between the "lived-in" landscape and the low bowl of the sacred landscape centered upon the ancestral place of power, in clear sight of both, was in itself an expression of the power and influence of the dead chiefs.

Rich assemblages of grave goods accompanied the chieftains to the afterworld. One of the most impressive was the Bush Barrow lozenge, a magnificent diamond-shaped plate of thin sheet gold 18 centimeters (7 inches) across. Finely decorated with distinctive patterns of incised lines, it is generally interpreted as an ornamental breast plate—an imposing mark of status. In the 1980s, Archie Thom and two colleagues claimed that the lozenge was a sophisticated astronomical observing instrument. By holding the plate horizontally and lining it up in the correct orientation, the various markings could have been used to indicate the sunrise and sunset positions on significant epoch dates in the "megalithic calendar" that Archie Thom's father, the Scottish engineer Alexander Thom, had proposed. It could be used in a similar manner to determine significant rising and setting points of the moon. The claim appeared to vindicate his father's theories.

But attractive as the idea seemed, problems emerged when it was examined in detail. For one thing, there would be various practical difficulties using such a device, not least in determining its correct orientation. The most serious problem, however, is that the directions supposedly marked by the patterns on the lozenge do not really fit very well. Several of the alignments actually fall between the markings, while many of the markings do not fit any of the alignments at all. The fact that the markings actually form a regular and symmetrical design (while the astronomical targets are not regular) argues strongly in favor of their being purely decorative rather than astronomically functional. And as if this were not enough, other lozenges exist in nearby burials with a similar form of decoration but different dimensions. Why should only this one function additionally as a calendrical device?

By the 1990s it had become clear that the other evidence supporting the idea of a "megalithic calendar" did not stand up to critical evaluation. However, the most direct blow to the calendrical interpretation of the Bush Barrow lozenge was delivered, ironically, when the historian John North attempted to interpret the lozenge independently as part of his own astronomical interpretations of prehistoric monuments and artifacts in southern England. Vehemently criticizing the existing astronomical interpretation, he proposed an equally complex but entirely different one of his own, thereby showing how easy it was to do so and in the process undermining confidence in both theories.

The example of the Bush Barrow lozenge demonstrates very clearly the dangers of trying to mould the evidence to fit a favored theory rather than letting the evidence speak for itself. Most likely the lozenge was simply a decorative artifact. It is impressive nonetheless and can be recognized as a considerable technological achievement without recourse to sophisticated calendars and astronomy.

**See also:**
"Megalithic" Calendar; Methodology; Thom, Alexander (1894–1985).
Nebra Disc; Stonehenge.

**References and further reading**

Darvill, Timothy, and Caroline Malone, eds. *Megaliths from Antiquity,* 347–348. York, UK: Antiquity Publications, 2003.

North, John D. *Stonehenge: Neolithic Man and the Cosmos,* 508–518. London: HarperCollins, 1996.

Ruggles, Clive. *Astronomy in Prehistoric Britain and Ireland,* 139–140. New Haven: Yale University Press, 1999.

Souden, David. *Stonehenge: Mysteries of the Stones and Landscape,* 52–53. London: Collins and Brown/English Heritage, 1997.

Thom, Archibald S., J. M. D. Ker, and T. R. Burrows. "The Bush Barrow Gold Lozenge: Is It a Solar and Lunar Calendar for Stonehenge?" *Antiquity* 62 (1988), 492–502.

## Cacaxtla

The citadel of Cacaxtla is prominently located on a hilltop in the highlands of Mexico, in the state of Tlaxcala, about eighty kilometers (fifty miles) southeast of Mexico City. It dates to the Epiclassic period (c. C.E. 650–850), a time when great cities such as Teotihuacan had collapsed and independent highland kingdoms had begun to develop. It was the main seat of the rulers of a group known to archaeologists as the Olmeca-Xicalanca (not connected with earlier Olmecs on the Gulf Coast).

The most impressive feature of Cacaxtla for the modern visitor is its huge murals, colorful and dramatic, reflecting a mixture of stylistic influences. These bear witness to tumultuous times. One of them vividly depicts the aftermath of battle and sacrifice, with the victorious dark-skinned warriors shown in jaguar pelts and the defeated lowland army in bird costumes. In one scene, the defeated captain is draped with Venus symbols and is being publicly humiliated prior to his execution. The archaeoastronomer John Carlson has argued convincingly that this episode was one manifestation of a wider cult of warfare and ritual sacrifice related to the patterns of appearance of the planet Venus. A small chamber discovered in 1987 contained two stuccoed pillars, each painted with a life-sized figure rich in symbolism relating to blood, water, and the planet Venus. One is male and one is female, but they also have scorpion, bird, and jaguar features and wear *Venus skirts,* kilts ornamented with a huge depiction of the glyph known to represent Venus. They are thought to represent deities strongly related to Venus, who had a vital role in the process of turning blood into water through ritual sacrifice (feeding the gods with human blood would encourage them to reciprocate by providing rain), thus ensuring fertility and renewal. It is possible that this room was the place where some of the most important sacrifices took place.

About one and a half kilometers (one mile) to the west of Cacaxtla, on an adjacent and higher hilltop, is the ceremonial center of Xochitecatl, a much older site that was occupied from about 700 B.C.E. and in use until an eruption of the volcano Popocatepetl forced its abandonment around C.E. 150. (Later, at the time of the dominance of Cacaxtla, parts of it were reoccupied.) One of its most distinctive features is a spiral pyramid built around 700 B.C.E. But it is the rectangular Pyramid of the Flowers that is most remarkable. It contains more than thirty infant burials (yet only one adult) and over two thousand clay figures, the great majority representing females of all ages. The obvious conclusion is that the latter were votive offerings related to fertility rites. The infant burials evoke practices known in (later) Aztec times, when children were sacrificed at mountain shrines on particular calendar dates to petition for rain.

Ascending the temple's western staircase to the large central platform—140 by 100 meters (460 by 330 feet)—reveals a striking view to the east and, directly ahead, the 4,600 meter (15,000 foot) volcano La Malinche, which dominates this region. The deposits of clay figures were found near this spot, which is scarcely surprising. Rain and fertility are clearly associated, and clouds tend to loom over the mountain even when the sky is otherwise clear. The later Aztecs certainly associated mountains with rains, believing (some have argued) that mountains were hollow "houses" filled with water—a conviction that is likely to reflect a much longer-standing element of Mesoamerican worldview. In pre-conquest times La Malinche was known as Matlalcueye, the great female mountain of sustenance. It is fair to conclude that rain and fertility rites at Xochitecatl might have been associated with this mountain for many centuries into the past, and possibly as far back as the Middle Formative Period when the pyramids were first built.

The view from the platform of the Pyramid of the Flowers reveals one more remarkable fact. From here, the palace of Cacaxtla sits immediately below and directly in line with the volcano. Is it too far-fetched to suggest that Cacaxtla was deliberately placed in an alignment whose significance was already long established? Possibly not; and it is even possible that cult practices related to this alignment have survived into modern times. Each year on the festival of St. Michael, September 29, the people of the village of San Miguel del Milagro, close to Xochitecatl, set out on a pilgrimage to the archangel's shrine. Chanting to celebrate the feast begins at sunrise. Around this day, as viewed from the Pyramid of the Flowers, the sun rises behind the summit of La Malinche, along the alignment of monuments and mountain. According to Carlson, these activities that still take place in a small Mexican village represent, in Christianized form, a living tradition of sacred geography, timed pilgrimage, and mountain veneration related to rain and fertility rites that extends well back into pre-conquest times. Such a continuity of tradition might seem unthinkable if we did not have good evidence that other fundamental aspects of Mesoamerican thought were remarkably durable—in particular the calendar itself.

Threading together in this way strands of evidence of different types—from archaeology, history, archaeoastronomy, and modern ethnography—raises many methodological issues, and if done carelessly can lead to quite unsustainable conclusions. Yet if done carefully it can bring some remarkable insights.

**See also:**
Christianization of "Pagan" Festivals; Methodology; Pilgrimage; Sacred Geographies.
Aztec Sacred Geography; Mesoamerican Calendar Round; Venus in Mesoamerica.

**References and further reading**
Carlson, John B. "La Malinche and San Miguel: Pilgrimage and Sacrifice to the Mountains of Sustenance in the Mexican Altiplano." In John B. Carlson, ed. *Pilgrimage and the Ritual Landscape in Pre-Columbian America.* Washington, DC: Dumbarton Oaks, in press.
Foncerrada de Molina, Marta. *Cacaxtla: La Iconografía de los Olmeca-Xicalanca.* Mexico City: Universidad Nacional Autónoma de México, 1993. [In Spanish.]
Instituto Nacional de Antropología e Historia. *INAH center Tlaxcala.* http://www.inah.gob.mx/inah_ing/cein/htme/cein29.html.
Krupp, Edwin C. *Skywatchers, Shamans and Kings,* 262–266. New York: Wiley, 1997.
Ruggles, Clive, and Nicholas Saunders, eds. *Astronomies and Cultures,* 215–226, 265–267. Niwot: University Press of Colorado, 1993.
Šprajc, Ivan. *Orientaciones Astronómicas en la Arquitectura Prehispánica del Centro de México,* 181 ff. Mexico City: Instituto Nacional de Antropología e Historia (Colección Científica 427), 2001. [In Spanish.]

# Cahokia

The many huge earthworks or mounds that remain visible in the Mississippi and Ohio valleys form a conspicuous testimonial to the technical achievements of indigenous North Americans before the arrival of European settlers. Inevitably, a number of them have attracted interest in potential astronomical alignments, and it is scarcely surprising that this includes the great Mississippian site of Cahokia.

The Mississippian culture flourished from about C.E. 1000 to 1400 in the central Mississippi valley, where the fertile floodplains were ideal for growing maize and other staple crops. Cahokia was not only the main economic and political focus of this culture but also the largest pre-Columbian town—worthy in fact of being called a city—north of Mexico. At the height of its development it covered over fifteen square kilometers (six square miles) and had an estimated population of more than twenty thousand. The city itself was laid out on a roughly cardinally aligned grid. At its heart was a ceremonial center containing over a hundred earthworks—temple or house platforms together with burial mounds. Most of this "mound center" is now preserved in Cahokia Mounds State Historic Site near East St. Louis.

The rectangular, flat-topped earthwork known as Monks' Mound is at the center of the ceremonial grid. It is the largest earthen mound in the whole of the Americas, some thirty meters (a hundred feet) high and covering around seven hectares (seventeen acres). But it was the discovery in the early 1960s of rings of postholes marking the sites of five timber circles to the west of Monks' Mound that triggered intense astronomical interest in the site. By analogy to Woodhenge—a set of concentric, oval-shaped timber rings found near Stonehenge in England—the discoverers named them *woodhenges,* although it is dangerous to draw close similarities between superficially similar sites from completely different cultural contexts. The five original woodhenges at Cahokia, identified by extrapolating from strings of postholes forming circular arcs running through excavated areas, appear to have had diameters ranging from thirty-seven meters (120 feet) to seventy-one meters (233 feet). They were good approximations to true circles, with evenly spaced posts. Most remarkably, initial estimates of the numbers of posts in the five circles—twenty-four, thirty-six, forty-eight, sixty, and seventy-two—implied that numerology was vitally important, and that strict design principles were operating. Four of the five posthole circles overlap, suggesting that they represent successive constructions fulfilling a particular purpose. Since this time, a number of other postholes and posthole structures have been discovered at Cahokia.

The original suggestion made by archaeologist Warren Wittry in the early 1960s was that the various circles of timber posts functioned as devices for observing the rising position of the sun at the solstices and equinoxes, perhaps for the purpose of regulating an agricultural or ceremonial calendar. However, this involved the totally arbitrary selection of data: for instance, three posts—indistinguishable in any other way from all the remaining posts in one of the circles—were identified as putative solar foresights as viewed from a central post. Other corroborating evidence presented at the time—for instance, relating to the orientation of the ramps used in setting each relevant timber post in its place—has proved totally fallacious, as have various further astronomical speculations.

An obvious question to have asked at the outset was: why, if sunrise observation was the primary purpose, was a whole circle of posts needed? Indeed, circles of this size—unless supplemented by the use of distant horizon foresights—cannot define the relevant alignments at all precisely. It is possible that the circles had a cosmological function or meaning, but the publicity given to the unfounded astronomical speculations has undoubtedly hampered serious consideration of this issue.

Instead, Cahokia stands as an unfortunate example of bad practice—

uncritical spotting of alignments upon preconceived astronomical targets undertaken without due consideration to the broader interpretative context.

**See also:**
Cardinal Directions; Cosmology; Equinoxes; Methodology; Solstitial Directions.
Circles of Earth, Timber, and Stone; Hopewell Mounds.

**References and further reading**
Aveni, Anthony F. *Skywatchers*, 304–305. Austin: University of Texas Press, 2001.
Fowler, Melvin, ed. "The Ancient Sky Watchers of Cahokia: Woodhenges, Eclipses, and Cahokian Cosmology." *Wisconsin Archaeologist* 77 (3/4), 1996.
Krupp, Edwin C. *Skywatchers, Shamans and Kings*, 295–299. New York: Wiley, 1997.
Schaefer, Bradley E. "Case Studies of Three of the Most Famous Claimed Archaeoastronomical Alignments in North America." In Bryan Bates and Todd Bostwick, eds. *Proceedings of the Seventh "Oxford" International Conference on Archaeoastronomy*. In press.

# Calendars

The regular motions of celestial bodies can be used to keep track of the passage of time, and observations of different cycles in the sky have underlain many different types of calendars developed and used from early prehistory to the present. The moon is an obvious tool for this purpose: it is conspicuous in the sky, and its cycle of phases is easily tracked and of a convenient periodicity (between twenty-nine and thirty days). The sun is another obvious candidate: the annual passage of its rising and setting position along the horizon, and its changing arc through the sky, which gives rise to days of different lengths, is directly related to the seasonal year. The stars follow the same track through the sky day after day, the changing time of night of their appearance and disappearance are also correlated with the seasonal year. Even the planets follow regular cycles, more complex but nonetheless recognizable and readily observable.

The diversity of human calendars, and the complexity of many of them, arises partly because these natural cycles do not fit together neatly. The length of the lunar phase cycle (synodic month) is not a whole number of days, the length of the seasonal year is not a whole number of lunar phase cycles, and so on. Sometimes nature provides, by chance, a reasonably close fit—thus five synodic cycles of the planet Venus are very close to eight seasonal years—and some human cultures seem to have gone out of their way to identify such correlations. A prime example is the Maya, as is clear from the almanac known as the Dresden Codex. Others observe certain cycles in the heavens and seem to ignore the rest. It is scarcely surprising that there is a broad correlation between calendars and latitude, since the general appearance of the skies and of the celestial cycles depends upon it. For example, the annual variation in the length of the day, and in the horizon position of sunrise (or sunset), is much greater, and hence more obvious, at higher latitudes than near to the equator.

These are convenient categorizations for us and for the modern astronomer, but they may have made no sense to peoples in the past. To understand the nature of a calendar we must understand the needs it fulfilled—practical, ideological, and social. In some cases more than one distinct calendars fulfilling different needs may have run in parallel. We must also understand that a calendar may have been conceptualized within a very different framework of understanding from our own. It then comes as less of a surprise that calendars are commonly regulated using astronomical cycles observed in combination with a multitude of other natural cycles, many of which (to our way of thinking) are less regular or reliable.

One of the greatest pitfalls is to try to identify "stages" in the development of the modern Western calendar, seeing these as a progression along an inevitable path of calendrical development. This encourages a form of intellectual imperialism (or ethnocentrism) in which we attempt to measure the achievement of others in relation to our own. This stepped approach is easily refuted by a number of indigenous calendars recorded in modern times. The calendar of the Mursi of Ethiopia is particularly valuable in this respect. What we actually find, rather than a single progression, is considerable diversity combined with remarkable ingenuity and adaptability to local circumstances.

Another pitfall is to simplify calendrical developments in the distant past, postulating the existence of calendars in common use over great swaths of the prehistoric world and/or changing little over many centuries. Notorious examples of this kind of oversimplification are the "megalithic" calendar in Neolithic Britain and the "Celtic" calendar in Iron Age Europe. Another is the proposition that the modern Borana calendar represented a calendar that had been propagated, without variation, through two millennia since being encapsulated in a set of alignments at Namoratung'a in northern Kenya. Research has shown, instead, that calendars are very often adapted rapidly to changing circumstances.

On the other hand, broad calendrical principles can be preserved with remarkable consistency, as happened amongst the scattered islands of Polynesia. Yet even here, considerable local variations developed: even between individual islands and parts of islands within the Hawaiian group. There were variations, for example, in the naming of months, the timing of months within the seasonal year, and rules for inserting intercalary months. A similar degree of variation is evident between different city-states in Classical Greece and also within pre-Columbian Mesoamerica.

See also:
Ethnocentrism; Lunar and Luni-Solar Calendars; Space and Time, Ancient Perceptions of.
Ancient Egyptian Calendars; Borana Calendar; Celtic Calendar; Delphic Oracle; Dresden Codex; Gregorian Calendar; Hawaiian Calendar; Hopi Calendar and Worldview; Horizon Calendars of Central Mexico; Javanese Calendar; Julian Calendar; "Megalithic" Calendar; Mesoamerican Calendar Round; Mursi Calendar; Namoratung'a.
Heliacal Rise; Inferior Planets, Motions of; Lunar Phase Cycle; Solstices; Superior Planets, Motions of.

References and further reading
Aveni, Anthony F. *Empires of Time: Calendars, Clocks and Cultures.* New York: Basic Books, 1989.
McCready, Stuart, ed. *The Discovery of Time.* Naperville, IL: Sourcebooks, 2001.

# Callanish

In one of the most remote corners of Britain stands one of the country's most impressive megalithic monuments. The standing stones of Callanish (an anglicization of the Gaelic name, Calanais) can be found on the western shores of the Isle of Lewis in the Outer Hebrides off western Scotland. They occupy a commanding situation overlooking what is now a stark landscape of heathery peat bogs and, to the west, the sea inlet of East Loch Roag. The site, dating to the third millennium B.C.E., consists of a ring of tall menhirs—the ring is 13 meters (43 feet) across and the stones vary from 3 meters (10 feet) to 4 meters (13 feet) in height—surrounding a small, chambered tomb. Rows of four or five stones radiate out to the east, west, and south. Northwards, but displaced by about ten degrees to the east, runs a longer double row or avenue. Several smaller stone circles and settings of standing stones are found in the vicinity.

The site first achieved astronomical notoriety when a survey was published by Vice-Admiral Boyle Somerville in 1912. It included the first-ever suggestion that a megalithic monument might have been aligned upon the moon at an extreme rising or setting point in its 18.6-year cycle, known as the lunar node cycle. As Somerville pointed out, the avenue is aligned southwards in the direction of the most southerly possible setting point of the moon, which the moon can only reach every eighteenth or nineteenth year. This idea was elaborated in the 1960s and 1970s by Gerald Hawkins, the author of *Stonehenge Decoded,* and by the Scottish engineer Alexander Thom. The Greek historian Diodorus of Sicily, writing in the first century B.C.E., referred to a sacred precinct or circular temple in the island of the Hyperboreans where the moon appeared close to the earth and the god returned every nineteen years, and it has been suggested by some people, including the British archaeologist Aubrey Burl, that Diodorus referred to

*The standing stones of Callanish, Isle of Lewis, Scotland. (Corel Corp.)*

Callanish rather than Stonehenge, as was more generally assumed. (Both interpretations are problematic.)

Thom proposed that Callanish was one of several "lunar observatories" marking significant lunar rising and setting points very precisely (down to a small fraction of the moon's diameter) using distant foresights—in this case, the mountain of Clisham on Harris, some 26 kilometers (16 miles) away. However, there is no clear view of Clisham from the avenue—it is obscured both by the taller stones of the central ring and by a small rocky outcrop just to the south of the site. Thom's high-precision "lunar observatories" in general have not stood the test of time.

However, the failure of mid-twentieth-century attempts to populate the prehistoric past with an intellectual or priestly elite prepossessed with high-precision astronomical observations should not blind us to the basic relationship with the moon that exists at the Callanish stones, or with what this might actually have meant to the Neolithic populations of this area. The full or nearly full moon scraping uncommonly low over the hills to the south and then—as viewed along the avenue—passing behind and setting among the stones of the circle, casting them into silhouette, would have been a truly spectacular sight, one that could only have been seen once or twice in a generation. If this was intentional from the outset, then it would certainly explain the skewing of the avenue away from the meridian. It would also have ensured that this location was charged with tremendous sacred power at these special times.

It has been suggested that the standing stones of Callanish and the various smaller megalithic monuments in the surrounding area incorporated numerous alignments upon prominent horizon features and extreme lunar rising and setting positions. The Callanish stones, it was proposed, stood at the heart of a complex that encapsulated a variety of relationships between built monuments, prominent natural features in the landscape, and the motions of the moon. The general idea is not implausible: indigenous societies commonly organize sacred space to reflect cosmic relationships perceived in the wider visual setting, and the visible environment included the sky. But in the absence of corroborating evidence, it is almost impossible to argue convincingly for any particular scheme. We have no way of knowing which relationships actually were perceived as significant in the past, and any choice that we make is ultimately subjective. Whether the Callanish stones really represent a temple whose significance related to the moon appearing in a special way in every nineteenth year remains an open question.

**See also:**
Cardinal Directions; Somerville, Boyle (1864–1936); Thom, Alexander (1894–1985).
Stonehenge.
Meridian; Moon, Motions of.

**References and further reading**
Ashmore, Patrick. *Calanais: The Standing Stones.* Stornoway, Scotland: Urras nan Tursachan, 1995.
Burl, Aubrey. *From Carnac to Callanish: The Prehistoric Stone Rows and Avenues of Britain, Ireland and Brittany,* 63–65, 179–180. New Haven: Yale University Press, 1993.
———. *A Guide to the Stone Circles of Britain, Ireland and Brittany,* 148–152. New Haven: Yale University Press, 1995.
———. *The Stone Circles of Britain, Ireland and Brittany,* 202–207. New Haven: Yale University Press, 2000.
Ruggles, Clive. *Megalithic Astronomy: A New Archaeological and Statistical Study of 300 Western Scottish Sites,* 75–98. Oxford: British Archaeological Reports (British Series 123), 1984.
———. *Astronomy in Prehistoric Britain and Ireland,* 88–89, 134–136. New Haven: Yale University Press, 1999.
———, ed. *Archaeoastronomy in the 1990s,* 309–316. Loughborough, UK: Group D Publications, 1993.
———, ed. *Records in Stone: Papers in Memory of Alexander Thom,* 426–431. Cambridge: Cambridge University Press, 2002.
Ruggles, Clive, and Alasdair Whittle, eds. *Astronomy and Society in Britain during the Period 4000–1500 BC,* 63–110. Oxford: British Archaeological Reports (British Series 88), 1981.
Thom, Alexander. *Megalithic Lunar Observatories,* 68–69. Oxford: Oxford University Press, 1971.

*The Caracol at Chichen Itza, viewed from the west. (Courtesy of Clive Ruggles)*

## Caracol at Chichen Itza

An extraordinary feature of Mesoamerican astronomy, given the complexity and sophistication evident from the ethnohistory and written sources such as the Dresden Codex, is the apparent lack of any observatories or observing instruments apart from the cross-sticks depicted in various codices, widely interpreted as a naked-eye sighting device. One of the very few serious candidates for a building used as an observatory is the Caracol at Chichen Itza.

Chichen Itza is one of the most famous Maya cities. Like Uxmal, it flourished around C.E. 800, but its influence appears to have persisted while many other cities were abandoned and fell into ruin. During the Postclassic period (C.E. 900–1300), before Chichen Itza itself fell into ruin, its sphere of influence became greatly extended and the constructions forming its ceremonial center became bigger and more impressive than ever. Even so, the Caracol stands out as exceptional. The reason is not so much its size as its shape: it is round, while every other visible construction here is straight-sided. As such the Caracol is almost unique, and certainly uniquely preserved. The round tower was ascended by a curved staircase running inside its double outer walls, leading to a small upper chamber with windows on various sides—or so it is supposed, since less than half of it now survives.

The openings are little more than niches, which means that they provide restrictive sighting devices. They do look out just above the horizon, which has led numerous researchers to measure alignments and speculate on their precise function. This anomalous and unaesthetic building seems an ideal candidate for an observatory where Maya priests observed the heavens to regulate the calendar.

To the person seeking statistical verification that the tower was used for astronomical observations, the results are disappointing. There are no consistent alignments on the sun, moon, or planets. And while stellar observations might have been important, there are too many possibilities—too many bright stars—to place much credence on any apparent stellar alignments that have been found. Further progress must consider the cultural context. Given the close association of various nearby structures at Chichen Itza with the planet Venus, the claim that two of the three surviving windows were ideally placed for observing the furthest northerly and southerly settings of this planet deserves closer scrutiny. But in view of the total lack of cultural evidence attesting to its use as an observatory, the strong possibility remains that the Caracol had nothing to do with astronomical observations at all.

**See also:**
Dresden Codex; Kukulcan; Venus in Mesoamerica.

**References and further reading**
Aveni, Anthony F. *Stairways to the Stars: Skywatching in Three Great Ancient Cultures,* 135–138. New York: Wiley, 1997.
———. *Skywatchers,* 272–283. Austin: University of Texas Press, 2001.
Krupp, Edwin C., ed. *In Search of Ancient Astronomies,* 190–199. New York: Doubleday, 1977.

# Carahunge

In a mountainous region of southern Armenia, near the town of Sisian, is an impressive stone setting consisting of over 150 standing stones varying from about one meter (three feet) to 2.8 meters (nine feet) in height. Approximately fifty more stones have now fallen. Of Neolithic date, probably built no later than the third millennium B.C.E. and possibly considerably earlier, the site broadly resembles many better-known megalithic monuments on the Atlantic seaboard of Europe, and particularly in Britain, Ireland, and Brittany. Surrounding a central dolmen is an oval-shaped ring about thirty-five meters (115 feet) in diameter consisting of about forty stones. An avenue runs out from the ring to the northeast, and other rows run north and south.

The site has been interpreted as an astronomical observatory and acclaimed as an Armenian Stonehenge. A distinctive feature of Carahunge is that about eighty of the stones in the north-south rows contained a small cir-

*Part of the stone setting at Carahunge in Armenia. (www.carahunge.com)*

cular hole running through their upper section (although only about fifty of these stones survive intact). These curious holes have carefully smoothed edges and are around five centimeters (two inches) in diameter. Some are as much as twenty centimeters (eight inches) deep, opening out into wider depressions carved into each side of the stone. Three holes enter in one side of a stone, turn a right angle and then point directly upwards. An unresolved question is whether the holes, which seem remarkably unweathered if they are indeed prehistoric, were in fact added as a later feature.

A number of Armenian and Russian archaeoastronomers have investigated the possible use of these holes for observations of the sun and moon in prehistoric times. They have established, for example, that three or four of them are directed toward the point of sunrise on the summer solstice and another three or four toward the point of sunset on the same day. Other holes point in a variety of directions, all around the compass. They are also inclined at various angles to the horizontal, mostly up to about fifteen degrees; this means that most are directed toward points in the sky just above the local horizon. These facts raise the serious possibility that the holes were used for astronomical observations, whether contemporary with the construction of the original monument or later. It has even been suggested that the three right-angled holes contained mirrors and were used for zenith observation. To address these issues, a systematic study of the holes that remain in situ, paying careful attention to methodological issues, is urgently needed, because the site is unprotected and threatened by damage from sightseers and looters.

Inevitably there have been other claims—more speculative and less supportable—relating to the astronomical significance of the site. One is that it can be astronomically dated to the sixth millennium B.C.E. And direct comparisons with Stonehenge, which few now believe was an observatory, are less than helpful.

See also:
Astronomical Dating; Methodology.
Megalithic Monuments of Britain and Ireland; Stonehenge.

**References and further reading**
Bochkarev, Nikolai. "Ancient Armenian Astroarchaeological Monuments: Personal Impressions of Metsamor and Carahunge." In Mare Kõiva, Harry Mürk, and Izold Pustõlnik, eds. *Cultural Context from Archaeoastronomical Data and the Echoes of Cosmic Catastrophic Events*. Tallinn: Estonian Literary Museum and Tartu, Estonia: Tartu Observatory, in press.
Herouni, Paris. "The Prehistoric Stone Observatory Carahunge-Carenish." *Reports of NAS of Armenia* 4 (1998), 307–328. [In Russian.]
Vardanyan, Gurgan. *Carahunge: Armenia's Stonehenge.* http://www.carahunge.com/.

# Cardinal Directions

If we find monumental alignments pointing due north, east, south, or west, it is tempting to interpret them simply as "cardinally oriented," but we must be cautious in doing so. Why would such alignments arise?

Where architecture is intentionally aligned toward due north or south, disregarding the exceptional circumstance where this is because some significant place or landmark happens to be located in that direction, then it is astronomical in a broad sense. This is because north-south alignments follow the axis of symmetry of the daily motions of the stars in the night sky as they swing each day around the celestial poles. From some locations, certain important celestial bodies may be seen to "rise" or "set" in these directions, but this can only happen if the horizon is not flat, since all the celestial bodies move along on the level when they are due north or south. In general, north-south alignments relate to the daily motions of the heavens as a whole rather than to any particular celestial body or event. Three very different examples illustrate the reasons for keeping architecture "in tune with the cosmos" in this way: the Great North Road at Chaco Canyon—the so-called Chaco

Meridian; the Forbidden City in Beijing; and the Pyramids of Giza in Egypt.

Alignments to the east or west are more problematic. An alignment on equinoctial sunrise or sunset, for example, will be approximately toward due east or west, respectively, but was not necessarily perceived as "cardinal." Then again, the concept of "east" may be a much broader one. Christian churches are generally conceived as oriented eastwards, for a variety of liturgical reasons that dictated this to be the direction that Christian worshipers should face when praying. However, their actual orientations span a range spreading as much as several tens of degrees away from true east. If we have only archaeological evidence to go on, especially if we have no reason to assume any great precision, then it is difficult to see how we can distinguish an alignment conceived to be "easterly" from one upon the equinoctial rising sun, or indeed upon some other star or asterism that happens to rise in that general direction (such as, in the present day, Orion's belt).

As if this were not enough, there is a deeper problem. In assuming that east is universally conceived as a cardinal direction, we are making the assumption that every human community, past and present, would be inclined to identify as significant four directions that divide the horizon into four exactly equal parts. This assumption is unwarranted. It is not true for the Chorti Maya people of the Yucatan in Mexico, for whom the east and west are seen as the directions of sunrise and sunset on the day when the sun passes directly overhead. The north and south directions, as we have said, arise as a direct consequence of the motions of the stars in the night sky as they swing each day around the celestial poles—in other words, they are discernible by direct observation of the heavens in a way that applies equally well at any point on the earth (apart from the poles). On the other hand, the directions east and west as we conceive them are not marked in any direct and consistent way by astronomical referents. (There is no direct way, for example, of identifying a particular sunrise as equinoctial: to do this one must employ some device such as counting the number of days or dividing the horizon between the solstices.)

In short, because the designation of the four directions due north, south, east, and west as cardinal is so familiar to us, it is easy to think of their significance as an absolute, but this would be a mistake. There is no reason to suppose concepts such as "east" and "west" in any prehistoric society bore any direct relation to our concepts, or meant anything at all. And yet a common trait among historical and modern indigenous communities, particularly in the Americas, is to conceive the cosmos to be divided (horizontally) into four parts. This is known as a quadripartite cosmology. Within such a worldview, "east" might refer to what we would describe as a range of directions around due east. But there is no reason a quadripartite cosmology must divide the horizon into exactly equal quarters, or for each part to be centered upon one of "our" cardinal directions. Indeed, quadripartite cosmologies are commonly related to the sun, so that the easterly direction is not a complete quarter of the horizon but that part within which the sun rises at some time during the year.

**See also:**
Equinoxes; Solstitial Directions.
Casa Rinconada; Chaco Meridian; Chinese Astronomy; Church Orientations; Crucuno; Pyramids of Giza.
Celestial Spheres; Meridian.

**References and further reading**
Aveni, Anthony F. *Skywatchers,* 294–295. Austin: University of Texas Press, 2001.
Ruggles, Clive. *Astronomy in Prehistoric Britain and Ireland,* 148. New Haven: Yale University Press, 1999.

# Casa Rinconada

Casa Rinconada was the largest of the great kivas in Chaco Canyon. Kivas were subterranean circular buildings used as meeting places or for sacred ceremonies. In Chaco they were generally built with their roofs at ground level, within the central plazas of Great Houses such as Pueblo Bonito, where over thirty-five examples, great and small, are to be found. Casa Rinconada was unlike the other kivas in both these respects. It was built on its own, separately from any Great House, and while the dirt floor was 3.5 meters (twelve feet) below the surrounding ground level, its walls protruded above the ground. It measured twenty meters (sixty-six feet) across, and its vast roof (of which no trace remains) was supported by four vertical timber uprights placed near the center. There were entrances to the north and south, and around the outer wall were twenty-eight regularly spaced niches together with six larger, irregularly spaced ones.

During the 1970s, archaeoastronomers were apt to scrutinize ancient monuments anywhere in the world for potential solar and lunar alignments reflecting those that had been so famously proposed a decade earlier at Stonehenge in England; Casa Rinconada was analyzed closely in this regard. However, as these early investigators had to admit, the roof was a problem. Only through what seems to have been a single window in the north-northeast could sunlight (or moonlight) actually have entered the kiva, and the best they could come up with was that after sunrise at the summer solstice, a beam of sunlight entering through this window could have illuminated one of six niches that, for some unfathomable reason, were built slightly larger than the rest. However, the window in question has been reconstructed and its original position is uncertain, and even if it is in the original position, it seems that one of the four roof supports may have been in the way. All of

Casa Rinconada and its situation in Chaco Canyon, looking toward Pueblo Bonito on the other side of the valley (see also Chaco Canyon). (Courtesy of Clive Ruggles)

this serves to illustrate the dangers of naïve "alignment hunting."

It also obscures a more fundamental property of the orientation of Casa Rinconada. The kiva was cardinally aligned: its main axis (through the entrances) was due north-south, and it was also broadly symmetrical about the east-west axis. The intention, surely, was cosmological. Creation myths that have come down to modern indigenous peoples in the vicinity tell how the first kiva was circular to reflect the circle of the sky, and conceptual principles that result in a quartering of space in relation to the cardinal directions are commonplace among indigenous North American groups to this day. As archaeoastronomer Ray Williamson concluded back in 1982, Casa Rinconada "was meant to serve as an earthly image of the celestial realm; . . . it was not an observatory, but a ritual building whose structure reflects the central place astronomy had in ancient Pueblo religious practice" (Aveni 1982, p. 205).

It is even more intriguing that principles of cardinality seem to have extended into the landscape. In 1978, archaeologist John Fritz drew attention to the fact that three major Chaco structures, the Great House of Pueblo Alto to the north of the canyon, Casa Rinconada within it, and Tsin Kletzin, another Great House on the mesa to the south, lie on a north-south line. This, argued Fritz, was the main axis of the city. The two largest Chacoan buildings, Chetro Ketl and Pueblo Bonito, lie on either side of this axis, almost equidistant from it, and on an east-west line. Fritz's contention was that these two cardinal axes were of fundamental importance in the minds of the Chacoans, a theory that received a dramatic endorsement with the subsequent discovery of what has become known as the (Great) North Road, which extends the north-south axis physically northwards for at least fifty-five kilometers. Archaeologist Steve Lekson has extended this idea still further, arguing the existence of a "Chaco Meridian" that extended from Aztec, over eighty kilometers (fifty miles) north of Chaco, right down to Paquime in modern Mexico, a staggering 630 kilometers (390 miles) to the south.

An investigation in the 1980s of the orientations of the principal walls of Chacoan Great Houses showed that only four out of fourteen of them were aligned cardinally. However, three of the four lie on Fritz's axes: Pueblo Bonito (north-south) and Pueblo Alto and Tsin Kletzin (both east-west). Casa Rinconada itself was also cardinally aligned, as we have already seen. Only Chetro Ketl stands out as an exception, its principal wall oriented at an azimuth of about seventy degrees. The fact that all but one of the buildings situated on the cardinal axes were themselves cardinally aligned surely adds strength to the idea that the cardinal directions and Fritz's axes were of fundamental importance. (Conversely, all but one of the buildings not lying on these axes were not cardinally aligned.)

During its first half-century of existence, archaeoastronomy has matured considerably as a discipline, moving on from "alignment hunting" to addressing culturally relevant questions. In the same period, Casa Rinconada has been transformed from a building containing numerous potential solar and lunar alignments (an idea that never really resonated with the cultural evidence) to a central point within an astronomically referenced sacred landscape.

**See also:**
Archaeoastronomy; Cardinal Directions; Pilgrimage.
Chaco Canyon; Chaco Meridian; Fajada Butte Sun Dagger; Navajo Hogan; Pawnee Earth Lodge; Stonehenge.

**References and further reading**
Aveni, Anthony F., ed. *Archaeoastronomy in the New World*, 205–219. Cambridge: Cambridge University Press, 1982.
———. *World Archaeoastronomy*, 153–154, 365–376. Cambridge: Cambridge University Press, 1989.
Krupp, Edwin C. *Echoes of the Ancient Skies,* 231–236. Oxford: Oxford University Press, 1983.
Lekson, Stephen H. *The Chaco Meridian: Centers of Political Power in the Ancient Southwest,* 82–86. Walnut Creek, CA: AltaMira Press, 1999.
Noble, David G., ed. *New Light on Chaco Canyon*, 69–70. Santa Fe: School of American Research Press, 1984.
Redman, Charles, ed. *Social Archaeology: Beyond Substance and Dating,* 37–59. New York: Academic Press, 1978.

## Catastrophic Events

Unexpected events in past skies, such as eclipses or the appearance of comets and meteors, could cause fear and even social disruption. In particular, they could threaten the power of elites, which was often predicated on being able to control the celestial bodies and keep them regular and predictable.

It has been suggested by some astronomers that cometary activity might have been much greater at certain epochs in the past, resulting in periods when sightings of spectacular comets and meteors were much more common. One such period occurred during the second millennium B.C.E., causing considerably more "active" skies, and the unrest that these caused brought about significant social and political upheaval, together with abrupt changes in beliefs systems in many parts of the world. The astronomical arguments remain controversial, but even if one accepts them, there are problems on the archaeological side. For sure, major social and ideological changes affected many human societies all round the globe during the second millennium B.C.E., but one has to question whether this period was much different from any other era in the last few millennia in this respect. In any case, the impact of fearsome sights in the sky on any *particular* group of people would have depended greatly upon all the other, more

The Barringer Meteor Crater near Winslow, Arizona. By chance, the camera has caught a meteor in the sky above the crater. (Bryan Allen/Corbis)

localized problems that were of concern at the time. The sky was important, but it wasn't everything.

Since the 1990s, increased cometary activity in the past has engendered a great deal of interest, but largely for another reason: the increased probability of physical impacts from small meteorites hitting the earth. Though derived from the same astronomical cause, this is a fundamentally different issue in human terms. Meteoritic impacts have a direct physical outcome. Even a small meteoritic impact can have a devastating short-term effect on the climate, causing atmospheric dust, which leads to growth-free "cold summers" over vast areas.

And such effects did happen, which is clearly shown by evidence from dendrochronology (tree-ring dating). This dating technique is capable of assigning individual growth rings in particular timbers to a particular year. As a result, we know that during certain years in the last five millennia, there was virtually no significant tree growth anywhere in the northern hemisphere. The years with the most severe climatic downturns were 2345 B.C.E., 1628 B.C.E., 1159 B.C.E., and 536 C.E. When these events were first identified in the tree-ring record, the cause was suspected to be volcanic eruptions, but the Northern Irish dendrochronologist Mike Baillie has argued strongly against this. If one accepts that the cause of some or all of the climatic downturns was indeed a meteoritic impact, this does not mean such impacts were more likely in the past; on the contrary, they may be as likely today as ever.

Whatever did cause these climatic downturns, questions about their impact on the Earth's population at the time can only be addressed by carefully piecing together fragments of archaeological, historical, and environmental evidence. In the meantime, the popular wave of interest in meteoritic impacts that ensued in the late 1990s generated everything from theories of global catastrophe and revolutionary social change throughout the world in the mid-sixth century C.E. to totally unsustainable ideas that Stonehenge in England was built to observe and predict meteoritic impacts.

See also:
Comets, Novae, and Meteors; Lunar Eclipses; Power; Solar Eclipses. Stonehenge.

**References and further reading**
Bailey, Mark. "Recent Results in Cometary Astronomy: Implications for the Ancient Sky." *Vistas in Astronomy* 39 (1995), 647–671.
Bailey, Mark, Victor Clube, and Bill Napier. *The Origin of Comets*. London: Pergamon Press, 1990.
Baillie, Mike. *From Exodus to Arthur: Catastrophic Encounters with Comets*. London: Batsford, 1999.
Keys, David. *Catastrophe*. London: Century Books, 1999.
Peiser, Benny, Trevor Palmer, and Mark Bailey, eds. *Natural Catastrophes during Bronze Age Civilisations: Archaeological, Geological, Astronomical and Cultural Perspectives*. Oxford: Archaeopress (BAR International Series 728), 1998.
Steel, Duncan. *Rogue Asteroids and Doomsday Comets: The Search for the Million Megaton Menace That Threatens Life on Earth*. Chichester: Wiley, 1995.

# Celestial Poles
*See* Celestial Sphere.

# Celestial Sphere
The earth is surrounded by stars at various distances, but when the night sky is viewed on a clear night from anywhere on the earth, it appears that all the stars (and other celestial objects) are fixed onto a huge sphere surrounding the observer. From any given place at any given time, roughly half of the sphere is visible above the horizon, provided that there are clear views unobscured by things like trees or buildings. The other half is below the horizon and invisible.

The earth rotates in space once a day, and the stars are effectively motionless. (They are actually in motion with respect to one another, but generally the effects of these *proper motions* are so negligible that they will not affect the configurations of the stars in a way detectable with the naked eye over many tens of thousands of years or more.) This means that from the viewpoint of an observer on the earth, the celestial sphere appears to rotate around him or her once a day, with all the stars affixed to it. (The fact that the observer is not at the center of the earth is of little consequence except in the case of the moon, for which lunar parallax must be taken into account before we can determine the position of the moon in the sky, either now or in the past.)

If we are interested in how human societies perceive the sky, as opposed to what modern physics and astronomy can tell us about the universe, then it is how the sky appears rather than how it actually is that is important. Thinking in terms of the celestial sphere enables us to define a number of fundamental concepts that are extremely useful in discussing ancient skies.

For example, on this rotating sphere we can identify an equator and two poles. The latter are called the north and south celestial poles, and they are directly overhead as viewed from the north and south poles respectively on the earth. More generally, and most usefully, we can define lines of latitude (*declination*) and longitude (*right ascension*), which allow us to specify the position of any star on the sphere.

**See also:**
Declination; How the Sky Has Changed over the Centuries; Lunar Parallax; Precession.

**References and further reading**
Aveni, Anthony F. *Skywatchers,* 49–57. Austin: University of Texas Press, 2001.
Krupp, Edwin C. *Echoes of the Ancient Skies,* 3–6. Oxford: Oxford University Press, 1983.

# Celtic Calendar

The concept of a Celtic calendar that divided the year into eight precisely equal parts, marked by festivals on the solstices, equinoxes, and mid-quarter days, has influenced archaeoastronomy for decades and does so still, in that many people continue to perceive monumental alignments upon sunrise or sunset on the equinoxes or mid-quarter days as inherently significant. Certainly pre-Christian calendrical festivals existed, such as Imbolc, Beltaine, Lughnasa, and Samhain (celebrated in modern pagan traditions at the beginning of February, May, August, and November, respectively), and continued to be important, for example, in early medieval Ireland. However, we should not accept uncritically the notions that their precursors were both widespread within Iron Age Europe and precisely determined. The whole concept of Celtic culture is now generally felt to be problematic, with many archaeologists arguing that it has more to do with modern perceptions of ethnicity than with historical or archaeological evidence of any widespread conformity between cultures in the European Iron Age. Given these uncertainties, the idea of a precise and all-pervasive Celtic calendar must be treated with considerable caution.

Nonetheless, there are indications that a precise division of the year into eight parts, based on observations of the sun, might have existed in pre-Roman times. Chief among these is the Coligny calendar, a bronze tablet about 1.5 meters by 0.9 meters (five feet by three feet) in size (though found in fragments, some of which were missing) dating to the second century C.E. It is a public calendar covering five years, with dates and festivals marked, and it is luni-solar in character: a late remnant of an indigenous calendar from the pre-Roman world. Lunar months alternating between twenty-nine and thirty days, and even intercalary months, are marked; but so, too, are dates that recur once every three months or so, each mysteriously marked "PRINI LAG" or something similar. These dates occur at intervals of ninety-one, ninety-three, ninety-one, and ninety-two days—intervals of almost exactly one quarter of the year (which has 367 days in the Coligny calendar). It has been suggested that they actually mark Celtic quarter-day festivals.

Regardless of the "Celticity" of the indigenous calendar in this region of southern Gaul (the calendar was discovered about eighty kilometers [fifty miles] from Lyon), the vital question is whether it provides solid evidence for a pre-Roman calendar in which the main seasonal festivals were timed by counting off exact numbers of days in the solar year. In view of the fact that the indigenous calendar concerned was clearly lunar-based, this seems very unlikely. Perhaps the regular intervals were introduced as part of the process of making the existing calendar conform to the Roman system.

See also:
Christianization of "Pagan" Festivals; Lunar and Luni-Solar Calendars; "Megalithic" Calendar; Equinoxes; Mid-Quarter Days.
Beltany; Boyne Valley Tombs.
Solstices.

**References and further reading**
Collis, John. *The European Iron Age*. London: Routledge, 1997.
Cunliffe, Barry. *The Ancient Celts*. Oxford: Oxford University Press, 1997.
Gibson, Alex, and Derek Simpson, eds. *Prehistoric Ritual and Religion: Essays in Honour of Aubrey Burl*, 190–202. Stroud: Sutton, 1998.
Hutton, Ron. *The Stations of the Sun: A History of the Ritual Year in Britain*, 218–225, 327–331, 360–370. Oxford: Oxford University Press, 1996.
James, Simon. *The Atlantic Celts: Ancient People or Modern Invention?* Madison: University of Wisconsin Press, 1999.
King, John. *The Celtic Druids' Year: Seasonal Cycles of the Ancient Celts*. London: Blandford, 1994.

Le Contel, Jean-Michel, and Paul Verdier. *Un Calendrier Celtique: Le Calendrier Gaulois de Coligny.* Paris: Éditions Errance, 1997. [In French.]

McCluskey, Stephen C. *Astronomies and Cultures in Early Medieval Europe,* 54–60. Cambridge: Cambridge University Press, 1998.

Olmsted, Garrett. *The Gaulish Calendar.* Bonn: Habelt, 1992.

Ruggles, Clive. *Astronomy in Prehistoric Britain and Ireland,* 142, 159. New Haven: Yale University Press, 1999.

Ruggles, Clive, and Nicholas Saunders, eds. *Astronomies and Cultures,* 102–109. Niwot: University Press of Colorado, 1993.

# *Ceque* System

The capital of the Inca empire, Cusco (or Cuzco), was at the heart of a huge scheme of sacred geography. At the very center was the Temple of the Sun, known as the Coricancha ("golden enclosure"). The entire Inca empire was partitioned into four divisions known as *suyus,* a division that was so fundamental in Inca thought that the Inca name for their own empire was Tahuantinsuyu ("four parts"). The boundary lines separating the four *suyus* radiated out from Cusco, indeed from the Coricancha itself. This much is known from the accounts of several different chroniclers; but one of the most thorough and meticulous, Bernabé Cobo, tells us of the existence of no fewer than forty-one (or forty-two) such radial lines, and he records that they were known as *ceques* (or *zeq'e*).

The spiritual order of the Inca empire was based around shrines called *huacas* (or *wak'a*), sacred places for the worship of gods, prayers, and sacrifices. *Huacas* were often located at places that were in some way special or exceptional, such as caves, springs, mountain peaks, bends in rivers, or unusual rocks or trees. Cobo recorded well over three hundred *huacas* in the vicinity of Cusco, documenting each of them individually. Three hundred thirty-two of them were located on one of the *ceques,* with a number more that were not part of the system; none of the *ceques* contained fewer than three shrines, and a couple had as many as fifteen. The *ceques,* however, were more than just lines along which the sacred places lay. They were also linked in various ways to social relations and the division of labor. According to Cobo again, each *ceque* was given over to one of the three main social classes (those directly related, less directly related, or unrelated to the Inca ruler), and it was the responsibility of representatives of that social group to take care of its *huacas* and to carry out the appropriate ceremonial activities there. Protocols for assigning representatives of different family groups (*ayllus*) to communal work activities, such as the maintenance of irrigation canals, were also governed by the *ceque* system. In these and various other ways, the *ceque* system made many of the principles of Incaic social and political organization tangible in the landscape.

Cobo's accounts seem to indicate that the *ceques* were conceived as

*The Andean shrine of Kenko Grande, near Cusco in Peru, one of many sacred places or huacas lying on* ceque *lines. (Wolfgang Kaehler/CORBIS)*

straight lines diverging radially from the Coricancha, the symbolic center of the world, and extending out into the cosmos. This way of organizing things spatially—lines radiating out from a center—seems to extend well back into pre-Inca times, evident, for example, in the lines on the desert at Nasca. It is also evident in the *quipu*—recording devices consisting of knotted strings—which can be laid out in a radial fashion, and has even survived to the present day in the layout of some of the more remote and traditional Andean villages.

It has been suggested that some of the Cusco *ceques* were astronomically aligned. Cobo's account of various *huacas* includes hilltops where monuments were placed that marked the sun's arrival at significant dates, such as the times to sow crops. This leaves open the question of where the hilltop *huacas* were to be viewed from, but archaeoastronomical investigations suggest that in some cases, at least, the answer was "along the *ceques*." A related claim is that some of the *ceques* radiating out from the Coricancha hit the visible horizon at astronomically significant points such as the rising and setting of the sun at one of the solstices, something that is not surprising given that various chroniclers describe horizon observations of the sun from Cusco and describe pillars that were erected to mark these spots and to facilitate the observations. However, the evidence for the solstitial *ceques* is far from conclusive.

Anthropologist and historian Tom Zuidema has gone considerably further by suggesting that the *ceque* system was linked directly to an elaborate ritual calendar based neither upon solar observations nor upon lunar phase-cycle (synodic) months but upon *sidereal* months. A sidereal month, the time it takes for the moon to complete a circuit through the stars, is two days shorter than a synodic month, lasting 27.3 days. This may seem like a very esoteric concept, but it is a cycle that is particularly noticeable near the equator (where the celestial bodies rise and set almost vertically) by observing what stars the moon (whatever its phase) appears on a level with, for example, as it rises. This is done, for instance, by the Borana in east Africa.

Zuidema's interpretation was triggered by the fact that the number of *huacas* recorded by Cobo in the vicinity of Cusco, which he took to be 328, is exactly equal to the number of days in twelve sidereal months. Each *huaca*, then, according to Zuidema, corresponded to a day in the calendar. This might well be dismissed as speculative nonsense, a mere numerical coincidence, except that Cobo tells us that acts of worship (such as sacrifices) took place at the various shrines in a sequential order, proceeding from one to the next, working gradually round the Cusco horizon. What of the remaining thirty-seven days in each seasonal year? These, according to Zuidema, represented the period of time when the fields lay fallow between last harvest and first planting. These were "dead days" when no ceremonies took place at any *huaca*, and corresponded to the time when the Pleiades were invisible, being too close to the sun. The various strands of evidence in support of Zuidema's ritual calendar at Cusco are highly complex and hotly disputed, but the interpretation remains a highly intriguing one.

It has become evident, however, that we would be unwise to interpret the *ceques* too literally as a set of dead straight lines radiating out into the landscape from the center of Cusco. During the 1990s, archaeologist Brian Bauer conducted a field project around Cusco aiming to identify the actual locations of as many as possible of the original *huacas*. His conclusion was that the *ceques* were not straight at all, but actually zigzagged through the landscape. The *huacas*, according to Bauer, were defined by the innate power of places themselves, and they were subsequently grouped into *ceques*. They were certainly not placed on preconceived lines. In other words, the *huacas* defined the *ceques*, not the other way round.

It is possible, then, that some earlier work on *ceque* alignments has been misconceived. On the other hand, the non-straightness of the *ceques* does not detract from the fact that the whole landscape was evidently full of sacred and special places at which, for example, the sun was observed to rise or set on certain days of the year as seen from others. This is quite clear from the chroniclers' accounts. We might conclude that Incaic perceptions of straightness do not accord with ours. But it may simply be that the *ceques*, although often straight over relatively small distances that could easily be sighted along, became more sinuous as they stretched out over the wider landscape. The ideal may not have been matched by the reality, but perhaps this was not a critical concern in practice.

In sum, the *ceque* system was undoubtedly both a basic organizing principle of the Inca empire and a mechanism of political control. The extent to which it was also a manifestation of calendrical principles and practices is likely to be debated for a long time to come.

**See also:**
Antizenith Passage of the Sun; Cobo, Bernabé (1582–1657); Lunar and Luni-Solar Calendars; Sacred Geographies; Solstitial Directions.
Borana Calendar; Cusco Sun Pillars; Island of the Sun; Misminay; Nasca Lines and Figures; Quipu.

**References and further reading**
Aveni, Anthony F. *Stairways to the Stars: Skywatching in Three Great Ancient Cultures,* 147–176. New York: Wiley, 1997.
———. *Between the Lines: The Mystery of the Giant Ground Drawings of Ancient Nasca, Peru,* 124–135. Austin: University of Texas Press, 2000. [Published in the UK as *Nasca: The Eighth Wonder of the World.* London: British Museum Press, 2000.]
———. *Skywatchers,* 314–321. Austin: University of Texas Press, 2001.
Aveni, Anthony F., and Gary Urton, eds. *Ethnoastronomy and Archaeoastronomy in the American Tropics,* 203–229. New York: New York Academy of Sciences, 1982.
Bauer, Brian. *The Sacred Landscape of the Inca: The Cusco Ceque System.* Austin: University of Texas Press, 1998.
Bauer, Brian, and David Dearborn. *Astronomy and Empire in the Ancient Andes.* Austin: University of Texas Press, 1995.
D'Altroy, Terence. *The Incas,* 155–169. Oxford: Blackwell, 2002.
Selin, Helaine, ed. *Astronomy across Cultures,* 216–219. Dordrecht, Neth.: Kluwer, 2000.
Zuidema, R. Tom. *The Ceque System of Cuzco: The Social Organization of the Capital of the Inca.* Leiden: Brill, 1964.

# Chaco Canyon

The sight of the Great Houses of Chaco Canyon—vast multistory constructions containing scores, and in a few cases hundreds, of rooms built in a dry and inhospitable valley—is nothing short of extraordinary. These and other architectural edifices were built by the ancient Puebloan peoples who occupied the canyon and the surrounding area between about C.E. 950 and 1125. The largest Great House, Pueblo Bonito, covers an area of 1.8 hectares (4.5 acres) and contains seven hundred rooms arranged in an irregular semicircle round a central plaza. The building increases in height

*The extraordinary ruins of Pueblo Bonito in Chaco Canyon, viewed from the mesa top to the northwest. (Courtesy of Clive Ruggles)*

from a single story by the plaza up to an incredible five stories next to the outer wall. In the plaza itself were numerous kivas—underground chambers built for religious ceremonies—of various sizes, including two "great kivas" around fifteen meters (fifty feet) across.

The canyon has also been the focus for a variety of archaeoastronomical theories that, as much as anything, illustrate the changing focus of the discipline since its early development in the 1970s to the present day. For example, a hot issue in the 1970s was whether a rock-art pictograph painted on one of the canyon walls depicted a particular supernova explosion in C.E. 1054. Another discovery that attracted a great deal of popular attention was the "sun dagger"—an interplay of sunlight and shadow on two spiral petroglyphs high up on the isolated pinnacle of Fajada Butte at the eastern entrance to the valley. Finally, there have been the inevitable studies of orientations and alignments, which in the early days followed a paradigm that had been established in the context of British stone monuments: this was to search for alignments upon a predefined set of astronomical targets consisting of the horizon rising and setting positions of the sun at the solstices and equinoxes and of the moon at the lunar "standstills."

These early approaches tended to be motivated more strongly by an interest in the nature of the astronomical practice itself than by an interest in the culture within which the practice operated. More recently, there has been a closer convergence between the questions asked by archaeoastronomers and the broader social issues of greater interest to archaeologists at large.

Archaeologically, one of the most intriguing questions about Chaco is why social organization and architectural technology should have reached such impressive heights in such an improbable place, and how they were supported. It is clear that a huge and well-organized infrastructure was needed, for example, in transporting tens of thousands of large timber beams for the construction of floors and roofs over distances of a hundred kilometers (sixty miles) or more from the nearest forests. A key issue is the size of the resident population. It seems unthinkable that the scant food sources available could have supported a population that filled all the rooms of the Great Houses, and this, together with other factors such as the preponderance of kivas, invites the suggestion that the canyon was first and foremost a religious center, with a relatively modest number of permanent residents. If so, then it was likely controlled by an elite who exerted both ideological and political control. One suggestion is that pilgrims are likely to have converged here in considerable numbers at particular times, and that these times would have been astronomically determined.

In recent years, convincing evidence has emerged that the houses, kivas, roads, stairways, and even middens around Chaco formed a carefully planned "ritual landscape" centered upon the Great Houses of the canyon itself. Within this scheme of things, it is perhaps not surprising that a number of buildings seem to have been deliberately oriented along one or the other of the cardinal directions, or else display a strong degree of symmetry about their north-south and/or east-west axes. Similar principles seem to have extended to the layout of buildings within the landscape, with several of the principal ones lying on an east-west axis and a north-south axis, which cross in the valley. When these axes were first discovered, many dismissed them as fortuitous rather than deliberate, but this changed with the subsequent discovery of the (Great) North Road, which connects with the north-south axis and runs due north from Chaco for more than fifty kilometers.

Astronomy, then, was part of the conceptual framework that lay behind the location and design of many of the principal Chaco constructions—at least on a basic level, since the north-south direction is ultimately defined in relation to the daily rotation of the celestial sphere about the celestial pole. This may not sound too exciting to a modern astronomer looking for astronomical "sophistication" (as defined in his or her own terms), but it is extremely interesting from a cultural perspective, because it suggests a degree of continuity of worldview extending down to the present day. The principles that defined the spatial layout of Chaco buildings and their landscape were apparently derived from a quadripartite cosmology—a perceived division of the world into four quarters around some central place. Such divisions have persisted as a characteristic feature of indigenous North American worldviews up to modern times.

**See also:**
Casa Rinconada; Chaco Meridian; Chaco Superova Pictograph; Fajada Butte Sun Dagger.
Celestial Sphere; Diurnal Motion.

**References and further reading**
Lekson, Stephen H. *Great Pueblo Architecture of Chaco Canyon, New Mexico.* Albuquerque: University of New Mexico Press, 1987.
Lekson, Stephen H., John R. Stein, and Simon J. Ortiz. *Chaco Canyon: A Center and Its World,* 45–80. Santa Fe: Museum of New Mexico Press, 1994.
Malville, J. McKim, and Nancy J. Malville. "Pilgrimage and Astronomy in Chaco Canyon, New Mexico." In D. P. Dubey, ed. *Pilgrimage Studies: The Power of Sacred Places,* 206–241. Allahabad, India: Society of Pilgrimage Studies, 2000.
Noble, David G., ed. *New Light on Chaco Canyon.* Santa Fe: School of American Research Press, 1984.
Plog, Stephen. *Ancient Peoples of the American Southwest,* 96–111. London: Thames and Hudson, 1997.
Renfrew, Colin. "Production and Consumption in a Sacred Economy." *American Antiquity* 66 (2001), 14–25.
Romano, Guiliano, and Gustavo Traversari. *Colloquio Internazionale Archeologia e Astronomia,* 137–151. Rome: Giorgio Bretschneider Editore, 1991.

# Chaco Meridian

Chaco Canyon in New Mexico was at the heart of a bustling landscape. Not surprisingly for a major cultural center, trackways converged on it from the surrounding villages and settlements. However, some of these make little practical sense in terms of communication or the transportation of goods. Typically a few meters (fifteen–thirty feet) wide, these "roads" were created by digging out topsoil or clearing away surface rocks and were sometimes lined with stone or adobe kerbs. And they ran almost perfectly straight, sometimes for several kilometers, completely disregarding the topography. (In this sense they bear more than a passing resemblance to some of the Nasca lines in faraway Peru.) Sometimes they were multiple, with up to four roads running in parallel. When they approached the steep walls of the canyon itself, rather than deviating to find easier routes down into the valley, they descended directly via precarious steps carved into the living rock (supplemented where necessary by less durable devices such as ramps). At first, these roads were thought to connect the Great Houses of Chaco with outlying villages or sacred sites many tens of kilometers away, but many of the roads running out from the canyon (or from outlying sites and pointing back toward Chaco) stop short after a few kilometers—seemingly functioning as symbolic pointers only.

One, which has become known as the (Great) North Road, is much longer and particularly remarkable. This nine-meter-wide road runs northwards from Pueblo Alto, one of the earlier structures in the Chaco Canyon area (built about 1020 C.E. over a pre-existing Great House), continuing through the landscape for more than 55 kilometers (35 miles) before reaching another canyon (Kutz Canyon) at a site known as Twin Angels Pueblo. Allowing for a dog-leg westwards along the canyon floor, another stretch striking out northwards would have brought the North Road a total of eighty-five km (fifty-three miles) to Aztec, a major cultural center in the Four Corners area where construction began just as major construction activity ceased at Chaco: in the early twelfth century.

The orientation of the North Road, and especially the main extant stretch from about three kilometers (two miles) north of Pueblo Alto northward to the southern rim of Kutz Canyon, is impressively close to the true north-south direction—the meridian. The southernmost third of this fifty-kilometer (thirty-mile) section heads only half a degree west of north, whereupon it turns slightly and heads two degrees east of north for the remainder. The whole of this stretch of road therefore remains within a conceptual north-south corridor little more than one kilometer wide. What determined its orientation? Did it simply join together two already existing places? The chronological evidence seems to rule out this idea. Instead, the construction of the road seems to have coincided broadly with the time when construction activity was just commencing at Aztec (and Chaco was on the point of falling into sudden decline).

Is the alternative possible, then, that the orientation defined the location of Aztec? If so, we must first ask the question of whether the cardinal orientation of the road was deliberate. The layout and design of the structures in Chaco Canyon itself strongly suggest that the answer is yes: many of the Great Houses and kivas in the canyon, such as Casa Rinconada, were built on a cardinal orientation, and some of the principal ones were themselves placed on cardinal axes within the landscape. Yet the orientation of the road was not so precise that one has to postulate the use of anything but basic surveying techniques. More impressive is the principle—the strength of belief that determined not only that a number of minor pueblos should be situated along the road, but that a major new center should be constructed on this very axis, connected, as it were, back to the source. This principle is also demonstrated elsewhere in the landscape around Chaco, where several roads (conceptually, if not actually) connect later buildings to earlier ones. (Examples also occur elsewhere in the world. One is the Neolithic stone circle and henge at Avebury in England, where a stone avenue [the Kennet Avenue] ran for over two kilometers [1.5 miles], connecting the great henge to a centuries-older multiple-timber circle known as the Sanctuary).

Why was the north-south direction important? The answer suggested by the evidence from the layout of Chaco itself is that it was cosmologically significant—it was part of structuring the landscape according to the prevailing worldview. And the motivation for doing this, to judge from countless similar practices the world over, was to keep human action fundamentally in harmony with the cosmos and hence to ensure well-being and stability.

In the late 1990s, archaeologist Steve Lekson proposed a radical new theory, suggesting that the North Road was only part of a far longer construction that he terms the "Chaco Meridian." Its southern extension, he suggested, stretched southwards from Chaco Canyon for a staggering 630 kilometers (390 miles) into modern Mexico, to the most important regional center in the period C.E. 1250–1450: the city of Paquime. Lekson's argument proceeds as follows. The fact that the three major regional centers of Aztec, Chaco (that is, its principal north-south axis through Pueblo Alto), and Paquime lie exactly on a north-south line (to within four kilometers from east to west) seems an extraordinary coincidence, if coincidence it is. Now consider their chronology: building at Chaco took place during the period c. 950–1125; Aztec c. 1100–1275; and Paquime c. 1250–1450. As soon as one city was abandoned, it seems, another rose up in its place. The implications are clear. For some reason—perhaps an environmental disaster triggered by overexploitation of local resources—the controlling elite were forced to abandon Chaco in the early twelfth century. When this happened, having staked out a north-south line on which their new center had to lie and identified a suitable spot on that line, they subsequently relocated there. In the late thirteenth century there was an extended drought in the Four Corners area, and that the whole vicinity suffered extensive depopulation. So the elite moved again, but this time they followed the Meridian southwards—a long way south.

These ideas are challenging and controversial, but their social and cognitive ramifications are profound, in that they challenge two long-held notions concerning what is now the Southwest between C.E. 900 and 1450. The first is that the peoples of the Southwest were organized into a variety of political units essentially operating on a local scale, though communicating via networks of trade and exchange. But the sheer scale of the Chaco Meridian, and the idea that the controlling elite could have uprooted themselves so far north and then, a few generations later, several times as far in the opposite direction, seems to contradict this idea. Instead, it is possible that a vast region stretching from the Four Corners area to northern Mexico could have been economically and politically integrated, and furthermore, that this situation might have prevailed for a considerable time.

The second controversial notion is that social and economic change was driven purely by environmental factors. Not that there is any reason to doubt the conventional view that the two abandonments were driven by environmental catastrophes (although a number of questions surround the first, in particular). The point is that the locations of the new, alternative cities—first Aztec and then Paquime—were primarily determined by constraints existing within people's own minds. The manner in which worldview influences action in numerous other human cultures suggests the following explanation. Chaco was conceived as the very center of the cosmos. That center could not be shifted arbitrarily, since this would introduce further disorder into the world. Only by extending the world axis north and south could a new center, appropriately connected back to the old one and to the existing order, be established and thus continuity—and survival—be assured.

Whatever new evidence is brought to bear upon the Chaco Meridian in the future, Lekson's theory serves as a fine illustration of a much broader general principle: human action in the past was typically determined by a mix of factors, some of which seem to us wholly pragmatic and others purely conceptual. In the context of another worldview, what is "in the mind" and what we would view as "objectively real" influences in the external world are not separated but woven together.

**See also:**
Cardinal Directions.
Avebury; Casa Rinconada; Chaco Canyon; Nasca Lines and Figures.
Meridian.

**References and further reading**
Aveni, Anthony F., ed. *World Archaeoastronomy*, 365–376. Cambridge: Cambridge University Press, 1989.
Lekson, Stephen H. *The Chaco Meridian: Centers of Political Power in the Ancient Southwest*. Walnut Creek, CA: AltaMira Press, 1999.
Plog, Stephen. *Ancient Peoples of the American Southwest*, 110–111. London: Thames and Hudson, 1997.

## Chaco Supernova Pictograph

The walker exploring the trail northwestward up Chaco Canyon toward the ruins of Peñasco Blanco—one of several Great Houses built in the canyon by Anasazi peoples in the tenth and eleventh centuries—passes a number of notable rock-art panels, but one in particular is remarkable in its simplicity and power. At one point, high up the vertical sandstone cliff that forms the side of the canyon, some 6 meters (20 feet) off the ground, is a narrow overhang. And painted on the bottom of this are three straightforward designs: a (life-sized) hand, a ten-pointed star, and a crescent—surely a depiction of the crescent moon.

*The "supernova" petroglyph at Chaco Canyon, viewed from below. (Courtesy of Clive Ruggles)*

This site has achieved particular notoriety because it has been claimed to be a representation of a supernova—an exploding star, previously invisible to the naked eye, that suddenly grows spectacularly in brightness. The supernova in question is one that astronomers know to have occurred in C.E. 1054 and which was certainly recorded by the ancient Chinese. It is often referred to nowadays as the Crab supernova, because the aftermath of the explosion is visible to astronomers today as a cloud of gas in the constellation of Taurus known as the Crab Nebula. For a few weeks the new star was significantly brighter even than the planet Venus and visible in broad daylight.

Anyone who has seen Venus and the crescent moon close together, hanging low in a clear blue and otherwise empty twilight sky, will know just what a conspicuous sight this is. In 1054, though—in the early morning of July 5 (in today's Gregorian calendar)—an even more extraordinary sight would have greeted anyone sheltering by the cliff looking eastwards into the predawn sky. On this morning, the crescent moon would have appeared right next to the supernova, and the suggestion is that this is precisely what the Chaco pictograph represents. (It is a pictograph rather than a petroglyph because it is painted onto, rather than carved out of, the rock.)

A number of factors seem to support this argument. First, the date fits with the archaeological evidence of when Anasazi activity reached such ex-

The situation of the "supernova" petroglyph on the bottom of a rock overhang. (Courtesy of Clive Ruggles)

traordinary heights in this apparently inhospitable valley, with the construction of numerous Great Houses and kivas. Second, the juxtaposition of the crescent moon and star in the pictograph fits the actual configuration visible in the sky on that July morning in 1054. Third, several other rock-art sites in the western United States have been interpreted as depictions of the same event. And finally, there is independent evidence of the ancient Chacoans' interest in astronomy, manifested most famously in the interplay of sunlight and shadow on the "sun dagger" petroglyph on Fajada Butte at the eastern entrance to the canyon.

Nonetheless, even if we accept that the crescent is an unmistakable representation of the crescent moon, is the star necessarily the Crab supernova? The answer, of course, is no. The "star and crescent" is one of the most pervasive astronomical symbols among humankind, particularly in the Islamic world. It is usually taken to represent the crescent moon and Venus, which—aside from transient phenomena such as supernovae—is the brightest object in the sky apart from the sun and moon. Eye-catching conjunctions of the crescent moon and Venus occur at quite frequent intervals (several times during each eight-year period that it takes Venus to return to essentially the same place relative to the sun and stars). Could the juxtaposed star and crescent symbols in the Chaco pictograph simply be depicting one of these conjunctions of the crescent moon and Venus? They clearly could: just because one rather attractive and exciting interpretation seems to fit the evidence, we cannot accept this as incontrovertible proof if there remains at least one equally plausible, though slightly more mundane, alternative.

Another uncomfortable question is: what is the hand doing there? It has been argued that it is merely a symbol showing that the place is sacred, but its presence might also suggest the uncomfortable possibility that these three icons belonged together as a triplet, conveying symbolic meanings perhaps, rather than (two of them but not the other) representing something "literal" in the sky.

What characterizes all these arguments is that they are based on speculations that pay no attention to the cultural context. In fact, all of them are rendered futile by an observation made by a Zuni ethnographer almost as soon as the supernova interpretation was originally proposed, back in 1975. The modern Zuni have sun-watching stations for calendrical observations, and traditionally mark them using a pictograph with four elements: a crescent, a hand, a star, and a sundisk. Although more worn and less visible, there is indeed a sundisk accompanying the other symbols on the overhang. This, surely, was a sun-watching station: astronomical certainly, but with nothing at all to do with the Crab supernova.

**See also:**
Comets, Novae, and Meteors; Cosmology; Star and Crescent Symbol; Symbols.

Casa Rinconada; Chaco Canyon; Chaco Meridian; Chinese Astronomy. Inferior Planets, Motions of.

**References and further reading**
Aveni, Anthony F., ed. *World Archaeoastronomy,* 144–145. Cambridge: Cambridge University Press, 1989.
Brandt, John, and Ray Williamson. "The 1054 Supernova and Native American Rock Art." *Archaeoastronomy* 1 (supplement to *Journal for the History of Astronomy* 10) (1979), S1–38.
Frommert, Hartmut, and Chrisine Kronberg. *Supernova 1054—Creation of the Crab Nebula.* http://seds.lpl.arizona.edu/messier/more/m001_sn.html.
Krupp, Edwin C., ed. *In Search of Ancient Astronomies,* 137–141. New York: Doubleday, 1977.
Mitton, Simon. *The Crab Nebula,* 15–29. New York: Charles Scribner's Sons, 1978.
Plog, Stephen. *Ancient Peoples of the American Southwest,* 100–101. London: Thames and Hudson, 1997.
Schaafsma, Polly. *Rock Art in New Mexico* (rev. ed.). Santa Fe: Museum of New Mexico Press, 1992.

# Chinese Astronomy

The political history of ancient China was no less turbulent than that of many other parts of the world. From about 4000 B.C.E. onwards it developed from a multitude of farming villages into warring local chiefdoms, and these gradually confederated into larger competing states ruled by powerful dynasties. The best known of these are the Shang and Zhou, which successively controlled ever greater swathes of eastern China during the second and first millennia B.C.E. Yet despite all this social turmoil, a characteristic cultural tradition emerged that is instantly recognizable in arts, crafts, and architecture. One reason is that it arose in geographical isolation from other developing cultures farther west. In 221 B.C.E., China was unified, thus forming one of the world's greatest empires. Further great dynasties, such as the Han, T'ang, Song, Yuan (when China was under Mongul control), Ming, and finally the Qing, endured until the Chinese revolution of C.E. 1911. Imperial China is characterized at various stages by impressive and highly distinctive developments in scholarship, literature, and art.

Many aspects of Chinese astronomy, from ancient times to the later imperial dynasties, are equally distinctive. Some of the earliest Chinese records of astronomical events, for example, are found on divinatory devices known as *oracle bones* that date to the Shang and Zhou dynasties. These were animal bones or turtle shells heated with a hot needle until cracks appeared in their surface. The shape of the cracks provided prognostications used by kings to assist in affairs of state. Fortunately for modern scholars, the prognostication was recorded on the bone itself, together with the eventual outcome (tending to confirm that the prognostication was correct). Some oracle

*One of the collection of astronomical instruments on the roof of the Old Observatory in Beijing, China. (Courtesy of Clive Ruggles)*

bones relate to astronomical events, including both solar and lunar eclipses, and form the earliest Chinese records of such phenomena.

In the ensuing centuries and millennia of China's imperial era, the recording of occurrences in the skies became systematized. Yet in China, unlike (for example) in ancient Babylonia, a clear distinction was maintained between what we might see as scientific (astronomical) and divinatory (astrological) traditions. The first was predicated on the regularity of calendrical cycles, which could be established through careful observation and recording and then reliably predicted. The latter concerned unpredictable phenomena, which had divinatory significance: these were omens to be noted and acted upon. There were even separate practitioners: experts in calendrical methods and experts in celestial patterns (whom we might choose to categorize as astrologers). Both were highly valued state employees. The goal of the state astronomers, and their principal raison d'être, was to produce ever more detailed and accurate almanacs that would reinforce (perceptions of) the emperor's control of order in the cosmos. The purpose of the astrologers, on the other hand, was to warn the emperor of portents in the skies and aid him in making crucial decisions. Rulers felt they had a mandate from heaven, but if they ignored signs of approval or disapproval and failed to act accordingly, consequences could be dire. The two sets of records—calendrical and astrological—combine to bequeath us an unparalleled data set of observations of astronomical events, both regular and recurrent (such as planetary motions) and transient (one-off), such as the appearance of comets, novae, meteor showers, aurorae, sunspots, and of course eclipses.

Occasionally, a form of celestial phenomenon would pass from the domain of divination to that of calendrics when its underlying regularity was discovered. The classic example of this in the Chinese case is that of lunar eclipses. Around the year 0, the Chinese discovered a cycle of ten years and 334 days (135 lunations), which, like the longer Saros cycle discovered by the Babylonians, defines a period after which a lunar eclipse will tend to recur. With this discovery lunar eclipses essentially became predictable, and while they continued to be recorded, they ceased from this point to have any significance as omens.

A landmark in ancient Chinese astronomy was the development of a distinctive reference system that enabled specialists to map stars, to note the positions of the sun, moon, planets, or transient phenomena, and to reckon time at night. This was the system of twenty-eight *xiu* or *lunar lodges*. These are best conceptualized as twenty-eight regions into which the sky was divided like the segments of an orange and which were identified by convenient stars and asterisms situated within them. If the *xiu* had been of equal width, the moon would have moved from one to the next in each twenty-four-hour period. However, the widths varied considerably, and the

Chest from the tomb of Marquis Yi of Zeng (late fifth century B.C.E.) featuring a representation of the asterisms marking the twenty-eight lunar mansions. (Instructional Resources Corporation)

moon could spend two or three nights in a very wide *xiu*, but only a few hours in a narrow one.

In Imperial China, astronomy was inextricably linked to the supreme power of the emperor, who was seen as uniquely able to harness the sacred forces of the heavens and preserve the balance and harmony of the cosmos in the service of the people. This power was reflected (and the emperor's right to rule thereby reinforced) both in the spatial layout of the palace and in the timing of appropriate rituals. From the Ming dynasty (fourteenth century) onwards, the imperial palace—otherwise known as the Forbidden City—was in the heart of Beijing and was approached along the meridian, directly toward the north celestial pole, which was seen as the very heart of the heavens. The Emperor moved around the sacred capital on prescribed paths at different seasons, leaving the palace so as to undertake rituals at various nearby shrines at the appropriate times. Two of the most important ceremonies took place at the solstices, when the emperor was required to actively assist in the transitions between the seasons. At winter solstice, he ascended the Round Mound, the highest part of the Temple of Heaven complex to the south of the Forbidden City, climbing up from the south and publicly paying homage by himself facing the center of the heavens in the north. At other times, subjects permitted to visit the emperor in the Forbidden City were required, similarly, to approach his throne—the center of the earthly realm—from due south.

Chinese astronomers established an impressive record in the development and construction of astronomical instruments. One of the most remarkable is a giant gnomon, in the form of a brick tower, built by the astronomer Guo Shou Jing in C.E. 1276. This enabled Guo to establish the length of the tropical (seasonal) year with an error of only twenty-six seconds. A collection of large Bronze instruments that were used for mapping and timekeeping can now be found on the roof of the Old Observatory in Beijing.

However, it is as well to remember the context in which these achievements took place. There was certainly a demand for ever greater astronomical and calendrical precision, but it resulted ultimately from the perceived need for the emperor to maintain cosmic harmony. In other words, the reasons astronomy was maintained at such a high level for so long in Imperial China were ultimately ideological and political rather than purely scientific, at least as we would see it.

See also:
Astrology; Comets, Novae, and Meteors; Eclipse Records and the Earth's Rotation; Lunar Eclipses; Power; Solar Eclipses.
Babylonian Astronomy and Astrology.
Cardinal Directions; Celestial Sphere; Meridian; Solstices.

**References and further reading**
Aveni, Anthony F., ed. *Empires of Time: Calendars, Clocks and Cultures,* 305–322. New York: Basic Books, 1989.
———, ed. *World Archaeoastronomy,* 55–75. Cambridge: Cambridge University Press, 1989.
Harley, J. Brian, and David Woodward, eds. *The History of Cartography, Volume Two, Book Two: Cartography in the Traditional East and Southeast Asian Societies,* 511–578. Chicago: University of Chicago Press, 1994.
Hoskin, Michael, ed. *Cambridge Illustrated History of Astronomy,* 48–49. Cambridge: Cambridge University Press, 1997.
Krupp, Edwin C. *Echoes of the Ancient Skies,* 196–198, 259–263. Oxford: Oxford University Press, 1983.
———. *Skywatchers, Shamans and Kings,* 271–274. New York: Wiley, 1997.
Needham, Joseph. *Science and Civilisation in China, Vol. 3: The Sciences of the Heavens and Earth.* New York: Cambridge University Press, 1959.
Ruggles, Clive, and Nicholas Saunders, eds. *Astronomies and Cultures,* 32–66. Niwot: University Press of Colorado, 1993.
Selin, Helaine, ed. *Astronomy across Cultures,* 423–454. Dordrecht: Kluwer, 2000.
Thurston, Hugh. *Early Astronomy,* 84–109. Berlin: Springer-Verlag, 1994.
Walker, Christopher, ed. *Astronomy before the Telescope,* 245–268. London: British Museum Press, 1996.
Xu Zhentao, David Pankenier, and Jiang Yaotiao. *East Asian Archaeoastronomy: Historical Records of Astronomical Observations of China, Japan and Korea.* Amsterdam: Gordon and Breach, 2000.

# Christianization of "Pagan" Festivals

Wherever a ruling elite seeks to impose or stimulate a change in the dominant religious beliefs of the populace, as when controlling a new population following a victory in war, they may tear down old places of worship and build different ones on the same sites, replace existing sacred myths and stories with ones that reflect the new ideology, and introduce novel rites and ceremonies in the hope of eliminating the existing ones. This is as true of the spread of Christianity as of any other religion. Throughout history, new Christian churches have been placed on the sites of "pagan" temples. This process is particularly evident in the Republic of Georgia, where the conversion to Christianity occurred as early as the fourth century C.E., and archaeologists excavating under early churches are wont to discover pre-Christian temples built several centuries earlier. Where indigenous religious festivals were timed in relation to the calendar, or tied to particular astronomical observations, it made sense to schedule the new Christian festivals to coincide with them, thereby upstaging them. A well-known example of this is the timing of Christmas to coincide with pre-existing winter solstice festivals; likewise, the feast of St. John is scheduled close to the summer solstice.

The result was the transformation of pagan festivals into Christian ones, but the original meaning was not always lost. In Mexico, for example, the feast day of St. Michael, as observed in the village of San Miguel del Milagro in the state of Tlaxcala, preserves some aspects of ancient observances relating to mountain and fertility gods.

It has been supposed that the timing of certain Christian feasts on mid-quarter days reflects the Christianization within Europe of earlier Celtic festivals dividing the year into eight equal parts. The best known of these is the feast of All Saints, which takes place on November 1, on the traditional date attributed to the Celtic festival of Samhain. The notion of a Celtic calendar that was widespread through western Europe in the Iron Age and provided an accurate division of the year into eight equal parts is problematic. Yet there is no doubt that a significant pagan "start of winter" festival somehow became transformed into a Christian festival. Its main association now, in many parts of the world, is with the dead, but this probably came later.

**See also:**
Celtic Calendar; Mid-Quarter Days.
Cacaxtla; Mithraism.
Solstices.

**References and further reading**
Hutton, Ron. *The Pagan Religions of the Ancient British Isles*, 247–341. Oxford: Blackwell, 1991.
———. *The Stations of the Sun: A History of the Ritual Year in Britain*, 360–370. Oxford: Oxford University Press, 1996.

McCluskey, Stephen C. *Astronomies and Cultures in Early Medieval Europe*, 60–76. Cambridge: Cambridge University Press, 1998.

Ruggles, Clive, and Nicholas Saunders, eds. *Astronomies and Cultures*, 100–123. Niwot: University Press of Colorado, 1993.

## Church Orientations

Christian churches generally point eastwards. Liturgical traditions dating from medieval times associate the direction east—the rising place of the heavenly bodies and particularly of the sun—with the resurrection of Christ and the dawning of the "day of eternity" for righteous souls. Worshipers therefore face, symbolically, their eventual home in paradise.

"East," however, evidently did not necessarily mean "due east." The orientations of many churches actually deviate from true east by considerable amounts. In modern times church builders were often constrained by the space available within densely populated towns and cities, so this is scarcely surprising, but in earlier centuries and in rural settings such limitations seldom existed. Given the care and attention that were afforded to many other aspects of a church's construction, it seems inconceivable that their orientations were merely poor attempts at facing due east. The question, then, is how the direction in which a given church should face was determined in practice. What motives and procedures led to the resulting orientations is seldom clear from surviving historical accounts or liturgical texts, and the question has fascinated a number of scholars in recent years.

There have been various suggestions, the main ones of which nearly all relate in some way to the rising sun. One is that churches simply needed to face within the eastern quarter of the horizon; the details were not important. A second is that they needed to face within the solar arc—toward a point on the horizon where the sun would rise at some time in the year—but again, the actual position within this range was not important. A third suggestion is that they faced the rising sun on the day when the foundations were set out. A fourth: they faced sunrise on the feast day of the saint to whom the church was dedicated. A fifth: they faced sunrise on the equinox. And a sixth possibility: they faced sunrise at Easter. The list is not exhaustive, and it is not at all improbable that different traditions and practices held sway in different places and at different times.

Occasionally, we have historical evidence of a particular practice. For instance, there is clear support for the third suggestion in seventeenth and eighteenth century England. Historian Sir Henry Chauncy, in a book describing the historical antiquities of Hertfordshire written at the end of the seventeenth century, reported a common practice of aligning the foundations in the direction of sunrise. Over a century later, the poet William Wordsworth graphically described a similar practice far away in the county of Westmoreland. Those who were charged with building Rydal Chapel kept vigil through a whole night, waiting for the sun to appear, at which time they carefully laid out the foundation stones and thus ensured that the building—and especially the high altar—would be correctly placed. The night concerned was not chosen arbitrarily: it was the night preceding the feast day of the church's patron saint.

Nevertheless, it is possible that Chauncy, Wordsworth, and others were merely expressing a romanticized notion rather than hard evidence. For that, we need to measure the orientations of particular churches and compare them with sunrise on the day when construction began, or with sunrise on the patronal feast day. Sadly, the date when construction started was rarely recorded; what mattered was the date of consecration of the completed church. Identifying the patronal feast day can also give us problems, since we do not always know the saint to whom the church was originally dedicated. Even where the church is still in use, the modern patron may not be the same as the original one.

In most cases, the only evidence remaining is from the orientations of the churches themselves. Here we must be careful, since so many churches have been partially rebuilt, extended, or altered since their original construction. It is certainly necessary to pay strict attention to historical records of construction phases where they are available. How can the orientations themselves help us choose between the six possibilities already mentioned? The answer is that if we observe a large enough group of churches conforming to a common practice, then the resulting spreads of orientations should be distinctively different. At all but the most northerly latitudes (above about 55°N), general orientation within the solar arc would result in a narrower range of orientations than general orientation within the eastern quarter, but both practices would simply result in a fairly even spread of orientations within this range, perhaps concentrated toward the middle. Orientation on sunrise on the day of laying the foundations would also result in a spread of orientations throughout the solar arc, but concentrated toward the ends if the foundation dates were spread evenly through the year. This is because the change in the sun's rising position from day to day is much smaller close to the solstices.

In order to investigate the remaining possibilities, we must determine more precisely the dates on which the sun rose at a particular spot on the horizon. For this, it is necessary to take into account the altitude of the horizon and to convert azimuths (orientations) into declinations. Orientation of sunrise on saints' days would result in sharp spikes corresponding to sunrise on a few specific dates in the year. (Here we encounter another complication: the dates in many cases would have been determined according to the Julian calendar rather than the modern Gregorian calendar.) Orientation on equinoctial sunrise should result in a single, reasonably sharp concentration of declinations, but quite possibly significantly offset from the true (astronomical) equinox, both because of the use of the Julian calendar and depending upon how the equinox was determined in practice. Easter sunrise orientation, finally, would result in a wider spread of orientations, yet much narrower than the whole solar range, since it would correspond to sunrise over the four- or five-week period following the vernal equinox (again, reckoned in the Julian calendar for earlier churches) where Easter could fall.

In recent years, some studies of the orientation of churches of various ages in eastern Europe, Austria, Italy, and Great Britain have supported the idea of orientation upon sunrise on patron saints' feast days. There is also evidence of solstitial and equinoctial orientation among thirteenth- and fourteenth-century churches in Hungary and elsewhere. The significance of these dates in medieval times is unquestionable: as early as the fourth century, feasts marking the births of Christ and John the Baptist had been established at the two solstices, together with feasts marking their respective conceptions (the Feast of the Annunciation and the Feast of the Visitation) at the two equinoxes. Other surveys, however, have concluded that patronal orientations never even existed. It seems that problems of data selection, together with a lack of rigorous statistics, may lie at the heart of these inconsistencies.

A recent systematic study of the orientations of medieval village churches in central England, by the American archaeoastronomer Stephen McCluskey, has revealed a complex picture that does not fit cleanly any of the six models. McCluskey restricted his sample to around a hundred churches that were dedicated to four particular patrons: Mary, John the Baptist, All Saints, and Andrew. The related feast days would be the spring equinox, the summer solstice, November 1, and November 30. While the orientations of the churches all fit comfortably within the solar arc—facing the direction of sunrise at some time in the year—they did not correlate with the four relevant saints' days. Instead, the spread of orientations corresponded broadly to the date range from mid-February to mid-April (or in the late summer and autumn, the period from early August to early October) in the Julian calendar. This spread is too narrow to fit the "any time of year" model, yet too wide to fit the "Easter" model. There is no obvious, simple explanation, and it seems likely that a number of different traditions and practices were operating, even within this confined group of churches. However, certain preferences do stand out. For example, All Saints churches were preferentially oriented upon sunrise on the (Julian) spring equinox, something that may be explained in liturgical texts connecting the day of eternity—"god's eighth day"—with (strictly?) eastward orientation, and with all the saints. Another, unexpected, trend is that the naves of churches of St. John the Baptist tend to point *westward* toward sunset on the feast of the conception of this saint on the autumnal equinox. Was this their defining characteristic?

Despite the patchiness and evident complexity of the data, there seems little doubt that astronomical observations, and particularly observations of the rising sun, were commonly used to determine the orientations of medieval churches all over Europe. Such practices may have continued well into early modern times, especially in rural areas. The investigations of such issues to date have only touched the tip of the iceberg: the evidence is extensive and remains largely unexplored.

See also:
Cardinal Directions; Equinoxes; Methodology; Nissen, Heinrich (1839–1912); Solstitial Directions.
Gregorian Calendar; Julian Calendar.
Altitude; Azimuth; Declination; Sun, Motions of.

**References and further reading**
Ali, Jason, and Peter Cunich. "The Orientation of Churches: Some New Evidence." *Antiquaries Journal* 81 (2001), 155–193.
Aveni, Anthony F., ed. *World Archaeoastronomy*, 430–440. Cambridge: Cambridge University Press, 1989.
Benson, Hugh. "Church Orientations and Patronal Festivals." *Antiquaries Journal* 36 (1956), 205–213.
Campion, Nicholas, ed. *The Inspiration of Astronomical Phenomena: Proceedings of the Fourth Conference on the Inspiration of Astronomical Phenomena, Magdalen College, Oxford, 3–9 August 2003*, 209–224. Bristol: Cinnabar Books, 2005.
Hoskin, Michael. *Tombs, Temples and Their Orientations*, 7–8. Bognor Regis, UK: Ocarina Books, 2001.
Jaschek, Carlos, and F. Atrio Barandela, eds. *Actas del IV Congreso de la SEAC «Astronomía en la Cultura» [Proceedings of the IV SEAC Meeting "Astronomy and Culture"]*, 149–155. Salamanca, Spain: Universidad de Salamanca, 1997. [Article is in English.]
McCluskey, Stephen C. "Calendric Cycles, the Eighth Day of the World, and the Orientation of English Churches." In Clive Ruggles and Gary Urton, eds. *Cultural Astronomy in New World Cosmologies*, in press.
Morris, Richard. *Churches in the Landscape*, 208–209. London: Dent, 1989.
Zenner, Marie-Therese, ed. *Villard's Legacy: Studies in Medieval Technology, Science and Art in Memory of Jean Gimpel*, 197–210. Aldershot, UK: Ashgate, 2004.

# Circles of Earth, Timber, and Stone

The practice of building circular, or at least roughly circular, enclosures and monuments was widespread in Europe in the Neolithic and Bronze Age. In Britain, for example, some of the earliest farming communities in the fourth millennium B.C.E. built earthworks known as *causewayed enclosures*. These

*A set of concentric timber ovals at Woodhenge, Wiltshire, England, viewed along the major axis in the direction of midsummer sunrise. The positions of the original timber posts are marked by concrete pillars. (Courtesy of Clive Ruggles)*

are mostly fairly circular, although some are highly irregular, and most of the examples are in southern England. They seem to have had a range of purposes, but they were almost certainly places where people from small surrounding communities came together. Later in the Neolithic, more regular circular earthworks known as *henges* started to appear, as did circles of upright timbers and circles of standing stones. Henges are generally more regular (better approximations of true circles) than causewayed enclosures. They consist of a ditch and exterior bank, usually with one or two entrances, and vary in size from a few meters across to the large *henge enclosures* or *superhenges* such as Avebury in Wiltshire, southern England. The large henge at Avebury is some 350 meters (1,150 feet) across, had a ditch some nine meters (thirty feet) deep and a bank eight meters (twenty-five feet) high, and surrounds an entire village. (Ironically, the name *henge* was coined for a type of monument that was like Stonehenge but without the stones; the name *Stonehenge* means "hanging stones" in Old English. The irony is that, by the generally accepted definition we have just given, Stonehenge is not actually a henge; its earthen enclosure has the bank on the interior of the ditch.)

In trying to understand the purpose and meaning of this type of monument, it is less helpful than was once thought to classify them primarily by the type of materials used, that is, as earthen circles, timber circles, or stone circles. This is because these do not form three separate categories—far from it. Excavations suggest that many circles of uprights were probably sculpted first in wood and then repeated in stone, a good example being Temple Wood in Argyll, western Scotland. Many henges containing stone circles have survived into modern times (e.g., the Ring of Brodgar and Avebury), although the sequence of chronological development is seldom clear without excavation. Others contained timber circles, and Woodhenge, situated some three kilometers (two miles) east-northeast of Stonehenge, contained no fewer than six concentric ellipses (some have suggested that they represent the postholes of a roofed building, but this has been the subject of intensive debate). Finally, Stonehenge itself represents an earthen circle with an interior ring of timber uprights, later superseded by the famous constructions in stone.

Stone circles (whether built as such or as elements of more complex sites) survive relatively well and the remains of many of them are still conspicuous in today's landscape. Timber uprights, on the other hand, have long since rotted away, perhaps leaving postholes that might one day be discovered by excavation. Many earthen ditches have long since filled in and banks been ploughed away, only to be revealed, where we are lucky, in cropmarks visible from aerial photographs. It is a reasonable supposition that the number of timber and earth circles built in Neolithic and Bronze Age Britain was at least comparable with, and possibly considerably greater than, the number of stone circles.

Why build a circle? On the face of it, a circle is the simplest way to mark off an area of sacred space, but there may be a great deal more to it than that. Important cognitive principles surely lay behind such monuments and helped determine how they were integrated into the landscape. The later prehistoric circles of earth, timber, and stone were surely places where people gathered, very likely for formal or ritualistic activities or observances. Perhaps, in some cases, only a privileged few were allowed to take part. The presence of apparently formal approaches to some sites, such as the earthen avenue that leads into Stonehenge and the stone avenues leading into Avebury, suggests that the experiences of the participants were carefully orchestrated. The permeability of the boundary is clearly important: there would have been a world of difference between undertaking a private observance inside a large henge with a high outer bank, cut off from the wider world and hidden from outside view, and giving an open performance inside a stone circle, feeling fully a part of that wider world. The British archaeologist Richard Bradley has identified a number of examples where circles seem to have been "closed off" over time, perhaps as part of a wider process of making both the places themselves and what went on there less generally accessible, more limited to a privileged few.

One possibility, for which there are plenty of historical and ethnographic parallels, is that many of these circles might have been conceived as *central places,* conceptual centers of the world. It is even possible that the circular shape might have been intended, at least initially, to be a reflection of the visible horizon, and that some circles might have been perceived as a microcosm encapsulating and reflecting the properties of the wider cosmos in various ways. Following this idea, a number of British archaeologists have explored various ways in which the properties of the surrounding world might have been reflected at these sites. The possibilities include the shape of a ring reflecting the surrounding topography, the shapes of stones in stone circles reflecting the shapes of hills behind them, and alignments (e.g., of entrances) toward sources of water. Unfortunately there are so many possibilities that it is difficult not to be entirely subjective.

Nonetheless, it is by this route that the alignment studies undertaken by archaeoastronomers for many years have begun to dovetail within the agenda of archaeologists interested in wider issues of cognition and cosmology. For astronomers, it is stone circles that have attracted most attention, since they leave readily surveyable remains above the ground and (subject to vegetation cover) alignments that can still be directly viewed. Nonetheless, some important astronomical features of timber circles have been noted, such as the orientation of Woodhenge, which, in common with Stonehenge itself and another henge in the general vicinity, Coneybury (which has now been obliterated), faces northeastwards toward midsummer sunrise.

See also:
Alignment Studies; Cosmology; Sacred Geographies.
Avebury; Brodgar, Ring of; Cahokia; Stone Circles; Stonehenge; Thornborough.

**References and further reading**
Bradley, Richard. *The Significance of Monuments.* London: Routledge, 1998.
Burl, Aubrey. *The Stone Circles of Britain, Ireland and Brittany.* New Haven: Yale University Press, 2000.
Gibson, Alex. *Stonehenge and Timber Circles.* Mount Pleasant, SC: Tempus, 1998.
Harding, Jan. *Henge Monuments of the British Isles.* Mount Pleasant, SC: Tempus, 2003.
Mercer, Roger. *Causewayed Enclosures.* Princes Risborough, UK: Shire, 1990.
Oswald, Alastair, Carolyn Dyer, and Martyn Barber. *The Creation of Monuments: Neolithic Causewayed Enclosures in the British Isles.* Swindon, UK: English Heritage, 2001.

## Circumpolar Stars

If you are in the northern hemisphere, then some stars that are close to the north celestial pole will never rise or set but simply circle around the pole once a day. These are known as circumpolar stars. Roughly speaking, these will be the stars whose declination, when added to your latitude on the earth, comes to more than 90°. For example, at latitude 40°N, all those stars with a declination greater than about 50° will circle around in the sky. The exact figure will depend upon the altitude of your northern horizon.

The same is true if you are in the southern hemisphere, except that the south celestial pole is now the relevant one, and the declination of the star (negative) plus your latitude (also counted as negative for the purpose) must now add up to less than −90°. In other words, at latitude 30°S, the circumpolar stars will be those with a declination less than about −60°.

**See also:**
Altitude; Celestial Sphere; Declination; Heliacal Rise.

**References and further reading**
Krupp, Edwin C. *Echoes of the Ancient Skies,* 5, 104. Oxford: Oxford University Press, 1983.

# Clava Cairns

Neolithic passage tombs, of which the most famous and impressive examples include Newgrange in Ireland and Maes Howe in the Scottish Orkney Islands, are found extensively in Ireland and also in significant numbers in the more northerly parts of Scotland. In a cluster around the northern Scottish city of Inverness are found a few more modest examples of the genre, which dating evidence suggests were constructed at a relatively late date—around the year 2000 B.C.E. The so-called Clava cairns derive their name from a concentration of seven such monuments at Balnuaran of Clava, on the southern banks of the river Nairn some eight kilometers (five miles) east of Inverness. They comprise two types: the passage tombs (or *passage graves*) themselves and *ring cairns,* cairns surrounded by a circular raised bank but without a passage. Circles of standing stones surround several examples of both types.

Orientation was clearly an important consideration when a Clava cairn was being constructed. The orientations of the passages, where they exist, fall without exception within a quarter of the compass between west-southwest and south-southeast. The encompassing stone circles tend to have their stones graded in height with the tallest in the southwest. The southwesterly preference in the orientations of these monuments appears to follow a dominant tradition very similar to the one that controlled the broadly contemporary recumbent stone circles found farther to the south and east.

The most evident concentration of cairns at Balnuaran of Clava comprises two passage tombs and a single ring cairn, all surrounded by stone circles. The two passage tombs are arranged so that their passages are directly in line with each other. The ring cairn, although placed between them, is slightly off to one side and does not obscure the alignment. Alexander Thom visited the site as part of his extensive campaign in the 1950s and 1960s to survey many hundreds of Scottish megalithic monuments. He noted that the two passages were not only directly in line with each other, but also closely aligned upon midwinter sunset. Twenty years later the archaeologist Aubrey Burl completed a study of the Clava cairn orientations and concluded that, while the solstitial alignment was not generally repeated at other sites, the whole group had a consistent, broader pattern of orientation related to the moon. The simplest interpretation is that it related to the midsummer full moon. In this respect, again, the Clava cairns seemed to have much in common with the nearby recumbent stone circles. As it happens, the direction of midwinter sunset falls within the broader lunar range, and the obvious conclusion from considering the monuments systematically as a group was that their significance was in fact lunar, with the specifically solar alignment at Balnuaran of Clava itself being a fortuitous occurrence within this range. The Clava cairns stood for almost twenty years as a clear example of how studying a group of monuments as a whole can modify, and constrain, "one-off" interpretations of individual sites.

All this changed in the late 1990s when another British archaeologist, Richard Bradley, commenced excavations at Balnuaran of Clava. One of his principal conclusions was that structural risks had been taken in order to conform to prevailing norms that (to our view) have no practical value—indeed, that seem to run counter to common sense. These included grading the heights not only of the surrounding stone circle but of the kerbstones and other stones within the cairn that governed its structural integrity: the tallest were placed on the southwestern side, something that made the overlying cairn structure inherently rather unstable. This confirms the importance of orientation but also hints at a much richer set of prevailing symbolic or aesthetic principles that governed the construction of these monuments.

Another extraordinary fact revealed by the excavations was that the various stones used in the construction of the cairns were often not, structurally speaking, best suited for the job. The choice of stone often seems to have had more to do with color than with (what we would see as) practical necessities such as size, shape, or strength. Most intriguing of all, stones of certain colors predominated in different parts of the cairns—with, broadly speaking, a preference for white stones facing sunrise in the east but red stones facing sunset in the west. Changes in color seem to have been related to the directions of solstitial sunrise and sunset. It is here that Bradley's discoveries provide a direct challenge to the lunar conclusion: they suggest that the predominant symbolism at the site was solar. This, combined with the very fact that the solar alignment is so precise—the setting midwinter sun shone down the full length of the passage—argue strongly that the solar symbolism, and particularly the solstitial alignment, were deliberate.

The wider issue that this conclusion raises is methodological. Every aligned structure must point somewhere, and we must seek corroborating evidence if we wish to increase our degree of belief that any particular alignment upon a specific astronomical target was actually intentional or, at the least, came to mean something to certain people in the past. There are two main ways to do this. The first is to investigate whether the alignment is repeated at similar monuments in the locality. The second is to seek a broader range of contextual evidence relating to the case in question: evidence that could inform us about the broader symbolic principles prevailing in that particular instance. The example of Balnuaran of Clava shows the potential of both methods, but also shows that they can be in direct conflict, and raises the question of how we can best achieve a reconciliation in such circumstances. One solution may be to postulate that this site, evidently the most complex and sophisticated in the Clava cairn tradition, incorporated a layer of symbolism relating to the sun that was additional to the more commonplace, and more basic, relationship to the moon.

**See also:**
Cosmology; Methodology; Solstitial Alignments; Solstitial Directions; Thom, Alexander (1894–1985).
Maes Howe; Newgrange; Recumbent Stone Circles.

**References and further reading**
Bradley, Richard. *The Good Stones: A New Investigation of the Clava Cairns.* Edinburgh: Society of Antiquaries of Scotland (Monograph Series 17), 2000.
Burl, Aubrey. *The Stone Circles of Britain, Ireland and Brittany,* 233–242. New Haven: Yale University Press, 2000.
Larsson, Lars, and Berta Stjernquist, eds. *The World-View of Prehistoric Man,* 123–135. Stockholm: Swedish Academy of Sciences, 1998.
Mithen, Steven, ed. *Creativity in Human Evolution and Prehistory,* 227–240. London: Routledge, 1998.
Ruggles, Clive. *Astronomy in Prehistoric Britain and Ireland,* 130–131. New Haven: Yale University Press, 1999.
Ruggles, Clive, and Alasdair Whittle, eds. *Astronomy and Society in Britain during the Period 4000–1500 BC,* 257–65. Oxford: British Archaeological Reports (British Series 88), 1981.
Trevarthen, David. "Illuminating the Monuments: Observation and Speculation on the Structure and Function of the Cairns at Balnuaran of Clava." *Cambridge Archaeological Journal* 10 (2000), 295–315.

# Cobo, Bernabé (1582–1657)

Our knowledge of the Inca empire as it was shortly after the time of first contact with the Europeans owes a great deal to ethnohistoric evidence, the first-hand accounts of a number of chroniclers in the first two or three generations following the conquest. Some of these were early settlers—priests, administrators, soldiers, and the like—and others were natives or of mixed blood.

Bernabé Cobo was a Spanish-born Jesuit priest and missionary who settled in Peru in 1599. He spent thirty years on various postings in South America, mostly in southern Peru and northern Bolivia, before moving to Mexico in 1630. During this time he observed and meticulously recorded a wealth of information about animals, plants, and people and their customs. In 1650 he returned to spend the last few years of his life in Peru.

Cobo is best known for his multivolume work *History of the New World,* eventually completed in 1653. Although it appeared over a century after the Spanish conqueror Pizarro first entered Cusco in 1533 and included a good deal of secondary material taken from earlier chronicles, some now lost, it provides one of the most detailed and wide-ranging accounts of practices and beliefs in the Inca world and is generally considered one of the most reliable. But for Cobo, we would know little if anything of the elaborate system of *ceque* lines that surround the Inca capital of Cusco, conceptual lines that defined and controlled social divisions as well as sacred space.

**See also:**
*Ceque* System; Cusco Sun Pillars; Island of the Sun.

**References and further reading**
D'Altroy, Terence. *The Incas.* Oxford: Blackwell, 2002.
Hamilton, Roland, ed. and trans. *History of the Inca Empire* [a translation of books 11 and 12 of Cobo's *Historia del Nuevo Mundo,* 1653]. Austin: University of Texas Press, 1979.
———. *Inca Religion and Customs* [a translation of books 13 and 14 of Cobo's *Historia del Nuevo Mundo,* 1653]. Austin: University of Texas Press, 1990.
Urton, Gary. *Inca Myths,* 28–32. Austin: University of Texas Press, and London: British Museum Press, 1999.

# Coffin Lids

Unlike those of Babylonia, the astronomical records of ancient Egypt, if they once existed, have generally not survived. It is possible that they were recorded on papyrus rather than on clay tablets, as was the Babylonian case. So-called pyramid texts, found inside tombs and coffins dating back as far as Old Kingdom times (c. 2600–2100 B.C.E.), and the younger "coffin texts" of the First Intermediate Period and the Middle Kingdom, do list various celestial bodies that represent deities, but they are in no way records of observations; their primary concern is with the passage of the soul to eternal life. The architecture of temples and pyramids in Old Kingdom times gives us some insights into how astronomical considerations affected and determined their construction. However, our main source of information about the development of astronomy and calendrics during Middle Kingdom Egypt comes from wooden coffin lids.

A modest number—about twenty—of such lids survive, dating mainly from the Ninth and Twelfth Dynasties (c. 2160–1780 B.C.E.). Painted scenes preserved on the inside incorporate arrays of hieroglyphs, typically forty rows long by thirteen columns wide. Each box in the array contains a name, and these were, for the most part, repeated along diagonals. The overall effect is of a checkerboard pattern with diagonal stripes, but there are numerous glitches—where a diagonal alignment of symbols jumps by a column, for example.

What was their purpose? The remarkable broad consistency between them, despite individual differences, suggests an elegant answer that makes sense in view of our broader knowledge of ancient Egyptian calendars and clocks. Each box in the grid contains the name of a star or star group known as a *decan*. In a typical table, three rows contain each of the names that appear elsewhere: these are a list of the thirty-six available decans. The remaining thirty-seven rows, containing the nearly diagonal grid, with its occasional glitches, represent the thirty-six weeks of the Egyptian civil calendar—each ten days long—together with the *epagomenal* period of five additional days that made up the year. For any given week, the names reading across the row represent the decans whose rising marked the successive "hours" during the night (these divisions of the night, unlike our hours, varied in length). A week later, each will rise roughly an "hour" earlier, which explains the diagonal pattern. However, since the position of the stars is irregular, it would not have been possible to choose decans that were evenly spaced across the sky, and this helps to explain the glitches.

Although they are often called *diagonal calendars* or *diagonal star clocks,* these were not instruments as such, but arrays of tabulated data. They helped the priest determine the time of night for nocturnal observances. For any given week in the civil year, the user could read off the stars whose appearance in the east would mark a particular hour in the night. We must also bear in mind that the data tables we find on the coffin lids were not actually functional—except, presumably, in the sense that the deceased would take them to use in the afterlife. The tables that actually served as priestly aids have not survived. In any event, each of them must have become obsolete relatively rapidly, because of the slippage of the civil calendar relative to the seasonal year by one day in every four years, and by a whole ten-day "week" after only forty.

The coffin-lid calendars were probably copies of the tables that the priests actually used. It is likely that these copies were made by artists who were not astronomical experts (unlike the creators of the actual tables). Perhaps they made mistakes in the transcription. Indeed, this may well explain other anomalies in some of the coffin-lid tables that do not seem to have an astronomical explanation. Reasonable as this sounds, it creates a headache for the modern interpreter. Do particular features reflect astronomical reality or were they merely errors in the transcription? To what extent did accuracy actually matter? It is hardly surprising that a definitive identification of all of the decans has eluded modern scholars and continues to do so at the present time.

See also:
Ancient Egyptian Calendars; Egyptian Temples and Tombs.
Star Rising and Setting Positions.

**References and further reading**
Faulkner, Raymond. *The Ancient Egyptian Pyramid Texts*. Oxford: Oxford University Press, 1969.
———. *The Ancient Egyptian Coffin Texts*. Oxford: Aris and Phillips, 2004. (Originally published in three volumes, 1973, 1977, and 1978.)
Krupp, Edwin C., ed. *In Search of Ancient Astronomies,* 208–211. New York: Doubleday, 1977.
Neugebauer, Otto. *The Exact Sciences in Antiquity,* 70–91. Princeton: Princeton University Press, 1951. (2nd ed., Providence: Brown University Press, 1957, pp. 71–96; further corrected ed., New York: Dover, 1969, pp. 71–96.)
Neugebauer, Otto, and Richard A. Parker. *Egyptian Astronomical Texts, I: The Early Decans*. Providence: Brown University Press, 1960.
Walker, Christopher, ed. *Astronomy before the Telescope,* 37–38. London: British Museum Press, 1996.
Wallin, Patrik. "Celestial Cycles: Astronomical Concepts of Regeneration in the Ancient Egyptian Coffin Texts." Unpublished Ph.D. thesis, Uppsala University, Sweden, 2002.

## Cognitive Archaeology

In order to probe the nature of astronomy in prehistoric cultures we must turn to the material evidence available in the archaeological record. Yet it is technological rather than cognitive achievement that is most obviously recorded there, in the form of artifacts such as stone axes and metal weapons, and constructions such as flint mines and megalithic monuments. Hence the traditional three-age system (Stone, Bronze, Iron) for categorizing periods in prehistory. Since the 1960s, environmental archaeology has provided a range of evidence pertaining to climate, landscape, and vegetation, as well as health and diet. Although this evidence has helped reveal many details of human subsistence activity in relation to the environment, such as developments in animal husbandry and agriculture, the emphasis is, once again, primarily technological. Social archaeology gained popularity in the 1970s, as people became more interested in political and economic relationships, structures, and spheres of influence. Archaeologists could make deductions from distributions of monuments and estimates of the work (and hence the labor organization) involved in their construction. But understanding prehistoric cognition always remained a more distant goal. Only in the 1990s, in the wake of intensive debates about archaeological theory and method that took place during the two previous decades, has cognitive archaeology begun to seem a feasible field of study.

We may be able to make specific deductions about the mental processes that made possible a particular technological achievement. For example, it seems clear that the layout of the sarsen circle and trilithons constructed at Stonehenge in around 2400 B.C.E. using massive stones weighing up to forty-five tons must have been carefully planned before each one was manually hauled overland to the site, a journey of over thirty kilometers involving a huge expenditure of labor. From this we can deduce that certain people, at least, possessed a level of numerical and geometrical skills that included the ability to produce a scaled-down plan or model, perhaps constructed in wood; the ability to deduce from that model the numbers, sizes, and shapes of stones actually required; the means to communicate the appropriate numbers and measurements to the place from which the sarsen blocks were being brought; and the ability to determine the place of erection of each stone.

But even this tells us nothing about the ideas and beliefs that led them to conceive of constructing the sarsen monument in the first place. For this, we need to develop a broader framework—a body of theory that suggests more general relationships between human thought and material remains, and more disciplined methods by which we can extrapolate from objects and patterns in the archaeological record to concepts in people's minds. This issue has been the subject of extensive debate among theoretical archaeologists, and there is no simple answer.

See also:
Symbols.
Stonehenge.

**References and further reading**
Johnson, Matthew. *Archaeological Theory: An Introduction*, 85–97. Oxford: Blackwell, 1999.
Renfrew, Colin, and Paul Bahn. *Archaeology: Theories, Methods and Practice*, 4th ed., 496–501. London: Thames and Hudson, 2004.
———, eds. *Archaeology: The Key Concepts*, 41–45. Abingdon: Routledge, 2005.
Renfrew, Colin, and Ezra Zubrow, eds. *The Ancient Mind: Elements of Cognitive Archaeology*. Cambridge: Cambridge University Press, 1994.

# Coligny Calendar
*See* Celtic Calendar.

*Cometary observations made by English astronomer Edmond Halley in 1682. Halley deduced that comets seen in 1531, 1607, and 1682 were in fact the same comet and correctly predicted its return, which happened in 1758. The comet was then named in his honor. (Library of Congress)*

## Comets, Novae, and Meteors

The majority of objects we see in the night sky move in regular and readily recognizable cycles, but a few phenomena are much more sporadic. These include meteors, or "shooting stars"; comets, which at their most spectacular are seen as bright, blurry patches with a long tail; and novae, new stars that appear where nothing was visible before.

Meteors are by far the most common of the three. On almost any night a patient observer can see a random shooting star fleetingly passing from one point in the sky to another. In fact, shooting stars are not stars at all but small particles of dust that fly into the earth's atmosphere (or, to put it another way, the earth flies into them), whereupon they burn up. When they are unusually large and close to the observer, they can appear as bright and colored fireballs. Occasionally a small fragment of rock, or a meteorite, will fall to earth; large meteorites are rarer still but can cause local damage or even environmental catastrophe. Meteors often occur in systematic bursts known as meteor showers, with scores or even hundreds visible in the space of a few hours. When watching these closely, an observer soon notes that all the shooting stars appear to radiate from the same point in the sky. It is as though the earth is heading at breakneck speed toward this spot, rushing past stationary points of light. This is, in fact, pretty much what *is* happening, since the earth's orbit is passing through a stream of material (although the material in the stream is itself in orbit). This means that many meteor showers occur regularly, around the same date each year, when the earth's orbit passes through the same region in space. However, varying conditions can mean that while in some years a given shower will be strikingly visible, in other years it will pass almost unnoticed.

Comets can be truly spectacular, their tails stretching far across the sky, and a few are so bright that they can be seen in daytime. Such occurrences are relatively rare, though. The best examples in the last two centuries were seen in 1811, 1843, 1861, 1882, and 1910. The best example in the later twentieth century (which, though impressive, did not compare with any of these) was Comet Hale-Bopp in 1997. The appearance of any particular comet, and how this varies from night to night, depends on many factors and differs considerably from one comet to another. A comet may take several weeks to reach its maximum brightness or do so extremely rapidly, and it may only be at its most striking for a day or two. Some comets recur regularly—the most famous being Comet Halley, which reappears roughly every seventy-six years—but these comets cannot always be relied upon to be conspicuous on every occasion (indeed, Halley passed almost unnoticed in 1986). In practice, then, the appearance of a bright comet to peoples in the past—even those who kept accurate records, such as the Babylonians and Chinese—was unpredictable.

While the majority of stars shine constantly with little perceptible variation for literally billions of years, some are variable and a few—known as novae—are liable to sudden outbursts that cause their brightness to increase considerably. However, the most incredible phenomena of this type are the so-called supernovae—stars that literally explode or implode and whose brightness, as a result, suddenly increases by a huge factor. When supernovae occur relatively close to us (in astronomical terms), then what before was invisible to the naked eye suddenly makes its appearance as a *guest star*, which, in some cases, can outshine everything in the sky apart from the moon and sun. There have been no great supernovae in the last four centuries and only around eight have been reliably recorded in historical times. The earliest, in the constellation of Centaurus, was recorded by Chinese astronomers in C.E. 185, although there is some dispute about whether this *was* actually a supernova. The most recent was Kepler's Supernova of 1604. It has been suggested that a guest star that appeared in 1054, and was certainly recorded by the Chinese and possibly by others in Asia and Europe, was also recorded in North America. However, the idea that a famous pictograph at Chaco Canyon was a depiction of this supernova does not withstand close scrutiny.

Any unexpected and conspicuous occurrence in the sky—including solar and lunar eclipses as well as rare meteorological phenomena—has the potential to cause panic and even social upheaval. The political control wielded by ruling elites often rests upon a spiritual power derived from their apparent links with the heavens. Unanticipated and unnerving celestial events can serve to undermine this power, sometimes catastrophically.

**See also:**
Catastrophic Events; Power.
Chaco Supernova Pictograph; Chinese Astronomy.

**References and further reading**
Brandt, John C., and Robert D. Chapman. *Introduction to Comets*, 1–54. Cambridge: Cambridge University Press, 2004.
Frommert, Hartmut, and Chrisine Kronberg. *Supernovae Observed in the Milky Way: Historical Supernovae.*
http://seds.lpl.arizona.edu/messier/more/mw_sn.html.
Ridpath, Ian, ed. *Norton's Star Atlas and Reference Handbook* (20th ed.), 88–94, 126–129. New York: Pi Press, 2004.
Stephenson, F. Richard, and David A. Green. *Historical Supernovae and Their Remnants.* Oxford: Oxford University Press, 2002.

## Compass and Clinometer Surveys

To determine the astronomical potential of a horizon point we need to measure its azimuth and altitude and then calculate the declination. Where we are concerned with possible alignments set up to within about half a degree or better (half a degree is roughly the diameter of the sun or moon), this is best done using a theodolite. However, when the accuracy required is not as great as this, a magnetic compass together with a clinometer may well be sufficient.

To be suitable for the task of determining the azimuth, a compass must be of the prismatic variety, enabling the user to view a distant object, a sighting mark (such as a vertical line), and the scale all at the same time. The measurement obtained is the magnetic bearing, or clockwise angle from magnetic north, as opposed to what is needed, which is the azimuth or clockwise angle from true north. To convert a magnetic bearing into a (true) azimuth, one must know the difference between true north and magnetic north. This difference—which is sometimes known as the magnetic declination, a term that can cause considerable confusion with (astronomical) declination—varies from place to place and time to time. Values for a particular place and time can often be obtained from topographical maps, or from various web sites on the internet. It is important to make sure that the correction is applied in the right direction: if magnetic north is to the east of true north, then the (true) azimuth will be greater than the magnetic bearing; if it is to the west then the (true) azimuth will be less.

A clinometer, or inclinometer, performs a similar task for measuring altitudes except that the zero point is determined simply by leveling a bubble before taking the reading, so that no further calibration is needed. Some manufacturers produce combined compass-clinometers as a single instrument, which are extremely useful for the archaeoastronomer.

Using a compass and clinometer has some clear advantages over using a theodolite. They are light and easily portable instruments, which is a major advantage if a site is difficult of access. There is no need for preliminary setting up and no need to take supplementary readings such as timed observations of the sun, which have the added problem of being dependent upon the weather (since a clear view of the sun is needed).

However, the compass also has some disadvantages. The main one is that the reading it gives can be affected by magnetic rocks or other magnetic materials close by (steel-rimmed glasses, for example). It is desirable to take various precautions, such as keeping metal objects well away from it, and measuring alignments in both directions wherever possible—the bearings should be 180 degrees apart. Best of all is to calibrate the readings from a particular location by measuring one or two reference points whose location, and hence true azimuth, can be determined independently from maps or digital topographic data. For this, one needs to know one's own location to within, typically, a few meters; this can be determined using another portable instrument, the GPS receiver.

Many archaeoastronomers prefer to use a theodolite, while others argue in favor of a compass and clinometer. Which instruments are best in any given instance actually depends upon the problem being addressed.

### See also:
Field Survey; GPS Surveys; Theodolite Surveys.
Altitude; Azimuth; Declination.

### References and further reading
Belmonte, Juan, and Michael Hoskin. *Reflejo del Cosmos*, 25–28. Madrid: Equipo Sirius, 2002. [In Spanish.]

Hoskin, Michael. *Tombs, Temples and Their Orientations*, 10–13. Bognor Regis, UK: Ocarina Books, 2001.

Ruggles, Clive. *Astronomy in Prehistoric Britain and Ireland*, 165. New Haven: Yale University Press, 1999.

# Constellation Maps on the Ground

It is often claimed, especially in popular accounts of archaeoastronomy, that human artifacts or constructions were laid out on the ground in the form of groups of stars in the sky. One such idea is that a group of temples at Angkor in Cambodia were laid out in the shape of the constellation

Draco. Another claim is that the Pyramids of Giza in Egypt were laid out in the form of the three stars of Orion's Belt. Groups of cup-like depressions resembling the Southern Cross and other constellations have been found on rock platforms in Aboriginal Australia.

Given that there is no independent evidence to support the supposition that any of these correspondences was intentional, we must ask how likely it is that they could have arisen by chance. (The patterns of stars in the sky, as opposed to their positions, have not changed significantly since earliest antiquity, so insecure dating is not a problem.) The chances of being able to fit a random set of points on the ground, such as buildings, to a set of stars in the sky might not seem very great. However, there are a large numbers of stars in the sky and a lot depends upon the size of the errors one is prepared to tolerate. And once the researcher selects some points and ignores others, the likelihood of a good fit is very greatly increased. A British television documentary broadcast in 1999 managed to fit the locations of selected major public buildings in New York City to the brightest stars of the constellation Leo with surprising ease. This correspondence clearly was not intentional.

We cannot dismiss outright the possibility that, at certain times in certain places in the past, things on the ground were configured to map out, represent, or reflect constellations in the sky. Anyone proposing this, however, must be able to describe and justify the selections made, provide an objective assessment of the goodness of fit, and determine the probability of the results obtained being fortuitous. They also face wider problems. First, there are few, if any, known cultural precedents among historical or modern indigenous communities for arrangements that map the stars; this will increase any archaeologist or anthropologist's inherent skepticism. Second, patterns visible archaeologically were often created in many stages, over a considerable period of time, whereas this idea implies that a single planning concept was adhered to. Finally, the use of some of the more sinuous Western constellations, such as Draco, must also arouse suspicions, since which collections of stars are perceived to form a constellation varies from one culture to another.

A particularly notorious example is the so-called Glastonbury Zodiac. This theory, which dates back to the 1920s, holds that features in the area surrounding the isolated hill of Glastonbury Tor, in Somerset, England—ancient field boundaries, paths, ditches, and banks—mimicked the shapes of zodiacal constellations. As soon as the evidence is examined in any detail, however, the arbitrary nature of the whole scheme becomes readily apparent. Those fragments of boundaries that best fit the theory are simply assumed to be the remnants of ancient boundaries, while others that do not are discarded. When we do this, we are forcing the evidence to fit the theory rather than using the evidence to test the theory, which any objective assessment soon shows to be unsupportable.

It is not out of the question that some sacred architecture in the past was laid out to reflect the configurations of groups of stars in the sky. But without due attention to the cultural context and a strict attention to methodology, investigations in this area can become no more than a modern game of fitting stars on a star chart to points on an archaeological map, with no cultural significance whatsoever.

**See also:**
Ley Lines; Methodology.
Aboriginal Astronomy; Angkor; Pawnee Star Chart; Pyramids of Giza.
How the Sky Has Changed over the Centuries.

**References and further reading**
Bauval, Robert, and Adrian Gilbert. *The Orion Mystery: Unlocking the Secrets of the Pyramids*. London: Heinemann, 1994.
Hancock, Graham, and Santha Faiia. *Heaven's Mirror*. London: Penguin Books, 1999.
Maltwood, Katherine. *A Guide to Glastonbury's Temple of the Stars*. London: James Clarke and Co., 1964. Originally published in 1929.
Williamson, Tom, and Liz Bellamy. *Ley Lines in Question*, 162–170. Tadworth: World's Work, 1983.

# Cosmology

A cosmology is a shared system of beliefs about the nature of the world—the cosmos—as it is perceived by a group of people. The term *worldview* is often used to mean much the same thing, as is (less frequently) the Spanish term *cosmovisión*.

How people perceive the world influences what they do and where they do it. Principles of cosmology may be reflected in domestic architecture and sacred buildings, the design of great monuments, city layouts, and even whole landscapes. Numerous examples in the modern world or historically documented illustrate this point, such as the design of the Navajo hogan, the Pawnee earth lodge, and the Yekuana roundhouse; the layout of the medieval Hindu city of Vijayanagara and of the imperial Forbidden City in Beijing; and the landscape around the Hopi village of Walpi.

Cosmology intimately involves astronomy, since the sky is an integral part of the world that people see around them. But cosmology is not restricted to astronomy. In many indigenous communities people freely associate things and events in the sky with those in the terrestrial world, in spirit worlds, or in a perceived world populated with beings that, from a Western scientific point of view, are imaginary and fantastical. Objects and happenings in the sky are also seen as intimately connected with actions and occurrences in the realm of human relations. It is people in the Western world who are exceptional in this respect, separating, in accordance with the Linnaean tradition, the perceived world into different categories that can be analyzed using different branches of science such as astronomy, geology, biology, and the social sciences. (And even in the Western world, despite modern science, astrology still continues to have a strong influence.) Today, the separation of the world of everyday concerns from the domain of the sky is exacerbated by the city lights, which prevent much of the modern population ever (at least, in the normal course of things) from seeing a really dark night sky.

For many indigenous peoples, on the other hand, and (as we now suspect) for virtually every human community way back into prehistory, the sky is an important if not critical part of the world in which they live, and it is crucial to keep human activity in harmony with it. Thus, for example, the traditional seasonal progression of the Lakota through the Black Hills of South Dakota is kept in tune with the path of the sun as it moves through certain constellations. The constellations themselves are directly associated with particular landmarks.

The perceived relationships among people, land, and sky can often guide human action in what we would see as a pragmatic way. The Lakota are one example of this, since their annual cycle of movement through the landscape following the buffalo was necessary for subsistence. Two rather different examples are the Caterpillar Jaguar constellation of the Barasana people of the Colombian Amazon and the star Marpeankurrk in Aboriginal Australia.

Archaeologists can seek to identify cosmological relationships in the archaeological record—in the orientations of buildings and monuments, for example—and alignment studies within archaeoastronomy form an important part of this process. However, spotting associations that might have had cosmological significance is far from a straightforward task, especially for the prehistorian, who may have little corroborating evidence other than the existence of an alignment itself. One problem, clearly, is that any alignment could have arisen fortuitously. Just because a structure may be aligned, say, upon a prominent (as it might seem to the modern investigator) hill or the midsummer sunrise does not mean that this was either intentional or meaningful—either to the builders or to those who used the place subsequently. Another problem is that we cannot prejudge the types of association most likely to have been significant (the possibilities are almost endless), let alone the actual meaning(s) that might have been ascribed to them. We have already intimated that prominence is a subjective concept, and one has to avoid the dangers of a checklist of possibly meaningful types of association that is formulated in the context of a modern worldview but may be totally inapplicable in the context of another. This is particularly the case in seeking astronomical associations such as alignments, where all too often an investigator will approach a site with a ready-made toolkit of astronomical targets.

In fact, it is when cosmological considerations act counter to pragmatics that they may be most striking in the archaeological record. An excellent example of this is the passage tomb at Balnuaran of Clava in Invernessshire, Scotland, where an apparent cosmological necessity, the need to have taller kerbstones on the southwestern side (following a local tradition evident in the nearby recumbent stone circles) seems to have resulted in structural principles being compromised so that the whole cairn was rendered unstable. In other cases, the arguments about whether an evidently intentional alignment reflects cosmological, or simply pragmatic, concerns, may be more difficult to resolve. Thus, the fact that many Iron Age roundhouses in Britain are oriented predominantly east- and southeastwards can be interpreted in two different ways: first, that it is the result of a pragmatic consideration, an alignment that allows the warming rays of the morning sun to shine into the house; second, that the orientation was dictated by cosmological principles, paralleling many modern examples such as the Navajo hogans and Pawnee earth lodges already mentioned. And these two explanations are not necessarily mutually exclusive.

#### See also:
Alignment Studies; Archaeoastronomy; Astrology; Landscape; Methodology; Orientation; Sacred Geographies; Space and Time, Ancient Perceptions of.
Aboriginal Astronomy; Barasana "Caterpillar Jaguar" Constellation; Chinese Astronomy; Clava Cairns; Hopi Calendar and Worldview; Iron Age Roundhouses; Lakota Sacred Geography; Navajo Hogan; Pawnee Earth Lodge; Recumbent Stone Circles; Yekuana Roundhouses.

#### References and further reading
Krupp, Edwin C. *Skywatchers, Shamans and Kings*, 15–42. New York: Wiley, 1997.
Ruggles, Clive, and Nicholas Saunders, eds. *Astronomies and Cultures*, 1–31, 139–162, 254–255. Niwot: University Press of Colorado, 1993.

## "Cosmovisión"
*See* Cosmology.

## Crucifixion of Christ
Can biblical, historical, and astronomical data be combined to produce an exact date for the crucifixion? The question has intrigued Christian scholars right back to Sir Isaac Newton and continues to spark fierce debate from time to time. The Bible places the crucifixion within the period when Pontius Pilate was procurator of Judaea, C.E. 26 to 36, and other historical constraints are generally agreed to rule out years earlier than C.E. 29 and

later than C.E. 34, with C.E. 29 and 34 themselves being unlikely. The biblical accounts also make it clear that the crucifixion took place on the afternoon before the Sabbath began, that is a Friday afternoon, and on the day before (or possibly on the day of) the feast of the Passover, that is on the Jewish date 14 Nisan or 15 Nisan. But in which year?

Attempting to answer this question involves correlating two calendars that are very different in nature. The Julian calendar of the Romans ran in a fixed and pre-determined way, tied to a solar year of 365-and-a-quarter days in length and ignoring the actual heavenly bodies. The (Gregorian) calendar used in the Western world today is directly descended from the Julian one. The Jewish calendar, on the other hand, was a lunar one and was, at least in the first century C.E., always regulated by direct observations of the moon.

In the Jewish calendar, the new day begins at sunset, with the Sabbath having fallen unfailingly on every seventh day since Biblical times. The new month begins when the new crescent moon first appears in the evening sky. A major problem, then, in trying to date the crucifixion is that we have to estimate the likelihood that the new crescent moon would have first been seen on a given evening or whether it would not have been noticed until the next. Another problem for us is that an intercalary month needed to be inserted periodically in order to keep the Jewish calendar in step with the seasonal year, and in the first century C.E. this was also done empirically. Although in theory the main trigger for inserting an extra month before Nisan was that Passover should not precede the vernal equinox, it is unclear how the equinox was itself determined (and to what accuracy), and in practice a variety of pragmatic criteria seem to have been used, including the availability (depending upon the weather) of first fruits and sacrificial lambs.

Given all these uncertainties, we cannot obtain a clear-cut answer to the critical question, which is whether 14 Nisan or 15 Nisan fell on a Thursday-Friday (the Jewish day running from sunset Thursday to sunset Friday) in any given year, and hence whether one or the other might be a candidate for the date of the crucifixion. However, the years C.E. 31 and 32 seem to be nonstarters. Newton himself favored April 23, C.E. 34, but historical arguments weigh against this. The clear favorites are generally April 7, C.E. 30 and April 3, C.E. 33. One view favors the C.E. 33 date on the grounds that (as astronomical calculations show) a lunar eclipse occurred on that very day. This seems to accord with the prophecy (Acts 2:20) that "the moon [shall be turned] into blood before that great and notable day of the Lord comes," which some interpret as describing an actual event at the time of Christ's death. On the other hand, the lunar eclipse was only partial and in its final stages when the moon rose: the moon would not have turned red for this reason. Furthermore, this and other references in the Bible to the moon turning to blood clearly refer to Judgment Day rather than to the crucifixion.

A final difficulty is that if the crucifixion really took place on a Friday, then it did not allow Jesus to spend three days and three nights in the tomb before Sunday morning. If one entertains the possibility that the crucifixion actually happened on a Wednesday or Thursday, then some of the eliminated years re-enter the frame.

In the end, it is more likely to be historical arguments than astronomical ones that will tip the balance in this issue.

**See also:**
Equinoxes; Lunar and Luni-Solar Calendars; Lunar Eclipses.
Gregorian Calendar; Julian Calendar.

**References and further reading**
Humphreys, C.J., and W. G. Waddington. "Dating the Crucifixion." *Nature* 306 (1983), 743–746.
———. "Crucifixion Date." *Nature* 348 (1990), 684.
Pratt, John. "Newton's Date for the Crucifixion." *Quarterly Journal of the Royal Astronomical Society* 32 (1991), 301–304.
Schaefer, Bradley E. "Lunar Visibility and the Crucifixion." *Quarterly Journal of the Royal Astronomical Society* 31 (1990), 53–67.

# Crucuno

Crucuno is one of a number of megalithic enclosures—settings of standing stones or boulders arranged to enclose a space—found in two clusters in Brittany, northwest France. One group is found toward the far northwest coast, around the modern city of Brest, and the other is found on the south coast of the Département of Morbihan, around Carnac. Often grouped together under the term *cromlechs*, some of these enclosures are very roughly circular or oval, while others are shaped like a horseshoe, a barrel, or a letter "D"; a few are closer to a rectangle or square. Crucuno, situated some five kilometers (three miles) northwest of Carnac, is certainly the most impressive of these. The twenty-two stones form the remains of what appears to have been an almost perfect rectangle, 33.2 meters by 24.9 meters (108.8 by 81.6 feet), oriented in the cardinal directions, with the longer sides east-west. Although the site has been largely reconstructed, the reconstruction appears to have been done faithfully, so that the layout can be trusted.

Amazingly, this simple structure combines geometrical perfection with cardinal orientation and astronomical alignment. In addition to being a near-perfect rectangle, the lengths of its sides were in the ratio 3:4. And in addition to the sides being oriented in the cardinal directions, the diagonals are oriented toward the rising and setting positions of the sun at the two solstices.

Yet we should guard against overinterpretation. The ratio of the sides does not necessarily imply that the builders were aware of the Pythagorean rule whereby the length of the diagonal is also a whole number of units (5).

*Part of the megalithic rectangle at Crucuno, Brittany, France. (Courtesy of Clive Ruggles)*

It could be just that, for aesthetic or other reasons, they chose to have the side lengths in the proportion 3:4 instead of, say, 2:3 or 4:5. Then there is the question of whether the solstitial orientation of the diagonals is fortuitous. Crucuno is located at a latitude (47.6°N) where, given a reasonably flat horizon, constructing a cardinally oriented 3:4 rectangle with its longer sides east-west will ensure the (approximate) solstitial orientation of the diagonals, whether or not this was intentional. On the other hand, perhaps the solstitial orientation of the diagonals was deliberate and the ratio of the lengths of the sides followed, unintentionally, as a result. Either way, it seems inconceivable that the latitude was deliberately chosen in order to achieve both solstitial orientations and Pythagorean proportions at once: the location of the site was surely determined in relation to the territory occupied by the people who built it, as is evident from its position within a tightly confined cluster of various types of megalithic enclosure in the Carnac area. Lastly, it is tempting to identify the east-west orientation as aligning upon the sunrise and sunset at the equinoxes, but this also makes several questionable assumptions, not least that the equinox as a concept meant something to people in prehistory.

The Scottish engineer Alexander Thom believed fervently that, in addition to a sophisticated knowledge of astronomy, "megalithic man" knew a good deal of geometry, including Pythagorean triangles, and used a precise unit of measurement, the "megalithic yard" of 0.829 meters (2.72 feet), in constructing monuments throughout the length and breadth of Britain, as well as in Brittany. The evidence he used to support this idea has now been largely refuted, but it is curious that the 3:4 ratio of side lengths is marked so clearly at Crucuno and that this implies a unit of 8.3 meters (27.2 feet), which is almost exactly ten of Thom's megalithic yards. The most probable explanation is that a unit of around this length tends to arise through local practices of pacing.

Most of the other recorded stone rectangles in Brittany are in a dilapidated state or have been removed entirely, but similarities have been noted with the station stones at Stonehenge. These four stones marked the corners of a rectangle whose side lengths were in the ratio 5:12, another Pythagorean combination yielding a diagonal of length thirteen units. The station stone rectangle is not cardinally oriented, but its shorter sides are themselves oriented upon midsummer sunrise and midwinter sunset, while the longer sides align roughly with an extreme rising position of the moon. This is similar enough to be intriguing, but different enough that it does not provide compelling corroborative evidence.

Nonetheless, despite the many problems and unresolved issues, the fact that people in the Neolithic or Bronze Age could have encapsulated four distinct characteristics—geometrical and numerological perfection, cardinal orientation, and astronomical alignments—in such a simple setting is truly remarkable. We do not have to accept Thom's megalithic yard at face value or postulate that the builders conceptualized Pythagorean geometry to appreciate this fact.

Nor do we have to assume that the builders chose their latitude in order to believe that the solstitial alignments of the diagonals were fully intentional. True, Crucuno would not have "worked" if it had been located much farther north or south. It is also true that such ideal latitudes are few and far between. For example, a square enclosure with similar properties would need to be placed much farther north—around Carlisle in northern England (latitude 55.0°)—and a 3:4 rectangle with its longer sides north-south would have to be located around latitude 59.6°, somewhat to the north of the Orkney Islands in northern Scotland. But what seems to us a case of remarkably good luck could, for people living in the Carnac area and unaware of differences farther afield, simply have seemed an intrinsic property of nature. In encapsulating the interconnectedness of geometry and astronomy in this modest but powerful way, the Crucuno rectangle may have provided a clear confirmation of the unity and integrity of the cosmos.

**See also:**
Cardinal Directions; Equinoxes; Solstitial Directions; Thom, Alexander (1894–1985).
Stonehenge.

**References and further reading**

Burl, Aubrey. *Megalithic Brittany,* 133. London: Thames and Hudson, 1985.

———. *A Guide to the Stone Circles of Britain, Ireland and Brittany,* 254–255. New Haven: Yale University Press, 1995.

———. *The Stone Circles of Britain, Ireland and Brittany,* 331–348. New Haven: Yale University Press, 2000.

Giot, Pierre-Roland. *La Bretagne des Mégalithes,* 83. Rennes: Éditions Ouest-France, 1997. [In French.]

Le Cam, Gabriel. *Le Guide des Mégalithes du Morbihan,* 67. Spézet, France: Coop Breizh, 1999. [In French.]

Ruggles, Clive. *Astronomy in Prehistoric Britain and Ireland,* 82–83. New Haven: Yale University Press, 1999.

Thom, Alexander, and Archibald S. Thom. *Megalithic Remains in Britain and Brittany,* 19–20, 175–176. Oxford: Oxford University Press, 1978.

## Cumbrian Stone Circles

In and around the Lake District, in the county of Cumbria in the northwest of England, are a group of nine large stone circles in spectacular landscape settings with panoramic views of the surrounding hills. The British archaeologist Aubrey Burl has also noted four similar large circles scattered slightly farther afield—three across the Solway Firth in Dumfriesshire, southwest Scotland, and one across the Irish Sea in County Down, Northern Ireland—that also seem to form part of this distinctive group of monuments.

These great stone circles are impressive in scale, varying from some thirty meters (100 feet) to over a hundred meters (330 feet) in diameter. The largest, Long Meg and Her Daughters, contains sixty-six stones (originally there may have been up to ninety). Nowhere does the idea that stone circles were conceived as *microcosms,* sacred spaces relating to the wider world and used for ritual performances, seem more appropriate than at some of these majestic circles in their breathtaking settings.

Astronomical alignments have been noted at a number of these circles. A large outlier at Long Meg and Her Daughters (Long Meg itself) is located in the direction of midwinter sunset from the center of the adjacent ring. The entrance at Swinside is roughly in the direction of midwinter sunrise. One of the best known of these circles, Castlerigg, contains a number of low-precision alignments noted by Alexander Thom, who considered it "a symbolic observatory." Systematic studies of the group as a whole do back up these initial impressions to some extent, revealing an apparent obsession with the cardinal directions and the solstitial directions. This may well reflect the existence of quadripartite cosmologies not dissimilar to those so prevalent in pre-Columbian Mesoamerica and indigenous North America.

**See also:**
Cardinal Directions; Solstitial Alignments; Thom, Alexander (1894–1985).
Circles of Earth, Timber, and Stone; Navajo Cosmology; Pawnee Cosmology.

*Part of the stone circle at Castlerigg, Cumbria, with the hill of Blencathra behind. (Corel Corp.)*

**References and further reading**
Burl, Aubrey. *Great Stone Circles*, 32–45. New Haven: Yale University Press, 1999.
———. *The Stone Circles of Britain, Ireland and Brittany*, 103–126. New Haven: Yale University Press, 2000.
Ruggles, Clive. *Astronomy in Prehistoric Britain and Ireland*, 131–133. New Haven: Yale University Press, 1999.
———, ed. *Records in Stone: Papers in Memory of Alexander Thom*, 175–205. Cambridge: Cambridge University Press, 2002.

## Cursus Monuments

In addition to the large numbers of circles of earth, timber, and stone constructed in Britain and Ireland during the Neolithic and Bronze Age, there were also a variety of monumental constructions that were linear in form, including rows of free-standing timber uprights or standing stones. Two very different examples of the latter are the 2 kilometer- (1.5 mile-) long Kennet Avenue at Avebury and the many short rows of no more than six megaliths found in western Scotland and the west of Ireland. The earliest easily recognizable type of linear monument, however, is the *cursus*, a pair of roughly parallel ditches and banks, typically a few tens of meters (one hundred to three hundred feet) apart, that run for some distance, typically a few hundred meters. Some eighty examples are known, of which the longest is the Dorset cursus in southwest England, which runs up and down across the undulating chalk uplands of Cranborne Chase for a remarkable ten kilometers (seven miles).

The name *cursus* derives from the spurious characterization by William Stukeley in the eighteenth century of one such monument as a racetrack for chariots. Their actual purpose is far from clear. The dating evidence suggests that cursuses were constructed in the Early Neolithic, generally within the later part of the fourth millennium B.C.E., so that they predate many of the more conspicuous, and hence better-known, British prehistoric monuments. The so-called Stonehenge cursus, for example, which is located just 1 kilometer north of Stonehenge and runs roughly east-west for some 3 kilometers (2 miles), is scarcely visible on the ground but may well have existed for several centuries before the earliest discernible activity took place at Stonehenge itself, and it may predate the arrival of the stones by as much as a whole millennium.

Cursus monuments are found in many parts of Britain with possible examples farther afield, although the main concentration is in southern England. Around twenty are known in the Upper Thames Valley alone. They represent some of the earliest monuments that impacted the British landscape in a way that is still visible today (though in most cases they are visible only from cropmarks evident in aerial photographs). They are also among the most baffling. Were they pathways or processional ways? If so, it seems strange that in most cases the ends, or *terminals,* were not open: instead, the banks along the sides curved or bent round, closing them off. It is debated whether the ditches and banks were sufficient to form a physical barrier, or whether the monuments could have been entered or crossed from almost anywhere. There is some evidence to support the assertion that they were cleared of tall vegetation, or at least built through cleared or partially cleared areas.

Did cursus monuments mark out formalized, or ritualistic, rites of passage? Some, like the Stonehenge cursus, joined two high points in the landscape, and it has been speculated that they provided some sort of symbolic link between two settlement areas. On the other hand, at Rudston in Yorkshire four different cursuses converge upon (or diverge from) a point in the valley through the chalk wolds, a natural focus for settlement through the ages, seemingly connecting it to the surrounding higher ground in various directions. Yet again, many of the Thames valley cursuses are situated on a single gravel terrace, their orientations ensuring that there were no sudden changes in elevation along their length.

Many cursuses also seem to encapsulate a variety of symbolic relationships with the wider visual environment. A number appear to be linked with streams and watercourses, while others are aligned with local hilltops. Yet others are astronomically aligned. Standing at what was once the northeastern terminal of the Dorset cursus, on Bottlebush Down, prehistoric viewers looking along the cursus to the southwest would have seen a conspicuous long barrow, built right across the cursus where it crossed the horizon. Around the shortest day of the year, the sun would have set directly behind this barrow. Although this segment of the Dorset cursus is virtually impossible to make out from the ground, the terminal can be identified and the barrow on Gussage Cow Down remains conspicuous. Another solstitially aligned cursus, the Dorchester cursus in Oxfordshire, has fared even less well: it has been completely destroyed by a combination of quarrying and the construction of a bypass.

It cannot be proved statistically that these solstitial alignments were deliberate. They are not repeated at most cursuses. Even at the Dorset cursus itself, only one segment is solstitially aligned: there were several other segments, built in more than one stage. Even if it was done intentionally, we can never know exactly why it was important to the people who built these particular cursuses to align them upon the midwinter sun. Some have suggested that there was a desire to keep them in harmony with nature, to appear to be part of the functioning of nature in a broad sense. And yet in some other cases a clear opportunity for a solstitial alignment seems to have been passed over. At Cleaven Dyke in Perthshire, central Scotland, is a two kilometer- (over one mile-) long earthen bank running between two ditches, generally considered to be a variant cursus monument. From the vicinity of its northwestern terminus a prominent hill is clearly visible, and this would have lined up approximately with the midsummer setting sun. If astronomical orientation really was important, it is hard to conceive that such an obvious alignment, with solstitial sunset coinciding with a prominent hill, would have been shunned. The cursus was not, however, oriented in this direction but over ten degrees differently. Some other suggested solar alignments are also problematic. The Stonehenge cursus, for instance, is often said to be aligned upon equinoctial sunrise and sunset, but there are a number of problems with this, as with supposed equinoctial alignments generally.

A different type of astronomical alignment is found at Thornborough in Yorkshire. The cursus here is associated with circular earthworks that were later superseded by three large henge monuments. To the northeast it was aligned upon midsummer sunrise, but to the southwest it was aligned upon the setting of the stars of Orion's belt. Since there are a good many bright stars in the sky, and their rising and setting positions vary considerably over the centuries owing to precession, it is possible to fit stars to almost any alignment, given the uncertainty in the archaeological dating of most prehistoric monuments. This one would certainly not be considered significant were it not for the fact that Orion's belt alignments are found repeatedly in different features at the site and fit with other archaeological evidence suggesting that the site was a pilgrimage center where ceremonies were held in early autumn. Orion's belt could have been important because its annual reappearance in the morning sky—its heliacal rising—served to forewarn potential participants.

In sum, cursus monuments do not appear to have any common ritualistic or symbolic purpose that is reflected in their relationships to the wider landscape or sky. Yet we can see tantalizing indications that both might have been important in particular cases.

See also:
Equinoxes.
Avebury; Circles of Earth, Timber, and Stone; Short Stone Rows; Thornborough.
Precession; Star Rising and Setting Positions.

**References and further reading**
Barclay, Alastair, and Jan Harding, eds. *Pathways and Ceremonies: The Cursus Monuments of Britain and Ireland.* Oxford: Oxbow Books, 1999.
Barclay, Gordon, and Gordon Maxwell. *The Cleaven Dyke and Littleour: Monuments in the Neolithic of Tayside,* 49–52. Edinburgh: Society of Antiquaries of Scotland, 1998.
Bradley, Richard. *Altering the Earth,* 50–62. Edinburgh: Society of Antiquaries of Scotland, 1993.
Ruggles, Clive. *Astronomy in Prehistoric Britain and Ireland,* 127–129. New Haven: Yale University Press, 1999.

## Cusco Sun Pillars

The Inca elite claimed kinship to the sun, and this needed to be constantly reinforced. Public rituals were regulated by the annual motions of the sun along the horizon, the most important of these being the festival of Inti Raymi, which took place at the time of the June solstice. This is still enacted today in the form of a pageant held annually at the ruins of the fortress of Sacsahuaman (Saqsaywaman) in the hills immediately outside the city of Cusco (Cuzco), once the capital of the entire Inca empire.

Several chroniclers, including Bernabé Cobo, give accounts of pillars that were erected on the horizon around Cusco in order to mark sunrise or sunset on key dates in the year. These eye-witness accounts span the whole of the century immediately following the European conquest and differ in many details, not least the number of pillars and their locations. There are, however, independent and largely consistent accounts of four pillars ("small towers") placed on a hill to the west of the city to mark planting times. Garcilaso de la Vega, the son of a Spanish captain and an Inca princess, described four sets of four pillars, each set marking one of the four solstitial di-

The Inti Raymi solstice pageant at Sacsahuaman, outside Cusco in Peru, photographed in 1984. (Courtesy of Clive Ruggles)

rections. In each case, according to Garcilaso, the critical rising or setting position of the sun was flanked by two smaller towers, themselves flanked by two larger towers to guard them.

Given the importance of the sun in Inca religion, it is scarcely surprising that horizon observations of the sun might have been the dominant seasonal markers and calendrical determinants, but some chroniclers' accounts also speak of a lunar calendar and distinctive ceremonials and rituals (such as llama sacrifices) being carried out each month. How the lunar and solar aspects of the calendar were resolved is not clear, although Garcilaso maintains that the solstitial pillars were used for this purpose. There are also some grounds for suggesting that pillars were used to observe lunar as well as solar risings and settings.

Within a generation after the conquest the pillars fell into disuse, although some may have survived for more than a century. No material traces remain today, though archaeologists and archaeoastronomers have tried to identify where some of them stood. Archaeoastronomer Anthony Aveni has identified what he feels is the likely location of a set of four pillars to the west of Cusco, on a hill called Cerro Picchu, suggesting that they marked sunset not on either of the solstices but on the dates of solar antizenith passage. However, the evidence is equivocal and his conclusions are hotly disputed.

**See also:**
Antizenith Passage of the Sun; Cobo, Bernabé (1582–1657); Lunar and Luni-Solar Calendars; Solstitial Directions.
*Ceque* System; Island of the Sun.
Lunar Phase Cycle; Solstices.

**References and further reading**
Aveni, Anthony F. *Stairways to the Stars: Skywatching in Three Great Ancient Cultures,* 161–165. New York: Wiley, 1997.
———. *Between the Lines: The Mystery of the Giant Ground Drawings of Ancient Nasca, Peru,* 130–135. Austin: University of Texas Press, 2000. (Published in the UK as *Nasca: The Eighth Wonder of the World.* London: British Museum Press, 2000.)
———. *Skywatchers,* 312–316. Austin: University of Texas Press, 2001.
Aveni, Anthony F., and Gary Urton, eds. *Ethnoastronomy and Archaeoastronomy in the American Tropics,* 203–229. New York: New York Academy of Sciences, 1982.
Bauer, Brian. *The Sacred Landscape of the Inca: The Cusco Ceque System.* Austin: University of Texas Press, 1998.
Bauer, Brian, and David Dearborn. *Astronomy and Empire in the Ancient Andes.* Austin: University of Texas Press, 1995.
Selin, Helaine, ed. *Astronomy across Cultures,* 203–206. Dordrecht, Neth.: Kluwer, 2000.

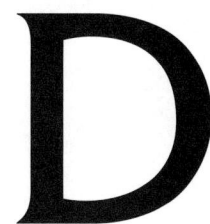

# Dating
*See* Astronomical Dating.

# December Solstice
*See* Solstices.

# Declination
Declination is the single most useful concept in archaeoastronomy. It basically means no more than "latitude" on the spinning celestial sphere, with the axis of spin defining the north and south poles. Every star has a particular declination and moves daily around a line of constant declination. This means that if you go to a place and identify a point of interest on the horizon—perhaps because an interesting archaeological feature points there—then all you have to do is to determine the declination of that horizon point, whereupon you can deduce what will rise and set there, and furthermore what would have risen and set there at any epoch in the past. The declination of a horizon point depends upon the azimuth and altitude of the point, together with the latitude of the observer. When calculating a declination, we need to make allowance for atmospheric refraction and, in the case of the moon, for lunar parallax.

The declination of the north celestial pole is 90° and that of the celestial equator is 0°. The declination of Polaris, for example, is 89° and that of the Pleiades is 24°. By convention, declinations south of the celestial equator are negative, so that the declination of Sirius, for example, is –17°, and that of Acrux, the brightest star in the Southern Cross, is –63°. The declination of the south celestial pole is –90°.

Confusingly, the word *declination* is also used to mean the difference between magnetic and true north. Equally confusingly, astronomers use the term *latitude* in the context of a different coordinate system on the celestial sphere: one related not to its spin axis but to the sun's apparent path around it—the ecliptic.

**See also:**
Archaeoastronomy; Field Survey.
Altitude; Azimuth; Celestial Sphere; Ecliptic; How the Sky Has Changed over the Centuries; Lunar Parallax; Precession; Refraction; Sun, Motions of.

**References and further reading**
Ruggles, Clive. *Astronomy in Prehistoric Britain and Ireland*, 18–19. New Haven: Yale University Press, 1999.

## Delphic Oracle

The oracle at Delphi, famous throughout ancient Greece from about the seventh century B.C.E., was (at least initially) only available for consultation on one day in the entire year. It was also set in a remote location in the foothills of Mount Parnassos, several days' journey from most ancient Greek cities. How, then, did pilgrims manage to arrive there at the appropriate time?

There might appear to be little difficulty, because the date was clearly specified as Apollo's birthday, the seventh day of the month of Bysios. The problem is that in ancient Greece, even as late as the fifth century B.C.E., there was no universal calendar. Instead, each urban center reckoned its own lunar calendar independently. In theory, observations of the first appearance of the new moon were used to determine the beginning of each month, but these were often subject to political interference, and the same was true of the insertion of intercalary months to keep the calendar in step with the seasonal year. Different cities had different sequences of months, different god and goddess cults, and different festival dates. The month of Bysios, for example, could fall at any time from (what in our calendar is) mid-January to mid-March. Certainly, no pilgrim relying on their own city's civic lunar calendar could be sure of arriving at the oracle at the appropriate time, perhaps within several weeks.

The key, it seems, was in observations of the stars. Their use as seasonal indicators had been known to Greek farmers since at least the eighth century B.C.E., as is clear from the writings of Hesiod. By the fifth century—following the development of peg-hole star calendars called *parapegmata*—some cities at least were using star sightings to regulate their own civic calendars. We also know from at least one *parapegma* that the relatively faint stars of the constellation Delphinus were recognized as a dolphin from at least the fifth century B.C.E. Finally, the cult title of the god (Apollo Delphinios) and the name of the site suggest an association with Delphinus.

Weaving these various strands together, it is not surprising to discover that the heliacal rise (first pre-dawn appearance) of Delphinus occurred to-

*The ruins of Delphi, Greece. (Corel Corp.)*

wards the end of December. If pilgrims used this event as the cue to commence the journey to the oracle, then they would have arrived in good time. What is more, this celestial event may well (at least initially) have defined the date when the oracle held forth. The key to this notion is Delphi's location within the landscape: it is situated in a natural bowl surrounded on three sides by towering cliffs. The high eastern horizon means that from the temple itself, the first appearance of Delphinus did not occur until about the end of January. According to tradition, Apollo left Delphi for a three-month period each year, which coincides (at least approximately) with the period when the constellation was too close to the sun and hence invisible. The consultation, then, would have taken place on the seventh day of the first lunar month in which the celestial dolphin had returned.

All this is strongly suggestive that the cult activities associated with the god Apollo Delphinios, and in particular the day when the oracle made its pronouncements, were defined by direct reference to the annual cycles of appearance of the constellation Delphinus. This direct connection may well have been used by pilgrims to know when to commence their journey to the oracle.

See also:
Hesiod (Eighth Century B.C.E.); Lunar and Luni-Solar Calendars; Pilgrimage.
Temple Alignments in Ancient Greece.
Heliacal Rise.

**References and further reading**
Flacelière, R. (trans. by D. Garman). *Greek Oracles,* 39. London: Elek, 1965.
McCready, Stuart, ed. *The Discovery of Time,* 74–79. Naperville, IL: Sourcebooks, 2001.
Parke, H.W. *The Delphic Oracle.* Oxford: Blackwell, 1956.
Salt, Alun, and Efrosyni Boutsikas. "Knowing When to Consult the Oracle at Delphi." *Antiquity* 79 (2005), in press.
Tuplin, C. J., and T. E. Rihll, eds. *Science and Mathematics in Ancient Greek Culture,* 112–132. Oxford: Oxford University Press, 2002.

## Diurnal Motion

*Diurnal* means "daily." Each day the celestial sphere turns once about an axis through two points known as the celestial poles. For an observer in the northern hemisphere, the north celestial pole is up in the sky while the south celestial pole is below the ground. The stars appear to turn around the north celestial pole, which defines the ("true") north direction. (In the southern hemisphere, the south celestial pole is up in the sky and defines the south direction.)

The diurnal motion is the most fundamental motion of the celestial bodies. For any place on earth (apart from the poles!) it defines the north-south axis known as the meridian, and hence any other direction in relation to this, regardless of the nature of the terrestrial landscape. Suppose, for example, that we find a group of monuments that are consistent in their orientation, and this has evidently been achieved over a wide area and despite a varied topography, so that it cannot be put down to prevailing winds, the local topography, or reference to distant landmarks. It follows that these orientations must be astronomical in the broadest sense, in that they can only have been achieved in relation to the diurnal motion of the celestial bodies.

**See also:**
Cardinal Directions; Orientation.
Celestial Sphere; Meridian.

**References and further reading**
Aveni, Anthony F. *Skywatchers,* 49–57. Austin: University of Texas Press, 2001.
Krupp, Edwin C. *Echoes of the Ancient Skies,* 3–6. Oxford: Oxford University Press, 1983.

## Dresden Codex

Most of our knowledge of Maya writing comes from monumental inscriptions in stone—public pronouncements carved into the walls of buildings or on stelae that stand like sentinels by a particular temple-pyramid. Sometimes textual commentaries were associated with striking carved friezes and vivid painted murals. Classic-period inscriptions and murals typically describe key events in the lives of kings and gods, but the later ones are increasingly concerned with warfare, conquest, and the sacrifice of captives. Many inscriptions contain calendrical dates, and these helped researchers establish the nature of the Maya calendar long before the remaining parts of the inscriptions began to be deciphered.

That these texts contained true hieroglyphic writing was not generally accepted until the later part of the twentieth century but is now beyond dispute. As we now know, syllabic glyphs were combined to produce phonetic expressions of words in a language conforming to strict principles of grammar and syntax, principles that have passed down recognizably into modern Maya languages such as Yucatec, spoken in the Yucatan peninsula of Mexico and in parts of northern Guatemala and Belize.

Ancient Maya script was without doubt the most highly developed writing system in the whole of pre-Columbian America. Yet much of it doubtless existed on perishable media and has been permanently lost. Hundreds of "books" in the form of folded strips of bark paper were burned to a cinder by the Spanish priests. From the Maya region only four of these codices survive, but these give us tremendous insights into learning and ritual. They are known, after the places where they were eventually discovered, respectively as the Dresden Codex, Madrid Codex, Paris Codex, and Grolier Codex. It is the first—the Dresden Codex—that holds the most fascination regarding astronomy. The whole book appears to be an astronomical (or rather astrological) almanac stuffed full of information about different celestial bodies. A vast amount of careful

*Page 20 of the Dresden Codex. It contains one complete almanac and parts of three others, featuring depictions of the Moon Goddess, Ix Chel. (Art Resource, NY)*

scholarship has been concerned with the detailed analysis of the contents of the codex. Among many other things, this work suggests that the Dresden Codex was actually a copy, probably made between the twelfth and fourteenth centuries C.E., of an original document as much as three or four hundred years older.

The Dresden Codex contains a number of tables that relate to particular celestial bodies. One of the most famous is the Venus table, which was first identified as such from the repeated appearance of the number sequence 236, 90, 250, and 8. These numbers seem to correspond to the canonical lengths of the appearance and disappearance of Venus during each synodic cycle. Another table records intervals between "danger periods" when solar eclipses might occur. It is also clear from the Dresden Codex that the Maya were obsessed with interlocking time cycles: they were well aware that five Venus cycles equaled eight years and also (nearly) 99 lunations; 46 *tzolkins* (260-day cycles) equaled 405 lunations, and so on. We would see these as coincidences of nature but to the Maya they represented natural rhythms of the cosmos.

The motivation for their interest in celestial cycles was, as we would see it, primarily astrological. They needed to be able to predict the motions of Venus, for example, so that they could judge the time when the omens would be best for waging war and capturing warriors for sacrifice to the gods. In this sense, the Dresden Codex and other books like it might have functioned both as astronomical tables and as divinatory almanacs.

See also:
Maya Long Count; Mesoamerican Calendar Round; Venus in Mesoamerica. Inferior Planets, Motions of.

**References and further reading**
Aveni, Anthony F. *Stairways to the Stars: Skywatching in Three Great Ancient Cultures*, 110–133. New York: Wiley, 1997.
———. *Skywatchers*, 169–207. Austin: University of Texas Press, 2001.
Coe, Michael D. *Breaking the Maya Code* (revised ed.). New York: Thames and Hudson, 1999.
Martin, Simon, and Nikolai Grube. *Chronicle of the Maya Kings and Queens: Deciphering the Dynasties of the Ancient Maya*. London: Thames and Hudson, 2000.
Thompson, J. Eric S. *A Commentary on the Dresden Codex: A Maya Hieroglyphic Book*. Philadelphia: American Philosophical Society, 1972.

# Drombeg

Drombeg is a carefully restored and well-preserved example of a south-west Irish axial stone circle, probably built in the early part of the second millennium B.C.E. Located near to the south coast of County Cork, a few kilometers from the village of Ros Ó gCairbre (Ross Carbery), it commands a

*The axial stone circle at Drombeg, Co. Cork, Ireland, viewed along the axial alignment toward midwinter sunset. (Courtesy of Clive Ruggles)*

wide view southward down toward the sea. It is relatively large as axial stone circles go: it is a little under ten meters in diameter and contains seventeen stones.

Drombeg is significant astronomically because of its axial alignment. As viewed along the axis from outside the circle to the northeast, looking in through the portals, the recumbent stone on the far side sits beneath a relatively close hill, less than a kilometer (half a mile) distant. In this hill, above the center of the recumbent stone, is a shallow but prominent notch. The axis and notch mark the setting position of the midwinter sun. The solstitial alignment of Drombeg was first recorded by Boyle Somerville in 1923 and has been much noted since. In fact, it is not exact: at around the time of the circle's construction the midwinter sun would have set somewhat to the left of the notch, only reaching it about two weeks before and after. Nowadays it is much closer. Nonetheless, even at the time of construction it would have been close enough to impress.

Much has been written about Drombeg, including speculations about other sunrises and sunsets that might have been observed from the circle at different times in the seasonal year. But everywhere has a horizon, and a circle of seventeen stones must of necessity form many alignments. Although it is not unreasonable to imagine Bronze Age people tracking the annual passage of the sun along the horizon, the archaeological record does not speak clearly about this. Even the intentionality of the main solstitial alignment is in some doubt. When Drombeg is considered in the wider context of the group of monuments of which it is a member, it is found that—despite the general consistency of the orientations of the axial stone circles toward the southwest or west—not a single other example is solstitially aligned. The obvious conclusion is that solstitial alignment was not important to the axial stone circle builders—even though the general design and orientation was—and that the solstitial alignment of this single example was probably fortuitous. On the other hand, however strong the framework of beliefs and traditions that governed and constrained the construction of one of these circles, it is always possible that the builders of this particular one had a special reason for going beyond the prevailing convention and incorporating a deliberate alignment upon sunset on the shortest day.

**See also:**
Alignment Studies; Somerville, Boyle (1864–1936).
Axial Stone Circles; Stone Circles.

**References and further reading**
Aveni, Anthony F., ed. *World Archaeoastronomy,* 470–482. Cambridge: Cambridge University Press, 1989.
Burl, Aubrey. *A Guide to the Stone Circles of Britain, Ireland and Brittany,* 218–219. New Haven: Yale University Press, 1995.
Ruggles, Clive. *Astronomy in Prehistoric Britain and Ireland,* 100–101. New Haven: Yale University Press, 1999.

# E

## Easter Island

Easter Island is the most isolated island on earth. Less than twenty-five kilometers (fifteen miles) across, it is well over 2,000 kilometers (1,300 miles) distant from the nearest habitable land in any direction. Yet when Europeans first chanced across it (Dutchman Jacob Roggeveen arrived there on Easter Sunday in 1722, hence the name), they found it inhabited. In fact, Rapa Nui (which is the Polynesian name for the island) had been colonized well over a millennium earlier, in about C.E. 400. Whether only a single canoe or fleet of canoes ever reached this remotest corner of Polynesia, or whether for a period there was further contact, we shall probably never know. What is certain is that the inhabitants of the island soon lost contact with their forebears and spent many centuries in complete isolation. Both archaeologists and ecologists have been bequeathed an incredibly valuable case study of people interacting with their environment within a closed system, and it is one that has some sobering lessons for the population of the planet as a whole. Certainly environmental disaster loomed toward the end, when the population had increased dramatically and all the forests had been destroyed.

Among the fascinations of Easter Island are the famous large statues or *moai*. Several hundred of them came to stand imposingly on platforms (*ahu*), mostly ringing the island and facing inland, but with a few in other locations. Scores more lie or stand in what resembles a frozen production line in the volcanic crater Rano Raraku, in various stages of production from quarrying to transportation. Why would an isolated population on a small island have put such a significant proportion of its human resources into producing them? One possibility is that they were seen as protective presences, but their full significance for the population of this tiny island remains largely a matter of speculation.

The island has long attracted astronomical attention. As far back as the 1960s, publications appeared describing *ahu* that were solstitially or equinoctially aligned. But since there are over 250 *ahu* on the island, the

*Ahu Nau Nau at Anakena, Easter Island, restored in 1978. (Courtesy of Clive Ruggles)*

question of bias in selection arose. Were the supposedly astronomically aligned *ahu* selected on the basis of a preconceived "toolkit" of targets? If so, it is possible that they had no statistical or cultural significance whatsoever. Attempts were made to resolve this question through systematic studies of *ahu* and *moai* orientations, both by an archaeologist (the Easter Island specialist William Mulloy) and later by an astronomer, William Liller. They provide only marginal evidence at best of any intentional orientations upon solstitial and equinoctial sunrise and sunset. The overriding tendency is for *ahu* to be situated close to the coast and oriented parallel to it, with the *moai* facing inland. Since such platforms are to be found all around the perimeter of the island, their orientations are scattered around the compass, with some falling inevitably within the solstitial and equinoctial ranges.

Perhaps the most noteworthy *ahu* from an astronomical point of view is one of a small minority that are not situated on the coast but a good way inland. Ahu Huri a Urenga is situated in a low saddle surrounded—unusually for Easter Island—by a hilly horizon with very little sea visible. Standing upon the platform is a distinctive and imposing *moai*. This is aligned not only upon the rising sun at the June solstice, but upon a prominent hill summit. Another hilltop marks equinoctial sunset. A number of apparently artificial depressions in boulders adjacent to the platform incorporate several solar alignments and hence, it has been claimed, could have functioned as a

*Ahu Nau Nau in its coastal setting. (Courtesy of Clive Ruggles)*

solar-ranging device. All this evidence still falls short of conclusive proof, but it does at least raise the possibility that natural features in the landscape were used to mark the solar rising or setting positions at different times of the year. This is a possibility that has been raised elsewhere in Polynesia, for example in the Nā Pali region of the Hawaiian island of Kauaʻi.

One of the most significant astronomical alignments to the inhabitants of Rapa Nui may actually have been a natural one. The village of Orongo was placed in an extraordinary situation on a knife-edge ridge overlooking the spectacular but treacherous crater of Rano Kau, with its marshy lake on one side and the ocean on the other. This sacred place was the ceremonial base for the extraordinary annual "birdman" ritual in which contestants had to swim two kilometers (over a mile) out to a small rock, wait there for the return of a migratory sea bird, and collect its first egg. As seen from Orongo, the sun rose behind the summit of Poike, a prominent volcanic hill on a far corner of the island, precisely at the winter (June) solstice. As the rising sun moved away from this peak, the spring ceremony approached. It is hard to believe that priests at Orongo could have failed to notice that the peak of Poike marked the limit of the sun's movement along the horizon, especially given that the word *poike*, according to the anthropologist and Maori scholar Sir Peter Buck (Te Rangi Hiroa), means "to be seen just above the horizon." However, the severe limitations of the topography make it unlikely that this was taken into account when the village was planned. Instead, the solstitial alignment of Poike from the crater-rim village may have been discovered later. We would see it as a coincidence, but it may well have been taken by the Easter Islanders as something that confirmed and reinforced the already evident sacredness of the place.

A remnant of Easter Island astronomy of a rather different nature may have been bequeathed to us in the form of Rongorongo script, a system of symbols used on Easter Island that seems to have provided a series of triggers for the reader rather than actual words or syllables. One fragment of Rongorongo has been interpreted as a lunar calendar.

> **See also:**
> Equinoxes; Methodology; Solstitial Directions.
> Nā Pali Chant; Navigation in Ancient Oceania; Polynesian Temple Platforms and Enclosures.
> Solstices.
>
> **References and further reading**
> Barthel, Thomas. *The Eighth Land: The Polynesian Discovery and Settlement of Easter Island.* Honolulu: University Press of Hawaiʻi, 1978.
> Fischer, Steven Roger, ed. *Easter Island Studies,* 122–127. Oxford: Oxbow Books, 1993.
> Flenley, John, and Paul Bahn. *The Enigmas of Easter Island.* Oxford: Oxford University Press, 2003.

Liller, William. *The Ancient Solar Observatories of Rapanui: The Archaeoastronomy of Easter Island.* Old Bridge, NJ: Cloud Mountain Press, 1993.

Mulloy, William T. "A Solstice-Oriented *Ahu* on Easter Island." *Archaeology and Physical Anthropology in Oceania* 10 (1975), 1–39.

Selin, Helaine, ed. *Astronomy across Cultures,* 148–152. Dordrecht, Neth.: Kluwer, 2000.

# Eclipse Records and the Earth's Rotation

Modern physics allows us to calculate, with extraordinary accuracy, the orbits of the earth about the sun and the moon about the earth. This enables scientists to predict the characteristics of solar (and lunar) eclipses for many centuries into the future, and in particular, to determine "paths of totality." These are narrow tracks across the surface of the planet defining those places where that rare and spectacular event—a total eclipse of the sun—will occur.

It is also possible to operate backwards in time and "predict" the occurrence of total solar eclipses in the past. Unfortunately, there are innate errors in this process that increase the further back (or forward) one tries to go. A significant effect is tidal friction, which has the effect of gradually slowing the rate of rotation of the earth. To illustrate the point, suppose that we made the assumption that the earth has rotated at the exactly same rate for the past two millennia. In fact it has slowed down slightly, and during that time it has actually rotated by one eighth of a turn more than we thought. The result will be that our eclipse track "predictions" for an eclipse occurring in the year 0 would be correct in latitude but some 45 degrees off in longitude, since the moon's shadow would actually have fallen further round the earth's surface than we expected.

It was first realized in the 1970s that this argument could be turned around. Eclipses were recorded by the ancient Chinese, Babylonians, Greeks, and Romans, as well as by astronomers in medieval Europe. As a result, there exist many scores of records of total solar eclipses (and other sun-moon-earth related astronomical events) dating between about 700 B.C.E. and C.E. 1600 with sufficient information on the timing of their occurrence to enable us to identify them with "predicted" eclipses. By comparing, through time, the difference in longitude between the "predictions" and the actual observations, it is possible to construct a remarkably accurate estimate of the rate of deceleration of the earth's rotation. Even so the modeling is not straightforward, and the main progress was made in the 1980s by the British historian of astronomy Richard Stephenson working together with astronomer Leslie Morrison.

This type of study uses archaeology and history—in the form of ancient eclipse records—to inform modern physics and astronomy. It has been named *applied historical astronomy* to distinguish it from archaeoastronomy, which does precisely the opposite. Archaeoastronomy uses modern astronomy to reconstruct the appearance of ancient skies with the ultimate aim of learning something about people's beliefs and practices in the past.

**See also:**
Archaeoastronomy; Lunar Eclipses; Solar Eclipses.
Babylonian Astronomy and Astrology; Chinese Astronomy; Swedish Rock Art.
Years B.C.E. and Years before 0.

**References and further reading**
Espenak, Fred. *Historical Values of Delta T.*
http://sunearth.gsfc.nasa.gov/eclipse/SEhelp/deltaT2.html.
Morrison, Leslie, and Richard Stephenson. "The Sands of Time and the Earth's Rotation." *Astronomy and Geophysics* 39 (1998), 5:8–5:13.
Stephenson, Richard. *Historical Eclipses and Earth's Rotation.* Cambridge: Cambridge University Press, 1997.

## Eclipses

*See* Lunar Eclipses, Solar Eclipses.

## Ecliptic

If the earth didn't rotate, and the atmosphere didn't scatter the light of the sun and obscure our view of the stars during the day, then as the earth moved slowly on its annual cycle around the sun we would see the sun moving slowly against the stars, completing a circuit around the celestial sphere once every year. This apparent path of the sun through the stars is known as the ecliptic.

Because the earth does rotate, the celestial sphere appears to spin around once every day. During each twenty-four-hour period, as it moves around with the stars, the sun progresses along the ecliptic by approximately one degree. If it were not for the atmosphere, we would see it by day among the stars. To many human communities it is important to know where among the stars the sun is. The zodiac familiar to people in the modern Western world—a band of twelve constellations through which the ecliptic runs—has its origins in ancient Babylonia. But many different "zodiacs" are possible, as for example among the Maya. The Lakota people of North Dakota traditionally keep their seasonal movement around the landscape in tune with the path of the sun through their own "zodiacal" constellations.

**See also:**
Lakota Sacred Geography.
Celestial Sphere; Obliquity of the Ecliptic.

**References and further reading**
Aveni, Anthony F. *Skywatchers,* 49–55, 200–205. Austin: University of Texas Press, 2001.

Goodman, Ronald. *Lakota Star Knowledge: Studies in Lakota Stellar Theology.* Rosebud, SD: Sinte Gleska College, 1990.

McCready, Stuart, ed. *The Discovery of Time,* 70–73. Naperville, IL: Sourcebooks, 2001.

# EDMs
*See* Theodolite Surveys.

# Egyptian Temples and Tombs

During the earliest dynasties in ancient Egypt, from about 3000 B.C.E. onwards, royal burials increased steadily in size and complexity. However, it is the construction of huge pyramids that for most people epitomizes ancient Egyptian burial customs and captures the imagination.

Large stone architecture first appeared during the Third Dynasty (twenty-seventh century B.C.E.) in the form of the 77-meter- (254 foot-)high stepped pyramid at Saqqara, some ten kilometers (six miles) south of modern Cairo. Pyramid building reached its peak (after some initial shortcomings) during the Fourth Dynasty (2613–2494 B.C.E.) with the construction of the famous Pyramids of Giza. Pyramids actually formed part of complexes, typically built on the edge of the desert plateau above the fertile river valley where they would dominate the surrounding landscape yet remain inaccessible to the masses: they were enclosed, together with other sacred structures, within a high wall. The precinct interior could only be reached by means of a covered causeway leading up from a separate temple in the valley, accessible by boat.

The subsequent Fifth Dynasty is particularly characterized by the construction of temples dedicated to the sun god, Ra (or Re). The pharaoh's power depended upon sun worship, and these "sun sanctuaries" followed the design of the mortuary complexes in having two enclosed precincts on different levels linked by a causeway. Six of the nine pharaohs of the Fifth Dynasty built temple complexes; the best preserved is that built by the sixth pharaoh, Neuserre, at Abu Ghurab.

All of this happened during Old Kingdom times, up to the mid-twenty-second century B.C.E. Ancient Egyptian civilization lasted for a further two millennia, its complex history including periods of political instability and social upheaval as well as two further periods of relative stability: the Middle Kingdom (mid-twenty-first century to mid-seventeenth century) and the New Kingdom (mid-sixteenth century to mid-eleventh century). Monumental tombs and temples proliferated in these later times, but they were generally more modest and not just the preserve of the kings themselves. However, the New Kingdom has left us some spectacular remains in the vicinity of its

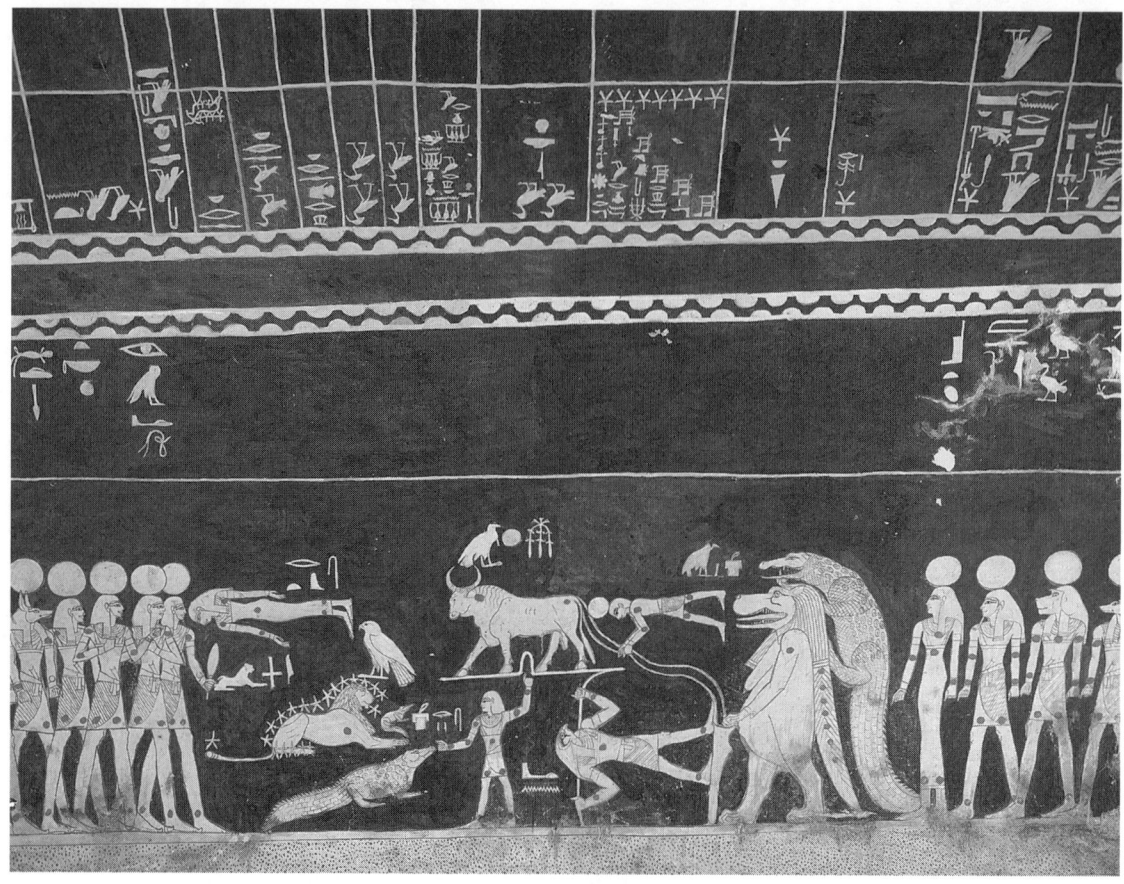

*Part of the painted ceiling of the burial chamber of the tomb of Seti I (New Kingdom, thirteenth century B.C.E.), listing and depicting the decans—stars and asterisms that marked ten-day calendrical periods. (Gianni Dagli Orti/Corbis)*

sacred capital, Thebes, some five hundred kilometers (three hundred miles) upriver to the south of Cairo, near modern Luxor. Amun (or Amon), the patron god of Thebes, had become identified with the existing sun god Ra, and the Great Temple of Amun-Ra at Karnak provided a spectacular setting at the heart of the city for public ceremonials relating to the sun god. On the opposite side of the river is the so-called Valley of the Kings, which contains over sixty underground pharaohs' tombs, including the famous (because it was discovered intact) tomb of Tutankhamun.

The most obvious clues to astronomical associations of temples and tombs are found in the inscriptions within them. A number of New Kingdom tombs and temples, for example, contain painted "astronomical ceilings," listing and depicting stars, constellations, and even planets. But why should these be placed inside tombs? The answer is that ancient Egyptians' understanding of the sky was framed within a worldview that bound together inextricably the gods, the otherworld *Duat*, the afterlife, and what was seen in the night sky. It had long been engrained in ancient Egyptian minds that the sun god Ra traveled nightly through *Duat,* the world beyond the horizon, on his journey to the eastern side where the sky goddess Nut would give birth to him once again. Similarly, most of the stars in the sky disappeared from view for a period of some weeks in the year, between their heliacal set and heliacal rise; during this time they too were understood to pass through *Duat.* Likewise, the deceased were required to pass through the twelve parts of the underworld in order to join the gods in the sky. As far back as Old Kingdom times, pharaohs' tombs contained sets of spells known as *pyramid texts* that were designed to ensure a safe passage. Circumpolar stars, on the other hand, were immortal: ever present in the night sky, they never crossed the horizon into *Duat,* never died, and were never reborn. It was these stars that the human soul, striving for immortality, endeavored to join.

These beliefs, and especially the deep importance attached to the north direction with its imperishable stars, may well have given rise to a practice during Old Kingdom times of aligning tombs and temples with the cardinal directions. Pyramids from Saqqara onwards were oriented to the cardinal points, and at the Pyramids of Giza this was achieved with remarkable precision. This practice of cardinal orientation also appears to have been followed, though not necessarily with such great precision, at Old Kingdom temples such as those at Abu Ghurab. Other celestial alignments have also been discovered at Old Kingdom temples and tombs. The great pyramid of Khufu, the largest of the three pyramids at Giza, contains two long shafts connecting the King's Chamber directly to the outside world. Thought for many years to be mere ventilation shafts, it was discovered in the 1960s that one was aligned upon the star Thuban in the north, the closest thing to a pole star at the time, and the other upon Orion's belt in the south. While it is possible that the alignments are fortuitous, it is more likely that they had a very real purpose relating to the journeys to the stars that the pharaoh would need to make in the afterlife.

Various New Kingdom temples seem to have been aligned with the sun. The central axis of the Great Temple of Amun-Ra is aligned toward winter solstice sunrise. The main enclosure at Karnak also contains several other temples with solstitial orientations. This is scarcely surprising, given that it was the center of a sun cult, but the reasons that particular importance was attached to the winter solstice are unknown. The main axis of another great New Kingdom temple, Abu Simbel, was aligned upon sunrise on (in our calendar) October 18 and February 22. (This temple was moved in the 1960s out of the way of the floodwaters created by the Aswan Dam, but the alignment was carefully preserved.) These dates were probably significant for calendrical reasons, but precisely what these were is hotly disputed.

A number of enthusiasts have investigated possible astronomical alignments at Egyptian temples and tombs over the years, going back to the pioneering efforts of the German scholar Heinrich Nissen and the British physicist Sir Norman Lockyer in the late nineteenth century. Not only solar but also stellar and even lunar alignments have been claimed, but there are inherent dangers in all this. The mere existence of an alignment does not prove that it was intentional, and the number of different structures in each temple complex provides many putative alignments. Only recently have systematic studies of temple and tomb orientations in ancient Egypt, sensibly interpreted in the light of the broader knowledge that has been obtained over the years from a range of archaeological evidence, begun to be undertaken. Preliminary results seem to indicate that temples in the Upper Egyptian Nile valley were predominantly topographically oriented, although some astronomical orientations were included as well, especially upon the winter solstice sun. According to another recent idea, the builders of Old Kingdom pyramids perfected (over time) their orientation and proportions so as to achieve impressive illumination effects at the summer solstice and equinoxes.

See also:
Astronomical Dating; Cardinal Directions; Lockyer, Sir Norman (1836–1920); Nissen, Heinrich (1839–1912).
Ancient Egyptian Calendars; Pyramids of Giza.
Circumpolar Stars; Heliacal Rise.

**References and further reading**
Belmonte, Juan. "Some Open Questions on the Egyptian Calendar: An Astronomer's View." *Trabajos de Egiptología [Papers on Ancient Egypt]* 2 (2003), 7–56.
Clagett, Marshall. *Ancient Egyptian Science: A Source Book, Vol. 2: Calendars, Clocks, and Astronomy*. Philadelphia: American Philosophical Society, 1995.
Fagan, Brian, ed. *The Oxford Companion to Archaeology*, 194–202. Oxford: Oxford University Press, 1996.
Hawkins, Gerald. *Beyond Stonehenge*, 193–218. New York: Harper and Row, 1973.
Hodson, F. R., ed. *The Place of Astronomy in the Ancient World*, 51–65. London: Royal Society, 1976.
Krupp, Edwin C., ed. *In Search of Ancient Astronomies*, 214–239. New York: Doubleday, 1977.
Lockyer, Norman. *The Dawn of Astronomy*. London: Cassell, 1894.
Neugebauer, Otto. *The Exact Sciences in Antiquity*, 70–91. Princeton: Princeton University Press, 1951. (2nd ed., Providence: Brown University Press, 1959, 71–96; further corrected ed., New York: Dover, 1969, 71–96.)
Neugebauer, Otto, and Richard A. Parker. *Egyptian Astronomical Texts, III: Decans, Planets, Constellations and Zodiacs*. Providence: Brown University Press, 1969.
Ruggles, Clive, ed. *Records in Stone: Papers in Memory of Alexander Thom*, 473–499. Cambridge: Cambridge University Press, 2002.
Selin, Helaine, ed. *Astronomy across Cultures*, 495–503. Dordrecht, Neth.: Kluwer, 2000.

Shaltout, Mosalam, and Juan Antonio Belmonte. "On the Orientation of Ancient Egyptian Temples: (1) Upper Egypt and Lower Nubia." *Journal for the History of Astronomy* 36 (2005), 273–298.

Walker, Christopher, ed. *Astronomy before the Telescope*, 32–37. London: British Museum Press, 1996.

# El Castillo
*See* Kukulcan.

# Elevation
*See* Altitude.

# Emu in the Sky

When the Europeans arrived in Australia at the end of the eighteenth century, they found a huge diversity of Aboriginal languages and traditions, much of which has been irretrievably lost in recent generations. Sky myths and stories were of great importance to many Aboriginal groups. Like other aspects of the Aboriginal cultural heritage, they varied greatly from one group to another, but some common elements do emerge from the fragments that have come down to us. A good example of this is the celestial emu.

The emu in the sky is a dark shape formed by dust clouds in the Milky Way. The clouds obscure the bright light behind them, and so appear as darker patches within the ribbon of diffuse light that encircles the sky. The emu's head is a particularly clear and compact black patch known to modern astronomers as the Coalsack, close to the Southern Cross. From here, a long neck runs along the Milky Way to a large body in the vicinity of Scorpius, with the hint of legs below. All this is completely invisible in modern light-polluted city skies where even the Milky Way itself can scarcely be seen. In remote areas, however, on a moonless night, it can be magnificent: particularly so when the emu "stands" on the horizon with its head halfway up toward the zenith. It is a huge and directly visible shape, looming darkly and imposingly behind the brilliant light of the foreground stars that fill the whole sky. For a Western observer unused to truly dark skies, recognizing the emu can come as a real shock, a moment of revelation that throws into a different perspective the Western predilection for forming constellations by joining stars with imaginary lines. This experience in itself serves as a reminder that different ways of perceiving things not only exist, but can be compelling and unforgettable once the mind comprehends them.

Earthly emus are unusual in that the male incubates the eggs and rears the young, and this was of great significance to male Aboriginal elders, who had the task of training boys in preparation for their initiation into manhood.

It is said that the changing appearance of the emu in the sky served to indicate the time (of year) when initiation ceremonies should be held. It has also been claimed that some of the rather distorted forms of the emu appearing in Australian rock art owe their appearance to the shape of the celestial emu.

The emu in the sky was recognized as such from one side of Australia to the other, although by no means universally or consistently. This has led some to propose that this particular celestial being may have a history that goes back thousands of years.

**See also:**
Aboriginal Astronomy.

**References and further reading**
Charles, Mary. *Winin: Why the Emu Cannot Fly*. Broome: Magabala Books Aboriginal Corporation, 2000. [Children's book.]
Johnson, Dianne. *Night Skies of Aboriginal Australia: A Noctuary*, 57. Sydney: Oceania Publications/University of Sydney, 1998.
Morieson, John. *The Night Sky of the Boorong*, 109–111. Melbourne: Unpublished master's thesis, University of Melbourne, 1996.

# Equinoxes

The word *equinox* is generally taken to refer to the days when, at every point on the earth, day and night are of equal length. But this definition of the equinox is a bit misleading. Since it gets light before the sun rises and remains light after the sun sets, the actual period of darkness at the equinox will be substantially less than twelve hours, the exact amount depending on latitude and how one defines the boundary between twilight and night. It is even misleading to say that the equinoxes are the days when the time between sunrise and sunset is the same as that between sunset and sunrise, because this definition assumes a flat horizon and the absence of atmospheric effects, particularly refraction. In practice, one cannot determine the equinox by measuring the length of time between sunrise and sunset.

From an astronomical perspective, the definition of the equinoxes is much clearer. They represent the positions in the earth's orbit where the axis joining the two poles leans neither toward nor away from the sun. These points are roughly a quarter of the way around from the two solstices, although not exactly, because the earth's orbit is elliptical rather than circular. It is also possible to express this technical definition from the perspective of an earth-based observer, using the concepts of the celestial sphere and declination. The equinoxes are the times when the sun, in its progress along its annual circuit through the stars known as the ecliptic, crosses the celestial equator, in other words, when its declination is zero. This occurs twice in each year, at the vernal, or spring, equinox, which generally falls on March 20 or 21 in the modern (Gregorian) calendar, and the autumnal equinox on September 22 or 23. (Of course, spring and autumn are transposed in the southern hemisphere.)

This concept of the equinox, in the sense in which astronomers use the term today, was a natural one for the astronomers of Hellenistic Greece, because they were attempting to develop *geometric models* that fit the available observational data. For example, Hipparchus in the second century B.C.E. developed a model of the motion of the sun around the earth in which the sun orbited in a circle at uniform speed but the earth was displaced from the center of the circle. If we were in Hipparchus's shoes, approaching the problem geometrically, then we might naturally progress as follows. First, draw a circle to represent the sun's orbit around the earth. Then divide it into quarter-points (the solstices and equinoxes). Then consider how these could best be determined through observation. Then measure the time intervals between them (leading to the discovery that they are not uniform). Finally, try to find a way of modifying the model to fit the observations. The point of this exercise is to show that, while the equinoxes do enter this particular thought process quite naturally, this is only because we are approaching the problem in a geometrical way. What if, in common with many other human societies in the past, especially the prehistoric past, as well as in the indigenous present, we were *not* inclined to try to understand or explain the cosmos using geometrical models? In other words, how relevant is the concept of the equinoxes likely to be among other human cultures in general?

To add a further complication, when people use the term *equinox* in a cultural context, they commonly mean not the true equinox (as just defined) but the halfway point between the two solstices. Unfortunately, this, like the "night and day" definition, is problematic in practice. For a start, one might mean the halfway point in space or the halfway point in time. The first of these might be identified, say, by observing the rising points of the sun at the two solstices and then marking out the day when it rises halfway between them. The second would involve counting the number of days between the adjacent summer and winter solstices, then counting half that number of days. In the case of the spatial halfway point, unless one is close to the equator, the sun does not rise vertically but at an angle. This means that, unless the horizon is smooth and flat, the resulting "equinoctial sunrise" could be as much as several degrees (and the resulting "equinox" many days) away from the true one. The temporal halfway point is less problematic, in that it is independent of place, thus making it easier for us to spot if things were consistently aligned upon sunrise (or sunset) on that date. The engineer Alexander Thom did believe that this event was extensively known in prehistoric Britain and marked by monumental alignments; for this reason the day halfway in time between the solstices is sometimes referred to as the *Thom equinox*. Because the earth does not orbit the sun at a constant speed, it actually occurs one or two days later in March, and earlier in September, than the true equinox, at a time when the sun's declination is about +0.5 degrees.

When monumental alignments are found facing close to east and west, such as the two passages at the passage tomb of Knowth in the Boyne Valley of Ireland, it is often assumed that they are equinoctial in the "halfway between the solstices" sense. It is certainly easier to argue on pragmatic grounds that the "halfway" rather than true equinox was the one marked in prehistoric times, in that the halfway point (whether in space or time) could have been identified using direct observations from the place concerned. However, this explanation leaves open the question of how this might have been achieved in practice, and to what precision. Identifying the Thom equinox, for example, requires an efficient system of recording or memorizing numbers of days up to at least 180 and presupposes that the solstices can themselves be defined to pinpoint precision—which is not self-evident, as is seen at the Bronze Age site of Brainport Bay in Argyll, Scotland. This is quite different from suggesting, for example, that past communities in temperate latitudes may have divided the seasonal year into two main portions, the start of summer and winter halves being recognized and celebrated at times that happen to approximate the equinoxes. The point is whether the equinox existed as a meaningful concept for them, and hence a meaningful concept for us in trying to explain, for example, monumental alignments to the east or west.

This question leads to another, fundamental question. Even if certain human societies in the past had the technical means to accurately determine the halfway point between the solstices, either in space or time, are they likely to have been motivated to do so? Historical and anthropological evidence suggests that the answer in most cases is no. To judge from historical and ethnographic examples, the norm is to identify dates that are significant in a local context, such as in a ceremonial calendar related to seasonal agricultural activities. Ceremonial significance might also be attached, say, to the day on which the sun rises or sets in line with a significant feature in the visible landscape, such as a sacred mountain: an example of this occurs at Cacaxtla in Mexico. The point is that space and time are generally not conceived in the abstract, as in the modern scientific tradition, but in relation to physically perceived objects and events. Now, the solstices are physically discernible (to whatever precision) as the days when the lengths of day and night are at their longest or shortest. In addition, the solstitial rising and setting points of the sun have a concrete significance in any landscape as the limits of the sun's motion, and the boundaries of those parts of the horizon where the sun can rise or set. The equinoxes, and the positions of equinoctial sunrise and sunset, on the other hand, have no inherent significance. Division into two equal parts only tends to strike the modern investigator as an obvious way of subdividing up the arc of horizon where the sun rises, or the time between the solstices, because he or she perceives space and time in an abstract way in the first place.

The fact that the equinox has, nonetheless, acquired crucial liturgical importance within the Christian world in connection with the timing of Easter is attributable to the roots of that tradition in the Classical world. The difficulties of recognizing and marking the equinox in medieval times were considerable, and this is reflected in the process and practice of orienting churches.

But outside this context the evidence from around the world that people in the past were directly interested in the equinox (however defined) and intentional equinoctial alignments is rather thin. For instance, if we look impartially at the patterns of orientation of local groups of European prehistoric temples and tombs, rather than just picking out individual examples, then the evidence suggests a clear interest in the motions of the sun, manifested in many different ways in different places and times, but no clear preference for equinoctial alignments. In Mesoamerica, to take another example, the easterly direction was itself of fundamental importance, and convincing evidence exists for horizon calendars marking sunrise at different dates in the year against topographic landmarks. However, there is little or no direct evidence of precise equinoctial alignments. Despite this, some researchers continue to argue that topographic landmarks could have been used to mark off equal divisions of the year into quarters or eighths.

When trying to interpret the orientations of ancient structures, it has become commonplace to apply a "toolkit" of potential astronomical targets in which sunrise and sunset on the solstices and equinoxes invariably appear at the top of the list. One problem is that the term "equinoctial" is often poorly defined and used all too often simply as a convenient label for east-west alignments. More seriously, it is usually considered self-evident (or else implicitly assumed) that the equinox was a meaningful concept, whatever the human society being investigated, and hence that sunrise or sunset on that date was a possible target for orientations. In fact, outside the framework of the modern Western scientific tradition (which has its roots in the Greek geometrical tradition), it is far from self-evident in most cases that the equinox was of any significance at all.

The equinoxes, perhaps more than any other astronomical concept, demonstrate the dangers of applying Western concepts too readily and uncritically in interpreting the material remains of human cultures in the past. It is essential to deconstruct concepts that are particular to our own mindset if we want to understand actions that arise as a result of the mindsets of others.

See also:
Cardinal Directions; "Megalithic" Calendar; Solstitial Directions; Space and Time, Ancient Perceptions of; Thom, Alexander (1894–1985).

Boyne Valley Tombs; Brainport Bay; Cacaxtla; Church Orientations; Gregorian Calendar; Horizon Calendars of Central Mexico.
Declination; Ecliptic; Solstices.

**References and further reading**
Ruggles, Clive. "Whose Equinox?" *Archaeoastronomy* 22 (supplement to *Journal for the History of Astronomy* 28 [1997]), S45–50.
———. *Astronomy in Prehistoric Britain and Ireland,* 148–151. New Haven: Yale University Press, 1999.

## Ethnoastronomy

Paralleling the definition of archaeoastronomy, one can define ethnoastronomy as the study of beliefs and practices concerning the sky among modern peoples, and particularly among indigenous communities, and the uses to which people's knowledge of the skies are put. The term *ethnoastronomy* seems to have been coined in 1973, not long after the term *archaeoastronomy,* first appearing in the title of a review article by Elizabeth C. Baity, "Archaeoastronomy and Ethnoastronomy So Far," in the journal *Current Anthropology*. There is no clear dividing line between archaeoastronomy and ethnoastronomy, and many would prefer simply to combine the two fields under one heading, such as *cultural astronomy*.

See also:
Archaeoastronomy.

**References and further reading**
Chamberlain, Von Del, John Carlson, and Jane Young, eds. *Songs from the Sky: Indigenous Astronomical and Cosmological Traditions of the World.* Bognor Regis, UK: Ocarina Books, and College Park, MD: Center for Archaeoastronomy, 2005.
Ruggles, Clive, and Nicholas Saunders, eds. *Astronomies and Cultures,* 1–31. Niwot: University Press of Colorado, 1993.
Selin, Helaine, ed. *Astronomy across Cultures.* Dordrecht, Neth.: Kluwer, 2000.

## Ethnocentrism

Ethnocentrism is the tendency to create a privileged view of our own (modern Western) culture and hence to project our own ways of comprehending things onto the people we are interested in. Clearly, this is something to be avoided if we are trying to understand other ways of understanding the world, which is the broader agenda into which archaeoastronomy fits.

In its most extreme form, ethnocentrism manifests itself as the tendency to judge the "achievements" of a past culture as if they can be measured on a linear scale with ourselves at the pinnacle. It is very easy to fall into this trap quite innocently, simply by praising some intellectual achievement of a past culture. Yet if a time machine took us to meet face to face with the people of whom we were speaking, they would surely find this practice immensely patronizing: from their point of view it would be we who are clearly failing to comprehend their way of viewing the world. Very likely *we* would be seen to have progressed only a short way up *their* scale of achievement.

A problematic issue arises. To reach what people in the modern Western world, at least, would consider to be the most reasonable, reliable, and sustainable conclusions on the basis of the evidence available—for example, whether an astronomical alignment at a monument was likely to have been deliberate and, if so, what it meant to the people who created it—we must be scientific in the broadest sense. This is another way of saying that we must select the evidence fairly and consider it objectively, rather than just taking into account the evidence that fits our favorite theory and ignoring the rest. However, some argue that in order to avoid ethnocentrism we must not accord "our science" a privileged place in analysis and must even strive *not* to be objective. This problem is relevant in archaeology, and indeed in many of the social sciences as a whole: it is a form of *cultural relativism* that has caused a good deal of confusion and aggravation. The resolution of this apparent paradox is simple. Even when we are studying other worldviews, and the context in which they operated, we must seek to comprehend them using our own mindset. Although we should avoid ethnocentric interpretations, we must still be "scientific" in our broad approach to interpreting the evidence.

**See also:**
Methodology; Nationalism.

**References and further reading**
Aveni, Anthony F. *Skywatchers*, 6–7. Austin: University of Texas Press, 2001.
Ruggles, Clive. *Astronomy in Prehistoric Britain and Ireland*, 80. New Haven: Yale University Press, 1999.

# Extinction

If you attempt to watch a star setting in the middle of the night, you may well be in for a disappointment. Unless the western horizon is a high one, as the star approaches it, it will gradually become dimmer and (depending upon a number of factors including, of course, the initial brightness of the star) may well disappear completely. Even in a location where the skies are little affected by modern light pollution, this will still be the case. The cause is *atmospheric extinction*—the absorption and dispersion of the starlight as it passes through the earth's atmosphere on the last leg of its long journey to the human observer on earth. When a star appears close to the horizon, its light has to pass through many times as much atmosphere as when it is high in the sky. The practical effect is that surprisingly few stars—even bright ones—are actually visible at the moment of rising or setting, even in remote places on the darkest of nights.

The effect of extinction is that a star will only be visible above a certain altitude, known as the *extinction angle*. That angle will depend upon the brightness of the star and upon atmospheric conditions. At high latitudes, where the celestial bodies rise and set at relatively shallow angles, extinction adds another confounding factor to any attempt to date a monument on the basis of postulated stellar alignments.

See also:
Astronomical Dating.
Altitude; Star Rising and Setting Positions.

**References and further reading**
Aveni, Anthony F. *Skywatchers,* 105–107. Austin: University of Texas Press, 2001.
Ruggles, Clive, ed. *Archaeoastronomy in the 1990s,* 157, 163. Loughborough, UK: Group D Publications, 1993.
Schaefer, Bradley E. "Atmospheric Extinction Effects on Stellar Alignments." *Archaeoastronomy* 10 (supplement to *Journal for the History of Astronomy* 17 [1986]), S32–S42.

# F

## Fajada Butte Sun Dagger

Fajada Butte is an isolated, 135 meter- (440-foot-)high column of rock at the southern entrance of Chaco Canyon in New Mexico. In a precarious location on a narrow ledge are three stone slabs between 2 and 3 meters (7–10 feet) in height, with narrow gaps between them, which lean against the cliff. On the cliff immediately behind the slabs are two spiral petroglyphs. For a few days around summer solstice, the larger spiral is directly lit by sunlight for just eighteen minutes. Shortly before noon, a spot of light appears right at the top of the spiral, quickly grows into a vertical dagger of light passing directly through the center, and then passes downward to disappear near the bottom of the spiral. On days before and after the solstice the dagger appears for longer and to the right of the center, the displacement being detectable only two or three days away from the solstice itself. The smaller spiral manifests a similar effect at the equinoxes. At winter solstice the two light daggers frame the larger spiral by appearing down each side of it.

This spectacle achieved considerable fame following its discovery in the 1970s and remains for many the best-known manifestation of ancient astronomy in the United States. Yet fierce debates surrounded it for many years, centering upon the question of whether the sun dagger was actually significant to, or even noticed by, the person or people who carved the spiral petroglyphs. It is not contentious to suggest that this inaccessible spot may well have had a sacred significance for ancient Puebloans. Neither is it difficult to envisage a sun priest coming here to conduct a lonely ritual, deriving cosmic power and making suitable offerings when the sunlight showed that the time was right. This is not so different from practices recorded in later, historic Pueblo times. But was it really a sun shrine? Did the sun dagger phenomenon even exist when the spirals were first carved? The slabs have shifted slightly since the 1970s, probably because of the damaging effects of the large numbers of visitors resulting from the publicity the initial discovery received, and before the site was closed off to the public. This shifting raises the possibility that shifting also could have occurred in

*Fajada Butte viewed from the east at sunrise. The petroglyphs upon which the sun dagger falls are located high up the Butte on a narrow ledge. (Courtesy of Clive Ruggles)*

the centuries between the period when the site was in use and its discovery. We can take nothing for granted.

Even ignoring this possibility, one much-debated question is whether the three slabs were placed in position or had simply fallen there naturally. Either way, placing the spirals to get the light daggers striking them in the correct way at the correct times would have been a formidable task involving long days studying the patterns of sunlight and shadows against the half-hidden rock face. In the absence of any independent evidence, how can we judge whether this feat was actually achieved? The question must be turned about to focus upon our own methodology. It becomes a question of how likely we are to interpret wholly fortuitous phenomena as deliberate. Given rock art in a restricted location, how likely are we to be able to spot totally unintentional but nonetheless impressive light and shadow phenomena occurring at what *we* would take to be significant times?

In the decade following the initial discovery, the team who had first uncovered the sun dagger phenomenon carried out further research at the site. As a result, they extended their hypothesis to suggest that the Fajada Butte petroglyphs were significant in relation not only to sunlight but to the light of the moon. At the time of moonrise, there would have been a sharp division between parts of the spirals lit up by moonlight and those still in shadow. But the motions of the moon are more complex than that of the sun, varying over an 18.6-year cycle known as the lunar node cycle. Lunar dagger configurations in relation to the spirals were duly identified, and the conclusion was reached that the ancient Puebloans knew and recorded the longer-term motions of the moon. This conclusion was subsequently reinforced by studies of alignments of various buildings in Chaco Canyon, which seemed to show significant alignments upon limiting moonrise and moonset positions that occur only around the times of *lunar standstills,* and these only occur once in every lunar node cycle. The alignment evidence proved highly controversial, both because of selection effects in the data and the fact that independent historical or ethnographic evidence of an interest in the lunar standstills as a concept is lacking anywhere in pre-Columbian or indigenous America. Ironically, what arose as an additional (lunar) elaboration of the original (solar) light-and-shadow idea turned out for many critics to undermine it, since it raised serious concerns about how easy it is to see significance in things that are, in fact, purely fortuitous.

The question remains: by focusing on the solstitial and equinoctial light dagger phenomena, might we be ignoring equally impressive but different phenomena that occur at numerous other times of day and year? If so, then we should question whether the solstitial and equinoctial ones were really intentional. This is an issue that has been extremely difficult to resolve until recently, with the development of a new visualization tool, which, at the time of writing, is on display at the Adler Planetarium in Chicago. This is a computer reconstruction of the Chaco petroglyph in three dimensions. It allows the user to set the time to any day in the year and any time of day, and to watch the light and shadow effects. Using such a tool, the user can reach an informed conclusion about whether the solstitial sun daggers are really so special.

The consensus that has emerged after the many debates is that the daggers that appear at the summer and winter solstices could well have been deliberate, and that the site may well have been a sacred place used as a sun shrine. This conclusion is considerably strengthened by a number of other "solstice daggers" that have now been found at other sites in the U.S. southwest. The lunar alignments at Fajada Butte, however, are almost certainly fortuitous, and the equinoctial ones remain open to question.

**See also:**
Equinoxes; Methodology.
Casa Rinconada; Chaco Meridian; Is Paras; Newgrange.
Moon, Motions of; Solstices.

**References and further reading**
Aveni, Anthony F., ed. *Archaeoastronomy in the New World,* 169–181. Cambridge: Cambridge University Press, 1982.
Carlson, John B., and W. James Judge, eds. *Astronomy and Ceremony in the Prehistoric Southwest,* 16–18, 71–88. Albuquerque: Maxwell Museum of Anthropology,1987.
Fountain, John W., and Rolf M. Sinclair, eds. *Current Studies in Archaeoas-*

*tronomy: Conversations across Time and Space,* 101–120. Durham, NC: Carolina Academic Press, 2005.

Krupp, Edwin C. *Echoes of the Ancient Skies: The Astronomy of Lost Civilizations,* 152–156. Oxford: Oxford University Press, 1983.

Romano, Giuliano, and Gustavo Traversari, eds. *Colloquio Internazionale Archeologia e Astronomia,* 137–150. Rome: Giorgio Bretschneider Editore, 1991.

Sofaer, Anna, Volker Zinser, and Rolf M. Sinclair. "A Unique Solar Marking Construct." *Science* 206 (1979), 283–285.

Zeilik, Michael. "A Reassessment of the Fajada Butte Solar Marker." *Archaeoastronomy* 9 (supplement to *Journal for the History of Astronomy* 16 [1985]), S69–S85.

## Field Survey

The goal of archaeoastronomical field surveys is to assess the astronomical potential of a set of archaeological or material remains. Often we are interested in structural alignments, asking where they point to on the horizon and what rises or sets there, or would have risen or set there at some point in the past. Sometimes we are interested in the interplay of sunlight and shadow. We might be interested to know, for example, the particular times of the year and day when sunlight suddenly streamed into the dark interior of a prehistoric tomb such as Newgrange in Ireland, or when streaks of sunlight or shadow fell in distinctive ways across the spiral rock carvings on Fajada Butte in Chaco Canyon.

It is, of course, possible to go to an archaeological site and simply observe what happens, for example, at a sunrise or sunset. Being there in this way has the advantage that you can assess the visual impact of a phenomenon first-hand and in context, and maybe spot alignments or light-and-shadow effects that might otherwise have been missed. For those lucky enough to be able to revisit a place on many different occasions, special effects only visible at certain times may suddenly show up. Some important discoveries have certainly been made in this way, including the famous midwinter sunrise alignment at Newgrange. However, there are dangers. An obvious one is that a site and its landscape may have altered significantly since earlier times: a monument may have been constructed and reconstructed in several stages or may have suffered damage more recently. Because of this, effects that are spotted now may have appeared rather different, if they appeared at all, in the past. A more subtle danger arises because people are inevitably drawn to a site on what they perceive as special days, such as the solstices. If you spot an eye-catching effect on a day such as this, it is easy to forget that equally eye-catching effects might also be present on other days, thereby decreasing the likelihood that the solstitial effects were actually intentional and meaningful to the original builders. Yet another problem is that the rising and setting position of the celestial bodies may have been different then from now. How great a problem this is depends upon the astronomical body or phenomenon in question, the precision of the alignment or effect, and the time span that has passed since it was (ostensibly) set up.

For professional work, a set of methods and techniques has been established by archaeoastronomers over the years. These form an essential part of the broader study of archaeoastronomy, just as excavation and other field techniques form a central part of archaeology as a whole.

For assessing alignments, the standard procedure is to determine the azimuth and altitude of the relevant point or range of points on the horizon using an instrument such as a theodolite or a magnetic compass and clinometer, depending upon the accuracy required. From this we are able to determine the declination, which provides an indication of what rises and sets there, or would have risen or set there at any given era in the past. In particular, it is not necessary to wait for a particular day such as the solstice in order to determine where the sun would have risen and set on that day, or on any other day in the year. Depending upon the accuracy required, it may also be possible to determine horizon data directly from large-scale topographic maps or by analyzing digital topographic data using computer software such as Geographical Information Systems (GIS). This approach is also useful in helping to overcome practical problems such as the obscuration of the horizon by modern buildings or trees, or even the bad weather that might beset a survey. There is also great potential for using computer technology to combine three-dimensional reconstructions of sites, monuments, and landscapes with visualizations of the whole sky; this is a technique that remains to be fully exploited.

For light-and-shadow effects, various forms of simulation are possible, including three-dimentional computer reconstructions. For example, at the time of writing, a simulation of a "sun dagger" that crosses one of the spiral petroglyphs at Chaco Canyon, where the user can set the day of the year and the time of day, can be viewed in the cultural astronomy gallery at the Adler Planetarium in Chicago.

It is important to be rigorous in presenting the results of field surveys so that others can objectively assess the evidence. A photograph of a pretty sunset in a horizon notch behind a standing monument is generally of little value because it is unquantitative: there can often be great flexibility in choosing a suitable spot for taking the photograph, and one often has only the photographer's word as to when the picture was taken. Scaled horizon diagrams and associated site plans are as essential to the archaeoastronomer as carefully drawn and scaled excavation plans are to the archaeologist.

A vital, though less tangible, aspect of field procedure is selection of data. The mere existence of an astronomical alignment does not prove that it was intentional or meaningful in the past. This means that we must pay attention (some would say, strict attention) to selecting data fairly. It proves nothing to go to a site and, wittingly or unwittingly, choose the alignments that look promising astronomically while ignoring the rest. Selection criteria form an essential part of the methodology that underlies all archaeoastronomical fieldwork.

**See also:**
Archaeoastronomy; Compass and Clinometer Surveys; Methodology; Theodolite Surveys.
Fajada Butte Sun Dagger; Newgrange.
Altitude; Azimuth; Declination; Precision and Accuracy; Solstices.

**References and further reading**
Aveni, Anthony F. *Skywatchers,* 120–124. Austin: University of Texas Press, 2001.
Ruggles, Clive. *Astronomy in Prehistoric Britain and Ireland,* 164–171. New Haven: Yale University Press, 1999.

# Fiskerton

The site of Fiskerton in Lincolnshire, in the fenlands of eastern England, is one of a number of wooden trackways or causeways built during the Late Bronze Age and Iron Age (late second and first millennium BC) across marshes, shallow lakes or rivers at various locations in Britain and continental Europe. Whether we might wish to classify them as "bridges" (connecting dry land) or "jetties" (for example, leading out onto platforms over open water) is not always clear from the available evidence, but what they have in common may have to do less with this (to us) pragmatic aspect of their function as with their strong association with votive deposits connected with water. Underneath and along both sides of the Fiskerton causeway, which was supported by two rows of upright posts, were scattered literally hundreds of what appear to be votive offerings, including precious weapons such as daggers and swords as well as more mundane items of metalwork such as hammers.

This is interesting enough in itself, but what sets Fiskerton apart is evidence that the causeway underwent major refurbishment at regular intervals. One of the advantages of ancient timber artifacts and constructions, where they remain to be examined by the modern archaeologist, is that they are susceptible to dendrochronology (tree-ring dating). This is a technique that can specify the date of felling to a particular year, and even season. Thus we know that the Fiskerton causeway was periodically maintained by adding new posts, and that the felling dates span a period of over a century, from 457 B.C.E. to 339 B.C.E. It appears that the felling took place predominantly in the winter or early spring, and that few if any timbers were seasoned before use—in other words, they were put into use within a year. Remarkably, the causeway uprights seem to have been felled—and, by implication, placed—at regular intervals of between sixteen and eighteen years.

Given the evident sacred significance of the site, an obvious possibility is that the regular episodes of major refurbishment—likely accompanied by due ritualistic observances—were astronomically regulated. Archaeologist Andrew Chamberlain has examined the possibilities in detail. A prime candidate would seem to be the Saros cycle. This represents a period after which the sun, moon, and earth return to almost exactly the same configuration in space. The upshot is that, at any time, the moon's phase and position in the sky are more or less identical to what they were eighteen years and eleven days before. In particular, if a lunar eclipse occurs once, then another is likely to occur eighteen years and eleven days later. Chamberlain concluded that the major episodes of construction were indeed correlated with total lunar eclipses—those associated with two particular Saros series (nos. 46 and 47).

The implication here is that the people who used Fiskerton could predict eclipses as opposed to merely reacting to them. Unfortunately, this appealing conclusion may not be as clear-cut as it seems. For one thing, the fact that one lunar eclipse may be prominently visible does not necessarily mean that the next one in the given Saros series will be prominent as well; the successor may, for example, occur below the horizon during the daytime. Second, many different Saros cycles, running concurrently, produce eclipses from one year to the next. Third, some lunar eclipses will be missed on cloudy nights. In the absence of any form of counting or recording, the occurrence of total lunar eclipses observed from any given location could not seem anything but entirely irregular. There are some short-term cycles that, once recognized, will yield partial predictive success, but only after very careful recording maintained systematically over several generations (as among the ancient Babylonians) can longer-term cycles such as the Saros become evident. And even then, the Saros is not the only cycle that might be recognized: the ancient Chinese, for example, identified a shorter one of 10 years and 334 days, while the Maya discovered a longer one of 32 years and 272 days. While we should not underestimate the capacity of people in Iron Age Britain to keep records had they been so motivated, we have no direct evidence of this.

It remains possible that some of the bursts of construction at Fiskerton were carried out in response to lunar eclipses, and indeed there is some evidence from other European trackway sites of correlations with various lunar eclipses that might support this conclusion. However, in this case the apparent regularity of the interval between the bursts of activity at Fiskerton would have to be fortuitous, since people would have had no way to distinguish only those eclipses from particular Saros series. Further evidence and statistical analysis may clarify some of these issues.

Despite all these imponderables, Fiskerton has a wider significance. One of the most important things about the sky as a cultural resource is that the regular cycles of the celestial bodies provide a way keeping the timing of various activities in tune with nature. Yet archaeoastronomers are seldom faced with evidence that bears directly on this, since the time resolution of most archaeological data is so imprecise. (Some have tried to use the astronomy to do the dating, but this is dangerous because it can so easily result in circular argument.) Even the best radiocarbon dating can seldom tie down a particular action to much better than the nearest century, and often the uncertainties are much greater than this. Fiskerton represents a rare example of a prehistoric site—certainly the first in Britain—where the timing of several successive bursts of activity can be tied down to the very year, if not the season. This enables us to ask not only whether they were correlated with the regular cycles of the heavens, but also whether they may even have had some connection to actual celestial events such as particular eclipses. This sets an important precedent for other places where we are lucky enough to be able to recover timber artifacts and monuments.

**See also:**
Astronomical Dating; Lunar Eclipses.
Babylonian Astronomy and Astrology; Chinese Astronomy; Dresden Codex.

**References and further reading**
Field, Naomi, and Michael Parker Pearson, eds. *Fiskerton: An Iron Age Timber Causeway with Iron Age and Roman Votive Offerings*. Oxford: Oxbow Books, 2003.
Catney, Steve and David Start, eds. *Time and Tide: The Archaeology of the Witham Valley*, 16–32. Heckington, UK: Witham Valley Archaeological Research Committee, 2003.
Pryor, Francis. *Britain B.C: Life in Britain before the Romans*, 283–286. London: HarperCollins, 2003.

# Forbidden City
*See* Chinese Astronomy.

# Giza
*See* Pyramids of Giza.

# Governor's Palace at Uxmal

Uxmal is situated in the Puuc Hills region of northern Yucatan. One of the largest Classic Maya cities, it reached its apex around C.E. 800, then was suddenly abandoned, a fate shared by many other Maya centers during the ninth century. The various buildings that form its ceremonial center are straight-sided and consistently aligned (on a grid rotated by nine degrees clockwise from the cardinal directions), but with a single glaring exception. The so-called Palace of the Governor (a traditional name not necessarily related to its original purpose)—one of the most splendid buildings on the site, an elongated rectangle in shape with a beautifully decorated façade containing seven entrances—is noticeably skewed from the common grid (by just under twenty degrees).

The Governor's Palace looks towards the distant (now ruined) pyramid of Cehtzuc, a little under five kilometers (three miles) away. (There is considerable confusion in the early literature, since Cehtzuc was originally misidentified as a different ruin, Nohpat.) This direction—east-southeast, or more precisely, an azimuth of 118 degrees—corresponds to the maximum southerly rising point of the planet Venus at around the time the temple was built. Another possibility, argued strongly by some scholars such as the Slovenian archaeologist Ivan Šprajc, is that the Venus alignment actually operated in the opposite direction, from Cehtzuc toward the Governor's Palace.

If this were a prehistoric monument that we were investigating in the absence of historical or documentary evidence, then little credence would be placed on such an alignment. The planetary motions are notoriously complex, and their extreme rising or setting points are subject to complicated patterns of change operating on various time scales. Furthermore, the extremes for all the planets are all fairly close together on the horizon and can

The front of the Governor's Palace at Uxmal, aligned upon the most southerly rising point of Venus. (Courtesy of Clive Ruggles)

be difficult to distinguish, both from one another and from the sun and moon. Why should we believe that this Venus alignment (in whichever direction) was deliberate? The most direct answer is the abundance of Venus glyphs within the carved frieze on the front of the building—over three hundred of them. To one able to read Maya writing, the building fairly screams "Venus." A recent reinterpretation of the hieroglyphic *throne inscription* above the central doorway has reinforced this conclusion by arguing that it depicts Maya zodiacal constellations and includes reference to Venus.

See also:
Methodology.
Dresden Codex; Venus in Mesoamerica.
Inferior Planets, Motions of.

**References and further reading**
Aveni, Anthony F., ed. *Archaeoastronomy in Pre-Columbian America*, 163–190. Austin: University of Texas Press, 1975.
Aveni, Anthony F. *Stairways to the Stars: Skywatching in Three Great Ancient Cultures*, 139–142. New York: Wiley, 1997.
Aveni, Anthony F. *Skywatchers*, 283–288. Austin: University of Texas Press, 2001.
Bricker, Harvey M., and Victoria R. Bricker. "Astronomical References in the Throne Inscription of the Palace of the Governor at Uxmal." *Cambridge Archaeological Journal* 6 (1996), 191–229.

Kowalski, Jeff Karl. *The House of the Governor: A Maya Palace of Uxmal, Yucatan, Mexico*. Norman and London: University of Oklahoma Press, 1987.

Krupp, Edwin C., ed. *In Search of Ancient Astronomies*, 199–202. New York: Doubleday, 1977.

Ruggles, Clive, ed. *Archaeoastronomy in the 1990s*, 270–277. Loughborough, UK: Group D Publications, 1993.

# GPS Surveys

Global Positioning System (GPS) receivers use signals from satellites to determine their position on the ground. Given a reasonably clear view of the sky, a standard GPS receiver can usually locate itself in the horizontal to within a few meters. This information is clearly useful in locating or relocating sites of interest, especially in remote areas, but it also has some uses in the course of archaeoastronomical field survey. Sometimes we wish to know the azimuth of a feature on a distant horizon, such as a mountain peak, in order to assess its potential astronomical significance. Then again, we might wish to know the azimuth of a reference point such as a church spire in order to calibrate a compass and clinometer survey. Given that the distant point is readily identifiable on a topographic map, one way to do this is to determine its coordinate position from the map, and to determine the coordinate position of the observing point using a GPS receiver. The azimuth of the distant point can then be deduced.

Standard GPS receivers are not accurate enough to be used for measuring the azimuth between two points (that is, determining the orientation of an alignment) unless the points in question align, in turn, with a distant landmark whose azimuth can be determined with the aid of a map, as already described. Trying to determine the azimuth of one point from another simply by placing a GPS receiver at each of them would, given a ten meter uncertainty in the position of either, require the two points to be something like two kilometers (one and a half miles) apart in order to achieve half-degree accuracy in the result. Differential GPS receivers can be a good deal more accurate, perhaps determining a location to within one or two meters, but even then the two points would need to be several hundred meters (up to half a mile) apart for this technique to be of any use.

Carrier-phase GPS receivers relate to standard GPS receivers as FM radio does to AM radio. Although still prohibitively expensive for most users, they have the potential to transform field survey techniques generally in the near future. A carrier-phase GPS receiver has the capacity to determine a position to within a centimeter or two if not better. This should dispense with the need for heavy surveying equipment such as a theodolite in many cases, although even this sort of GPS receiver will experience difficulties if there is not a clear view of the sky, for example because of buildings or trees.

**See also:**
Compass and Clinometer Surveys; Field Survey; Theodolite Surveys. Azimuth; Declination.

**References and further reading**
Ruggles, Clive, Frank Prendergast, and Tom Ray, eds. *Astronomy, Cosmology and Landscape*, 179–181. Bognor Regis, UK: Ocarina Books, 2001.

## Grand Menhir Brisé

The landscape around Carnac on the south coast of Brittany, northwestern France, is so full of megalithic monuments that one almost grows blasé about them. There are huge passage tombs (also referred to as passage graves) and other types of burial monument; great multiple alignments of standing stones several hundreds of meters (over a thousand feet) in length; stone circles and variants, known locally as *cromlechs;* and numerous individual standing stones, often themselves of considerable size (some are several meters [over 15 feet] high). Le Grand Menhir Brisé (the Great Broken Menhir) is situated close to the modern town of Locmariaquer, adjacent to two burial mounds of considerable size: a passage tomb known as the Table des Marchand (Merchants' Table), and a 160 meter- (close to five-hundred-foot-) long tumulus known by its Breton name Er Grah, much of which was destroyed in the mid-twentieth century to make room for a visitors' parking lot, but which has now been carefully restored.

The Great Menhir stands out as by far the largest single standing stone in the area, indeed the largest standing stone in Europe; incredibly, it weighed well over three hundred tons and measured some 20.5 meters (67 feet) from its base to its tip. It does not stand now: at some stage it broke, the top part falling one way and the lower part in the opposite direction. The top part split into three on impact, so that the stone now lies in four huge pieces. The fact that the whole menhir did not topple over in the same direction implies that it could not have fallen while being erected, nor could it have been deliberately pushed over. The only viable conclusion seems to be that it was shaken from side to side by an earthquake: in other words, it fell naturally.

The Scottish engineer Alexander Thom, famous for his interpretation of the British megalithic sites as solar and lunar "observatories" of considerable precision, interpreted the menhir as a "universal" lunar foresight that was used to track the changing rising and setting positions of the moon. The rising point of the moon moves up and down the eastern horizon (and similarly the setting point on the western horizon) between limits that are reached once every month. These limits themselves vary over a cycle of 18.6

*The largest two pieces of Le Grand Menhir Brisé viewed from the west, with the dolmen known as La Table des Marchand visible in the distance. (Courtesy of Clive Ruggles)*

years, so that in every nineteenth year the range of rising (and setting) positions of the moon is at its widest, while nine and a half years later it is at its narrowest. The outer and inner "limits of the limits," where the moon rises and sets at times that are (rather misleadingly) referred to as its major and minor "standstills," form eight lunar horizon targets that, if known, would have helped people keep track of the 18.6-year cycle. All eight horizon targets were, according to Thom, accurately marked by the Great Menhir. This was achieved by carefully placing backsights some distance away in each of the eight opposite directions, to mark places where people could stand and observe the rising or setting of the moon in relation to the Great Menhir in the distance.

Unfortunately, there are many problems with this interpretation. How could the positions of the backsights have been fixed without observing programs lasting many generations? How useful would the "inner" target markers have been in practice? (The moon passes these positions twice every month!) What happened when a critical observation was missed owing to bad weather? Given these and many other concerns, people were quick to ask whether Thom's evidence could really support such an idea. One problem that soon emerged was that the alleged backsights formed a motley mix of prehistoric monuments, and in four out of eight cases were probably not genuine prehistoric monuments at all. Yet the landscape in this area is strewn with impressive monuments. Wouldn't it be quite easy to find an equally convincing backsight in almost any direction from the Great Menhir? The answer, it is now generally agreed, is: very likely indeed.

The importance of this example is methodological. How could Thom have been so selective with the evidence without realizing it? As we know from the accounts of those who worked with him, he merely set out to identify suitable backsights in each of the eight directions he considered significant. Why didn't he at least try some other, randomly chosen, directions as a control? The probable answer is that he was too convinced by his own theory—that Neolithic people observed the motions of the sun and moon to great precision—to feel it necessary. He was not alone in this. It is natural to want to find evidence that supports a favorite theory, especially if it is one that has taken a lifetime to develop. Yet it is precisely for this reason that correct field methodology, which ensures that one gives due consideration to *all* the evidence, is so critically important.

In fact, The Grand Menhir is now known to have stood at the end of an alignment of more than a dozen stones, varying in height. Most of the smaller stones in the row were subsequently felled and some at least were then reused in the construction of tombs. A split carving shows without any trace of doubt that one former menhir—possibly one of this row—was split asunder, with one end ending up as the capstone at the Table des Marchand and another as the capstone at the tomb of Gavrinis several kilometers away across a strait (it is now on an island). Not only does the very early date for menhir construction that this implies drive a final nail in the coffin of the lunar foresight hypothesis (since the lunar targets shift slightly over the centuries owing to the changing obliquity of the ecliptic). More importantly, it shows the potential complexity of the archaeological and chronological context, and the vital importance of taking this into account when formulating astronomical theories.

**See also:**
Methodology; Thom, Alexander (1894–1985).
Crucuno.
Moon, Motions of; Obliquity of the Ecliptic.

**References and further reading**
Burl, Aubrey. *Megalithic Brittany,* 134–137. London: Thames and Hudson, 1985.
———. *From Carnac to Callanish: the Prehistoric Stone Rows and Avenues of Britain, Ireland and Brittany,* 131–146, 153–156. New Haven: Yale University Press, 1993.
———. *A Guide to the Stone Circles of Britain, Ireland and Brittany,* 250–261. New Haven: Yale University Press, 1995.
———. *The Stone Circles of Britain, Ireland and Brittany,* 331–348. New

Haven: Yale University Press, 2000.
Hadingham, Evan. *Circles and Standing Stones,* 163–167. London: Heinemann, 1975.
———. "The Lunar Observatory Hypothesis at Carnac: A Reconsideration." *Antiquity* 55 (1981), 35–42.
Hornsey, Richard. "The Grand Menhir Brisé: Megalithic Success or Failure?" *Oxford Journal of Archaeology* 6 (2) (1987), 185–217.
Patton, Mark. *Statements in Stone: Monuments and Society in Neolithic Brittany.* London: Routledge, 1993.
Politzer, Anie, and Michel Politzer. *Des Mégalithes et des Hommes.* Spézet, France: Coop Breizh, 2004. [In French.]
Ruggles, Clive. *Astronomy in Prehistoric Britain and Ireland,* 34–35. New Haven: Yale University Press, 1999.
Thom, Alexander, and Archibald S. Thom. *Megalithic Remains in Britain and Brittany,* 98–110. Oxford: Oxford University Press, 1978.

## "Green" Archaeoastronomy

This term denotes an approach in archaeoastronomy that is primarily concerned with developing rigorous procedures in studying the possible astronomical alignment of monumental structures. It involves developing strict criteria for data selection and fieldwork methodology, and the formal statistical analysis of the results. This approach emerged in Britain during the 1970s in the attempt to move alignment studies forward and as a response to the huge controversy between archaeologists and astronomers caused by the conclusions of Alexander Thom.

**See also:**
Thom, Alexander (1894–1985).
Alignment Studies; Archaeoastronomy; "Brown" Archaeoastronomy; Statistical Analysis.

**References and further reading**
Aveni, Anthony F., ed. *World Archaeoastronomy,* 3–12. Cambridge: Cambridge University Press, 1989.

## Gregorian Calendar

In 1582 Pope Gregory XIII introduced a dramatic calendrical reform to overcome the problem with the existing Julian calendar: it was getting gradually further and further out of step with the seasonal ("Tropical") year. The problem would not have occurred if the mean length of the year were exactly 365 days and 6 hours, but it is in fact somewhat shorter—365 days, 5 hours, 48 minutes, and 46 seconds. By the sixteenth century, the Julian calendar was running ten days behind the true solar year.

The solution to the problem was two-pronged. First, the pope issued a decree that the day following October 4, 1582, would be October 15. This brought the calendar back in step with the seasonal year. Second, steps were taken to prevent a significant error from accumulating in the future. A scheme was introduced whereby in each subsequent period of four hundred years, three years that would have been leap years under the "every fourth year" rule now would not be. This makes the mean length of the calendrical year 365 days, 5 hours, 49 minutes, and 20 seconds, which is only 34 seconds longer than the true (Tropical) year and means that the accumulated error will not reach a full day again for some 2,500 years. The specific scheme adopted was that century years not divisible by four hundred would not now be leap years, that is, 1700, 1800, 1900, 2100, etc.

Calendrical reform took place immediately in Catholic countries, but not surprisingly, in Protestant countries the pope's decree was ignored. The Gregorian calendar was not adopted in Britain and its colonies until 1752, by which time the error had increased to eleven days, and in Russia until the Bolshevik revolution of 1917. In Ethiopia, the Julian calendar still remains in use, and the timing of Easter for Orthodox Christians in eastern Europe still follows the Julian calendar.

Although it bears no direct relationship to ancient calendars, the Gregorian calendar is related closely to the seasonal year and so provides a convenient frame of reference when discussing the timing of seasonal activities, astronomical observations, or the dates when the sun rose or set in line with temples or buildings. If we are interested in whether the orientation of certain temples or buildings had a calendrical significance, we do have an advantage where we know something of the calendar involved. An obvious case in point is the orientation of medieval churches, where we must be especially careful to distinguish between the Gregorian calendar as our point of reference and the Julian calendar that was being used by the builders.

See also:
Church Orientations; Julian Calendar.

**References and further reading**
Aveni, Anthony. *Empires of Time: Calendars, Clocks and Cultures,* 116–118. New York: Basic Books, 1989.

## Group E Structures

Uaxactún is one of scores of Maya architectural complexes located in the Petén, a forested lowland region in northern Guatemala. From modest beginnings in the first millennium B.C.E. it grew to become one of the most important Maya cities, the focus of a city-state that thrived for several centuries before finally collapsing around C.E. 900.

A rather peculiar group of structures at Uaxactún, known as Group E, was identified as far back as the 1920s as potentially having astronomical significance. The structures lie on opposite (western and eastern) sides of a plaza located on the eastern side of the ceremonial center of the city. (There are also structures to the north and south of the plaza, but these seem to be of secondary significance.) In around C.E. 200 an eight meter- (twenty-six foot-) high pyramid was built on the western side of the plaza, known to archaeologists as Structure E-VII-sub. Later, this pyramid was covered by a fourteen meter- (forty-five foot-) high pyramid, Structure E-VII. Across the plaza on the eastern side was an elongated platform running lengthwise from north to south, upon which there were three buildings ("temples" E-I, E-II, and E-III), evenly spaced. Viewed from the western pyramid, the three buildings to the east are approximately in line with sunrise at the summer solstice, equinoxes, and winter solstice, respectively. This was one of the earliest astronomical alignments to be recognized in Maya architecture, and the site became renowned as a "solar observatory."

A detailed investigation in the 1980s by archaeoastronomer Anthony Aveni and architect Horst Hartung confirmed that the solstitial and equinoctial alignments were precise, provided that the observer was standing about 3.5 meters (11 feet) up the western pyramid. This, together with various strands of archaeological evidence, led them to propose that the whole structure evolved in three stages. At first, E-VII-sub was a truncated pyramid standing only to about 3.5 meters (11 feet) high, and the eastern mound was a simple rectangular platform. An observer standing on the western pyramid would have been able to track the sun rising along the horizon behind—and just about level with—the eastern platform. The length of the platform just spanned the solar rising range. At some stage, the three "temples" were added. These provided points of reference against which precise observations of the solstices and equinoxes could now be made. When E-VII-sub was completed to its full height, such observations could still be made by standing at the appropriate level on the eastern staircase, but all this changed when a new, larger pyramid was built over the top. An observer on this new pyramid would see the eastern horizon well above the eastern temples, and the precise alignments would no longer be functional.

A skeptic examining the Group E alignments might well ask whether they could have arisen fortuitously. One approach to this problem would be to look for similar structures elsewhere, to see if they systematically incorporate similar alignments. Such an approach has been used extensively in Europe, both to confirm suspected megalithic alignments and—as at Drombeg stone circle in Ireland—to isolate megalithic alignments that seem convincing in themselves but become less so when the site concerned is considered as one of a group. In fact, over fifty architecturally similar structures are now known to exist at other Maya sites, mostly concentrated in the Petén within about a hundred kilometers (sixty miles) of Uaxactún: they are so similar to Group E that they have become known as *E-group structures*.

These are broadly characterized by an eastern platform in the shape of an elongated rectangle supporting three equally spaced buildings in a roughly north-south line. There are several variants, including some with three separate constructions on the eastern side. Many of these, while clearly preserving the general form, are so irregular that they could not reproduce the solstitial and equinoctial alignments found at Uaxactún, at least in anything like a similar way.

One's immediate reaction might be to dismiss the Uaxactún alignments as fortuitous. However, we have a plausible chronological development at Uaxactún itself obtained by combining the archaeological and alignment evidence. This suggests another interpretation: that the other E-groups were *nonfunctional copies* of the Uaxactún observatory. Just as in the later stages at Uaxactún itself, they preserved a ritual significance that had its origin in real observations but no longer needed to be confirmed or reinforced by repeating those same observations. Unfortunately, as dating evidence has slowly emerged, it no longer seems that Uaxactún Group E was a particularly early example of the genre. Many other E-groups were built at around the same time—the transition between the Late Preclassic Period and Early Classic Period around C.E. 200—and at least two seem to have been considerably earlier. Even more recently, other *pseudo E-structures* have been discovered that date from the Late Classic period (as late as the seventh century C.E.) but appear to incorporate functional solstitial alignments. A more sustainable argument, then, is that while Maya rulers in the northern Petén over a considerable time span desired to have one of these ritualistic complexes, the actual solar observations mattered more to some than to others.

If solar observations were carried out at Uaxactún and some other E-groups, what then was their purpose? One possibility is that they related—at least originally—to an empirical calendar based upon direct observations of the horizon rising position of the sun. Such a calendar existed in this region, it has been suggested, before an invasion from the north in about C.E. 400 brought greater emphasis on the more abstract 260-day cycle that came to characterize calendars throughout Mesoamerica. Recent statistical analysis of the alignments incorporated in the Group-E-type assemblages does support the conclusion that many of them incorporated sunrise alignments on dates marking calendrically significant intervals. Even so, it may be misleading to portray even the functional E-groups as observatories, since they were primarily ritual complexes. It seems more likely that, where solar observations actually took place, they served as much as anything to regulate calendrically related rituals or ceremonials. One of these may well have been the ball game—a sacred game played all over Mesoamerica and claimed by some scholars to have been played at the equinoxes, since they symbolized the time when the forces of nature were in balance. A great many of the E-groups are found close to a ball court—an arena where such games were played. In many cases, though, any association between the sun and the timing of a ceremony such as the ball game did not need to be reaffirmed by making actual observations.

**See also:**
Alignment Studies; Equinoxes; Solstitial Directions.
Drombeg; Maya Long Count; Mesoamerican Calendar Round.

**References and further reading**
Aveni, Anthony F. *Skywatchers*, 288–293. Austin: University of Texas Press, 2001.
Aveni, Anthony F., ed. *World Archaeoastronomy*, 441–461. Cambridge: Cambridge University Press, 1989.
Aveni, Anthony F., Anne S. Dowd, and Benjamin Vining. "Maya Calendar Reform? Evidence from Orientations of Specialized Architectural Assemblages." *Latin American Antiquity* 14 (2003), 159–178.
Aylesworth, Grant R. "Astronomical Interpretations of Ancient Maya E-Group Architectural Complexes." *Archaeoastronomy: The Journal of Astronomy in Culture* 18 (2004), 34–66.
Whittington, E. Michael, ed. *The Sport of Life and Death: The Mesoamerican Ballgame*, 42–45. London: Thames and Hudson, 2001.

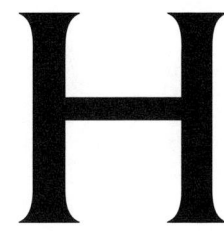

## Ha'amonga-a-Maui

As ancient Polynesians voyaged across the Pacific, discovering and colonizing a great variety of islands, the cultures they established developed in different ways according to the environment and the available natural resources. At one end of the scale were sand and coral atolls where little but coconuts could be made to grow; lifestyle there was frugal, and many of these atolls were settled for only a short period before being abandoned. By contrast, in large and fertile island groups such as Tonga, the Society Islands, and the Hawaiian chain, populations thrived, and complex social hierarchies and powerful chiefdoms developed. The Tongan Islands, in the heartlands of ancient Polynesia, are unusual in that, never having been overthrown by European invaders, the pre-contact social structure has in essence persisted without severe cultural disruption right up to the present day.

On the largest Tongan island, Tongatapu, is found Oceania's most famous archaeoastronomical artifact: the coral trilithon known as *Ha'amonga-a-Maui*. According to tradition, its construction was ordered by Tu'itatui, the eleventh sacred ruler of Tonga, in around C.E. 1200. It consists of three large coral monoliths, two standing and one placed as a lintel across the top, and bears an uncanny resemblance to the sarsen trilithons at Stonehenge in England. Nothing quite like it is known elsewhere in Polynesia. It is aligned along its length upon the rising position of the sun on the June solstice, and ceremonies to mark the occasion still take place.

Despite the superficial similarity between *Ha'amonga-a-Maui* and the Stonehenge trilithons, and the existence of solstitial alignments at both sites, it would be plainly ridiculous to posit a direct connection between these two sites on opposite sides of the world and separated in time by some three millennia. Even so, it might seem impressive that human thought could develop in such similar ways at such different places and times. However, there is absolutely no historical evidence to back up the assertion that the Tongan alignment was significant at the time of its construction. The present tradition dates no further back than 1967, when the King of Tonga, Taufa'ahau

Tupou IV, himself an amateur astronomer, began to take an interest in the alignment of the monument.

In fact, valuable insights about ancient Tongan astronomy may be found by focusing less on monumental alignments and examining instead a broader range of customs and traditions, including contemporary ones. In Tonga, many cultural practices have been unusually well preserved, though constantly subject to modern influences. Stories and poetry frequently refer to topographic features in the landscape and describe the motions of objects in the sky. Designs on bark-cloth and incised markings on war clubs often incorporate symbols of the heavenly bodies. Dance movements can have sacred and cosmic significance. As all this reminds us, knowledge of celestial phenomena formed an integral element of worldview in Tonga as well as throughout Oceania and among human societies everywhere.

> **See also:**
> Cosmology; Solstitial Directions.
> Navigation in Ancient Oceania; Polynesian and Micronesian Astronomy; Stonehenge.
>
> **References and further reading**
> Bellwood, Peter. *The Polynesians: Prehistory of an Island People* (rev. ed.), 69–72. London: Thames and Hudson, 1987.
> Gifford, Edward W. *Tongan Society.* Honolulu: Bishop Museum Press, 1929.
> Hodson, F. R., ed. *The Place of Astronomy in the Ancient World,* 137. London: Royal Society, 1976.
> Kaeppler, Adrienne L., and H. Arlo Nimmo, eds. *Directions in Pacific Traditional Literature: Essays in Honor of Katharine Luomala,* 195–216. Honolulu: Bishop Museum Press, 1976.
> Kirch, Patrick V. *On the Road of the Winds: An Archaeological History of the Polynesian Islands before European Contact,* 219–230. Berkeley: University of California Press, 2000.
> Liller, William. *The Ancient Solar Observatories of Rapanui: The Archaeoastronomy of Easter Island,* 48–49. Old Bridge, NJ: Cloud Mountain Press, 1993.
> Ruggles, Clive, ed. *Archaeoastronomy in the 1990s,* 132–133. Loughborough, UK: Group D Publications, 1993.
> Selin, Helaine, ed. *Astronomy across Cultures,* 137–139. Dordrecht, Neth.: Kluwer, 2000.

# Hawaiian Calendar

The Hawaiian islands were settled, probably in around C.E. 400, by Polynesians who had voyaged northwards across the Pacific for some 3,000 kilometers (2,000 miles) from central Polynesia. This new island chain provided a substantial land mass and a rich and fertile environment in which, over the subsequent centuries, agriculture developed and intensified, and the population thrived and multiplied. During the last two centuries or so before European contact, powerful social hierarchies developed, controlled by high chiefs (*ali'i*), with lesser chiefs, priests, and other specialists, and commoners (fishermen and farmers) being assigned to progressively lower ranks. There may also have been a lowest social class of slaves and outcasts. Everyday life, whatever a person's rank in society, was pervaded by ritual observances and taboos tied to a strict calendar. This calendar derived from an earlier prototype, variants of which were in use all over Polynesia by the time of European contact.

What we know of the Hawaiian calendar derives largely from three native historians, David Malo, Samuel Kamakau, and Kepelino, whose accounts were first recorded in the mid-nineteenth century, in the decades following the arrival of the Christian missionaries and the virtual eradication of indigenous religious practices. While broadly consistent, the three descriptions of the calendar differ markedly in many details. This variation not only indicates that there were significant differences from one island to another at around the time of European contact, but also suggests that the implementation of the calendar in practice, while following common principles, operated on a local basis. In view of the broad similarities between the recorded Hawaiian calendars and those recorded elsewhere in Polynesia, the general principles were probably of considerable antiquity.

The Hawaiian year, termed the *makahiki,* was divided into twelve months according to the phase of the moon. Most month names were used on more than one island and followed broadly the same sequence, but their timing in the seasonal year varied considerably from one Hawaiian island to another. The month *Iki-iki,* for example, was said to have occurred in March on O'ahu, in May on the big island of Hawai'i, in July on Kaua'i, and in August on Moloka'i. The main division of the year was always into equal summer and winter halves, but the two seasons were recorded as containing different sets of six months on different islands. All this might be taken as evidence that no additional (intercalary) month was ever inserted to keep the month sequence in step with the seasons, so that the sequence of months moved around in the solar year. Indeed, there is no record of an intercalary month being inserted, or of what procedure was followed in order to know when to do so. On the other hand, if there were no intercalation, then the timing of events that *were* fixed in the seasonal year (such as the beginning of summer) would progress on average by one month in every second or third year, and there is no evidence of this happening either.

The Hawaiians did make use of observations of astronomical events fixed in the solar year. Piecing together fragments of the available accounts, it appears that the beginning of the summer and winter halves of the year were marked by two major annual events involving the Pleiades. The beginning of winter, in about November, was marked by the appearance of the Pleiades in the eastern sky just after sunset as opposed to rising later in the night (loosely, their acronical rise). The beginning of summer, in about May, was marked by the first appearance of the Pleiades in the eastern sky before dawn, after not having been visible at all for several weeks (their heliacal rise). The Hawaiian term for the new year is the same as that for the year itself, *makahiki,* and this appears to be a contraction of *makali'i-hiki,* which means the rising of the Pleiades.

Each month consisted (nominally) of thirty nights—Hawaiians counted the nights rather than the days—whose names, like those of the months themselves, differed somewhat from island to island or place to place. According to Malo, the day when the new crescent moon was first sighted was known as *Hilo,* meaning "twisted," because slender and twisted was how the moon appeared, and he goes on to name the other twenty-nine nights. There is no explicit reference to the last night, *Muku,* being skipped in approximately every second month, but this must have happened in order to keep the months correlated with the lunar phases. Nonetheless, it is quite clear that a rich set of prognostications was attached to the various nights of the month. For example, according to Kepelino, potatoes, bananas, and gourds planted on the day following *Hilo* will thrive, while those planted on *Ku-kolu,* the fifth night, will "just shoot up like coconuts" and be useless. Many native Hawaiians today still keep track of the phase of the moon and will only carry out certain activities, such as planting crops, on the correct day or days in the month. People's personalities were also seen as related to the night of the month on which they were born. Before the arrival of the Europeans, each month contained four *tapu* (taboo) periods dedicated to the worship of particular gods, when appropriate ritual observances took place. Each was imposed at sunset and lifted at dawn two or three days later.

The term *makahiki* not only referred to the new year and the year as a whole, but also to a period starting at new year and lasting, by most accounts, for some three or four months. It was a sort of prolonged harvest festival, during which normal labors were suspended, as was the regular monthly cycle of ritual observances and taboos associated with them. Instead, a number of rites took place that were connected with the god Lono—who was particularly associated with (dryland) agriculture, fertility, and medicine—as well as with other gods in the Hawaiian pantheon. When Captain Cook first arrived on the Big Island of Hawai'i, he was received and revered as Lono himself, but when he was forced to return just eight days after his departure, in order to repair a mast, he was killed. The deification of Captain Cook almost certainly owed a great deal to the timing of his appearance in relation to the calendar, which may also have had a bearing on his subsequent demise. The first arrival occurred, by good chance, during the *makahiki* festival when the mythical return of Lono was annually re-enacted. The second happened inopportunely, after the *makahiki* period had finished. At this time Lono was meant to have departed, having fulfilled his annual task of regenerating nature, leaving the *aliʻi* to rule in the normal way until the following year. His unexpected return constituted a threat to the perceived order of things and he met a ritualistic death. But this interpretation remains fiercely debated. The opposite view is that Cook was primarily revered as a visiting chief and met his end in a skirmish that had a very down-to-earth explanation—it resulted from his attempts to take an *aliʻi* hostage in return for a stolen boat.

A wider issue of considerable significance here is how outsiders should best try to comprehend, and hence to respect, indigenous traditions. Is it not demeaning to native Hawaiians, as the Sri Lankan anthropologist Gananath Obeyesekere has argued, to assume that they were incapable of recognizing a real threat and dealing with this threat in a "common-sense" manner by attacking the perpetrator? Or is it more demeaning, as the American anthropologist Marshall Sahlins has argued, to fail to recognize and respect the worldview within which Hawaiian religious beliefs and calendrical rituals operated, rather than trying to rate it against our own rationality?

See also:
Lunar and Luni-Solar Calendars.
Kumulipo; Polynesian and Micronesian Astronomy; Polynesian Temple Platforms and Enclosures.
Heliacal Rise; Lunar Phase Cycle.

**References and further reading**
Beckwith, Martha W. *Kepelino's Traditions of Hawaii,* 82–96. Honolulu: Bishop Museum Press, 1932.
———. *Hawaiian Mythology,* 31–41. Honolulu: University of Hawai'i Press, 1970.
Johnson, Rubellite K. *The Kumulipo Mind: A Global Heritage,* 217–222. Honolulu: Anoai Press, 2000.
Kamakau, Samuel M. *The Works of the People of Old,* 13–19. Honolulu: Bishop Museum Press, 1976.
Makemson, Maud. *The Morning Star Rises: An Account of Polynesian Astronomy,* 122–129. New Haven: Yale University Press, 1941.
Malo, David. *Hawaiian Antiquities (Moʻolelo Hawaiʻi)* (2nd ed.), 30–36, 141–153. Honolulu: Bishop Museum Press, 1951.
Obeyesekere, Gananath. *The Apotheosis of Captain Cook.* Princeton: Princeton University Press, 1997.
Sahlins, Marshall. *Islands of History.* Chicago: University of Chicago Press, 1985.
———. *How "Natives" Think.* Chicago: University of Chicago Press, 1995.
Selin, Helaine, ed. *Astronomy across Cultures,* 113–117. Dordrecht, Neth.: Kluwer, 2000.
Valeri, Valerio. *Kingship and Sacrifice: Ritual and Society in Ancient Hawaii,* 12–15, 200–233. Chicago: University of Chicago Press, 1985.

# Heliacal Rise

As the celestial sphere rotates, every star moves constantly around the line of a particular "latitude" or *declination*. This means that from any place on earth, every star that is not circumpolar will rise and set in the same positions, day in and day out, year in and year out (although this is not the case on longer time scales). However, since the (diurnal) rotation period is slightly less than a day—approximately four minutes shorter—this means than the *time* of rising or setting will not remain the same but will be approximately four minutes earlier each day.

For any given place and any given star (apart from circumpolar ones), there will be a time of year, typically lasting a few weeks, when that star is not visible in the night sky at all, because it is rising at approximately the same time as the sun and setting at about the same time as the sun. In other words, it is up in the sky during the day, when it cannot be seen, and below the horizon at night. Gradually, however, the rising time will progress earlier and earlier each day until there comes a day when the star can be seen briefly in the morning sky before the sky brightens too much prior to sunrise. This is known as the *heliacal rise*. After this, the star will rise progressively earlier, getting ever higher in the sky before it is lost in the pre-dawn twilight. Eventually, the star will reach a point, typically five months or so after the date of heliacal rise, where it crosses the sky from east to west during the night and can just be seen to set before the sky brightens. This first visible setting is often referred to as the *acronical set,* although the term strictly applies when the star sets at sunrise, an event that will already have passed unnoticed some days earlier. At around the same date, the star's rising time will have moved back through the night and be approaching the previous sunset. The last time the star can be seen to rise before this occurrence becomes lost in the evening twilight is often referred to (again loosely) as the *acronical rise*. During the ensuing months the star will be rising in daylight but setting ever earlier during the night. Finally, the day will come when it can only be seen briefly in the evening sky after sunset before itself setting, after which it will once again become invisible. This is known as the *heliacal set*.

This general series of events—disappearance, heliacal rise, acronical rise/set, heliacal set, and then disappearance again—holds for all stars apart from circumpolar ones, which never rise or set. On the other hand, details such as the length of disappearance, and whether acronical rise occurs before or after acronical set, will depend upon the brightness of the star, the latitude of the place, the declination of the star, and even atmospheric conditions (since these can affect the visibility of a star in the evening or morning twilight). Nonetheless, for any particular star and place of observation, the approximate dates of the various events can be reliably specified.

The terms *heliacal rise* and *heliacal set* are also applied to planets and have the same meaning of "first appearance in the pre-dawn sky" and "last appearance in the evening sky," respectively. However, the apparent motions of the planets are more complex than those of the stars, and those of inferior and superior planets are fundamentally different from each other.

**See also:**
Celestial Sphere; Circumpolar Stars; Declination; Diurnal Motion; How the Sky Has Changed over the Centuries; Inferior Planets, Motions of; Precession; Superior Planets, Motions of.

**References and further reading**
Aveni, Anthony F. *Stairways to the Stars: Skywatching in Three Great Ancient Cultures,* 46–48. New York: Wiley, 1997.
———. *Skywatchers,* 110–113. Austin: University of Texas Press, 2001.
Schaefer, Bradley E. "Heliacal Rise Phenomena." *Archaeoastronomy* 11 (supplement to *Journal for the History of Astronomy* 18) (1987), S19–S33.
———. "Astronomy and the Limits of Vision." *Vistas in Astronomy* 36 (1993), 311–361.

# Heliacal Set
*See* Heliacal Rise.

# Henge Monuments
*See* Circles of Earth, Timber, and Stone.

# Hesiod (Eighth Century B.C.E.)

Ancient Greece is well known as the place where the modern science of astronomy has its roots. This resulted from the blending together, in the Hellenistic era that followed the conquests of Alexander the Great in the latter part of the fourth century B.C.E., of two earlier traditions: a Babylonian tradition of careful empirical observation and a Greek tradition of cosmological and philosophical theorizing about the cause of observed phenomena. It is easy to forget that the classical world not only contained famous philosophers and astronomers—Plato, Aristotle, Aristarchus, Apollonius, and Hipparchus—but also contained common people, such as farmers, who made use of observations of the sun, moon, and stars in going about their everyday activities, as they had done for many generations.

Hesiod was a farmer who lived in central Greece in the eighth century B.C.E. He was also a poet, and his epic *Works and Days* was written to be chanted or sung. It was passed on for many generations and eventually written down three centuries later, whence it has come down to us. Thanks to Hesiod we know something of the ways in which his contemporaries regulated their activities through the seasons by watching the skies, thereby helping to overcome the vicissitudes of the climate. *Works and Days* contains a series of tips about signs to watch for and what to do in response to them—a sort of farmer's almanac that could be easily memorized and recited. Tips such as those contained in Hesiod's poem had doubtless been carefully accumulated and passed down over many generations.

The *Works* part of *Works and Days* refers to many seasonal astronomical indicators that make good sense from a modern perspective. For example:

> At the time when the Pleiades, the daughters of Atlas, are rising,
> begin your harvest, and plow again when they are setting.
> The Pleiades are hidden for forty nights and forty days,
> and then, as the turn of the year reaches that point
> they show again, at the time you first sharpen your iron.
> (*Works and Days* 383–387.)

Descriptions such as this one of the changing appearance of the Pleiades refer to observable events in the skies that are tied to the seasons, and give good practical advice. So useful are these types of observation—and in particular observations of the heliacal rise (first pre-dawn appearance) of stars and asterisms—as seasonal indicators, that similar practices have evolved, and persisted, as an aid to subsistence all over the world, from prehistoric times right up to the present day.

The *Days* part, in contrast to the *Works,* contains prognostications associated with the phase of the moon that, from a modern scientific perspective, represent no more than irrational superstition. However, it is important to realize that, to the farmers of the time, there was no fundamental distinction between the two parts: they would have viewed them on equal terms. This illustrates an important general principle. In attempting to interpret ancient practices relating to astronomy, we can rarely separate what, from a modern perspective, might appear to be "rational" behavior with pragmatic outcomes from "irrational" actions that are only explicable if one knows something about the prevalent systems of beliefs.

**See also:**
Javanese Calendar; Temple Alignments in Ancient Greece.
Heliacal Rise.

**References and further reading**
Aveni, Anthony. *Empires of Time: Calendars, Clocks and Cultures,* 41–51. New York: Basic Books, 1989.
Lattimore, Richmond, trans. *Hesiod.* Ann Arbor: University of Michigan Press, 1959.

# Hopewell Mounds

The custom of mound building was widespread in eastern North America prior to European contact, and it was long-enduring. Earthen burial mounds were built as far back as the fourth millennium B.C.E. The tradition spread and developed until reaching a culmination that lasted from the late second millennium B.C.E. until well into the mid-second millennium C.E. It was characterized by the construction of huge earthworks typically several hundred meters (more than a thousand feet) across—comparable in size to the largest British henges, such as Avebury. These earthworks are found down the Mississippi valley from Wisconsin to Louisiana, as well as eastwards into Ohio in the north and Alabama in the south. A distinctive phenomenon within the mound-building tradition, particularly from the late first millennium C.E. onwards, was the construction of effigy mounds—earthworks in the form of birds, animals, (occasionally) people, and more abstract designs. Serpent Mound in southern Ohio is one of the best-known examples of these.

The number of mounds built by ancient native North Americans is staggering: over fifteen thousand examples have been documented within the state of Wisconsin alone. Sadly, many have been obliterated and others only survive as almost imperceptible bumps in ploughed fields. Still others have met more distinctive fates—such as the enclosure in the shape of a conjoined circle and octagon at Newark in Ohio, which has been landscaped into a golf course. The largest were on a grand scale, and in just a few cases one can still gain a sense of how impressive they must originally have been.

The peoples who adopted the practice of mound building were many and various, each doubtless turning the practice to their own purposes. These peoples occupied a variety of woodland and valley environments and, over the centuries, developed new modes of subsistence. One of the clearest transitions was the adoption of maize as a staple food in the late first millennium C.E. Archaeologists identify a number of distinctive cultural assemblages: Poverty Point, named after the earliest known large ceremonial earthwork in the lower Mississippi valley, dating to around 1200 B.C.E.; Adena, in the Ohio valley in the mid-first millennium B.C.E.; Hopewell, which flourished later in the same area; and Mississippian, in the first part of the second millennium C.E., its most famous site being the great center of Cahokia, sited close to modern St. Louis.

The Hopewell assemblage is characterized by enormous enclosures in the shape of circles, squares, and other clear-cut geometrical shapes, generally thought to have been built between about 200 B.C.E. and C.E. 400. These are often found in association and even connected together, as at Newark. From an engineering perspective they are awe-inspiring, which has led to much speculation about other aspects of their construction, such as the use of precise measurement units and the incorporation of astronomical alignments. The latter include a number of solstitial alignments: for example, along the diagonal axis of several square earthworks, such as Hopeton, Anderson, Dunlap, Hopewell, and Mound City. Even the Serpent Mound—belonging to the later effigy mound tradition—has its head pointing at least roughly toward midsummer sunset. Various alignments upon potentially significant horizon rising and setting positions of the moon have also been claimed. The best known of these is the axis of the circle-octagon at Newark, which is aligned upon the most northerly rising position of the moon.

One thing is certain: these sites were not astronomical observatories. They served a variety of purposes. The geometrical earthworks demarcate large spaces, and it is generally assumed that they functioned, among other things, as ceremonial centers. One suggestion is that rituals related to burying the dead changed over time into grander ceremonials aimed at ensuring the continuance of world order and seasonal renewal. This explanation certainly fits the archaeological evidence of burial mounds gradually becoming transformed into large ceremonial earthworks. If so, it reflects a modification in the way people expressed and acted upon their beliefs relating to death, ancestors, and regeneration.

If these great earthworks were focuses for ceremonial activity preserving the world order, then they would surely reflect that perceived world order in their overall design. But how do we go about trying to understand this in any detail? One possibility is to examine modern indigenous worldviews in the region, on the basis that some aspects of early indigenous American cosmological beliefs may have survived through to modern times. For example, in the case of Wisconsin effigy mounds it has been suggested, by comparison with modern indigenous beliefs there, that the bird-shaped mounds were representations of powerful spirits inhabiting the upperworld, while land and water animals represented spirits of the lowerworld. Clusters of effigy mounds, then, both reflected—and preserved—the balance between the two realms of nature, earth and sky. (Other modern groups conceive of three realms—sky, earth, and a watery underworld—but the same general principles apply.)

Hopewell mounds are more distant in time from the indigenous present. Yet, just as modern traditional houses of many native North American groups reflect the fundamental perceived division of the (horizontal) world into four quarters, demarcated either by the cardinal or solstitial directions, so many of the Hopewell earthworks have a fourfold symmetry and are cardinally or solstitially aligned. More speculatively, it has been suggested that circles and squares (or octagons) might themselves represent the earth and sky. When conjoined, as at Newark, they would have represented the whole universe in microcosm, defining spaces in which ceremonies or other activi-

*Part of the Serpent Mound, a solstitially oriented effigy mound in southern Ohio. (Courtesy of Clive Ruggles)*

ties could be performed that were suited to each realm, dedicated to keeping each in its natural balance with the other.

However close to the truth such speculations might be, it is clear that any astronomical alignments that were intentionally built into the Hopewell mounds formed part of a much wider set of symbolic associations that helped to affirm the perceived order of things. Solstitial alignments, if deliberate, were not very precise. But if their purpose was merely to demarcate the four divisions of the cosmos, they did not need to be. The horizon rising and setting positions of the moon, on the other hand, have no place in modern native American thought, which is one reason why many archaeoastronomers remain unconvinced that such alignments were intentional. The other is that the alignments do not seem to occur in any consistent way.

### See also:
Solstitial Directions.
Avebury; Cahokia; Navajo Hogan; Pawnee Earth Lodge.
Moon, Motions of.

### References and further reading
Birmingham, Robert, and Leslie Eisenberg. *Indian Mounds of Wisconsin*. Madison: University of Wisconsin Press, 2000.

Carr, Christopher, and D. Troy Case, eds. *Gathering Hopewell: Society, Ritual, and Ritual Interaction*. New York: Kluwer, 2004.

Hively, Ray, and Robert Horn. "The Newark Earthworks." *Archaeoastronomy* 4 (supplement to *Journal for the History of Astronomy* 13) (1982), S1–20.

Mainfort, Robert, and Lynne Sullivan, eds. *Ancient Earthen Enclosures of the Eastern Woodlands*. Gainesville: University Press of Florida, 1998.

Romain, William F. *Mysteries of the Hopewell*. Akron: University of Akron Press, 2000.

Squier, Ephraim, and Edwin Davis. *Ancient Monuments of the Mississippi Valley*. Washington: Smithsonian Institution Press, 1998. (Originally published in 1848.)

## Hopi Calendar and Worldview

The Hopi people of Arizona remained almost undisturbed by European influences until about 1870. A number of traditional Hopi villages were situated on the very edges of mesa tops overlooking the surrounding barren plains, and Walpi, located particularly precariously on the top of a narrow mesa with precipitous cliffs on three sides, has remained relatively unspoiled. The inhabitants of Walpi are known to have observed an elaborate ceremonial calendar. The entire year was punctuated with ceremonies whose purpose was to "assure vital equilibrium, both social and individual, and conciliate the supernatural powers in order to obtain rain, good harvests, good health, and peace" (Ortiz 1979, p. 564). Some were carried out in public and some in private. Some were of considerable duration: a celebration lasting nine days occurred at the time of winter solstice. The calendar was regulated by carefully tracking the horizon rising position of the sun against various distant landmarks from different observing positions in and around the village.

The Hopi calendar, as practiced at Walpi and other traditional villages, provides a fine example of a seasonal calendar regulated by horizon sun observations. First recorded by the ethnographer Alexander Stephen in the 1890s, it has attracted a good deal of attention because it achieved an accuracy that many commentators found remarkable, generally keeping within two or three days of the "true" solar year. As a result, it has sometimes been portrayed as a classic example of the use of horizon sun observations to regulate crop planting and other subsistence activities, and so it is. Yet this only represents one aspect of the whole picture. The elaborate ceremonials ("ritual performances") that accompanied the various seasonal subsistence tasks were a vital part of an annual round of activities that—although *we* might try to break them up into those that were more "sacred" and those that were more "secular" or pragmatic in character—represented to the Hopi an integrated way of harmonizing human actions with cycles of events in the natural world. The calendar and all its associated ceremonies, in other words, had a key role in reaffirming the natural (cosmic) order.

This cosmic order also had spatial characteristics. These are reflected in the way people perceived the horizon and the whole landscape around them, and attributed meaning to different places. Working with the Hopi in the 1970s, historian Stephen McCluskey found out that the points on the horizon behind which the sun rises and sets at the solstices are themselves sacred places. Some of them are visited at the appropriate times of the year, when decorated prayer sticks and other offerings to the sun are placed on shrines. The place of midwinter sunrise in the southeast is the house of the sun, out of which the sun is said to come eating from a red stone bowl. The place of midsummer sunset in the northwest is the house of *Huzruing wuhti,* "hard being woman," who is associated with hard substances such as shells, corals, and turquoise. The sun stands directly above her house before descending into it through a hatchway in the roof (Ruggles 1993, 40).

Observations of the sun rising and setting at the solstices themselves are actually superfluous to the ceremonial calendar, since the sun's day-to-day movement at these times is minuscule and there are no suitable foresights that would aid precision. The solstitial directions are important for a different reason: they mark the four principal directions that are sacred in the Hopi worldview, fundamental axes that divide the world into four parts centered upon a particular village. These directions are not conceived as geometrical abstractions but as empirical realities with a variety of symbolic associations. They result in a conceptual quartering of the world, or *quadripartite cosmology,* a type of worldview that is found, in different variants, among other indigenous American groups such as the Navajo and Pawnee.

Each of the mesa-top villages, then, had an associated "sacred geography" in which particular places in the landscape had specific meanings, many were sacred, and the village itself stood at the center of things. This view of the world was derived through experience and constantly reinforced in myth and practice. It contrasts absolutely with the Western view of land as a resource to be exploited.

The Hopi calendar did (and does) not simply regulate crop-planting activities but constantly reaffirms the structure and correct functioning of the Hopi cosmos.

**See also:**
Cosmology; Landscape; Solstitial Directions.
Navajo Cosmology; Pawnee Cosmology.

**References and further reading**
Aveni, Anthony F., ed. *Archaeoastronomy in the New World,* 31–57. Cambridge: Cambridge University Press, 1982.
McCluskey, Stephen C. "The Astronomy of the Hopi Indians." *Journal for the History of Astronomy* 8 (1977), 174–195.
———. "Calendars and Symbolism: Functions of Observation in Hopi Astronomy." *Archaeoastronomy* 15 (supplement to *Journal for the History of Astronomy* 21) (1990), S1–16.

Ortiz, Alfonso, ed. *Handbook of North American Indians, Volume 9: Southwest*, 524–532, 564–580. Washington, DC: Smithsonian Institution, 1979.

Ruggles, Clive, ed. *Archaeoastronomy in the 1990s*, 33–44. Loughborough, UK: Group D Publications, 1993.

## Horizon Calendars of Central Mexico

Since the emergence of archaeoastronomy in the 1970s, there have been numerous studies of the orientations of temple-pyramids and other constructions in the central Mexican highlands and throughout Mesoamerica. Particular attention has been paid to the Valley of Mexico, in which peoples had dwelt for millennia before the Aztec capital, Tenochtitlan, was eventually founded at what is now the center of Mexico City. Numerous instances have been identified of orientation upon sunrise at the solstices and on the day of solar zenith passage, as well as on other calendrically significant dates.

It has also been proposed that *horizon calendars* were developed by direct observation of the movement of sunrise on successive days along mountainous horizons full of natural foresights and reference points. Some researchers have argued that surprisingly many sunrise alignments correspond to dates separated from the solstices by intervals of 20, 52, 65, or 73 days—numbers that are also encapsulated in the intermeshing cycles of the Mesoamerican Calendar Round. Among the independent strands of evidence in support of this idea is the fact that the first appearance of the noonday sun in the Xochicalco zenith tube occurs exactly fifty-two days before the June solstice, and the final appearance fifty-two days after. These successive fifty-two-day intervals are particularly intriguing because the interval between them (at Xochicalco, the length of the period of darkness before the sun starts to appear at noon again) is more or less 260 days, the length of the calendrical cycle known as the *tonalpohualli*.

A more controversial suggestion is that horizon calendars were tied to concepts of space, with intervals in time corresponding to horizon sunrise positions separated by multiples of an angular unit of around 4.5 degrees. However it is more likely that the true significance of dates encapsulated in intentional sunrise alignments had to do with a much richer set of perceived connections—between temples, the perceived dwelling places of the gods, the correct place and timing of tributes to ensure seasonal renewal, and so on. In other words, the horizon calendars only made sense in the context of complex schemes of sacred geography.

One idea is that horizon calendars might have been developed in the Valley of Mexico at a very early date and may have helped stimulate the development of the Mesoamerican Calendar Round that is well known from historical and documentary evidence. One of the main proponents of this idea is the Austrian-born anthropologist Johanna Broda, who argues that an early horizon calendar existed at Cuicuilco, a large Preclassic temple-pyramid dating from the first millennium B.C.E. whose ruins are now lost within the suburbs of Mexico City. This temple, argues Broda, was built on an observation point where some of the natural features of the eastern horizon happened to mark—and others came to define—significant calendrical dates.

There are three strands to the argument. First, the main peaks on the eastern horizon correlate particularly well with sunrise on calendar dates that were significant in the later Mesoamerican calendar and some of which continue to be significant to the present day. Second, there are coincidences between these sunrise dates and the dates of the calendrical rituals performed at shrines erected on the slopes of some of the relevant mountains. And finally, the visual lines that connect Cuicuilco with the mountain peaks on the eastern horizon pass through a number of historically significant settlements, sacred mountains, and shrines, implying a network of *sacred lines* in the landscape that remained significant, and continued to be respected, over a long period. Broda's scheme has since been reassessed and modified by the Slovenian archaeoastronomer Ivan Šprajc, but he has confirmed the idea that calendrically significant intervals were marked at Cuicuilco.

This type of investigation raises a number of methodological questions that are more familiar from "green" archaeoastronomy and reassessments of British ley lines. Yet it rests on much firmer ground because of the abundance of historical evidence concerning sacred geography in ancient Mesoamerica.

**See also:**
"Green" Archaeoastronomy; Ley Lines; Methodology; Sacred Geographies. Aztec Sacred Geography; Mesoamerican Calendar Round; Zenith Tubes.

**References and further reading**
Iwaniszewski, Stanisław, Arnold Lebeuf, Andrzej Wierciński and Mariusz Ziółkowski, eds. *Time and Astronomy at the Meeting of Two Worlds*, 497–512. Warsaw: Centrum Studiów Latynoamerykańskich, 1994.
Romano, Giuliano, and Gustavo Traversari, eds. *Colloquio Internazionale Archeologia e Astronomia*, 15–22, 123–129. Rome: Giorgio Bretschneider Editore, 1991.
Ruggles, Clive, and Nicholas Saunders, eds. *Astronomies and Cultures*, 270–285. Niwot, CO: University Press of Colorado, 1993.
Šprajc, Ivan. "Astronomical Alignments at Teotihuacan, Mexico." *Latin American Antiquity*, 11 (2000), 403–415.
———. *Orientaciones Astronómicas en la Arquitectura Prehispánica del Centro de México*, 172f, 201f, 258f. Mexico City: Instituto Nacional de Antropología e Historia (Colección Científica 427), 2001. [In Spanish.]

# How the Sky Has Changed over the Centuries

The actual positions of the distant stars in space relative to one another have only changed by minuscule amounts over many millennia. This means that the distinctive patterns of the constellations visible in today's skies around the world have not changed significantly since early prehistory. On the other hand, the rising and setting positions of particular stars as viewed from a given spot, although constant day after day and year after year, do change significantly on a time scale of centuries. On this time scale, the whole mantle of stars shifts on the celestial sphere owing to a phenomenon called *precession* (short for the *precession of the equinoxes*). This shift has important methodological implications whenever we try to interpret apparent alignments upon stellar targets at ancient structures whose date can only be determined archaeologically to within a span of a few centuries.

The rising and setting positions of the sun, moon, and planets are not affected by precession but do nonetheless change by a smaller amount owing to a distinct phenomenon known as the change in the obliquity of the ecliptic. For example, in an equatorial location, the difference between the horizon position where the sun rises or sets at one of the solstices now and the corresponding position in 2000 B.C.E. is about the same as sun's own diameter. (In temperate zones the difference is slightly greater, because the sun rises or sets at a shallower angle rather than almost vertically.)

Some astronomers have argued that the appearance of spectacular comets and meteors occurred much more frequently at certain epochs in the past, for example during the Bronze Age. Since this would also have resulted in an increased probability of meteoritic impacts, the idea is usually discussed in the context of *catastrophism*, the idea that major natural catastrophes could have affected humanity in the past.

> **See also:**
> Astronomical Dating; Catastrophic Events; Comets, Novae, and Meteors; Methodology; Solstitial Directions.
> Celestial Sphere; Obliquity of the Ecliptic; Precession; Solstices; Star Rising and Setting Positions.
>
> **References and further reading**
> Peiser, Benny, Trevor Palmer, and Mark Bailey, eds. *Natural Catastrophes during Bronze Age Civilisations: Archaeological, Geological, Astronomical and Cultural Perspectives.* Oxford: Archaeopress (BAR International Series 728), 1998.
> Ridpath, Ian, ed. *Norton's Star Atlas and Reference Handbook* (20th ed.), 3–6. New York: Pi Press, 2004.
> Ruggles, Clive. *Astronomy in Prehistoric Britain and Ireland,* 57. New Haven: Yale University Press, 1999.

# I

## Inferior Planets, Motions of

The planets Venus and Mercury revolve around the sun on orbits inside that of the earth. And because they are on closer orbits, they move faster. Suppose for a moment that the earth simply vanished, leaving a sentient space monster orbiting the sun in its place, its head upward to the north. For want of anything better to do, the monster decides to ignore everything else in the sky but the sun and Venus, keeping its face toward the sun and watching Venus gradually progressing around it. At intervals it sees Venus at its closest approach come rattling past between it and the sun, then pull rapidly away to the right. The time when Venus passes the sun (only rarely will it pass directly in front of it) is known as *inferior conjunction*. As Venus continues to arc round the sun, its rightward motion decreases, but it continues to get farther away (and hence it also gets fainter). At some point, well before it has gained on the earth-monster by a quarter of an orbit, it will be at its furthest angle rightward from the sun, known technically as the point of *greatest elongation*. Thereafter Venus appears to move leftward toward the sun again, but ever more slowly, as it heads round to the far side of the sun from the earth-monster. Eventually, it passes behind the sun, an event known as *superior conjunction,* and then gradually emerges to the left. The sequence is now repeated in reverse and in mirror image, the planet increasing in proximity and brightness, gradually at first, then reaching its greatest elongation to the left of the sun, and finally moving rapidly once again toward inferior conjunction.

The sequence of events for an earth-based observer reflects those just described, except that we view them from the surface of a spinning planet. An observer in the northern hemisphere perceives the celestial sphere to be turning to the right. If Venus is to the right of the sun, it rises earlier, passes across the sky to the right (west) of the sun, and sets earlier. The directions are reversed for an observer in the southern hemisphere, whose perceptions can be modeled by turning the space monster upside down.

Just a few days after inferior conjunction, then, Venus suddenly appears in the eastern sky before sunrise. This event as known as the *heliacal rise*—a similar concept to the heliacal rise of stars. The planet is already bright, and in a few more days it dominates the dawn sky, easily the brightest object apart from the sun and moon. As it draws rapidly away westward from the sun, it rises earlier and climbs higher in the sky by sunrise, until at greatest (western) elongation it can be visible in the pre-dawn sky for as much as three hours. This occurs about seventy-one days after inferior conjunction. Subsequently, it draws closer to the sun once again but much more gradually, steadily losing brightness as its distance from the earth increases. After a further 180 days or so, it disappears from the pre-dawn sky. This time—around superior conjunction—it remains hidden for much longer, around fifty days on average. The cycle is then repeated, in reverse, in the evening sky. Toward the end of this period it dominates the western sky after sunset. Then it rapidly disappears from the evening sky (this is known as the *heliacal set*). This is followed by inferior conjunction, and the cycle is repeated.

It may seem strange at first that Venus can only ever be seen during a short period around dawn or dusk. It never appears in the middle of the night, which would imply that it was opposite to the sun in the sky—that the earth was between the sun and Venus. However, since Venus is on a closer orbit to the sun, the earth can never be between Venus and the sun.

The period between successive occurrences of the same configuration (the *synodic period*) of Venus is 584 days, the average periods of visibility and invisibility being 263, 50, 263 and 8 days. It is an accident of nature that five Venus synodic cycles equate to almost exactly eight years, the difference being less than three days. This means that whatever Venus's appearance on a particular date, it will be almost exactly the same eight years (minus two days) later. It also means that the motions of Venus (over eight years) can be correlated with the seasons. Both of these facts had a particular significance in pre-Columbian Mesoamerica.

Mercury is much closer to the sun and orbits it in a much shorter time, but its apparent motions follow a similar pattern. Its synodic period is just 116 days, the average periods of visibility and invisibility being 38, 35, 38 and 5 days (with quite wide variations).

See also:
Star and Crescent Symbol.
Dresden Codex; Mesoamerican Calendar Round.
Celestial Sphere; Heliacal Rise; Superior Planets, Motions of.

**References and further reading**
Aveni, Anthony F. *Conversing with the Planets*, 24–33. New York: Times Books, 1992.
———. *Stairways to the Stars: Skywatching in Three Great Ancient Cultures*, 40–46. New York: Wiley, 1997.
———. *Skywatchers*, 80–94. Austin: University of Texas Press, 2001.

# Inuit Cosmology

The Inuit peoples, who live (mainly) in the extreme north of Canada, occupy one of the most hostile and challenging environments in the world. Their communities, scattered over eight thousand kilometers (five thousand miles) from the eastern tip of Siberia eastward to the west coast of Greenland, seem to an outsider to represent outposts in a vast, icy wilderness. Yet to the Inuit themselves the Arctic tundra is a homeland—the central part of which now forms the Canadian territory of Nunavut—that for countless generations has provided a variety of resources deriving from both land and sea animals and is navigable by kayak and dog-drawn sled.

The survival of human communities in such an inhospitable environment seems, to most outsiders, nothing short of miraculous. How survival was achieved varied from place to place, according to the resources available. Thus for the central Inuit, living around Hudson Bay, subsistence needs had to be satisfied in very different ways during the two seasons of the year. Winters were traditionally spent in fixed coastal settlements, with sea mammals the main source of food and other essentials. Summers were traditionally spent in temporary camps while hunters followed the caribou herds and other land animals. This basic dichotomy is reflected in central Inuit people's conception of the structure of the world. So, too, are gender qualities. For example, women are regarded as belonging more to the sea, sea mammals, and the winter season, while men belong more to the land, land animals, and summer. Such principles determine the materials of which, say, hunting (male) and sewing (female) tools are made; they can also influence the orientations of male and female burials. Similar principles also extend to a number of perceived Other Worlds: the Land of the Moon Spirit, Birdland, and Belowsea Land. Belowsea Land, for example, is ruled by Sea Woman, while the night sky is a reflection of terra firma, where the male moon spirit drives his sled across smooth ice (clear sky) or through trickier snow fields (clouds).

As for numerous other indigenous peoples, sky myth and symbolism formed an integral part of Inuit understanding of the workings of the cosmos—knowledge that ensured well-being and, ultimately, survival. What makes the Inuit case particularly interesting is that they are one of the very few sets of human communities living at very high latitudes (the only others are in northern Scandinavia and parts of Siberia). Here, the appearance and behavior of the celestial bodies is distinctive in several ways. Many Inuit communities lie within the Arctic Circle, which means that there is a period around the summer solstice when the sun never sets, and around the winter solstice when it never rises. During the hours (and—around the winter solstice each year for those living north of the Arctic circle—the weeks) of darkness, whenever the skies are clear, the stars are seen to pass around the sky in circles only shallowly inclined to the horizontal, a great many of them never disappearing below the horizon.

The Sun and Moon Spirits are prominent in Inuit cosmology; the sun is female and the moon male. Yet the sun's role in Inuit myth is limited; by far the more important figure is the moon, her brother. Moon Man was widely seen as a benevolent and approachable spirit, a direct help in maintaining human life. This is scarcely surprising, since it is the principal luminary during the long, dark winter nights, especially prominent during those years (depending upon the lunar node cycle) when the winter moon would circle above the horizon each month for several days on end.

Observances to mark the winter solstice, or (above the Arctic Circle) to mark the first brief noonday appearance of the sun after the "great darkness," were certainly important. Great festivities lasting for many days have been recorded in Greenland, but in many other places they were rather subdued, since this is one of the most difficult times of the year. Children (symbolic of renewal) were often to the fore: in one custom, they would smile at the newly appeared sun, but only with one side of their face. This was to show that while warmer weather was now assured, the coldest part of the winter was yet to come. The summer solstice, in contrast, apparently was of little or no importance. Summer festivals recorded by early ethnographers tended to take place later in the season or as the first terrestrial signs (such as the formation of ice floes) heralded the onset of winter.

Traditionally, Inuit communities have named relatively few bright stars and constellations. They saw the majority generally as spirits of the dead or as "holes in the sky." Ursa Major is one distinctive group of stars that was widely recognized; it was seen variously as a herd of caribou or as a single animal. It was used for navigating and for marking time. Specific meanings, and the stories associated with them, most often attached to those brighter stars that were seen to set and rise again. Thus Sirius, which only appeared low in the southern sky in the middle of winter, and flickered brightly in different colors owing to atmospheric effects, was known by some Inuit peoples as the "fox star," and was seen as a red fox and a white fox fighting to get into the same foxhole. The annual patterns of appearance and disappearance of these stars were also used to mark times of the year.

The exceptional nature of the terrestrial environment in which they live has strongly influenced Inuit knowledge and beliefs, but so too has their exceptional sky. The extent to which their distinctive view of the positions and motions of the celestial bodies has given rise to characteristic aspects of Inuit cosmology is an issue about which little more may be knowable, but it highlights the wider cultural significance of the shreds of evidence that have survived concerning traditional Inuit sky knowledge.

**See also:**
Cosmology.
Heliacal Rise; Moon, Motions of; Solstices.

**References and further reading**
MacDonald, John. *The Arctic Sky: Inuit Astronomy, Star Lore and Legend.* Toronto: Royal Ontario Museum and Iqaluit: Nunavut Research Institute, 1998.
Ruggles, Clive, ed. *Archaeoastronomy in the 1990s,* 59–68. Loughborough, UK: Group D Publications, 1993.

# Iron-Age Roundhouses

Unlike the preceding two thousand years, which remain conspicuous in the British landscape by way of their monumental tombs and temples while settlement evidence is sparse, the archaeological record of the late second and first millennium B.C.E. in Britain—conventionally labeled as the Middle Bronze Age to the Late Iron Age—is characterized by settlements. There were isolated farmsteads; villages, both open and fortified; and hill forts, some of which housed communities of several hundred people. The dominant form of domestic architecture during this period was the roundhouse, and several thousand of these have been uncovered by archaeologists over the years. They were generally of moderate size, typically between about eight meters (twenty-five feet) and fifteen meters (fifty feet) across. A ring of upright wooden posts or planks, either within the walls or forming a separate interior ring, supported a conical roof, which was probably thatched with straw or reeds and may often have been covered in turf. Inside this roof space, smoke from a central fire could accumulate before gradually permeating out through the thatch.

The roundhouse provided good and effective shelter from the British climate, and it is scarcely surprising that there was just a single doorway. But it may seem odd that the direction the entrance faced was far from random. At first, roundhouses faced predominantly southward, but from c. 1200 B.C.E. onward the great majority of roundhouse doorways faced generally toward the east or southeast. Why? An obvious practical explanation is that by placing the entrance in this direction the roundhouse interior would be sheltered from the prevailing westerly winds and could be warmed by letting the early morning sunlight enter the house. Yet if westerly winds were the major factor, might there not be greater variation in orientation? It also seems curious that little attempt was made to avoid the elements in the actual siting of buildings and settlements. Furthermore, similar orientation practices seem to extend more widely—for example, to hill fort entrances, where the prevailing wind would hardly have been a consideration.

The "obvious" explanation was increasingly challenged in the 1990s. Detailed analyses of the azimuths of roundhouse entrances showed that there are particular concentrations around due east (between about 85 degrees and 95 degrees) and similarly around southeast (between about 125 degrees and 135 degrees). There was a significant drop between the two as well as a sharp drop outside the whole range (85 degrees to 135 degrees). The second of the two ranges corresponds to the position of midwinter sunrise, prompting suggestions that the solstitial sun was the intended target, and a natural extension of the argument was to suggest that the easterly concentration had something to do with the equinox. But why should the equinox have been a significant target at all, since it does not represent a physical "station of the sun," only the half-way point between its two extreme rising positions? The position of winter solstice sunrise, on the other hand, is certainly tangible, in that it represents the limit of the part of the horizon where the sun can rise, separating it from the southern quarter, which the sun only ever passes over. But even so, why should the direction of sunrise at different times in the year have had any bearing upon the way people oriented their houses and other structures?

Vital clues are provided by practices still known among certain modern indigenous peoples, together with broader archaeological evidence concerning the spatial distribution of different activities within the Iron Age roundhouses themselves. To take one modern example, hogans, traditional houses of the Navajo, face eastwards toward the sacred mountain of the east, for reasons that have to do with keeping life in harmony with a cosmos perceived to be divided into four quarters. The earth lodges of the Pawnee provide another example where the dominant practice of entrance orientation—again, toward the east—derives from the dominant worldview. A more general principle is at work in these cases, of which entrance orientations form just one part: these modern roundhouses serve as models of the world—microcosms—designed as such so that people can live their lives at one with the cosmos.

Recent work provides compelling evidence that there were many symbolic divisions of space and activity within Iron-Age roundhouses in relation to their overall design and orientation: for example, between living, eating, and sleeping; between preparing and eating food; between male and female activities; and so on. Everyday activities that we would see as mundane and unconsequential were, it seems, strictly enacted in accordance with a prevailing worldview, as is the case among the modern indigenous peoples just mentioned. Within this context, it is scarcely surprising that the entrance orientations should have been heavily influenced by cosmological considerations.

Nonetheless, considerations that seem to us altogether more pragmatic may also have played their part. As in the case of the Pawnee earth lodge, the practical benefits of houses facing the warmth of the early morning sun could certainly have been recognized. The separation between behavior that we would see as having clear practical ends and what we might choose to describe as symbolic or even irrational did not exist in the minds of prehistoric people. There was no clear separation between special rites that appeased the cosmic powers and those mundane activities that filled and maintained life from hour to hour and day to day. In this sense, there was a sacred aspect to the very houses in which they lived out their daily lives; the orientation of those houses in relation to the rising sun is just one manifestation of this.

**See also:**
Cosmology; Equinoxes; Solstitial Alignments.
Navajo Cosmology; Navajo Hogan; Pawnee Cosmology; Pawnee Earth Lodge.
Azimuth; Solstices.

**References and further reading**
Champion, Timothy, and John Collis, eds. *The Iron Age in Britain and Ireland: Recent Trends*, 117–132. Sheffield, UK: Sheffield Academic Press, 1996.
Gwilt, Adam, and Colin Haselgrove, eds. *Reconstructing Iron Age Societies*, 87–95. Oxford: Oxbow Books, 1997.
Hill, J. D. *Ritual and Rubbish in the Iron Age of Wessex*, ch. 11. Oxford: Tempus Reparatum (BAR International Series 602), 1995.
Hunter, John, and Ian Ralston, eds. *The Archaeology of Britain*, 113–134. London: Routledge, 1999.
Parker Pearson, Michael and Colin Richards, eds. *Architecture and Order: Approaches to Social Space*, 47–54. London: Routledge, 1994.
Parker Pearson, Michael, Niall Sharples, and Jim Symonds. *South Uist: Archaeology and History of a Hebridean Island*, 69–79. Stroud, UK: Tempus, 2004.
Pryor, Francis. *Britain B.C.*, 320–331. London: HarperCollins, 2003.
Ruggles, Clive. *Astronomy in Prehistoric Britain and Ireland*, 153. New Haven: Yale University Press, 1999.

# Is Paras

Is Paras is one of an extraordinary type of prehistoric monument known as *nuraghi* that is found in copious numbers on the Mediterranean island of Sardinia. Located near to the modern town of Isili, Is Paras was built during the Bronze Age, probably around the middle of the second millennium B.C.E. Despite some damage to the exterior, it contains one of the most impressive examples of the central chamber characteristic of all nuraghi, and this is preserved intact. Circular in cross-section at ground level, measuring over 6.5 meters (21 feet) across, the walls rise vertically for several meters before beginning to close in. From this point, layer upon layer of stone blocks are gradually corbelled inwards, each supporting the next so as to form a high vaulted roof. Although it is built entirely of dry stone, it rises to an incredible 11.5 meters (over 37 feet), making it the tallest example known.

*Nuraghe Is Paras, Sardinia, viewed around noon close to the summer solstice. The dagger of light on the back wall is visible through the entrance. (Courtesy of Clive Ruggles)*

It has been suggested that the tower functioned as part of a system of inter-nuraghe alignments marking solar and lunar events. Viewed from another nuraghe, Nueddas, a few kilometers away, it is a prominent skyline feature and marks the position of midsummer sunset. However, what makes Is Paras of particular interest is a distinctive light-and-shadow phenomenon that also seems to relate to the summer solstice.

The apex of the vaulted roof is unusual in having a small round opening, about forty centimeters (sixteen inches) across. A stone, possibly used to cover the hole, was discovered on top of the tower by excavators in the late 1990s, but it could be moved by one or two people and is too small to provide a permanent cover. While it is possible that a larger, more permanent stone was used at one time to cover the hole, there is no sign of it now. It is therefore a strong possibility, though by no means a certainty, that the hole was designed to be uncovered at least at certain times.

When the sun is high in the sky, sunlight enters the chamber and casts a dagger of light onto the chamber wall. This only happens around the middle of the day, and since the sun is toward the south at this time, the dagger appears on the northern wall, moving in a "U"-shaped curve down and to the left until local noon is reached, whereupon it starts to move upwards again, continuing to the left. At noon on the summer solstice, the sun is as high as it ever reaches in the sky and the noontime dagger reaches its lowest point of all. This is on the very lowest layer of stones, within two centimeters of the floor. For a period of about twenty minutes on this day, it moves across on this level before starting to rise up again.

This example epitomizes the problems of interpreting a "one-off" phenomenon. Nothing like it has been discovered at any other nuraghe, despite the fact that there are several thousand of them, and this must cast doubt upon its authenticity. Could it simply be a coincidence? Was the hole permanently covered? And yet the fact that the light dagger reaches so close to the floor without actually touching it—surely an incredible coincidence if unintentional—seems to argue otherwise. Without further contextual evidence we may never know whether this phenomenon was deliberately intended, or if it was deliberate, what it actually meant to the people who built and used this tower back in the Bronze Age.

See also:
Methodology; Solstitial Alignments.
Fajada Butte Sun Dagger; Nuraghi.
Solstices.

**References and further reading**
Belmonte, Juan Antonio, and Michael Hoskin. *Reflejo del Cosmos*, 185–188. Madrid: Equipo Sirius, 2002. [In Spanish.]
Hoskin, Michael. *Tombs, Temples and Their Orientations*, 183–185. Bognor Regis, UK: Ocarina Books, 2001.
Zedda, Mauro Peppino. *I Nuraghi tra Archeologia e Astronomia*, 24–34, 55–56. Cagliari: Agorà Nuragica, 2004. [In Italian.]

# Islamic Astronomy

Scholars in the Islamic world were responsible for providing a vital bridge connecting ancient Babylonian and Greek astronomy to modern scientific astronomy. But for this bridge, the traditions of thought that had led to the development of mathematical astronomy in Hellenistic Greece (after the conquests of Alexander the Great in the fourth century B.C.E. had brought the Babylonian and Greek traditions into direct contact) would have been all but severed. In any case, several centuries had passed by the time (in the ninth century C.E.) that serious quantities of old astronomical treatises written in Greek began to be retrieved, translated into Arabic, and thence communicated around the Muslim world. There they helped give birth to one of the world's richest and most sophisticated astronomical traditions. Early elements of this tradition (including translations and re-translations of the ancient Greek texts) subsequently passed into Christian Europe, where they contributed to the development of ideas within the European Renaissance. (This legacy is most obvious today in the fact that a number of modern star names, such as Aldebaran, Alnilam, Alnitak, and Altair—as well as techni-

*Muslims bow their heads toward Mecca in prayer. Mecca, in present-day Saudi Arabia, is the birthplace of Muhammad, the prophet upon whose teachings Islam is based. (PhotoDisc/Getty Images)*

cal terms, such as azimuth, zenith and nadir—are Arabic in origin.) However, Islamic astronomy continued to develop in its own right until around the beginning of the sixteenth century, and Europeans remained largely ignorant of it. In fact, developments in Europe and the Muslim world remained largely independent until the nineteenth century.

Astronomy and astronomers generally thrived in the Islamic world. The distinctive nature of the astronomy that they studied and practiced derived from a particular mixture of traditions: indigenous folk astronomies, Babylonian and Greek ideas drawn from ancient texts, and elements of Persian and Indian astronomy absorbed as the Muslim world stretched eastwards. An example of this mixed intellectual heritage is the set of stars and asterisms known as *anwā'* that identified twenty-eight lunar stations or "mansions"; these helped identify the position of the moon on successive nights and hence to track its monthly course through the stars. The general concept may well have been adopted from Hindu astronomy, while the chosen reference stars derived from Bedouin knowledge of heliacal and acronical events. Muslim astronomers advanced mathematical astronomy in various ways, including studies of the motions of the sun, moon, and planets, and catalogues of stars and their positions, tabulated in substantial documents known as *zījes*. They were also responsible for the construction of many fine portable astronomical instruments and made a number of important technical advances in the design of the astrolabe and the quadrant.

Coexisting alongside what we would recognize as scientific astronomy was a tradition of folk astronomy that did not involve any attempt to make systematic observations or predictions based on mathematical models. What it did do was to regulate both agricultural activities and religious ones using straightforward observations of the skies. While mathematical astronomy could achieve the same ends, and indeed do so with a great deal more precision, it was the preserve of just a few specialists; folk astronomy, on the other hand, was accessible to all, and thus had far wider social implications since it determined everyday practice for ordinary people.

Prior to the widespread adoption of the Muslim faith, different cultural traditions within what became the Islamic world would have tended to produce a mish-mash of local seasonal calendars and/or astronomical rules of thumb molded by the particular subsistence requirements of a given community. The rapid exchange of ideas within the Islamic world itself brought about a degree of rationalization: thus a broadly consistent set of *anwā'* became widely used for divination and were increasingly adopted as a means of time keeping. Nonetheless, what makes it possible to speak meaningfully of "Islamic folk astronomy" is the fact that very particular requirements would have been imposed upon all practicing Muslims by sacred Islamic law. Folk astronomies all over the Muslim world adapted to satisfy these common needs.

The first essential was to observe religious festivals on the right days, and particularly to correctly define such critical dates as the beginning and end of the month-long fast in the holy month of Ramadan. The formal Islamic calendar is lunar, and consists of only twelve synodic (phase-cycle) months, since intercalary months are forbidden by the *Quran:* this means that it is eleven days shorter than the seasonal year. Whatever calendars were in local use for seasonal reckoning, it was necessary for liturgical purposes to determine the start of each new month by direct observations of the new crescent moon. People with good eyesight might be sent to watch the western sky on the critical evenings, and various solutions had to be found when the critical part of the sky was obscured by cloud. It is scarcely surprising that the star and crescent symbol, depicting the crescent moon, has become a global symbol of the Islamic faith.

An imperative for any Muslim is to observe the five daily prayers, which must take place within set intervals during the day and night. The intervals in question are specified according to the daily motions of the sun: in relation to the lengths of shadows when the sun is up; in relation to certain visible phenomena (such as the redness of the sky) at twilight; and in relation to set time intervals during the night (the evening prayer, specifically, should preferably be completed before a third of the night has passed). This means that the actual prayer times vary both according to latitude and longitude, and at any given place they change from day to day with the varying length of daylight. This makes the problem of determining the correct prayer times nontrivial, and before tables based on scientific data became available, the use of sundials and gnomons was common, with *anwā'* used to estimate the time of night.

Folk astronomy had a third and rather different application, though to something no less vital in the practice of the Muslim faith. This was to help determine the sacred direction, or *qibla*. One of the most basic necessities for any Muslim is to know the direction of Mecca, in order to determine the correct bodily orientation during prayers and other activities. Formally this has nothing to do with astronomy, but (again, before Muslim mathematicians had produced tabulations of the *qibla* for different latitudes and longitudes) astronomical observations were extensively used to provide the best available approximations, some of which were rather better than others. Typically they used the horizon rising or setting point of a particular bright star or the rising or setting arc of the sun. This seemed only natural as the base of the Kaaba itself—the cube-shaped stone structure at the center of the Great Mosque in Mecca that forms the sacred center of the Muslim world—had been known as far back as the seventh century to have its longer axis oriented upon the rising of the star Canopus and its minor axis aligned (roughly) in line with midsummer sunrise and midwinter sunset.

Historically, the assumed *qibla* is most evident in the orientation of mosques and in the layouts of some cities. In some cases, historical accounts attest to how the sacred direction was actually determined. Thus, we know that the earliest mosques in Iraq were built with their prayer walls facing midwinter sunset in order to face the northeastern wall of the Kaaba, while Egyptian ones were built with their prayer walls facing midwinter sunrise in an attempt to be parallel with the Kaaba's northwestern wall. However, in many cases we can only speculate about the methods used to determine the *qibla*. Since the errors were sometimes considerable, anyone trying to reconstruct the motivation behind mosque orientations in the absence of anything but the orientations themselves would have a hard time fathoming that the intention was always to orient them toward Mecca.

**See also:**
Lunar and Luni-Solar Calendars; Orientation; Star and Crescent Symbol. Heliacal Rise; Lunar Phase Cycle; Solstitial Directions.

**References and further reading**
Belmonte Avilés, Juan Antonio. *Tiempo y Religión: Una Historia Sagrada del Calendario*, ch. 6. Madrid: Ediciones del Orto, 2005. [In Spanish.]

Chamberlain, Von Del, John Carlson, and Jane Young, eds. *Songs from the Sky: Indigenous Astronomical and Cosmological Traditions of the World*, 26–31. Bognor Regis: Ocarina Books, and College Park, MD: Center for Archaeoastronomy, 2005.

Hoskin, Michael, ed. *Cambridge Illustrated History of Astronomy*, 50–63. Cambridge: Cambridge University Press, 1997.

McCluskey, Stephen C. *Astronomies and Cultures in Early Medieval Europe*, 165–187. Cambridge: Cambridge University Press, 1998.

Ruggles, Clive, and Nicholas Saunders, eds. *Astronomies and Cultures*, 124–138. Niwot: University Press of Colorado, 1993.

Schaefer, Bradley E. "Lunar Crescent Visibility." *Quarterly Journal of the Royal Astronomical Society* 37 (1996), 759–768.

Selin, Helaine, ed. *Astronomy across Cultures*, 468–471, 585–650. Dordrecht, Neth.: Kluwer, 2000.

Walker, Christopher, ed. *Astronomy before the Telescope*, 143–174. London: British Museum Press, 1996.

## Island of the Sun

The Inca empire dominated a huge swath of the Andean region running down the west coast of South America for almost a century before the arrival of the Europeans in 1532. At its height, it stretched for no less than 4,000 kilometers (2,500 miles) from north to south, an area that extends from the northern border of Ecuador down to Santiago, Chile. The Inca rulers rose to dominance remarkably quickly and ruled their vast and diverse state using bureaucratic and rigidly hierarchical procedures to organize labor and redistribute food and other raw materials. One mechanism of conquest was subsuming the religions and ideologies of captured peoples within a state-controlled cult. It was dedicated to three principal deities, among whom Inti, the Sun god, was prominent. The ruling Inca elite claimed to be descended from Inti himself, and in this way the dominant ideology they had imposed served also as a mechanism of political control, putting their own right to rule beyond question.

In one myth recounted by the chronicler Bernabé Cobo, the "true dwelling place of the sun" was said to be a large crag on the island in the huge inland lake of Titicaca, on the border between modern Peru and Bolivia. According to the legend, the ancient peoples had been without light for many days. "Finally, the people of the Island of Titicaca saw the Sun come up one morning out of that crag with extraordinary radiance. For this reason they believed that the true dwelling place of the Sun was that crag, or at least that crag was the most delightful thing in the world for the Sun" (Hamilton 1990, pp. 91–92). This myth evidently had pre-Inca origins, but under Inca control the island became a destination for pilgrimages to the sun's place of origin. A sanctuary was built around the sacred crag and access was tightly controlled. Common pilgrims were only permitted if they

*Compound containing the Intihuatana, or "sun stone," occupying a dominant position at the sacred heart of the Inca city of Machu Picchu. (Courtesy of Clive Ruggles)*

followed various strict protocols and donated appropriate offerings. Even then, they were not allowed near the sacred crag, but had to watch the sunrise ceremony from a distance. By combining elements of Cobo's and other accounts with archaeological examinations of the remains of the sanctuary and archaeoastronomical measurements to determine the rising direction of the sun at different times, archaeologist Brian Bauer and archaeoastronomer David Dearborn have deduced that the sunrise-watching ceremony almost certainly took place in June or July, perhaps at the June solstice itself.

The fact that the sunrise ceremony at the sanctuary on the Island of the Sun is historically attested allows us to answer questions that we would not be able to answer if we had to rely upon archaeological evidence alone. Compare, for example, later British prehistoric sites such as Brainport Bay, a solstitially aligned set of platforms and structures, where we would dearly like to know who made sunrise observations and for what purpose; Stonehenge, where—although it is evident that only a select few could observe solstitial sunrise or sunset from within the huge stones of the sarsen circle—we can only speculate about the observers' social status and purpose, and whether larger crowds were allowed to spectate from a distance; and the Thornborough henges, which appear from the archaeological evidence to have been a pilgrimage destination, and we are keen to ask when in the year people came here and what the place signified to them.

**See also:**
Cobo, Bernabé (1582–1657); Pilgrimage.
Brainport Bay; *Ceque* System; Cusco Sun Pillars; Stonehenge; Thornborough. Solstices.

**References and further reading**
D'Altroy, Terence. *The Incas,* 141–176. Oxford: Blackwell, 2002.
Dearborn, David, M. T. Seddon, and Brian Bauer. "The Sanctuary of Titicaca, Where the Sun Returns to Earth." *Latin American Antiquity* 9 (1998), 240–258.
Hamilton, Roland, ed. and trans. *Inca Religion and Customs* [a translation of books 13 and 14 of Cobo's *Historia del Nuevo Mundo,* 1653]. Austin: University of Texas Press, 1990.
Selin, Helaine, ed. *Astronomy across Cultures,* 200–201, 206–211. Dordrecht, Neth.: Kluwer, 2000.
Urton, Gary. *Inca Myths,* 34–37, 54. Austin: University of Texas Press, and London: British Museum Press, 1999.

## Javanese Calendar

To many generations of rice farmers in rural Java, Indonesia, it was not the stars of Ursa Major that formed the plough, but the stars of Orion. Here, close to the equator, the constellation appears on its side and Orion's belt plus three of the four outer stars (excluding Betelgeuse) were seen to resemble a traditional Javanese plough (*Weluku*): they constituted the constellation *Bintang Weluku*.

The changing appearance of this constellation (together with the Pleiades) over the year was used to provide various rules of thumb to regulate the different seasonal activities associated with inundated rice cultivation. The beginning of the new agricultural year was marked by the plough's first appearance in the pre-dawn sky (heliacal rise). This happened around the time of the June solstice. At this time the celestial plough was upright, just like earthly ploughs when in use. Three months later, with *Bintang Weluku* rising progressively earlier each night, came the onset of the rainy season and the time to prepare the tools and check the water channels. When the plough eventually rose at sunset (acronical rise) and was once again upright, it was the time for the women to sow the rice in the nursery and for the men to start plowing the fields. About a month later, at around the time of the December solstice, the process began of harrowing the fields (men) and transplanting the seedlings into the fields (women). (Planting was also regulated by the phase of the moon.) By now, *Bintang Weluku* was appearing higher and higher in the sky after sunset, and the end of this period was marked by its culmination (appearing at its highest point in the sky) at dusk. Four months later, when the rain had stopped and the rice had ripened, the harvest began. This season was marked by the celestial plough appearing progressively lower in the western sky after sunset. At this time it was perceived as upside down, "like a farmer's plough when the work is done." Soon afterwards, it disappeared completely (heliacal set).

The workings of this pragmatic stellar calendar are reminiscent of Hesiod's words of advice to Greek farmers in the eighth century B.C.E. Numerous other practices known from many different parts of the world also regulate seasonal farming activities (both growing crops and managing livestock) by reference to the observed behavior of prominent stars or asterisms.

*Bintang Weluku* remained in use until around the late nineteenth century and is still prominent in folk memory, despite the co-existence of two other methods of dividing the year. The first involved numerical (i.e., non-astronomical) cycles such as five- and seven-day weeks and 210-day years, which derived from a mix of Islamic and Hindu traditions. This system was institutionalized and also used for divination. The other calendar, known as the *pranotomongso* (or *pranatamongsa*), was based on measurements of the length of the sun's shadow at noon, regulated since the seventeenth century by using an accurate gnomon device known as a *bencet*. This calendar divided the year into twelve unequal divisions, although only four were widely used in practice. These were demarcated by the solstices and the days of solar zenith passage.

**See also:**
Hesiod (Eighth Century B.C.E.); Orion; Solstices; Zenith Passage of the Sun. Heliacal Rise.

**References and further reading**
Chamberlain, Von Del, John Carlson, and Jane Young, eds. *Songs from the Sky: Indigenous Astronomical and Cosmological Traditions of the World*, 320–335. Bognor Regis, UK: Ocarina Books, and College Park, MD: Center for Archaeoastronomy, 2005.
Selin, Helaine, ed. *Astronomy across Cultures*, 371–384. Dordrecht, Neth.: Kluwer, 2000.

## Julian Calendar

The new civic calendar introduced by Julius Caesar in 45 B.C.E. was revolutionary in that it ignored the moon. Earlier calendars in the Roman world had been based on the lunar phase cycle and had gotten badly out of step with the seasonal year. Remarkably, two key elements of our own modern calendar date all the way back to Caesar's innovations. The first is its division into twelve months, each with a fixed number of days, chosen so as to add up to the number of whole days in the seasonal year (365). The word *month* in this context is in fact a complete misnomer, since the divisions of our year are completely independent of the cycle of lunar phases. The second innovation was the concept of a leap year. By introducing an additional day once every four years, the average length of a year in the Julian calendar became 365.25 days, very close to the true length of the year.

The Julian calendar did not escape teething troubles. For over three decades an additional day was mistakenly inserted every third instead of every fourth year, necessitating a retrospective adjustment. Only by C.E. 8

was the pattern established that remains familiar today, whereby every fourth year (whenever the year is divisible by four) is a leap year.

The Julian calendar survived intact for over 1500 years in the Christian world, but (with the exception of Ethiopia) was eventually superseded by the more accurate Gregorian calendar. This happened at different times in different countries according to whether the dominant tradition was Catholic, Protestant, or Orthodox. Those who study calendrically related phenomena in medieval times, for example the orientations of medieval churches, have to be critically aware of the differences between the two calendars.

**See also:**
Lunar and Luni-Solar Calendars.
Church Orientations; Gregorian Calendar; Roman Astronomy and Astrology.

**References and further reading**
Aveni, Anthony. *Empires of Time: Calendars, Clocks and Cultures*, 111–115. New York: Basic Books, 1989.
McCready, Stuart, ed. *The Discovery of Time*, 88–89. Naperville, IL: Sourcebooks, 2001.

# June Solstice
*See* Solstices.

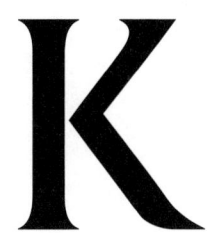

## Khipu
*See* Quipu.

## Kintraw

This megalithic monument on the west coast of Argyll, Scotland, consists of a single standing stone four meters (thirteen feet) high situated between two round cairns. It probably dates to the late third or early second millennium B.C.E. It is well known because it was the place where, in the 1970s, an archaeological excavation was first used to test an astronomical theory deduced from alignment studies.

The horizon all around this site is relatively close except toward the southwest, where there is a distant view toward the Isle of Jura, some 45 kilometers (28 miles) away. Here two adjacent mountain peaks stand out prominently with a deep, sharp notch between them. As viewed from the standing stone, this notch aligns precisely with the midwinter sunset. Close to the winter solstice, the top part of the setting sun would reappear briefly in the bottom of the notch, but at the solstice itself it would be slightly too low, and only then would it disappear from view.

When Alexander Thom discovered this alignment, he also discovered a problem. The notch is just hidden from the standing stone itself by an intervening ridge; only a 2.5-meter (8-foot) giant would be able to see it from the standing stone. Undeterred, Thom followed the alignment back toward the northeast where it crosses a deep gorge and then climbs a steep hillside. Halfway up this precipitous slope is a narrow ledge, which would have provided the perfect spot from which to observe the setting sun. From here, the standing stone appears directly beneath the deep notch. But there was no obvious sign of human activity here, just two rounded boulders. Was the ledge simply a natural feature? Here was the perfect opportunity for an archaeological test of an astronomical theory: if the ledge could be shown by excavation to be artificial, then this would vindicate the idea that it was used as an observing platform.

Archaeologist Euan MacKie rose to the challenge in 1970 and 1971. Unfortunately, his excavations did not reveal any clear and direct signs of human activity. However, they did show that the flat part of the ledge was covered with rounded pebbles, which he suggested had been carefully placed there to form a viewing platform. The problem was that they might have arrived naturally, rolling down the hill and coming to rest behind the two larger boulders. Analyses were undertaken of the orientations of the pebbles, in the hope of distinguishing between the two possibilities, but no clear conclusion could be reached. Thus MacKie's excavations proved inconclusive.

One of the general questions that arises from Thom's theory of highly precise solstitial sightlines is how they could have been set up in the first place. One of the most plausible ideas is that a row of people, stretching perpendicularly to the sightline, watched the sunset on days leading up to the solstice. Each day, fewer or them would see the sun reappear in the notch, as the sun's setting path sank slightly in the sky (the opposite would be true at a notch where the sun rose or set at the summer solstice). Each evening a temporary marker would be set up to mark the limiting position where the sun could be seen to reappear in the notch. Day by day, this position would move sideways, by less and less as the solstice approached, until the limit was eventually reached and the position began to move back in the opposite direction. For the summer solstice, the limit would mark the spot where the tip of the sun would only appear in the notch on the solstice itself; for the winter solstice, as at Kintraw, it would mark the spot from which the sun would only disappear on the solstice itself. This, then, would be the spot where a permanent marker would be erected.

There are two serious problems with this procedure at Kintraw. First, there is no room to set out a line of people as just described, because the platform is situated on the side of a precipitous hillside above a deep gorge. Second, the intervening ridge blocks the view of the distant notch from those positions where one would have had to stand in order to watch the sunset on the days before and after the solstice. There are also more general problems that bring into question the whole idea that solstitial sightlines of pinpoint precision could have been established at British megalithic sites such as Kintraw, Ballochroy, or the Ring of Brodgar. An obvious imponderable would have been the British weather: the setting-up procedure would only work given a reasonable run of clear evenings, something that would have been far from guaranteed even given that the weather in western Scotland in later prehistoric times was somewhat different from now. Another is the possibility of significant day-to-day variations in atmospheric *refraction* due to changing atmospheric conditions, which would alter the apparent altitude of the setting sun.

The reappearance of the setting sun in the notch at Kintraw for a period around the December solstice would undoubtedly have formed a spectacular sight. Certainly, it may have been deeply significant to the people who constructed (not necessarily at the same time) the cairns and monolith: the larger cairn, for example, had a setting of stones resembling an entrance (known as a *false portal*) facing in this general direction. Perhaps no greater precision than this was needed. By trying too hard to fit prehistoric people into the mould of ancient "scientific" astronomers, and by focusing too closely on the possible existence of alignments of pinpoint precision, we risk failing to grasp more useful insights into the part that astronomy really played in prehistoric burial and ceremonial practices.

See also:
Science or Symbolism?; Solstitial Alignments; Thom, Alexander (1894–1985). Ballochroy; Brodgar, Ring of; Megalithic Monuments of Britain and Ireland. Refraction.

**References and further reading**
Ruggles, Clive. *Astronomy in Prehistoric Britain and Ireland*, 25–29. New Haven: Yale University Press, 1999.

# Kukulcan

The pyramid of Kukulcan, also known as El Castillo, is one of the most impressive constructions at the Maya site of Chichen Itza. Some consider its innate calendrical significance to be expressed in the numerology of its four stairways, one on each side, each of which rises in ninety-one steps to the top platform; adding the final step into the temple on the top makes 365 steps in all. The pyramid is not cardinally oriented but skewed by about twenty-one degrees clockwise, and archaeoastronomers have investigated the possible reasons for this as part of systematic studies of the orientations of Maya temples and city plans. However, the phenomenon that has captured the public imagination was apparently discovered by chance by a caretaker at the site in the late 1920s.

Twice each year, at around the time of the equinoxes, the late afternoon sun slants across the stepped terraces of the northwestern (actually NNW) corner of the pyramid and casts a sinuous shadow onto one balustrade of the northern (actually NNE) staircase. At the bottom of this balustrade is a large stone serpent's head, and the shadow appears to complete the body of the serpent, undulating back up the balustrade. The shadow is visible for an hour or two, as sunset approaches.

Whether the phenomenon that has become known as the *equinox hierophany* was intentionally orchestrated by the builders is highly questionable. Given the overall fourfold symmetry of the monument, it seems incongruous that the northern staircase should be favored in this way. However,

*Crowds of people assemble at Kukulcan (El Castillo) to view the equinox hierophany in 1988. (Richard A. Cooke/Corbis)*

the pyramid does overlay an earlier version, and this one only had a single stairway which was on the northern (actually NNE) side. Furthermore, there is good evidence to relate the pyramid to Kukulcan, a plumed serpent (feathered rattlesnake) god prominent in the Mesoamerican pantheon, known later to the Aztecs as Quetzalcoatl. There is also good evidence to relate Quetzalcoatl to the planet Venus. A platform to the north of the Kukulcan pyramid has wall carvings containing a number of Venus symbols, and as a result has become known as the Platform of Venus. This has led some to suggest that the timing of the so-called equinox hierophany may have had more to do with the motions of Venus than the equinox.

Be this as it may, the site has certainly become the focus for modern pilgrimage. The hierophany is now witnessed by tens of thousands of visitors, which has become interesting as a modern sociological phenomenon in itself.

**See also:**
Equinoxes; Pilgrimage.
Caracol at Chichen Itza; Venus in Mesoamerica.

**References and further reading**
Aveni, Anthony F. *Skywatchers,* 298–300. Austin: University of Texas Press, 2001.
Carlson, John B. "Pilgrimage and the Equinox 'Serpent of Light and

*The serpent's head at the foot of north stair of the Kukulcan pyramid. Its body "appears" in shadow at the time of the equinox. (Courtesy of Clive Ruggles)*

Shadow' Phenomenon at the Castillo, Chichén Itzá, Yucatán." *Archaeoastronomy: The Journal of Astronomy in Culture,* 14 (1) (1999), 136–152.

Krupp, Edwin C. *Echoes of the Ancient Skies,* 298–299. Oxford: Oxford University Press, 1983.

———. *Skywatchers, Shamans and Kings,* 267–270. New York: Wiley, 1997.

# Kumukahi

Cape Kumukahi, in the southeastern corner of the Big Island of Hawai'i, forms the bleak, easternmost extremity of the Hawaiian island chain. Lava flows have pushed across this flat plain over the centuries, most recently in 1960, when the molten rock stopped just short of the lighthouse at the end of the road out across the cape—something regarded by many as a conciliatory gesture by the volcano goddess Pele. East of the lighthouse, close to the coast, is a rugged landscape formed of large irregular chunks of black lava.

This is an unlikely setting for astronomical observations, yet it is one of the very few places mentioned explicitly in this regard in Hawaiian oral history. Two traditions have come down to us. The first, recorded by the ethnographer Nathaniel Emerson in 1909 as a footnote in a book about sacred songs of the *hula,* refers to a "pillar of stone," itself called Kumukahi, and a "monolith," to the south of it, called Makahoni. According to Emerson:

> In summer the sun in its northern excursion inclined, as the Hawaiians noted, to the side of Kumukahi, while in the season of cool weather, called Makalii, it swung in the opposite direction and passed over to Makahoni. The people of Puna accordingly said, "The sun has passed over to Makahoni," or "The sun has passed over to Kumukahi," as the case might be. (Emerson 1997 [1909], p.197)

Ethnographer Martha Beckwith recorded a slightly different legend. By this account, a red stone at the extreme end of the cape represents the god Kumukahi, a god with healing powers who can take the form of a plover. Two further stones represent his wives, who "manipulate the seasons by pushing the sun back and forth between them at the two solstices" (Beckwith 1970 [1940], p.119).

The observations described in the myths are hardly of great sophistication. The significance of this example is that the Kumukahi pillars constitute one of very few calendrical markers whose location is identified in Polynesian oral traditions with absolute certainty. Without these legends, this barren corner of the Hawaiian islands would surely be one of the last places a modern archaeoastronomer might think to look for evidence of ancient astronomical observations. The visible landscape is formed entirely in relatively modern times, a substantial part of it within living memory; such landmarks as there are were formed by natural processes rather than human intervention. The stones and pillars of which the legends speak are not easy to single out from among the many natural lava pillars in the area, if indeed they are still standing. Even with the help of the legends, the exact nature of the observations, and hence of the alignments that may still exist, is not entirely clear. Nonetheless, a combination of archaeology and archaeoastronomy may yet manage to identify the Kumukahi pillars with a fair degree of confidence.

One cannot visit Kumukahi without wondering what the ancient Hawaiians could possibly have been doing in such a desolate spot. Yet this was undoubtedly a sacred place, featuring in a number of different songs and chants. We do not need to postulate the existence of more precise astronomical alignments in order to explain its sacred significance; we probably need look no further than the fact that it *was* an extremity, beyond which the boundless ocean extended away in the direction of the rising sun. In this sense, it may have something in common with remote Necker Island, at the opposite end of the chain.

See also:
Solstitial Directions.

Hawaiian Calendar; Necker Island; Polynesian Temple Platforms and Enclosures. Solstices.

**References and further reading**
Beckwith, Martha W. *Hawaiian Mythology* (repr.), 119. Honolulu: University of Hawai'i Press, 1970. Originally published in 1940 by Yale University Press, New Haven.
Emerson, Nathaniel B. *Unwritten Literature of Hawaii: The Sacred Songs of the Hula*. Honolulu: 'Ai Pōhaku Press, 1997. Originally published in 1909 by Government Printing Office, Washington, DC.
Hale'ole, S. N. *Ka Mo'olelo o Lā'ieikawai: The Hawaiian Romance of Lā'ieikawai,* trans. by Martha W. Beckwith, 627. Honolulu: First People's Productions, 1997. First published in translation in 1919 by the Smithsonian Institution, Washington, DC.

# Kumulipo

Just because a human society has not developed writing does not mean that it cannot generate literature. In many nonliterate cultures, certain stories—and particularly creation myths—were so sacred that they were carefully learned and passed on from generation to generation, strictly verbatim, and thereby preserved intact for many decades or even centuries. Recital or performance, sacred in itself, was typically a formal affair and subject to strict protocols. The audience was often restricted to those of sufficient rank or those suitably initiated in some way.

A fine example of such oral literature is the Hawaiian prayer chant known as the *Kumulipo*. Passed down through several generations of high chiefs (*ali'i*), it came into the possession of the last native ruler of Hawai'i, King David Kalakaua, after which it was recorded and translated into English. It originated as far back as the mid-eighteenth century, a longevity that is all the more remarkable given its considerable length—over two thousand lines. It was composed for the *ali'i* Ka-'Ī-i-mamao, and clearly served as a legitimization of his rank and power. Its purpose was to demonstrate how he (and, by extrapolation, his descendants) was directly descended from a string of culture heroes and ultimately from the gods themselves. The *Kumulipo* was, according to later Hawaiian informants, the very chant that was recited to Captain Cook during a ceremony at Hikiau heiau on the Big Island of Hawai'i, when he was being received as the god Lono.

The *Kumulipo* combines a creation story and description of the world with genealogies that recount (or, as an outsider would see it, establish) the lineage of the chief. It is in the form of a poem, although substantial chunks consist of no more than lists of names. Although genealogical prayer chants of this nature are not exceptional in Polynesia, the *Kumulipo* describes various natural phenomena in particular detail. It is, in this sense, an excellent natural history. The heavens are included, of course, and the work actually begins with a reference to the sun, moon, and Pleiades. Some have claimed that a much richer body of astronomical knowledge is represented in the form of hidden, double meanings, which were undoubtedly included. Hawaiian is a language ripe for punning; subtle word play, allusions, and different levels of meaning were an essential characteristic of sacred chants. The difficulty lies in accurately reconstructing them. Unfortunately, much was lost in the early transcriptions of Hawaiian, which ignored subtle but vital differences in vowel sounds that have only relatively recently been represented using appropriate diacritical marks. The underlying, hidden meanings of old texts are extremely difficult to reconstruct, even for native Hawaiian scholars, and can vary widely in their interpretations. In this regard, the *Kumulipo* is no exception.

The genealogical lists in the *Kumulipo* certainly include known star names. One native Hawaiian scholar, Rubellite Kawena Johnson, has argued that the chant describes the heavenly cycles in considerable detail using various numerological encodings and devices. Others, however, argue that the particular ordering of and relationships between the natural phenomena described have more to do with the requirements of good rhythmic balance. However this debate is ultimately resolved, the *Kumulipo* is interesting enough as an example of a creation story that not only describes the natural world and its human inhabitants but tries to explain it.

**See also:**
Hawaiian Calendar; Nā Pali Chant.

**References and further reading**
Beckwith, Martha W. *The Kumulipo: A Hawaiian Creation Chant* (repr.). Honolulu: University of Hawai'i Press, 1972.
Johnson, Rubellite K. *Kumulipo: Hawaiian Hymn of Creation, Volume I.* Honolulu: Topgallant Press, 1981.
———. *The Kumulipo Mind: A Global Heritage.* Honolulu: Anoai Press, 2000.
Sahlins, Marshall. *Islands of History,* 109–110. Chicago: University of Chicago Press, 1985.
Selin, Helaine, ed. *Astronomy across Cultures,* 93–94. Dordrect, Neth.: Kluwer, 2000.

# L

## Lakota Sacred Geography

The traditional homeland of the Lakota (Sioux) is the Black Hills, an isolated oval-shaped area of mountains on the border between South Dakota and Wyoming. The mountains stretch some 180 kilometers (120 miles) from north to south and 80 kilometers (50 miles) east to west, and are surrounded by flat plains. The name *Lakota* means "Friends of the Earth," and the Lakota, traditionally a nomadic people, kept their lives in harmony with the cosmos by tuning their seasonal path through the landscape to the annual cycle of the sun through the stars. The basis of their subsistence was the buffalo (*tatanka*), an animal they considered to embody the power of the sun. As they followed the herds through the landscape, they received spiritual instruction from the stars. Although their traditional way of life was virtually eliminated during the nineteenth and twentieth centuries, as were the extensive buffalo herds that used to roam this country, the Lakota are now sharing with outsiders elements of their tradition as they try to rebuild it.

The Lakota perceive a direct connection between certain places in the landscape and certain groups of stars in the sky. Thus Harney Peak, the highest mountain in the Black Hills, was associated with Seven Little Girls (the Pleiades), and isolated Devil's Peak was connected with Bear's Lodge (part of Gemini). During the year, the Lakota traditionally moved from their winter camps south of the Black Hills northward, toward and into the range. This journey, following the seasonal movements of the buffalo through the landscape, also followed the path of the sun through the constellations, and an integral part of the process was the performance of the appropriate sacred rites. On one level, this annual journey, regulated by observations of astronomical phenomena, simply served to ensure the successful exploitation of a vital food resource. However, this outsiders' perspective fails utterly to comprehend anything of the meaning of these practices to the people who undertook them. From the Lakota standpoint, the whole journey was a ceremony—a sort of ritual pilgrimage—that first and foremost preserved the balance of life and ensured peace and harmony in the cosmos in general.

A few details serve to give an idea of the complexities involved. The spring journey away from the winter encampments commenced when the sun entered the *Cansana Ipusye* constellation (parts of Aries and Triangulum). The same phrase, *Cansana Ipusye,* is used by the Lakota to describe the Dried Willow, the Sacred Pipe in which it was smoked, and the wooden spoon in which shamans would carry live coals (representing the sun) as the journey commenced. At the same time (around the spring equinox), the sacred powers in the sky would use the celestial wooden spoon (Ursa Major) to carry the sun to the celestial pipe (Aries/Triangulum).

When the sun arrived at the Pleiades, the people would camp at Harney Peak and perform ceremonies welcoming back the Thunders. After this, the sun moved toward the center of "the great circle of stars" (Rigel, Sirius, Procyon, Pollux, Castor, Menkalinan, Capella, and the Pleiades) while the people moved to *Pe Sla,* a "bare hill" in the middle of the Black Hills. Emerging out of the celestial great circle are the head (Hyades), backbone (Orion's belt), ribs (Betelgeuse/Bellatrix and Rigel) and tail (Sirius) of a buffalo, the animal that symbolized all life. Likewise, the Black Hills were seen as the place out of which all life was annually renewed on earth. For this reason, rites took place there at this time to welcome back life: water was poured on the earth to feed the plants, seeds were scattered for birds, and buffalo tongues were offered for meat-eating animals.

As the summer solstice approached, the sun approached the Bear's Lodge (*Mato Tipila*) and the people arrived at the Devil's Tower. *Mato Tipila* was created by the central figure in Lakota oral tradition, Fallen Star, to protect two children threatened by bears. The constellation is located on one side of, but nonetheless within, the great circle of stars, whereas its earthly equivalent, Devil's Tower, is in fact some one hundred kilometers (sixty miles) northwest of the Black Hills. This configuration seems contradictory from a Western perspective, serving to remind us how wrong we could be if we tried to reconstruct elements of Lakota worldview from partial historical information without having modern informants. We might easily try to enforce too literal an interpretation of the mapping between sky and earth, and dismiss the idea that the Devil's Tower could have been part of the scheme. It was at Devil's Tower that, in July or August, the annual Sun Dance took place. Contrary to popular belief, this was a solemn ritual, the last celebration of the season devoted to the sun, buffalo, and the Black Hills.

The Lakota, like many indigenous peoples, ascribe spiritual meaning to the whole landscape and to particular places within it. What makes their case particularly interesting, though, is the added dimension of timing. Cosmic harmony is preserved by being in the right place at the right time and performing the appropriate rites. The terrestrial world is connected to the spirit world both in space and time, and the key to this connection is the sky. Not only are places in the landscape associated with particular asterisms, but the time to be there is prescribed by reference to the sun's passage through the stars. This connection between the terrestrial and the spiritual also has the pragmatic effect (from a Western perspective) of making sure that the buffalo can be successfully hunted. But for the Lakota, the buffalo are not simply a resource to be exploited. They are part of the whole circle of life, who say "take my flesh with gratitude" (Chamberlain et al. 2005, p. 145), and humans have to return this gift by doing all they can to preserve the cosmic order.

**See also:**
Landscape; Orion; Sacred Geographies.

**References and further reading**
Chamberlain, Von Del, John Carlson, and Jane Young, eds. *Songs from the Sky: Indigenous Astronomical and Cosmological Traditions of the World*, 140–146. Bognor Regis, UK: Ocarina Books and College Park, MD: Center for Archaeoastronomy, 2005.
DeMallie, Raymond J., and Douglas R. Parks, eds. *Sioux Indian Religion*. Norman: University of Oklahoma Press, 1987.
Goodman, Ronald. *Lakota Star Knowledge: Studies in Lakota Stellar Theology*. Rosebud, SD: Sinte Gleska College, 1990.
Mizrach, Steven. *Lakota Ethnoastronomy*. http://www.fiu.edu/~mizrachs/lakota.htm.
Sundstrom, Linea. "Mirror of Heaven: Cross-cultural Transference of the Sacred Geography of the Black Hills." *World Archaeology* 28 (2) (1996), 177–189.

# Land of the Rising Sun

In modern Japan, ancient traditions coexist with the many emanations of modern Western culture. In centuries past, too, Japanese culture continually absorbed influences from abroad—namely from mainland Asia—and yet retained a distinct identity. Even the Japanese language, when it was first written down in the middle of the first millennium C.E., used kanji characters imported from the Chinese.

For a millennium and a half, from the fourth century until the end of World War II, the principles of Shinto religion forged unbreakable links between cosmology, political structure, and the sun. Successive emperors traced their ancestry directly back to the Sun Goddess, helping to forge a national identity linked with the sun that is still evident in the national flag. Shinto persisted despite the strong influence of Buddhist traditions from China that have coexisted in this island nation from the sixth century onwards. This coexistence involved some remarkable compromises, for instance, in locating and aligning temples and palaces. In the Shinto tradition this would be done with respect to places of spiritual power in the landscape, whereas in the Chinese tradition they would typically be aligned cardinally, reflecting the principle that spiritual and imperial power derived from the north celestial pole as the pivot of the heavens, and ensuring that the emperor would be approached, like the celestial pole itself, from the south. The plan of ancient Kyoto, for example, built in C.E. 794, with its palace complex approached from due south, reflected such principles every bit as faithfully as Beijing itself.

Aside from institutionalized astronomy, elements of star lore and folk calendars have been transmitted through countless generations of ordinary people, and some have persisted in rural areas despite the introduction of the Gregorian calendar. For example, the changing patterns of appearance of the Pleiades (Subaru), Hyades, and Orion's belt provided a succession of seasonal rules of thumb for rice farmers. One of these was that when Subaru—which resembled a collection of rice seedlings—set progressively earlier in the evening sky in the spring, this indicated the time for planting the actual rice seeds in the ground.

Archaeoastronomy is in its infancy in Japan. The Asuka plain, to the east of Osaka, contains several tombs (*kofun*) of high-status individuals that were erected in the seventh and early eighth centuries. Two of them, only about a kilometer (half mile) apart—Takamatsu Zuka Kofun, excavated in 1972, and Kitora Kofun, probed in 1998 using a miniature camera—contain paintings with strong astronomical associations. The ceilings depict the twenty-eight lunar "mansions" (known in Japan as *shuku*) and other constellations, while the walls show the animal gods associated with each of the cardinal directions. The two tombs demonstrate clear but nonetheless different Chinese and Korean influences. The region also contains a number of granite megaliths carved in the shape of human figures. These are of uncertain origin, date, and purpose, and while some have attracted archaeoastronomical interest, interpretations are far more speculative.

**See also:**
Archaeoastronomy; Orion; Power.
Chinese Astronomy and Astrology.
Celestial Sphere; Heliacal Rise.

**References and further reading**
Krupp, Edwin C. *Skywatchers, Shamans and Kings*, 196–207. New York: Wiley, 1997.
Renshaw, Steven, and Saori Ihara. "Archaeoastronomy and Astronomy in Culture in Japan: Paving the Way to Interdisciplinary Study." *Archaeoastronomy: The Journal of Astronomy in Culture* 14 (1) (1999), 59–88.
———. *Astronomy in Japan: Science, History, Culture.* http://www2.gol.com/users/stever/jastro.html.
Selin, Helaine, ed. *Astronomy across Cultures*, 385–407. Dordrecht, Neth. Kluwer, 2000.

Walker, Christopher, ed. *Astronomy before the Telescope*, 267–268. London: British Museum Press, 1996.

Xu Zhentao, David Pankenier, and Jiang Yaotiao. *East Asian Archaeoastronomy: Historical Records of Astronomical Observations of China, Japan and Korea.* Amsterdam: Gordon and Breach, 2000.

# Landscape

People in the past ordered the world of their experience, defining places within it and pathways through it, and in that way came to understand and control it. Vital as it was to exploit natural resources to the full in order to subsist, human activity in relation to the landscape was often structured according to symbolic or cosmological principles, forming what are known to archaeologists as *ritual landscapes* or *sacred geographies*. Examples of sacred geographies are many and varied among both historically recorded and modern indigenous peoples.

As is still the case for many indigenous peoples today, places may have become imbued with sacred or mystical significance because of their position within the landscape, as with the summits of high hills; because of the presence of prominent landmarks, such as caves, watering holes, or distinctive natural rocks; or because of remembered events that happened there. Some of these places may have been used for the construction of shrines or monuments, and others may have assumed sacred significance precisely because earlier constructions were built there. Sacred significance may be attached to lines and paths as well as particular places. For the inhabitants of the Hopi village of Walpi, the directions of sunrise and sunset at the solstices are themselves sacred, so that four sacred lines radiate out through the landscape from the village to the horizon. This configuration represents a very direct and obvious connection between the landscape and the sky, and also a relationship between space and time, in that each line is empowered by the actual occurrence of sunrise or sunset on the appropriate day.

Patterns of movement through the landscape may be vitally important as a way of reinforcing a group's understanding of the structure of the world and their place within it. Thus the Lakota people of South Dakota associate certain constellations with specific landmarks within the Black Hills. In accordance with their oral tradition, they undertake an annual progression through those hills that is understood to be related to, and seen to be in tune with, the path of the sun through the constellations. The Lakota example also illustrates how the needs of the sacred and the mundane—or, as we might see it, of ideology and pragmatism—need not necessarily be mutually exclusive or in conflict. The cycle of seasonal movements ensures that the Lakota follow the movements of the buffalo, a vital food source that they also saw as the very embodiment of the power of the sun.

The Lakota example also illustrates that no discussion of sacred landscapes is likely to be complete without a consideration of the sky. As soon as we start to think about how people perceived and conceived the world, and as soon as we start examining evidence of human action shaped by and undertaken in accordance with such perceptions, then we need to consider not only the land and sea but the totality of the visible environment within which those people lived. That includes the sky.

**See also:**
Cosmology; Monuments and Cosmology.
Aboriginal Astronomy; Aztec Sacred Geography; Hopi Calendar and Worldview; Lakota Sacred Geography.

**References and further reading**
Ashmore, Wendy, and Bernard Knapp, eds. *Archaeologies of Landscape: Contemporary Perspectives.* Oxford: Blackwell, 1999.
Barrett, John, Richard Bradley, and Martin Green. *Landscape, Monuments and Society.* Cambridge: Cambridge University Press, 1991.
Bradley, Richard. *An Archaeology of Natural Places.* London: Routledge, 2000.
Carmichael, David, Jane Hubert, Brian Reeves, and Audhild Schanche, eds. *Sacred Sites, Sacred Places.* London: Routledge, 1994.
Edmonds, Mark. *Ancestral Geographies of the Neolithic: Landscapes, Monuments, and Memory.* London: Routledge, 1999.
Goodman, Ronald. *Lakota Star Knowledge: Studies in Lakota Stellar Theology.* Rosebud, SD: Sinte Gleska University, 1990.
Renfrew, Colin, and Paul Bahn, eds. *Archaeology: The Key Concepts*, 156–159. Abingdon: Routledge, 2005.
Scarre, Chris, ed. *Monuments and Landscape in Atlantic Europe.* London: Routledge, 2002.
Tilley, Christopher. *A Phenomenology of Landscape: Places, Paths and Monuments.* Oxford: Berg, 1994.

# Ley Lines

In 1921, so the story goes, businessman Alfred Watkins had a revelation. Standing on a hillside in Herefordshire, England, overlooking an expanse of rural countryside containing a number of ancient features, he noticed that many of them seemed to lie on straight lines. A subsequent examination of Ordnance Survey maps revealed the apparent existence of numerous ancient straight trackways that formed a network of intersecting straight lines stretching from one end of Britain to the other, with ancient sites of various ages situated along them. Noticing that many of the trackways passed through places whose names contained the syllable *ley*, Watkins concluded that the word *ley* referred to the trackways themselves and named them *ley lines*. Drawing on the earlier astronomical work of Sir Norman Lockyer, he also concluded that some of the lines were oriented in the directions of sunrise and sunset at the solstices. By the following year he had published his first book, *Early British Trackways*; its sequel, *The*

*Old Straight Track,* published in 1925, attracted a wide following from a fascinated public.

Professional archaeologists, however, were highly skeptical. The idea that the whole of Britain could have been mapped out single-mindedly in prehistoric times seemed to them inherently unreasonable, and there seemed little in the way of hard evidence to support the theory. For a start, the suggested etymology of the term *ley* as referring to ancient tracks was highly questionable. More crucially, the ancient sites taken to define the lines were of widely different ages and types, and marker points actually included natural features such as large stones, trees, ponds, and mountain peaks. Furthermore, large numbers of equally plausible ancient sites and natural places did *not* lie on any of the proposed lines. We must ask the question: how easy is it, given a scatter of effectively random locations, to find straight lines joining some of them? It is possible to give a mathematical answer, but the informal answer is clear: surprisingly easy. The excavator of Stonehenge, Richard Atkinson, once demonstrated this using the locations of telephone boxes, which no one could argue had been laid out deliberately in straight lines, finding several convincing "telephone box leys." As to the question of the different ages of the sites forming supposed marker points, Watkins and his followers argued that newer buildings such as churches must have been constructed on the site of older markers. However, in most cases there was little or no independent evidence to support this, and in some cases the archaeological evidence stood in flagrant contradiction to the theory. A notorious example is a ley line joining Stonehenge (third millennium B.C.E.), Old Sarum (first millennium B.C.E.), and Salisbury cathedral (C.E. 1220) in Wiltshire. The site for Salisbury cathedral was on virgin marshland, actually chosen because a preferred site, several kilometers to the west, could not be secured.

Curiously, unbeknownst to Watkins and his followers in the 1920s, a similar idea had been developed in Germany around the same time, claiming that sacred places in that country were linked by *Heilige Linien* ("Holy lines"). This theory came, infamously, to be used in support of the Nazi political agenda. Back in Britain after the 1920s, though, ley lines soon faded into obscurity and would have been long forgotten but for their reemergence in the 1960s in a new guise. Watkins believed that his trackways were constructed by prehistoric surveyors physically sighting from one place to another, but the ley lines of the 1960s were conceived as lines of power, the paths of some form of spiritual force or energy accessible to our ancient ancestors but now lost to narrow-minded twentieth-century scientific thought. Public interest in ley lines mushroomed, and "ley hunting" became a hugely popular pastime, largely because someone with no professional training whatsoever could feel directly involved in the process of rediscovering the magical landscapes of the past. All that was required was an open mind, the energy to walk about in the landscape seeking potential sacred places, and the ability to feel the spiritual forces.

The dangerous aspect of this approach is that it abandons all attempts to examine the evidence objectively—to be scientific in any broad sense. The existence of the lines of force had to be accepted as an article of faith, to be backed up by subjective discoveries. Very few archaeologists bothered to waste their time on a topic that seemed to them a complete fiction, but this just served to confirm to the ley liners that the academics were both narrow-minded and arrogant. Meanwhile ley hunting spawned journals and books, sub-fields, experts, and doctrinal disputes; in short, it developed all the trappings of an alternative discipline, arrayed in competition against the academic establishment. The irony is that the ley hunters found themselves adopting the very paraphernalia of science that most of them were arguing so vociferously against.

For the most part, ley lines represent an unhappy episode now consigned to history. Yet humans do tend to construct lines in the landscape, conceptual or actual, a practice that has emerged in some human societies in the past. The best-known example is that of the Inca *ceques*, conceptually (though not actually) straight lines radiating out from the capital, Cusco, throughout the entire Inca empire. Sacred sites and places, *huacas*, lay on the *ceques*. There is even evidence that a few *ceques* (though by no means all) were oriented astronomically. In cases such as this we may wonder how ancient people perceived these lines, and whether in their minds they may indeed have been the loci of sacred power. In doing so, we may find ourselves, as anthropologists, investigating an innate human desire to discern, and hence to try to reinforce, a structure in the landscape that we can understand. The ley line phenomenon, at least in its later manifestation, may represent a more modern manifestation of this same innate desire.

**See also:**
Constellation Maps on the Ground; Lockyer, Sir Norman (1835–1920); Methodology; Nationalism.
*Ceque* System.
Solstices.

**References and further reading**
Burl, Aubrey. *Rings of Stone*, 80–82. London: Frances Lincoln, 1979.
Michell, John. *A Little History of Astro-Archaeology*, 58–65. London: Thames and Hudson, 1989.
Ruggles, Clive. *Astronomy in Prehistoric Britain and Ireland*, 3. New Haven: Yale University Press, 1999.
Watkins, Alfred. *The Old Straight Track*. London: Abacus, 1974. Originally published in 1925 by Methuen, London.
Williamson, Tom, and Liz Bellamy. *Ley Lines in Question*. Tadworth: World's Work, 1983.

*Sir Norman Lockyer. (Hulton-Deutsch Collection/CORBIS)*

# Lockyer, Sir Norman (1836–1920)

Sir Joseph Norman Lockyer was the first, at least in the English-speaking world, to introduce scientific method into the investigation of the possible astronomical significance of ancient monuments. He was a physicist, hailed for his discovery of helium through spectroscopic studies of the sun, and for many years the editor of the scientific journal *Nature*. In later life Lockyer became fascinated by possible solar and stellar alignments at Egyptian temples and surveyed several of them, publishing his results in 1894 in a book entitled *The Dawn of Astronomy*. He was responsible for the idea that some temples were carefully designed so that a beam of light from the sun, or possibly a bright star, would shine down a long narrow passage, illuminating a dark sanctuary in the interior on particular days. This idea would be applied more generally in very different contexts in later times, one of its most famous manifestations being at Newgrange in Ireland. Later Lockyer turned his attention to his native Britain, surveying several dozen megalithic monuments all over the British Isles and publishing the results in 1906 under the title *Stonehenge and Other British Stone Monuments Astronomically Considered*. This work inspired a generation of amateur archaeoastronomers, including the engineer Alexander Thom. Lockyer is less well known as the author of *Surveying for Archaeologists*, the first book on that topic, published in 1909.

> **See also:**
> Nissen, Heinrich (1839–1912); Somerville, Boyle (1864–1936); Thom, Alexander (1894–1985).
> Egyptian Temples and Tombs; Newgrange.
>
> **References and further reading**
> Lockyer, Norman. *The Dawn of Astronomy*. London: Cassell, 1894.
> ———. *Stonehenge and Other British Stone Monuments Astronomically Considered*. London: MacMillan, 1906.
> ———. *Surveying for Archaeologists*. London: MacMillan, 1909.

## Lunar and Luni-Solar Calendars

One lunar phase cycle, also known as a *lunation* or a *synodic month*, is completed on average every 29.53 days. It is one of the most obvious of all the celestial cycles and has formed the fundamental basis for reckoning time in a wide range of human communities to the present day. Recognition of the lunar phase cycle may extend well back into Palaeolithic times: markings on the 30,000-year-old Abri Blanchard bone have been identified as a possible lunar calendar. Yet the lunar phase cycle remains at the heart of some of the world's great calendar systems, including both the Jewish and Islamic calendars.

In the great majority of lunar calendars, the new month is taken to start following the new moon, with the first appearance of the thin crescent just after sunset. This particular point in the cycle is chosen because it is a clearly definable event; but even so there can be problems. Among the Mursi of southwestern Ethiopia, there is often disagreement as to whether the moon was really seen or not by someone who claims to have done so. Problems also arise if a critical evening is cloudy, and they are compounded if a single calendar needs to be consistently adopted over a wider area. The Islamic world faced this problem at an early stage, and at first Muslim astronomers used technical criteria to determine the theoretical visibility of the new moon. Nowadays an institutionalized algorithm is more widely adopted, in which the length of successive months alternates between twenty-nine and thirty days. However, it is still necessary to modify this from time to time.

In many calendars, each day in the month is given a name. The Greeks divided the month into three periods of ten days known as *decades,* while other cultures simply distinguish between the waxing and waning halves. A variety of portents may be associated with the particular days or periods in the cycle. One of the most intricate examples of this is the Hawaiian ritual calendar, in which each month contains a number of taboo periods. Lunar prognostications also live on in folk traditions in various parts of the world.

Problems occur as soon as any attempt is made to keep a lunar calendar in step with the seasonal year. Whether the number of months in a year is taken to be twelve or thirteen, the calendar will drift out of step with the seasons by more than a full month in only two or three years. Some lunar calendars, such as the Islamic calendar, simply proceed independently of the seasonal year. This calendar comprises twelve months and lasts 354 days, so that the start of each year moves back by eleven or twelve days in relation to the seasons. Thus, in 1994 the Islamic New Year occurred on June 11, while in 2004 it occurred on February 22.

Calendars that are based on the cycle of the lunar phases but take account of the seasonal (solar) year are known as *luni-solar.* In these calendars, a month is periodically added or subtracted, a process known as *intercalation.* It is not necessary to do this formally: the Mursi manage it through a process of institutionalized disagreement and retrospective correction. Generally, however, wherever people *do* need to know what month it is at any given time, definite procedures are required that determine the insertion or omission of an intercalary month.

One way is to incorporate other forms of astronomical observation that *are* directly linked to the seasonal year. Among the most common are the appearance and disappearance of certain stars or constellations in the dawn and dusk sky, known technically as their heliacal and acronical rise and set. Early ancient Egyptians, for example, used the heliacal rise of Sirius, which coincided with the beginning of the flood of the Nile, to decide whether to insert an intercalary month before the start of the new year. The Hawaiians and other Polynesians used the heliacal or sometimes the acronical rising of the Pleiades in similar ways. In ancient Greece, the practice of directly linking star appearances and disappearances with seasonal events or actions was documented as far back as the eighth century B.C.E. in the farmers' almanac written by Hesiod, and it may have preceded the development of lunar calendars. Certainly, star calendars, or *parapegmata,* existed by the fifth century B.C.E. These were stone tablets used like pegboards for tracking the 365 days in the year, and many of them were annotated with star configurations for each day. By this time lunar calendars existed too, and *parapegmata* could have provided any number of reference points—appropriate star observations—for keeping a lunar calendar in step with the seasonal year.

A different intercalation method came to light in Greece in the fifth century B.C.E. It emerged from a centuries-old Babylonian tradition that combined meticulous observation and recording with complex arithmetical calculations. The discovery was that adding seven intercalary months in every nineteen years kept a lunar calendar accurately in step with the seasonal year. The error is less than a fifth of a day in each cycle—just over a day in a century. This idea was propagated in Greece by Meton and bears his name—the Metonic cycle. The significance of the Metonic cycle is that it paved the way for empirical practices to be replaced by a set of mechanical rules. This development epitomizes the move toward a logical and mathematical way of understanding nature, and toward an abstract concept of time, which developed in ancient Greece and prevails in the Western world today.

Ironically, it is not clear that Meton's scheme caught on widely, and empirical intercalation schemes may have persisted in many Greek cities. Even more crude and subjective schemes of intercalation persisted in the Roman world, but the schemes became so dogged by political interference that the calendar got more than three months out of step with the seasonal year before it was eventually replaced in the Roman world by the solar-based Julian calendar.

**See also:**
Hesiod (Eighth Century B.C.E.).
Abri Blanchard Bone; Ancient Egyptian Calendars; Hawaiian Calendar; Islamic Astronomy; Julian Calendar; Mursi Calendar.
Heliacal Rise; Lunar Phase Cycle.

**References and further reading**
Aveni, Anthony. *Empires of Time: Calendars, Clocks and Cultures,* 106–115. New York: Basic Books, 1989.
McCready, Stuart, ed. *The Discovery of Time,* 57–87. Naperville, IL: Sourcebooks, 2001.

# Lunar Eclipses

When the moon is full, the side facing the earth is fully illuminated by sunlight, which means that it is also the side of the moon facing the sun. In other words, the earth is between the moon and the sun. More often than not, the earth doesn't pass exactly between them, but when this does happen—about once every sixth lunation—the moon passes through the earth's shadow (umbra). At such times an astronaut on the moon would see the earth passing across in front of the sun—a rather different experience from seeing a solar eclipse from the earth, since the earth's disc would appear about four times the size of the sun's.

The situation is actually more complicated than this example suggests, because the earth's shadow has an inner cone (the *umbra*) and an outer part called the *penumbra*. If our astronaut on the moon stood within the penumbra, he or she would see the equivalent of a partial solar eclipse. But if he or she stood within the umbra, the earth's disc could entirely cover the sun. Around the disc an eerie glow would be visible—orange, reddish, or brown—due to sunlight reflected in the earth's cloudy, dusty atmosphere.

But how does this appear from the earth? If the moon passes through the penumbra but misses the umbra—a so-called penumbral eclipse—then the moon's disc is slightly dimmed and the eclipse passes unnoticed to the naked eye. Such eclipses have no cultural impact and can be ignored. Umbral eclipses may themselves be partial or total, depending whether the entire moon passes into the umbra.

Lunar umbral eclipses, unlike solar ones, can be seen (weather permitting) from approximately half of the earth: from all places where the sun is below the horizon, it is night, and the moon—opposite to it in the sky—is up and visible. During a partial lunar eclipse—as well as in the stages leading up to a total one—part of the moon is in the earth's shadow and is dark. Sometimes a partially eclipsed moon could be mistaken for a normal crescent moon except that it is in the wrong position in the sky and, to the discerning eye its appearance is odd—the crescent may be at an unusual angle and have an odd shape. A total lunar eclipse, however, can be truly spectacular. As soon as no bright part remains to overpower the rest, the whole lunar disc reappears but now glows dimly with a deep red, brown, or orange color. "Totality" can last as long as one and three-quarter hours.

For modern city dwellers it is easy to miss even a total lunar eclipse, and for this reason we are inclined to think that for ancient peoples the social impact of a total *solar* eclipse—a truly rare event transforming day into night—would have been far greater. Yet anyone who relies upon the moon to illuminate a dark night or as the basis of their calendar regulating seasonal, subsistence, and ritual activities may be alarmed to watch the full moon being "eaten" away and then totally consumed. And such events are not rare: given clear skies and someone to wake us up when it happens, we would expect to see roughly one total lunar eclipse every three years on average.

To many peoples in the past, then, an eclipse of the moon was just as ominous, potentially frightening, and even calamitous, as a total solar eclipse. The Aztecs, certainly, were every bit as terrified of a lunar eclipse as of a solar one, as we know from the writings of the chronicler Fray Bernardino de Sahagún: when the moon's face darkened, women feared that their unborn children would be born lipless, noseless, or cross-eyed, or turned into mice. Maya, ancient and historical, variously saw the moon as being attacked, bitten, or eaten, and hence sick or dying, just as they saw the sun during a solar eclipse. The impression of the moon literally being eaten is reinforced by its blood-red color at the time of totality. Indeed, the moon being turned into blood was a common metaphor in medieval European chronicles.

The blood-red color was also taken as a portent of death. The *Anglo-Saxon Chronicles* record that in C.E. 734 the moon was as though drenched with blood; this lunar eclipse was taken to presage the death of the Venerable Bede in the following year. Similarly, it has been supposed that a prophecy that the moon would "turn to blood" at the crucifixion of Christ might refer to a lunar eclipse (although this is highly questionable). Lunar eclipses were certainly taken very widely as a bad omen. An unexpected lunar eclipse in 413 B.C.E., for example, led the Athenian forces to postpone their departure from Syracuse, which they had held under siege for two years during the Peloponnesian war. This delay resulted in total military defeat. Almost a millennium later, in C.E. 549, two Teutonic armies were preparing to wage battle in the Carpathian basin. It was nighttime. In their two camps, close but out of sight of each other, the chieftains and escorts on each side were likely drunk in their tents. Without warning, the soldiers of both armies—quite independently—simply fled from their camps, leaving their superiors with no choice but to conclude an armistice. The reason, it seems, was a total lunar eclipse.

The Maori had a less calamitous view. According to Maori myth, a woman named Rona was snatched up by the moon after insulting it when it disappeared behind a cloud while she was trying to fetch water at night. Rona can still be seen in the face of the moon. During a lunar eclipse, Rona was attacking the moon and trying to destroy it. The result of the conflict, however, was far from calamitous: the moon ultimately returned reinvigorated, young and beautiful.

Once people began to keep systematic records, the possibility emerged of predicting lunar eclipses. Since they can only occur at full moon, the challenge is to predict which full moons are the mostly likely candidates. There are a variety of cycles that, once recognized, provide the means to do this at various levels of reliability. The simplest in the short term involves regular counting in sixes and fives: if a lunar eclipse occurs in month zero, then another is most likely in months six, twelve, eighteen, etc.; might happen in months five, eleven, seventeen, etc.; but cannot happen in any other month. The best-known longer-term cycle is the so-called Saros cycle of eighteen years and eleven days (223 lunations) discovered by the Babylonians, but the Chinese used one of ten years and 334 days (135 lunations), and the Maya recognized one of thirty-two years and 272 days (405 lunations), a time period commensurate with their 260-day sacred calendar round known as the *tzolkin* (405 lunations = 46 tzolkins to within a fraction of a day). None of these cycles is failsafe, however. Some of the predicted eclipses will only be partial, and some may occur but not be visible because they happen during daylight. Those that occur high in the sky in the middle of the night, and last longest, will be the most conspicuous.

For many ancient peoples, the motivation for systematic astronomical observations—initially, at least—was divinatory. Thus ancient Chinese records of lunar eclipses date back to the Shang dynasty in the second half of the second millennium B.C.E., when they were recorded—together with associated prognostications and verifications—on turtle shells and animal bones that have become known as oracle bones. However, as lunar eclipses became more predictable, around the year 0 and beyond, they began to lose their significance as portents and to fall into the domain of calendrics. They simply became part of the framework of regular and reliable events whose cycles could be calculated using careful record-keeping and mathematics. (Solar eclipses, in contrast, remained unpredictable and so continued to be regarded as bad omens.) The Maya, on the other hand, seem to have been obsessed with establishing exact numerical relationships between natural and entirely manmade cycles, and in particular with the 260-day *tzolkin*. Each day of this sacred round had particular characteristics that were crucial in fixing rituals and making prognostications. In complete contrast to the ancient Chinese, the desire to incorporate what were effectively astrological considerations into their calendar drove the Maya to ever greater levels of astronomical achievement.

See also:
Astrology; Lunar and Luni-Solar Calendars; Power; Solar Eclipses.
Babylonian Astronomy and Astrology; Chinese Astronomy; Crucifixion of Christ; Dresden Codex; Fiskerton; Mesoamerican Calendar Round.
Lunar Phase Cycle.

### References and further reading
Aveni, Anthony F. *Stairways to the Stars: Skywatching in Three Great Ancient Cultures,* 33–37. New York: Wiley, 1997.
———. *Skywatchers,* 28–32, 173–184. Austin: University of Texas Press, 2001.
Aveni, Anthony F., ed. *World Archaeoastronomy,* 83–91, 389–415. Cambridge: Cambridge University Press, 1989.
Best, Elsdon. *The Astronomical Knowledge of the Maori,* 19–20. Wellington: Dominion Museum, 1922.
Espenak, Fred. *Lunar Eclipses of Historical Interest.* http://sunearth.gsfc.nasa.gov/eclipse/LEHistory/LEHistory.html.

———. *NASA/Goddard Space Flight Center Lunar Eclipse Page.* http://sun-earth.gsfc.nasa.gov/eclipse/lunar.html.

Ruggles, Clive, ed. *Archaeoastronomy in the 1990s,* 98–106. Loughborough, UK: Group D Publications, 1993.

Savage, Anne, transl. *The Anglo-Saxon Chronicles.* London: Pan Macmillan, 1984.

Schaefer, Bradley E. "Lunar Eclipses That Changed the World." *Sky and Telescope,* 84 (1992), 639-642.

Selin, Helaine, ed. *Astronomy across Cultures,* 174, 450. Dordrect, Neth.: Kluwer, 2000.

Stephenson, Richard, and David Clark. *Applications of Early Astronomical Records,* 10, 30–33. Bristol, UK: Adam Hilger, 1978.

Thurston, Hugh. *Early Astronomy.* Berlin: Springer-Verlag, 1994.

Xu Zhentao, David Pankenier, and Jiang Yaotiao. *East Asian Archaeoastronomy: Historical Records of Astronomical Observations of China, Japan and Korea,* 13–19, 61–106. Amsterdam: Gordon and Breach, 2000.

## Lunar Parallax

We use modern astronomy to determine the appearance of the sky in different parts of the world at different times in the past. The formulae we use are derived from mathematical models of the motions of the earth through space, as it orbits around the sun, spinning and wobbling under the gravitational influences of the sun and moon. One consequence of this is that when we use standard formulae to calculate the declination of a celestial body or event, we implicitly make the assumption that we are standing at the center of the earth. Although this sounds crazy, the distances of most celestial objects are so vast that it makes no practical difference to the archaeoastronomer studying naked-eye observations of the skies in the past. Only in the case of the moon is the difference important.

The lunar parallax is the difference between the position of the moon in the sky as viewed (theoretically) from the center of the earth and (actually) from somewhere on the earth's surface. If we are interested in the possibility of lunar alignments of a precision much greater than the size of the moon (about thirty arc minutes across), then we must take parallax into account. There are two ways of doing this. The usual way, which is adequate in most cases, is to adjust the "target" declination values that correspond to various lunar events, such as the moonrise or moonset at the major standstill limits. However, if we are interested, as was Alexander Thom, in possible alignments of very high precision—precise to a just few minutes of arc—then we must take account of the fact that the parallax correction varies slightly from place to place and from time to time. We can only do this by increasing by an appropriate amount, different in each case, the altitude of any given horizon point, and hence of the moon if it were rising or setting behind that point, in order to calculate where it would be seen in the sky, and hence what its declination would be if we were standing at the earth's center. The corrected declination, which can now be compared directly with the target declination, is known as the *geocentric lunar declination*.

**See also:**
Thom, Alexander (1894–1985).
Altitude; Declination; Moon, Motions of.

**References and further reading**
Ruggles, Clive. *Astronomy in Prehistoric Britain and Ireland,* 23, 60–61. New Haven: Yale University Press, 1999.

# Lunar Phase Cycle

The changing phases of the moon form the most obvious regular cycle of events in the night sky. This phase cycle of the moon, known by astronomers as the *synodic month* and also as a *lunation,* is between twenty-nine and thirty days long. The details depend upon the viewer's latitude on earth, the time of year, and upon the timing in relation to certain longer-term cycles of the moon, but generally the phase cycle adheres to the following pattern. After a period of no more than a few days when the moon is not seen at all, a thin crescent moon is seen low in the western sky just after sunset, following the sun downward. During the next few nights, the moon grows in phase night by night and also appears higher in the western sky night by night, taking longer until it sets. By the time half phase is reached, also known confusingly as *first quarter,* since this is a quarter of the way through the cycle, the moon appears close to the north-south meridian at dusk and sets approximately halfway through the night. During the next seven days it becomes gibbous. It now appears in the eastern sky at dusk, or may even be visible before dusk, taking most of the night to pass at first across the meridian and then set in the west in the early hours. Finally, the moon becomes full. The full moon rises around dusk and sets around dawn. Then, during the next seven days, while it decreases back to half phase, it rises progressively later each night and is still high in the western sky at dawn. By the time half phase *(third quarter)* is reached, it rises approximately halfway through the night and reaches the meridian around dawn. Then finally, as it diminishes back to a crescent, it rises progressively later and later, getting less and less far up into the eastern sky by dawn, until finally the thin crescent is seen only briefly just before sunrise.

The lunar phase cycle is likely to have been recognized by human communities well back into Palaeolithic times. One obvious practical consequence of the changing lunar phases is that at certain times in the month one can see to travel about at night, when hunting, for example. Seafaring communities could also become aware that the appearance of the moon is related to the tides. The fact that the length of the phase cycle is close to that of the female menstrual cycle has led to controversial suggestions that groups of women in hunter-gatherer communities would tend to synchronize their menstrual periods not only with one another but also with the phases of the moon. This, alongside the other effects, could lead to the regulation of food gathering, fishing, hunting, cooking, feasting, and sexual activity—all in accordance with the moon. And this cycle, in turn, would eventually become "institutionalized," reinforced through ritual practices and taboos related to the lunar phases, and encapsulated in myth. Although this sequence of events is speculative, examples abound of the influence of the lunar cycles on human activity. Indigenous peoples in the historical and modern world often tied ritual practices to the phases of the moon (the Hawaiian calendar is just one example), and many folk practices and beliefs relate to the lunar phase cycle, such as those found in the Baltic states of Lithuania and Latvia.

**See also:**
Hawaiian Calendar.
Meridian; Moon, Motions of.

**References and further reading**
Aveni, Anthony F. *Stairways to the Stars: Skywatching in Three Great Ancient Cultures*, 28–32. New York: Wiley, 1997.
———. *Skywatchers*, 67–71. Austin: University of Texas Press, 2001.
Knight, Chris. *Blood Relations*. New Haven: Yale University Press, 1991.

# Lunar "Standstills"
*See* Moon, Motions of.

## Maes Howe

The gently rolling landscape of the Orkney Islands, north of mainland Scotland, is home to a wealth of prehistoric monuments, few more impressive than the great passage tomb at Maes Howe. Probably built around 3000 B.C.E., it is the largest of its type in Scotland. Its dome-shaped mound is estimated to have been some 38 meters (125 feet) across and about eight meters (26 feet) high, and its huge central chamber, cruciform in shape and over 4.5 meters (15 feet) across in each direction, was probably more than 4 meters (13 feet) high originally. This was reached via a low passage over 15 meters (50 feet) long, which enters the mound from the southwest.

The passage at Maes Howe is aligned so that the light of the setting sun on the shortest day of the year shines along its entire length and onto the back wall of the main chamber. However, this solstitial alignment is far from exact. A similar phenomenon occurs daily from more than a month before the solstice until more than a month after it, and the same would have been true in prehistory. In fact, a most impressive light phenomenon occurs about twenty-one days (twenty-two to twenty-three days when the tomb was built) before and after the solstice. On these days the sun, having set behind the top of Ward Hill, on the adjacent island of Hoy, reappears briefly to the side of the hill a few minutes later, suddenly striking the back of the chamber for a second time.

Archaeologist Euan MacKie has argued that the tomb was not poorly aligned upon the solstice but very precisely aligned upon the sun's setting position twenty-two or twenty-three days before and after it. These dates, argued MacKie, were significant because they represented "epoch dates" in a calendar that divided the year into exactly sixteen equal parts of (on average) 22.8 days each. The existence of such a calendar in Neolithic Britain had been argued by the engineer Alexander Thom on the basis of a variety of supposed alignments at different megalithic monuments. However, it has subsequently been shown that if the data are selected fairly, there is no convincing evidence to support it. This fact, quite apart from a number of other

*Maes Howe passage tomb, Orkney Islands, Scotland. Behind the tomb are the hills of Hoy, which the entrance of the tomb faces. (Adam Woolfitt/Corbis)*

arguments, undermines the calendrical interpretation of Maes Howe.

On the other hand, it remains possible that the imprecise solstitial alignment was deliberate, and comparisons have inevitably been made between Maes Howe and the Irish passage tomb of Newgrange, which is also aligned upon the midwinter solstitial sun, though at sunrise rather than sunset. However, Maes Howe lacks anything like the famous "roof-box" that distinguishes Newgrange, and which allowed the sun to shine into that tomb even after the entrance had been blocked off. The entrance at Maes Howe was blocked off too, admittedly using a stone that was too short, and some have argued that this was to permit the continued entry of sunlight for a similar purpose, but this argument is controversial. And the alignment at Maes Howe is even less precise than that at Newgrange. Yet the fact that the alignment was imprecise does not mean that the tomb was badly constructed or unfit for its purpose. Some, indeed, have argued that the vagaries of the Orkney weather would have made this imprecision a necessity.

Yet to dwell too long on the precision of the alignment, or the actual interplay of sunlight within the tomb at or dates close to the solstice, may be missing the point. We may learn more of the purpose and meaning of the orientation of Maes Howe tomb by examining the archaeology of the living.

The Orkney Islands are rare in that they contain evidence of Neolithic settlements. One of these is the famous village at Skara Brae, with its stone houses huddled together as protection against the winds and weather, some containing still-visible central hearths and dressers made of stone. In the 1980s, excavations commenced at a newly discovered village at Barnhouse, only about one kilometer (one-half mile) from Maes Howe. They revealed a number of houses, cruciform in shape like the Maes Howe chamber, including a particularly large one remarkably similar in form to the nearby tomb. This discovery reinforces the oft-made suggestion that the design of tombs—houses for the dead—reflected that of houses for the living. One aspect of design is orientation, and tombs may well have echoed orientation practices common among houses for the living. In this way they did not only reflect what we would consider "practical" considerations; they also reflected principles that were symbolic or cosmological, though of no less importance to their builders. Such principles still operate among a number of modern indigenous peoples, good examples being the Navajo and Pawnee in the United States.

Neolithic houses in Orkney are characterized by a remarkable consistency in their spatial arrangement. A square hearth is always located in the center, a stone dresser is always found against the wall opposite the door, and rectangular stone boxes, generally assumed to be beds, were placed against the side walls. This uniformity provides ample evidence that activities in these houses were, indeed, governed and strongly constrained by a consistent set of ordering principles. These principles no doubt derived from how people perceived and understood the world, and served in their minds to keep everyday life in tune with the cosmos.

Studies of the orientations of houses at Barnhouse, together with other Neolithic villages in Orkney, show a marked preference for entrances in the southeast. Given other similarities between houses for the living and houses for the dead, it comes as little surprise that the many chambered tombs in the area were oriented in the same direction, Maes Howe itself being an exception. However, in many cases, the only indication of the orientation of an excavated house is the orientation of the hearth, since all other features have disappeared. We can extrapolate the orientations of the walls, but we cannot tell which wall contained the entrance. When studied systematically, the hearth orientations—and hence the wall-directions—cluster broadly but quite unequivocally around the intercardinal directions; the cardinal directions themselves were avoided. The four wall-directions also cluster around the directions of sunrise and sunset at the solstices and may have been intended to be oriented upon them, but this is uncertain. It is clear, though, that one of the basic ordering principles of the cosmos, which constrained the orientation of houses and, by extrapolation, tombs, was a quartering demarcated by the cardinal directions. This quartering recalls the *quadripartite cosmologies* common among indigenous American peoples such as the Navajo and Pawnee.

Given this context, the solstitial alignment at Maes Howe should surely be interpreted not as an indicator of an illusory prehistoric calendar of supposed remarkable precision, but in a more modest way, as a particularly fine reflection of prevailing views of the cosmos.

**See also:**
Cosmology; "Megalithic" Calendar; Solstitial Directions; Thom, Alexander (1894–1985).
Navajo Cosmology; Navajo Hogan; Newgrange; Pawnee Cosmology; Pawnee Earth Lodge.
Solstices.

**References and further reading**
Darvill, Timothy, and Caroline Malone, eds. *Megaliths from Antiquity,* 339–354. York: Antiquity Publications, 2003.
Fraser, David. *Land and Society in Neolithic Orkney* (2 vols.). Oxford: British Archaeological Reports (British Series, 117), 1983.
Parker Pearson, Michael, and Colin Richards, eds. *Architecture and Order: Approaches to Social Space,* 38–72. London: Routledge, 1994.
Renfrew, Colin, ed. *The Prehistory of Orkney* (repr.), 83–117, 305–316. Edinburgh: Edinburgh University Press, 1990.
Ritchie, Anna. *Prehistoric Orkney.* London: Batsford/Historic Scotland, 1995.
Ruggles, Clive. *Astronomy in Prehistoric Britain and Ireland,* 129, 158. New Haven: Yale University Press, 1999.
Sampson, Ross, ed. *The Social Archaeology of Houses,* 111–124. Edinburgh: Edinburgh University Press, 1990.

# Magellanic Clouds

The two Magellanic Clouds are the most extensive patches of fuzzy light (nebulae) in the sky, apart from the Milky Way itself, and can be seen from the southern hemisphere and tropics. Where the Milky Way sweeps past the south celestial pole, they are seen on the opposite side of the pole, the Small (or Lesser) Magellanic Cloud (SMC) more or less directly across from the Southern Cross, and the Large (or Greater) Magellanic Cloud (LMC) off to one side. Most of the LMC falls within the southern constellation of Dorado (the "goldfish"—or more strictly, the Hawaiian *mahi-mahi*). When the Southern Cross, which is within the Milky Way, appears high in the sky above the pole, the SMC will be seen close to the horizon below the pole, with the LMC over to its right, looking like detached pieces of the Milky Way itself. As nights progress, the Milky Way and the two Magellenic Clouds, together with all the individual stars and other objects in the sky, swing round the pole, just as the Great and Little Bear (Ursa Major and

Ursa Minor) and other stars do in the north, but in the opposite direction (i.e., clockwise).

The Magellanic Clouds are actually great clumps of stars, or *dwarf galaxies*, satellite appendages to our own large galaxy (the Milky Way). Each contains hundreds of millions of stars, but this is only a small fraction of the number of stars contained in the Milky Way itself.

Not surprisingly given their visual prominence in southern skies, the Magellanic Clouds frequently feature in southern-hemisphere cosmologies and myths. A number of different stories have been recorded among Australian Aboriginals. To one group in Arnhem Land, toward the northern tip of Northern Territory, Australia, the Magellanic Clouds represent the homes of two sisters. When only the SMC is visible, it is said that the elder of the two, who dwells in the LMC, journeys away. Her younger sister eventually persuades her to return in the wet season, when they can go out together to collect yams. They were also widely recognized in ancient Polynesia: to the ancient Hawaiians, for example, they were "white butterfly" and "dark butterfly." One story among the Khoisan bush-peoples of southwestern Africa starts with the culture hero ≠Gao (the ≠ symbol represents a palatal click, or the sound made when the tongue is pressed against the roof of the mouth and then quickly removed) standing on the LMC, looking around for game to hunt. He eventually spots three zebras (Orion's belt) but his arrow (Orion's sword), aimed at the middle one, falls short.

One could continue to recount such stories, but these three examples alone are sufficient to demonstrate the diverse ways in which different peoples can perceive the same objects in the skies, each colored by their own earthly environment and subsistence practices.

**See also:**
Aboriginal Astronomy; Polynesian and Micronesian Astronomy.
Celestial Sphere.

**References and further reading**
Pukui, Mary K., and Samuel H. Elbert. *Hawaiian Dictionary* (rev. ed.), 353. Honolulu: University of Hawai'i Press, 1986.
Selin, Helaine, ed. *Astronomy across Cultures*, 79–83, 187–188, 461. Dordrect, Neth.: Kluwer, 2000.

# Mangareva

Mangareva is the largest of the Gambier Islands at the southeastern corner of French Polynesia. South of this group lies nothing but clear ocean all the way to Antarctica. To the east lie just four small and scattered islands in the Pitcairn group, and then well over 2,000 kilometers (1,400 miles) beyond those, the remote Rapa Nui (Easter Island), the most isolated outpost of ancient Polynesia. In 1834, a Belgian priest by the name of Honoré Laval began working as a missionary on Mangareva. He remained on the island for almost four decades and compiled a detailed account of the indigenous peoples. Ironically, their traditional religious practices were simultaneously being relentlessly destroyed in the missionary zeal to convert them to Catholicism.

Laval describes the use of two tall, dressed stones set up side by side on a small mountain, for determining the solstice. When the sun rose between them, a "learned person" called Akaputu, seated on a flat stone in the middle of the village, announced that the sun had reached its resting place. The anthropologist and Maori scholar Sir Peter Buck (Te Rangi Hiroa), who visited the island a century later, coincidentally published his *Ethnology of Mangareva* in the same year Laval's account, which had languished in missionary archives in Belgium, was finally brought to publication. Buck documents four separate places where observations were made of the northern and southern limits of the course of the sun along the horizon. The place of observation, in one case at least, was a flat rock or stone on which the priest sat. From vantages close to the east coast, the changing rising position of the sun could be tracked against various islets visible on the reef to the east. Elsewhere, people took advantage of other foresights, such as natural landmarks on mountain ridges, and sometimes used the setting rather than the rising sun. In another case, the focus of attention was the shadow of a mountain at sunset: when this reached a certain stone, the winter (June) solstice had arrived. The practice of placing a pair of stones on the horizon to mark the limiting rising or setting position of the sun with maximum precision seems to have represented the culmination of this widespread tradition.

Archaeological and archaeoastronomical work by the American anthropologist Patrick Kirch has recently established, beyond any reasonable doubt, the location of one platform used to make solstitial observations as recorded by Laval. The platform at Atituiti Ruga is about twenty-three meters (seventy-five feet) square and contains a large, flat boulder at its center. From here, sunrise on summer (December) solstice was observed over Agakauitai, an island in the outer barrier reef that is visible on the sea horizon to the east. The shadow of the eastern peak of Auorotini (Mount Duff), which towers over the site to the north, passed across the platform shortly after local noon on the winter (June) solstice.

Why was it important to keep track of the sun? One possibility is that the solstices were used to mark the start of the new season or the new year. A side benefit of this would be to keep the calendar in phase with the seasons. The Mangarevan calendar had much in common with the calendars of most of the rest of Polynesia: it was a lunar calendar based on the phase cycle of the moon, days within the month were reckoned by nights of the moon, and it used month and day names of which variants appear more widely. As is the case with all lunar calendars, it would have been necessary to insert an intercalary month from time to time in order to keep the months in step with the seasons. Making observations of the solstices would be one way to do this.

However, even though the calendar is of a type familiar through Polynesia, clear evidence of careful observations of the solstices is nonexistent elsewhere, except for hints in Hawai'i. This raises the question: did a particular concern with the annual motions of the sun happen to develop in Mangareva but not in other places, or was it a wider Polynesian preoccupation that just happens to have survived long enough in this island to be reliably recorded?

**See also:**
Lunar and Luni-Solar Calendars; Solstitial Directions.
Easter Island; Hawaiian Calendar; Polynesian and Micronesian Astronomy.
Lunar Phase Cycle; Solstices.

**References and further reading**
Buck, Peter H. [Te Rangi Hiroa]. *Ethnology of Mangareva,* 414–415. Honolulu: Bishop Museum Press, 1938.
Kirch, Patrick V. "Solstice Observation in Mangareva, French Polynesia." *Archaeoastronomy: The Journal of Astronomy in Culture* 18 (2004), 1–19.
Laval, P. Honoré. *Mangareva: L'Histoire Ancienne d'un Peuple Polynésien,* 213–214. Braine-le-Comte, Belgium: Maison des Pères des Sacrés-Coeurs, 1938. [In French.]
Selin, Helaine, ed. *Astronomy across Cultures,* 142–143. Dordrect, Neth.: Kluwer, 2000.

# Maya Long Count

The Maya were the most enduring of the Mesoamerican civilizations. The origins of Maya society can be traced back to at least 400 B.C.E. in the lowlands of the Yucatan peninsula, when existing small villages—which, in the dry north, were built close to *cenotes,* sink holes full of vital water formed by the collapse of underground caves—started to develop into more sprawling towns with monumental structures at their centers. This process accelerated, with social elites beginning to gain power, during what is known as the Late Preclassic, or Late Formative, period, which lasted until about C.E. 250. In the Classic period that followed, the political structure of the Maya lowlands developed into a patchwork of city states—anywhere from about twenty-five to about sixty of them at different times—creating a continually shifting network of wars and alliances. Each political center was also a religious one, with ceremonial buildings and pyramids surrounding a large central plaza. But the ninth century brought change. Many of the largest Maya cities were suddenly abandoned for no clearly apparent reason. One of the exceptions was Chichen Itza, which endured and even reached its apex in the ensuing Postclassic period, from about C.E. 900–1300. Maya descendents, speaking dialects of the Maya language, continue to live in southern Mexico, Guatemala, and adjacent countries.

That the Maya peoples were prepossessed with astronomy is beyond dispute, although the motivation seems to have been primarily astrological. The Maya shared the general Mesoamerican concern with calendrical cycles, taking to tremendous lengths the quest to find coincidence between different numbers of cycles of different celestial bodies. However, one of their most incredible achievements was the adoption, development, and extensive use of the linear calendar known as the Long Count. Underlying this was a set of time units—*baktuns* (periods of four hundred years), *katuns* (twenty years), *tuns* (years of 360 days), *uinals* (twenty days) and *kins* (days)—and each progressive day was simply expressed as a number of each unit. For example, 9.17.19.13.16 meant nine *baktuns*, seventeen *katuns*, nineteen *tuns*, thirteen *uinals* and sixteen days.

One crucial prerequisite, without which Long Count dates could never have been recorded and, arguably, the Long Count is unlikely even to have been conceived, is a system for recording numbers using a fixed base. Modern (Arabic) numerals use a decimal (base ten) system, so that 8,945 signifies 5 units + 4 tens + 9 hundreds (10 x 10) + 8 thousands (10 x 10 x 10). The Maya did the same, except they used different symbols to express the digits and used a base of twenty. Thus, when they wrote (in their own vigesimal, rather than decimal, notation) 1.2.7.5 this signified 5 units + 7 twenties + 2 four-hundreds (20 x 20) + 1 eight-thousand (20 x 20 x 20)—in other words, the same number. (The use of base twenty makes sense: our base ten derives from counting using the fingers and thumbs, and people have twenty digits on their hands and feet.) To represent the individual numerical digits, the Maya used bars and dots. Each dot counted one and each bar five. Thus the digit seven was represented by a bar and two dots. When reading Maya numbers it is essential to remember that each set of bars and dots represents a single digit. In a base-twenty system, we need symbols for all the digits from one to nineteen, and so the largest, nineteen, was represented by three bars and four dots. Finally, in any fixed-base notation we need a symbol for zero. The Maya used a stylized nut, shell, or other variants. More esoteric ways of recording numbers also existed, such as representing each digit from 0 to 19 as a deity's head shown in profile with distinctive attributes.

Before any society can develop a fixed-base recording system for numbers, they must develop two concepts: first, the idea that the position of a digit can be used to signify value—that is, depending on its place, a digit can represent units, twenties, four-hundreds, etc.; second, the concept of zero in order to concoct a symbol to represent it. Once these bridges are crossed, two major advantages soon emerge. The first is the ease with which one can do simple arithmetic. It is a trivial matter to multiply a Maya numeral by twenty by simply adding a zero on the end, and other sums are easily solved using straightforward rules, just as with modern numbers; but they would be frightfully difficult using, say, Roman numerals. The second advantage is the ability to represent arbitrarily large numbers (since it is always possible to add further digits). It is this ability that aided the conceptualization of the long periods of time that underlie the Long Count.

The relationship between Long Count dates and base-twenty numbers fell out naturally, since each time unit was twenty times larger than the next smaller one, with the single exception of the *tun*, which is eighteen rather than twenty *uinals* in order to achieve a "year" not too different in length from the real one. In fact, all written records that have come down to us are calendrical, and in these the Maya used a modified vigesimal system to express all numbers, in which the third-to-last digit was 18, rather than 20, times the second-to-last one.

The significance for us of Long Count dates is that they permit us to date sites and artifacts. The significance for the ancient Maya was that they could achieve a grand conception of history. On the first point, most Maya calendrical inscriptions give the Long Count and Calendar Round (*haab* plus *tzolkin*) date. One of the main challenges is to find events, dated according to the Long Count, that can also be dated within our own (Gregorian) calendar to permit a correlation between the two. Astronomical events such as eclipses are one possibility; recorded events soon after the conquest are another. Once the correlation is determined any Long Count date can be converted to a Gregorian one. The issue of finding the definitive correlation has occupied Maya scholars for many decades.

On the second point, since the Long Count was a linear calendar, the Maya could conjure up key (mythical) historical events and assign them dates that seemed significant within the Long Count, such as the beginning of a new *baktun*. Even more fundamentally, they had to fix a *zero date* for the whole scheme—an assumed date of creation. The date chosen corresponds in the Gregorian calendar to about August 13, 3114 B.C.E. (depending upon the correlation). One explanation of why this particular date was chosen is that it was obtained by *retrogressing* the Calendar Round and other observed astronomical cycles in order to reach a date of key (perceived) astronomical significance.

The Long Count could also be extended into the future and provided a source of prognostication. It is commonly believed that the ancient Maya even predicted the end of the world, which would occur at the end of *baktun* 13, around December 21, 2012. However it now seems more likely that they considered the present era would contain 20 *baktuns*, or a complete *pictun* of 8,000 years, giving the world considerably longer.

A more sobering thought is that without the historical evidence of Maya inscriptions and the few books that only survive through fortunate accident, we would know little or nothing of the existence—let alone the distinctiveness or complexities—of Maya astronomy and calendrics. Although a few anomalous round towers, such as the Caracol at Chichen Itza, have been suggested as possible observatories, there is no convincing archaeological evidence of any Maya astronomical measuring instruments.

**See also:**
Calendars.
Caracol at Chichen Itza; Dresden Codex; Gregorian Calendar; Mesoamerican Calendar Round.

**References and further reading**
Aveni, Anthony F. *Stairways to the Stars: Skywatching in Three Great Ancient Cultures*, 125–127. New York: Wiley, 1997.
Aveni, Anthony F. *Skywatchers*, 136–139, 210–214. Austin: University of Texas Press, 2001.
Coe, Michael D. *The Maya* (4th ed.), 172–176. London: Thames and Hudson, 1987.
Tedlock, Barbara. "Maya Astronomy: What We Know and Why We Know It." *Archaeoastronomy: The Journal of Astronomy in Culture* 14 (1) (1999), 39–58.

## Medicine Wheels

The structures known as medicine wheels are distinctive configurations of boulders found across the Great Plains region of Canada and the United States. Over 130 examples are known; they are typically found in remote spots at high elevations. Among the most famous are Majorville Cairn in Alberta, Moose Mountain in Saskatchewan, and Big Horn in Wyoming. The archetypal medicine wheel is a few tens of meters (around a hundred feet) across and exhibits a characteristic "spoked wheel" structure, with an outer ring and a central cairn. In practice, however, no two are the same and the form varies widely. The supernatural and shamanic associations that have given rise to the term "medicine" are modern. Yet at sites such as Majorville, the large central cairn—striking both because of its size and because of the inclusion of a variety of colorful local stones topped with recent offerings—has evidently been cumulated over a very long period. Archaeological evidence corroborates this conclusion (the site was excavated in the 1970s) and demonstrates that there has been more or less continuous activity at the site for as long as five thousand years.

Medicine wheels were among the earliest indigenous structures in the Americas to be analyzed astronomically, at around the time of the rise of "megalithic astronomy" in Britain in the 1970s. Big Horn, for example, was found to have a plausible axis of symmetry oriented upon midsummer sunrise and excited much attention. This was followed by the discovery of alignments upon midsummer sunset and the rising positions of the stars Alde-

*Majorville medicine wheel, Alberta, Canada. The central cairn is an accumulation of boulders of many different types of brightly colored stone; the outer ring and some of the radial spokes are also visible. (Courtesy of Clive Ruggles)*

baran, Rigel, and Sirius. The idea that these alignments could have been intentional was backed up by the common-sense argument that the heliacal risings of the three stars, and of course the summer solstice sunrise and sunset, all occurred during the summer months, when the site would not have been covered with deep, drifting snow.

However, as astronomical interest in medicine wheels grew, the ease with which solar, stellar, and even lunar alignments could be identified—and replaced by new ones when errors were found in earlier calculations—became disconcerting. A more systematic study in the 1990s by the Canadian archaeoastronomer David Vogt revealed little overall consistency in patterns of orientation and astronomical alignment. The basic problem was that the alignment hunters had paid little attention to the cultural context.

Anthropological studies have suggested a variety of functions that medicine wheels may have served: as monuments to the dead, for example; as focuses for Sun Dance rituals; or as sacred places where those on solitary Vision Quests went to seek the patronage of guardian spirits. Such ideas do not rule out the possibility that solar and stellar alignments were incorporated into at least some medicine wheels as a vital part of their broader function; but it is difficult to see any need for high precision, and the ways in which alignments were achieved are likely to have been as many and various as the different forms of the medicine wheels themselves.

> **See also:**
> "Green" Archaeoastronomy; "Megalithic Astronomy"; Solstitial Alignments. Heliacal Rise.
>
> **References and further reading**
> Aveni, Anthony F. *Skywatchers,* 301–303. Austin: University of Texas Press, 2001.
> Chamberlain, Von Del, John Carlson, and Jane Young, eds. *Songs from the Sky: Indigenous Astronomical and Cosmological Traditions of the World,* 127–139. Bognor Regis, UK: Ocarina Books, and College Park, MD: Center for Archaeoastronomy, 2005.
> Hall, Robert. "Medicine Wheels, Sun Circles, and the Magic of World Center Shrines." *Plains Anthropologist* 30 (1985), 181–193.
> Krupp, Edwin C. *Echoes of the Ancient Skies,* 142–148. Oxford: Oxford University Press, 1983.
> ———. *Skywatchers, Shamans and Kings,* 217–220. New York: Wiley, 1997.
> Ruggles, Clive, and Nicholas Saunders, eds. *Astronomies and Cultures,* 163–201. Niwot: University Press of Colorado, 1993.
> Schaefer, Bradley E. "Case Studies of Three of the Most Famous Claimed Archaeoastronomical Alignments in North America." In Bryan Bates and Todd Bostwick, eds., *Proceedings of the Seventh "Oxford" International Conference on Archaeoastronomy.* In press.

## "Megalithic Astronomy"

This term, which is a chapter title in Alexander Thom's first book, came to be synonymous with Thom's ideas concerning the astronomical significance of the megalithic monuments of Britain. It is generally felt by archaeologists to be unhelpful (along with terms such as "megalithic mensuration," "megalithic geometry," and "megalithic man") because there was no "megalithic culture" as such. The prehistoric communities who built many of the megalithic monuments of Britain and Ireland also built structures in earth, timber, and other materials; it is just that the stone constructions have best withstood the passing of time so that many of them remain, conspicuous, in the modern landscape.

> **See also:**
> Thom, Alexander (1894–1985).
> Megalithic Monuments of Britain and Ireland.
>
> **References and further reading**
> Ruggles, Clive. *Astronomy in Prehistoric Britain and Ireland,* 1–81. New Haven: Yale University Press, 1999.

## "Megalithic" Calendar

Sir Norman Lockyer, writing early in the twentieth century, was one of the first to suggest that there was a widespread calendrical practice in Neolithic and Bronze Age Britain and Ireland that involved dividing the year into eight equal parts. This calendrical practice fed through into Celtic times two millennia later, and thence to medieval pagan and Christianized calendrical festivals. It has nothing to do with lunar months but instead involves counting the days from one of the solstices. The eight division points are the two solstices themselves, the two equinoxes, and the four mid-quarter days.

When Alexander Thom started to survey scores of British megalithic monuments several decades later, and to examine their astronomical potential, he noticed accumulations of alignments upon the position of sunrise or sunset at each of these eight dates. There was even some hint of an interest in the dates halfway in between them. On the basis of this evidence Thom proposed that there existed in Neolithic Britain a "megalithic calendar" dividing the year into sixteen equal parts, each twenty-two or twenty-three days long, demarcated by "epoch dates," which in the modern (Gregorian) calendar would correspond, to within a day or so, to June 21 (solstice), July 14, August 6 (mid-quarter day), August 28, September 20 (equinox), October 13, November 5 (mid-quarter day), November 28, December 21 (solstice), January 12, February 4 (mid-quarter day), February 27, March 22 (equinox), April 14, May 6 (mid-quarter-day), and May 29. While the sixteen-division calendar never gained wide acceptance, there was a great attraction in the idea that the eight-fold division of the year represented a thread of continuity stretching through the Celtic calendar and into the even more remote past. Indeed, mid-quarter day alignments have been found among stone circles, such as Beltany in County Donegal, Ireland; among Neolithic passage tombs dating back to the late fourth millennium B.C.E., such as Dowth (one of the large tombs in the Boyne valley) and Bryn Celli Ddu on the Isle of Anglesey (Ynys Môn) in Wales; and even in one Earlier Neolithic site (a U-shaped setting of timber posts) at Godmanchester in Cambridgeshire, England.

Yet this argument may be circular. There is a distinct danger of our picking out "calendrical" alignments because they seem interesting—because they form part of our list of astronomical targets deemed worthy of attention—while ignoring many other alignments that do not point at calendrical targets. Lockyer was certainly aware of existing ideas about the Celtic calendar, as was Thom. Reassessments of Thom's data, paying stricter attention to the fair selection of data, have failed to duplicate his results regarding the calendar, as have systematic studies of local groups of monuments. Added to this, the whole notion of a precise "Celtic" calendar, confidently assumed by Lockyer and Thom and their contemporaries, is now known to be highly problematic and questionable in itself.

Finally, one of the other criticisms of Thom's megalithic calendar was that he did not find calendrical sites as such (although the site at Brainport

Bay in Argyll, Scotland, was subsequently claimed as one), but rather a scatter of individual monuments with isolated alignments, each of which indicated sunrise or sunset on just one of the various calendrical epoch dates.

See also:
Celtic Calendar; Equinoxes; Lockyer, Sir Norman (1836–1920); Methodology; Mid-Quarter Days; Thom, Alexander (1894–1985).
Beltany; Boyne Valley Tombs; Brainport Bay; Bush Barrow Gold Lozenge; Gregorian Calendar; Maes Howe.
Solstices; Sun, Motions of.

References and further reading
Ruggles, Clive. *Astronomy in Prehistoric Britain and Ireland*, 49–55, 128–129. New Haven: Yale University Press, 1999.
Ruggles, Clive, ed. *Records in Stone: Papers in Memory of Alexander Thom*, 197–199, 225–229. Cambridge: Cambridge University Press, 2002.

## Megalithic Monuments of Britain and Ireland

All over Britain and Ireland, and particularly in the more remote northern and western areas of Britain and the western side of Ireland, the traveler is quite likely to encounter one of the hundreds—indeed, several thousands—of megalithic monuments (monuments built of large stones) that form a conspicuous legacy of the prehistoric past of the British Isles. Erected in the Neolithic period or the earlier part of the Bronze Age, between the later part of the fourth millennium B.C.E. and the later part of the second millennium B.C.E., this category of monuments includes huge passage tombs such as Newgrange in Ireland and Maes Howe in Scotland, and spectacular settings of standing stones such as Callanish in Scotland and Avebury and Stonehenge in southern England. It also includes a great many more modest tombs and temples, often found in remote fields and farmlands, as well as stone circles, short stone rows, and many single standing stones.

These alluring and mysterious monuments have attracted astronomical attention ever since the earliest antiquarians roamed the British Isles as far back as the seventeenth century. The first person to bring any semblance of a scientific method to this type of investigation was Sir Norman Lockyer at the beginning of the twentieth century. However, the most famous name in this regard is Alexander Thom who, from the 1950s to the 1970s, espoused the idea that they functioned as astronomical observatories and incorporated high-precision alignments upon the rising and setting positions of the sun, moon, and stars. These conclusions concerning "megalithic astronomy" caused controversy for many years, until a consensus among archaeologists and astronomers finally began to emerge toward the end of the 1980s.

From the archaeological point of view, it is important to realize that this category of monument does not reflect a particular cultural development

*The Stones of Stenness, Orkney Islands, Scotland. (Courtesy of Clive Ruggles)*

but rather a practice of construction using large stones that emerged among the early farmers of the Neolithic period and spread, developing all the time in various ways, until it finally faded out toward the end of the Bronze Age. It manifests itself in various forms of monument, doubtless with many different purposes. Some, like the great sarsen monument at Stonehenge, were unique, while others evidently represented regional traditions, since monuments of very similar form were found in considerable numbers in particular local areas. Examples of the latter are the wedge tombs of western Ireland and the recumbent stone circles of northeastern Scotland. Many of these regional practices appear to have emerged, and then faded away, within a fairly short period compared with the two millennia or so that the whole megalithic tradition persisted in Britain and Ireland. The whole tradition itself forms part of a wider megalithic tradition that is particularly prominent on the Atlantic fringes of western and northern Europe, although megalithic monuments are also found in other parts of Europe and indeed all over the world at such places as Carahunge in Armenia, Nabta Playa in Egypt, and Rujm el-Hiri in Israel. This wide distribution in itself helps to emphasize that we should not see the practice of building things with big stones as a key indicator of significant ethnic or cultural identities in the past.

All this helps to explain archaeologists' skepticism in the 1970s toward archaeoastronomical theories such as Thom's that implied the existence in Neolithic Britain of a "megalithic science" that was well developed, stable, and consistently practiced (by "megalithic man") from one end of the British Isles to the other. More recent evidence reveals much more complex and regionally specific developments in astronomy throughout the two millennia when megalithic monuments of various forms were being constructed. Important examples in this regard are the axial stone circles of southwest Ireland, the wedge tombs of western Ireland, and the Clava cairns and recumbent stone circles of northeast Scotland.

**See also:**
Lockyer, Sir Norman (1836–1920); "Megalithic Astronomy"; Thom, Alexander (1894–1985).
Avebury; Axial Stone Circles; Callanish; Carahunge; Clava Cairns; Maes Howe; Nabta Playa; Newgrange; Recumbent Stone Circles; Rujm el-Hiri; Short Stone Rows; Stone Circles; Stonehenge; Wedge Tombs.

**References and further reading**
Hunter, John, and Ian Ralston, eds. *The Archaeology of Britain*, 58–94. London: Routledge, 1999.
Ruggles, Clive. *Astronomy in Prehistoric Britain and Ireland*, 14–16. New Haven: Yale University Press, 1999.
Waddell, John. *The Prehistoric Archaeology of Ireland*, 57–178. Galway: Galway University Press, 1998.

## Megalithic "Observatories"

In the mid-twentieth century, the engineer Alexander Thom proposed that many British megalithic monuments functioned in prehistory as devices for observing the sun, moon, or stars. Actually, the horizon was the real observing instrument. Given a sufficiently distant horizon, and provided that the observing position is closely enough defined, then any clear, unmistakable horizon feature such as a jagged hilltop or a sharp notch between two steep slopes can be used to pinpoint a particular direction. From the right observing position, the notch or other feature can be used as a foresight marking the rising or setting point of a celestial body to remarkable precision, perhaps to as little as a few arc minutes. For instance, the difference in the rising or setting position of the sun on the solstice and a couple of days earlier or later is normally imperceptible. However, by choosing a suitable observing position and horizon notch, an oberserver could distinguish the solstice, Thom observed, on the basis of whether or not a brief flash of sunlight from the very edge of the sun's disc appeared in the notch. Many megalithic monuments, he claimed, marked both the place to stand and the direction in which to look. They could, in this sense, appropriately be labeled "observatories." Classic examples include the stone row at Ballochroy and the standing stone at Kintraw, both near the coast of Argyll, western Scotland.

Thom's ideas of high-precision alignments have not withstood the test of time. During the 1970s and 1980s many problems came to light, including day-to-day variations in atmospheric refraction, and extinction in the case of stars; and many difficulties were found with Thom's selection of data (see, for example, Brodgar, Ring of). Many archaeologists were deeply suspicious from the outset of the idea of "observatories" in prehistory. They did not necessarily dismiss out of hand the possibility of deliberate astronomical alignments; what disturbed them was the idea that monumental alignments were used primarily as observing devices. Surely, if the aim was merely to mark a rising or setting position, one could simply use a temporary marker such as a wooden stake, as is done by sun-watchers among the Mursi in Ethiopia. There would be no need to go to all the bother of erecting big stones. Given that big stones *were* used, it would seem more likely that they were encapsulating an alignment that was already well known, perhaps using it for symbolic or ceremonial purposes. This would be entirely consistent with the human tendency, identified by anthropologists, to encapsulate relationships of cosmological significance in monumental architecture. This sort of explanation also fits well with the lower-precision alignments that are more commonly indicated by research in archaeoastronomy today.

### See also:
Cosmology; Palaeoscience; Science or Symbolism?
Ballochroy; Brodgar, Ring of; Kintraw; Megalithic Monuments of Britain and Ireland; Mursi Calendar.
Extinction; Refraction.

### References and further reading
Ruggles, Clive. *Astronomy in Prehistoric Britain and Ireland*, 1–81. New Haven: Yale University Press, 1999.
Thom, Alexander. *Megalithic Lunar Observatories*. Oxford: Oxford University Press, 1971.

# "Megalithic Yard"
*See* Stone Circles.

# Meridian

The meridian is a line running directly overhead across the sky, joining the north point on the horizon to the south point. Technically, it is defined as the complete circle around the celestial sphere that also passes directly underneath, but this definition can be ignored for current purposes. Any observer on earth, unless they are standing at one of the two poles, has a meridian. Every celestial body crosses that meridian once each day, halfway in time between rising in the east and setting in the west, assuming that the eastern and western horizons are level or at a similar altitude; circumpolar stars, which never rise or set, cross it twice. The celestial north pole lies on the meridian for observers in the northern hemisphere, as does the celestial south pole for observers in the southern hemisphere. Outside the tropics, every celestial object reaches its highest altitude, or point of *culmination*, on the meridian. In particular, the sun crosses the meridian at local noon: at this time, the shadow of a vertical pole is shortest and points due north in the northern hemisphere or south in the southern hemisphere.

The cultural significance of the meridian is that it is an axis of symmetry around which all the celestial bodies swing evenly in the course of their diurnal motion. Anything aligned along this axis on the ground (i.e., due north–south) will appear naturally "in tune" with the symmetry of the heavens. Perhaps this was one reason the Great North Road leading into and out of Chaco Canyon was so closely aligned along the meridian. Meridianal alignments are also oriented, consequentially, upon the celestial pole, which is often seen as the hub of the heavens, the point that keeps the cosmos in place, or the point of connection with the upper world. Aligning earthly structures along the meridian can serve to confirm and reinforce the power of great leaders, whether living, as in the imperial Forbidden City in Beijing, or dead, as with those entombed in the Pyramids of Giza.

How, though, was the meridian determined in practice, especially when reasonable accuracy was desired? After dispensing with modern instruments, the simplest way in the northern hemisphere at the present time might be to orient directly upon Polaris, which would give true north to within a degree; but owing to precession no bright star marked the pole at most times in the past. One possibility is to use two chosen circumpolar stars to determine the north direction, using a plumb line to determine when one passed directly above the other. It has been suggested that, with the right choice of stars, this method would explain the extraordinary accuracy with which the Khufu Pyramid at Giza was cardinally aligned. The continued use of the same stars over subsequent decades would then have resulted in lower accuracy for subsequent pharaohs.

But there is a problem here. *We* can easily work out which two stars would give the best result because our modern instruments and computer programs tell us very accurately where true north actually is; people in the past did not, by definition, have such a yardstick. How could they have known which two stars were best? In fact, this was not an insoluble problem for them: one possibility is that they marked the direction defined by a vertically aligned pair of stars, then observed the same pair of stars at a different time of year and night when they were upside down (this would happen twelve hours after the first observation but would not generally be observable at that time because the sky would be light). The difference between the two directions would give an indication of the accuracy (in fact, the average of the two directions would be true north). Of course, this *is* only a suggestion, and the actual method used remains speculative; the point is that we have step out of our own shoes in order to recognize that there is a question to be answered.

See also:
Cardinal Directions.
Chaco Meridian; Chinese Astronomy; Pyramids of Giza.
Altitude; Celestial Sphere; Circumpolar Stars; Diurnal Motion; Precession.

**References and further reading**
Belmonte, Juan Antonio. "On the Orientation of Old Kingdom Egyptian Pyramids." *Archaeoastronomy* 26 (supplement to the *Journal for the History for Astronomy* 32) (2001), S1–S20.
Krupp, Edwin C. *Skywatchers, Shamans and Kings,* 271–274. New York: Wiley, 1997.
Lekson, Stephen H. *The Chaco Meridian: Centers of Political Power in the Ancient Southwest.* Walnut Creek, CA: AltaMira Press, 1999.
Spence, Kate. "Ancient Egyptian Chronology and the Orientation of Pyramids." *Nature* 408 (2000), 320–324.

# Mesoamerican Calendar Round

Ancient Mesoamerica—the region stretching from central Mexico to El Salvador and northern Honduras, including Guatemala and Belize—had a turbulent history. One city-state after another rose to power and then collapsed, often relatively suddenly and for no obvious cause. From the Olmec, who inhabited the hot and humid lowlands of the Mexican Gulf Coast as early as 1200–400 B.C.E., to the Aztec, whose powerful and in many ways cruel empire centered on the Valley of Mexico in the central highlands was at its height at the time of Spanish contact, the archaeological record of the region bears witness to numerous and varied polities that rose and fell.

Despite this backdrop of social turmoil, common aspects of Mesoamerican ceremonial centers (such as ball courts, the setting for a ritualistic ball game) bear witness to common practices and beliefs—in other words, a common worldview. An essential and apparently unassailable aspect of this worldview was the calendar. Its basic structure was totally distinctive—it is unlike any other calendar system the world has seen—yet extraordinarily consistent, and persistent, through ancient Mesoamerica. It consisted of two interconnected cycles. The first, relatively unexceptional in itself, comprised an endlessly repeating cycle of 365 days divided into eighteen months of twenty days plus five additional days.

It is the other cycle that is extraordinary and gives the Mesoamerican calendar its idiosyncratic character. Also endlessly repeating, it contained

260 days, organized around thirteen numbers and twenty day names, but with each running continuously, like two interlocking cogwheels. Thus, for the Maya (for whom the twenty day names were Imix, Ik, Akbal, Kan, Chicchan, Cimi, Manik, Lamat, Muluc, Oc, Chuen, Eb, Ben, Ix, Men, Cib, Caban, Etznab, Cauac, and Ahau), the day 1 Imix was followed by 2 Ik, 3 Akbal, etc., round to 13 Ben, but then followed 1 Ix, 2 Men, etc., round to 7 Ahau, then 8 Imix, 9 Ik, etc. Every possible combination of number and day name would be covered, but in what seems to us a very improbable order, before the commencement of the next 260-day cycle. The names of the two cycles, as well as the month and day names within them, varied from one Mesoamerican culture to another, but the cycles themselves were quite invariable. And they were themselves intermeshed, resulting in a *Calendar Round* that only returned to its starting point after slightly less than fifty-two years, in other words, exactly fifty-two rounds of the longer (365-day) cycle and seventy-three rounds of the shorter (260-day) one.

In central Mexico the Calendar Round was known as the Xuihmolpilli, or "binding of the years." The end of the cycle, when the years were bound, was a time of great fear. All fires were extinguished, houses were swept very clean and all rubbish removed. At an elaborate ceremony, commencing in the middle of the night (when the Pleiades crossed the meridian, according to one source), new fire was kindled in the heart of a sacrificed warrior, thus ensuring that the sun would reappear, at which point runners with torches would spread the new fire around the land. The great Templo Mayor in the center of the Aztec capital of Tenochtitlan was rebuilt six times, each rejuvenation involving the addition of an entirely new layer over the existing temple, thereby increasing its size substantially. It has been suggested that these episodes of rebuilding might well have been undertaken in preparation for the start of a new Calendar Round, although there is no direct evidence to support the idea. It is more generally accepted that they were ordered by successive new rulers.

The 260-day cycle was called the *tzolkin* in the Maya region and the *tonalpohualli* in central Mexico. Nobody knows its origins. While the 365-day cycle (*haab* for the Maya, *xihuitl* or *xiuhpohualli* for the Aztecs) was clearly related to the solar year, the *tzolkin* has no obvious astronomical derivation, although several different theories have been proposed. One suggestion is that it arose from observations of the zenith passage of the sun around the latitude of fifteen degrees (that is, at around the latitude of the Classic Maya site of Copan), where the interval between the two dates of zenith passage in a given year was, indeed, 260 days. However, the earliest inscriptions attesting to the 260-day calendar round, dating to the mid-first millennium B.C.E., come from much farther north, in the region of Oaxaca. We do know for certain that the 260-day cycle was of fundamental importance for divinatory purposes. Each day had its omens, good or bad, and these were used to regulate a variety of sacred rituals as well as more general activities. The *tzolkin* is still used for prognostications among Quiché Maya peoples to this day.

Ancient Mesoamericans were tremendously concerned with fitting together cycles of nature—or to be more precise, cycles of the celestial bodies—into neatly repeating longer cycles. All the while, the 260-day cycle, itself apparently unrelated to any cycle directly visible in the skies, endured and formed part of this process. Some larger cycles fell out naturally, owing to (what we would see as) coincidences of nature. The most important of these is the fact that eight synodic cycles of Venus are very close to five solar years. The Mesoamericans knew the synodic period of Venus to be 584 days, and by this reckoning five Venus cycles are exactly equivalent to eight *haab* or *xihuitl* years of 365 days. It follows that, since the number of 365-day years in two Calendar Rounds is divisible by eight, two Calendar Rounds must also be a whole number of Venus cycles: actually sixty-five of them. This coincidence of three cycles has been suggested as a reason that particular significance was attached to the end of every *second* Calendar Round.

Another astronomical coincidence is that three times the mean eclipse danger period, close to 173.3 days, which would be necessary knowledge for anyone hoping to predict eclipses, is equal to two 260-day cycles, so that the main danger of eclipses would recur around (every other occurrence of) the same three dates in the 260-day cycle. Yet another coincidence is the fact that the synodic period of the planet Mars is equal to exactly three rounds of 260 days. In the case of central Mexico, it is not always easy to pursue these ideas, since the threads of evidence are very indirect, but it is quite certain that the Maya took this process of cycle-matching to tremendous lengths: we know this from the Dresden Codex.

A number of religious texts both from before and from shortly after the conquest contain *cosmograms*—representations of the cosmos that can contain depictions of the gods, the four quarters of the world and their qualities, the path of the sun along the horizon, and the calendar, plus perceived interrelationships between them. A famous example in the Codex Féjérvary-Mayer, a pre-conquest document from central Mexico, contains a pattern of dots in the shape of a Maltese cross which represents the entire 260-day sacred round arranged in twenty groups of thirteen. In form, this resembles many of the pecked cross-circle designs found in and around a number of central Mexican sites. It is not surprising, then, that the numbers of small holes forming the basic elements of such designs seem to include, much more frequently than would be expected by chance, calendrically significant numbers such as twenty, eighteen, thirteen, and five.

Another way in which concepts of number, calendar, space, and time may have been tied together in ancient Mesoamerica is manifested in what has become known as *calendrical orientation*. Systematic studies of the orientations of buildings and city plans, both in central Mexico and in the Maya world, have revealed clear but curious patterns of orientation preference. (Many Maya cities, for example, have their street grids oriented a few degrees clockwise of the cardinal directions.) These preferences included sunrise and sunset on dates whose significance would have followed, as for many peoples worldwide, from directly observable extremes in the motions of the sun—the solstices and the days of zenith (and perhaps also anti-zenith) passage. However, there are also notable regional preferences for other days in the year. Some of these seem to indicate the existence of *horizon calendars* in which key dates were those counted off from the solstices in intervals such as twenty, fifty-two, sixty-five, and seventy-three days, numbers that also arise in the context of the Calendar Round.

However, the apparent coexistence of the Calendar Round and horizon calendars raises an issue. Calendars based on the movement of the sun up and down the horizon are implicitly tied to the true solar year, whereas the 365-day cycle of the Calendar Round that became established all over Mesoamerica would have slipped against this by one day in every four years, and the discrepancy would have become evident over time. Since our knowledge of the Calendar Round comes from documentary and historical sources, while the evidence for horizon calendars relies more on the statistics of alignments, it is not easy to resolve this issue. One possibility that has been suggested is that the horizon calendars were older developments, precursors of the Calendar Round. Another is that the formal, ceremonial 365-day calendar and the pragmatic horizon-based one existed in parallel, rather like what is thought (by most scholars) to have occurred with the civil and lunar calendars in ancient Egypt.

In most respects, the Mesoamerican calendar reflected the cycles of nature—particularly observable astronomical cycles—directly. The only apparent exception to this was the 260-cycle *tzolkin* or *tonalpohualli*: if this ever did have such a basis, it became lost in the mists of time. Yet this sacred cycle, with its complex set of associated prognostications, lay at the very heart of Mesoamerican worldview and calendrics.

See also:
Calendars; Zenith Tubes.
Ancient Egyptian Calendars; Aztec Sacred Geography; Cacaxtla; Dresden Codex; Mesoamerican Cross-Circle Designs; Teotihuacan Street Grid; Venus in Mesoamerica; Zenith Tubes.
Inferior Planets, Motions of.

**References and further reading**
Aveni, Anthony F. *Skywatchers*, 139–152. Austin: University of Texas Press, 2001.

Galindo Trejo, Jesús. *Arqueoastronomía en la América Antigua*. Madrid: Equipo Sirius, 1994. [In Spanish.]

Iwaniszewski, Stanisław, Arnold Lebeuf, Andrzej Wierciński and Mariusz Ziółkowski, eds. *Time and Astronomy at the Meeting of Two Worlds*, 15–24, 181–206. Warsaw: Centrum Studiów Latynoamerykańskich, 1994.

Romano, Giuliano, and Gustavo Traversari, eds. *Colloquio Internazionale Archeologia e Astronomia*, 15–22, 123–129. Rome: Giorgio Bretschneider Editore, 1991.

Selin, Helaine, ed. *Astronomy across Cultures*, 227–233. Dordrecht, Neth.: Kluwer, 2000.

Tedlock, Barbara. *Time and the Highland Maya* (rev. ed.). Albuquerque: University of New Mexico Press, 1982.

# Mesoamerican Cross-Circle Designs

Cross-circle designs, often referred to as *pecked cross-circles,* are found at several sites in the highlands of central Mexico and also as far south as the Maya city of Uaxactún, typically pecked into the stuccoed floors of buildings or carved as petroglyphs in rocks in the hills around cities. The greatest preponderance is in and around Teotihuacan, where over fifty are known. They generally consist of one, two, or three concentric circles intersecting four radial lines in the shape of a cross, though there are numerous variants on this design, including squares and Maltese crosses. They are typically around one meter (three feet) across, and made up of cuplike depressions about one centimeter (half an inch) in diameter. They have attracted a good deal of attention from archaeoastronomers, since they may have reflected aspects of astronomy and the calendar in three distinct ways.

The first way is in their numerology. Counts of the number of dots forming elements of cross-circle designs, such as on radial spokes between points where they cross the circles or on circle sectors between points where they cross radial spokes, reveal a preponderance of the numbers five, thirteen, eighteen, and twenty, which are significant as elements of the Mesoamerican calendar. Enough of them have a total count of 260 (the number of days in the sacred 260-day calendrical cycle) to suggest that this might have been significant too. They may have been symbolic representations of the calendar in some sense, an idea that seems less strange when we compare them with *cosmograms* found in some of the central Mexican codices. Yet ethnohistoric accounts also refer to the use of such designs, made up of cavities pecked into the floors of buildings, for a game called *patolli,* which involved moving pebbles around in the holes. Why should a gameboard have reflected the calendar? Perhaps because, in Aztec times at least, it had religious overtones.

The second way the cross-circle designs may have reflected aspects of

One of several cross-circle designs discovered in the 1980s adjacent to the Pyramid of the Sun at Teotihuacan, Mexico. One corner has been re-pecked to avoid a hole in the stucco floor, suggesting that the actual shape was less important than the number of holes. (Courtesy of Clive Ruggles)

astronomy and the calendar is in their orientation. Systematic studies suggest that the spokes of those found as geoglyphs on natural rocks tend (more often than would be expected by chance) to be oriented roughly in the solstitial directions. This would not be surprising if they were indeed cosmological symbols, reflecting the four corners of the perceived cosmos.

Finally, those carved on rock outcrops tend to be in hilly locations with clear views of distant horizons, particularly toward the east, once again suggesting a connection between landscape, calendar, and cosmos. Could some of them have marked observing points? Or, as has been suggested, were they used as surveyor's benchmarks? In the 1970s it was proposed that three of the cross-circles at Teotihuacan—one pecked into the floor of a building by the Street of the Dead and two carved into rocks on adjacent hills—defined two baselines that were used by surveyors in laying out the street grid. This idea was criticized on statistical grounds, because it depended upon the arbitrary selection of three cross-circles from fourteen then known at Teotihuacan (the total is now over fifty), and on archaeological grounds, because the central cross-circle, used in both baselines, was pecked into the floor of a building dating to a late phase, long after the street grid was set out. The statistical argument was criticized in its turn because of the established fact that cross-circles had different purposes. But we cannot use this as an excuse for simply selecting the data that fit a given hypotheses and ignoring the rest. The archaeological evidence seems to clinch the issue unless one appeals to the possibility (unproven) that the central pecked cross may have been cited on the position of an older one.

Be this as it may, it is quite clear that cross-circle designs had several different functions and meanings. One view is that they represent the propagation of a symbol whose form remained essentially unaltered but came to be used for different purposes. An alternative is that many if not all of these purposes were interrelated. It may seem strange to us to tie concepts of space and time together in such contorted ways, but the unfamiliar manner in which patterns of thought operate within other worldviews means we would certainly be wrong to close our minds altogether to such possibilities.

> See also:
> Mesoamerican Calendar Round; Teotihuacan Street Grid.
> 
> **References and further reading**
> Aveni, Anthony F. *Skywatchers,* 226–232, 329–334. Austin: University of Texas Press, 2001.
> Aveni, Anthony F. "Pecked Designs at Teotihuacan." *Journal for the History of Astronomy* 36 (2005), 31–47.
> Broda, Johanna, Stanisław Iwaniszewski and Lucretia Maupomé, eds. *Arqueoastronomía y Etnoastronomía en Mesoamérica,* 269–290. Mexico City: Universidad Nacional Autónoma de México, 1991. [In Spanish.]
> Ruggles, Clive, ed. *Records in Stone: Papers in Memory of Alexander Thom,* 442–472. Cambridge: Cambridge University Press, 2002.
> Ruggles, Clive, and Nicholas Saunders. "The Interpretation of the Pecked Cross Symbols at Teotihuacan." *Archaeoastronomy* 7 (supplement to *Journal for the History of Astronomy* 15) (1984), S101–S107.

# Meteors
*See* Comets, Novae, and Meteors.

# Methodology

Methodology concerns how we go about obtaining data and using those data to assess our theories. Whether we are considering archaeological evidence such as monumental alignments, or other forms of evidence relevant in archaeoastronomy such as ethnohistoric accounts, written documents, or a combination of these, we need agreed-upon procedures that give us confidence that we can reach the best, most reasonable, most reliable, and most sustainable conclusions on the basis of the evidence available. An accepted methodology is vital in any academic discipline, and it will constantly be under review: people will constantly strive to improve it.

One of the most fundamental methodological necessities is to be fair with the evidence. As a statistician involved in the debates about Alexander Thom's theories concerning the British megaliths once decreed: "Observe all there is and report all you observe." In practice it is seldom feasible to observe all there is, so fair ways of selecting (sampling) data become vital. "Fair selection of data" means doing so in a way that will not bias the eventual conclusion. In other words, it is vital that researchers not just consider and present the evidence that favors a favorite theory while ignoring or hiding the rest by accident or by design. This would occur if, in our enthusiasm to find data that fit a favored theory—for example, that ancient stone rows were aligned upon sunrise or sunset on particular days in the year—we kept searching until we found suitable candidates while ignoring many others that did not fit. Unless this principle is followed, no amount of complex statistical analysis will have any value whatsoever.

Nowhere are the dangers more evident than in popular claims that constructions on the ground, such as temples and pyramids, were laid out in the form of celestial constellations. A famous example is the suggestion that various temples at the ancient site of Angkor in Cambodia were laid out in the formation of the bright stars in the constellation Draco. If there are no firm criteria for selecting the buildings to be considered, then, given the large number of stars in the sky to select from, one can obtain surprisingly good fits totally fortuitously. One BBC television documentary, experimenting with this concept, managed to obtain a remarkably good fit between the layout of selected large buildings in New York City and the stars in Leo.

Each of the various different types of evidence about ancient astronomy presents its own methodological challenges. For instance, impressive light-and-shadow effects, occurring only when the sun or moon reach a particular configuration, are well known. Famous examples include the equinox hierophany at the pyramid of Kukulcan at Chichen Itza in Mexico, the sun dagger at Fajada Butte in Chaco Canyon, and the passage tomb at Newgrange in Ireland. Each of these, on the face of it, may have been intentional but might be the result of chance. As the American archaeoastronomer Anthony Aveni once said, "In the American southwest, the search for solar alignments [has been] conducted with a fervent passion not exceeded anywhere in the world. Many a weekend was spent by individuals in hot pursuit of daggers of light entering cracks and openings in caves and buildings—especially light patterns that crept across petroglyphs" (Ruggles 2002, p. 444). We have to devise ways of assessing the likelihood that an effect we observe might actually have been intentional. Where we have only the archaeological evidence to go on, this must inevitably involve some sort of statistical analysis.

In alignment studies there is a similar problem, stemming from the simple fact that every oriented building or structure must point somewhere, and there can be many different reasons for choosing a particular orientation. Often an ancient monument contains not just one obvious axis but many different alignments of buildings, stones, or other structures, and to pick and choose between them on the basis of a favored theory will lead to completely misleading conclusions. There are also many different horizon targets—many potential rising and setting points of celestial bodies that *we* might deem significant if we find alignments upon them. Added to this is a further problem, that the rising and setting positions of stars change over the centuries, owing to precession, and those of the sun, moon, and planets to a lesser extent, owing to the changing obliquity of the ecliptic. If, as is most often the case, we cannot date a structure at all accurately by archaeological means, then if we simply choose the best fit date, the possibilities are almost endless. If we allowed ourselves the fifteen brightest stars, for instance, and a range of five hundred years, we could fit a suitable star and date for one-third of all oriented structures. Clearly we have to be far more careful and critical.

What, then, can we do to reach a reasonable degree of confidence that any alignment, light-and-shadow phenomenon, or similar association that we spot in the archaeological record was actually intentional and meaningful? One possibility is to look for repeated occurrences. This approach has demonstrated the solar and, in some cases, lunar significance of many local groups of monuments in Britain and Europe, a classic example being the Scottish recumbent stone circles. On the other hand, repeated occurrences can only reveal prevailing practices that were widely and consistently adhered to. Another approach is to seek independent evidence that a particular alignment was deliberate. Archaeological excavations can sometimes reveal this, as has been the case at Balnuaran of Clava, near Inverness, Scotland. Here, it seems that the structural stability of a passage tomb was compromised in order to incorporate the correct astronomical alignment.

A fundamental methodological issue arises from the fact that people's behavior is not governed by unbending principles comparable to the physical laws that govern the universe. Even where the strictest doctrines dictated correct practice regarding, say, the construction of a Neolithic tomb, it is likely that some people "did their own thing." Treating archaeological data too much like data from physics, subjecting it to "scientific" testing, will not explain the actions of individuals in particular places and times, according to particular sets of circumstances. The stone circle at Drombeg in Ireland is a single solstitially oriented stone circle among a group of around fifty. Considering the group as a whole, one would be extremely inclined to put the solstitial alignment of Drombeg down to chance. However, it is certainly possible that the builders of this monument did have reasons for the alignment and that it was, in fact, deliberate. The crucial point is that one cannot use this argument to simply *assume* that the Drombeg alignment was deliberate; to do so would be to fall into the trap that we identified at the start, namely to choose the data that fits a favored theory (the solstitial alignment at Drombeg) while ignoring the rest (the alignments of the other forty-nine circles). Independent corroborating evidence, such as that obtained at Balnuaran of Clava, is needed.

Problems of data selection are not confined to archaeological evidence. Ethnographers may have the problem of distinguishing between informants' accounts that genuinely relate to practices that have been endemic for many generations and more recent inventions or intrusions. The situation is also different where we need to reconcile archaeological with historical and other forms of evidence such as iconography and written inscriptions. Thus a Venus alignment at the so-called Governor's Palace at the Maya city of Uxmal is a one-off that would not be given credence if we did not have a secure date for the alignment, extensive evidence for the importance of Venus in contemporary Maya culture, and many Venus glyphs on the building itself.

Another problem is that some alignments occur naturally, purely by chance, but may well have been afforded significance in the past. A good example is the Hawaiian island chain, which is quite closely aligned with June solstice sunset and December solstice sunrise. From an indigenous perspective, such alignments, once noticed, can seem very impressive, since they appear to demonstrate the integrity and harmony of the cosmos and the interconnectedness of land and sky. However, this possibility can present a headache for the modern archaeoastronomer seeking to prove that a particular astronomical meaning was attached to a natural alignment in ancient times, especially in the absence of corroborating evidence from history or ethnography. A great many alignments with plausible astronomical connections may well occur naturally in the landscape, and we have no knowledge as to which ones were actually "observed." Once again, we cannot use our own predilections to prejudge the issue.

In short, it is crucial not to try to fit a favorite theory to some evidence. We must select the data fairly and consider the possible alternative explanations. Unless we do this, we risk misleading not only everyone else but also ourselves. How best, in practice, to assess competing interpretations of the evidence at hand raises a series of questions. There is no simple answer to many of them. We must always keep an open mind and always be open both to new ideas and to new data.

See also:
Alignment Studies; Archaeoastronomy; Astronomical Dating; Constellation Maps on the Ground; Field Survey; Orientation; Statistical Analysis; Thom, Alexander (1894–1985).
Angkor; Bush Barrow Gold Lozenge; Clava Cairns; Drombeg; Easter Island; Fajada Butte Sun Dagger; Governor's Palace at Uxmal; Grand Menhir

Brisé; Is Paras; Kukulcan; Nā Pali Chant; Newgrange; Polynesian Temple Platforms and Enclosures; Recumbent Stone Circles.
Obliquity of the Ecliptic; Precession.

**References and further reading**
Ruggles, Clive. *Astronomy in Prehistoric Britain and Ireland*, 159–162. New Haven: Yale University Press, 1999.
———. "The General and the Specific." *Archaeoastronomy: The Journal of Astronomy in Culture* 15 (2000), 151–177.
Ruggles, Clive, ed., *Records in Stone: Papers in Memory of Alexander Thom*, 442–472. Cambridge: Cambridge University Press, 2002.

# Mid-Quarter Days

The solstices and equinoxes divide the year into four almost equal parts. If one further divides each of these parts into two, the additional dates are what are known variously as the *mid-quarter* or *cross-quarter days,* or sometimes as the *Scottish quarter days,* since in Scotland, as opposed to England, they—and not the solstices and equinoxes—were used as the basis of the legal division of the year into four. The dates are February 4, May 6, August 6, and November 5.

A number of later prehistoric monuments in Britain and Ireland are aligned upon sunrise or sunset close to one of the mid-quarter days, although whether these alignments were intended and hence reflect a widespread calendrical practice is a matter of considerable doubt. Perhaps the earliest example is a U-shaped setting of large timber posts constructed in the Earlier Neolithic (fifth century B.C.E.) at Godmanchester in Cambridgeshire, England, which was aligned upon sunrise in early May and early August. Others include the longer passage at the Neolithic passage tomb of Dowth, County Meath, Ireland (one of three large tombs in the Boyne valley), which is aligned upon sunset in early November and early February.

Four of the eight festivals traditionally associated with the Celtic calendar were also associated with mid-quarter days, namely Imbolc (early February), Beltaine (early May), Lughnasa (early August), and Samhain (early November), and these same dates found their way into pagan and Christianized calendrical festivals, most notably the feast of All Saints on November 1.

See also:
Christianization of "Pagan" Festivals; Celtic Calendar; Equinoxes; "Megalithic" Calendar; Solstices.
Boyne Valley Tombs.

**References and further reading**
Hutton, Ronald. *The Stations of the Sun: A History of the Ritual Year in Britain*. Oxford: Oxford University Press, 1996.
McCluskey, Stephen C. *Astronomies and Cultures in Early Medieval Europe*, 51–76. Cambridge: Cambridge University Press, 1998.

# Midsummer Sunrise

*See* Solstitial Alignments.

# Midsummer Sunset

*See* Solstitial Alignments.

# Midwinter Sunrise

*See* Solstitial Alignments.

# Midwinter Sunset

*See* Solstitial Alignments.

# Minoan Temples and Tombs

The Minoan civilization flourished in the Mediterranean island of Crete during the mid-second millennium B.C.E. and influenced later developments in the Aegean in various ways. It has been suggested, for example, that orientation practices in early and Classical Greece may derive in part from this more ancient culture. The rich Bronze Age heritage of Minoan Crete includes palaces, villas, and several peak sanctuaries—cult centers where archaeologists have uncovered thousands of clay figurines, apparently votive offerings left by pilgrims. Various solar and stellar alignments noted at peak sanctuaries such as Pyrgos and Petsophas hint at an intimate relationship between these temples and observations of the skies, as well as to the possible use of prominent horizon foresights.

Similar relationships may also have extended to tombs. At Armenoi, a huge Minoan cemetery on the western part of the island, are well over two hundred carefully constructed tombs consistently oriented eastwards within an arc that is best explained as relating to moonrise. Prominent on the skyline in this direction is Mount Vrysinas, a mountain that may well have had strong cultic associations with the moon, since a peak sanctuary located on its summit seems to have been dedicated to a moon goddess.

See also:
Pilgrimage.
Temple Alignments in Ancient Greece.

**References and further reading**
Blomberg, Mary, Peter Blomberg, and Göran Henriksson, eds. *Calendars, Symbols and Orientations: Legacies of Astronomy in Culture*, 127–134. Uppsala: Uppsala Astronomical Observatory, 2003.
Hoskin, Michael. *Tombs, Temples and Their Orientations*, 217–222. Bognor Regis, UK: Ocarina Books, 2001.
Jones, Donald W. *Peak Sanctuaries and Sacred Caves in Minoan Crete: A*

Comparison of Artifacts. Jonsered: Paul Åströms Förlag, 1999.

Peatfield, Alan A. D. "The Topography of Minoan Peak Sanctuaries." *Annual of the British School at Athens* 78 (1983), 273–279.

Ruggles, Clive, Frank Prendergast, and Tom Ray, eds. *Astronomy, Cosmology and Landscape*, 72–91. Bognor Regis, UK: Ocarina Books, 2001.

# Misminay

Misminay is a small village in the Peruvian Andes, geographically quite close to well-trodden tourist destinations, since it is situated almost midway between the city of Cusco and the famous ruins at Machu Picchu. It is, however, off the main tourist routes and thus relatively remote and difficult of access. In many ways it is unexceptional; what is unusual is that the astronomical knowledge and practices of the villagers are known in some detail, owing to the particular interest taken by anthropologist and ethnoastronomer Gary Urton in the 1970s.

The villagers recognize various objects in the sky and use several different types of astronomical observation to regulate their seasonal agricultural activities. One is to observe the annual cycles of appearance (*heliacal rise*) and disappearance (*heliacal set*) of different stars and asterisms. For example, there are two "celestial storehouses," the Pleiades and the five stars at the tail or hook of Scorpius. These are almost opposite in the sky, so one or the other is generally visible at any time in the night. In fact, the most significant event in the cycle, the heliacal rise of the Pleiades in June, does not directly signal the time to begin planting but instead heralds observations where prognostications are made about the year's crop yield on the basis of the perceived brightness of the stars. The times for planting and harvesting maize and potatoes are determined by another method: by looking at the changing rising position of the sun on the eastern horizon, and in particular, noting when it passes a given spot going in one or the other direction. The lunar phases are also said by some to be important in determining the best times for planting. In sum, the calendar is not regulated by simple observations of one type, but by a mix of observations of the heliacal risings and settings of stars, the changing rising position of the sun along the horizon, and (to some extent) the phases of the moon.

But there is a good deal more to Misminay astronomy than the agricultural calendar. The villagers' conception of the sky forms part of a worldview that affects many aspects of life, including the very layout of the village. Central to this worldview is the Milky Way, which at the latitude of Misminay (14°S) has a very distinctive cycle of motions about the heavens during each twenty-four-hour period. For several hours it appears to hang in the sky, arcing over from roughly the northeast to the southwest, dividing the sky into two parts. Then, relatively suddenly, it starts to tumble toward the horizon, falling and tipping until it seems to surround the observer, just above the horizon. Equally suddenly, it starts to rise again, climbing until it partitions the heavens in the orthogonal direction, from the northwest to the southeast. The sequence is then repeated in reverse. This pattern of motions only occurs at around this latitude and is difficult to visualize without the aid of a planetarium; but when seen it is extremely impressive. Only half of this diurnal cycle can be seen at any particular time of the year, since the other half occurs during daylight hours. This means that, for example, when the sky first gets dark during the rainy season at Misminay, from November to February, the Milky Way is seen arcing over from the northwest to the southeast; whereas during the dry season, May to August, it divides the sky the other way. For people living at this latitude and watching the skies through the year, these patterns of motion soon become apparent. A natural reaction is to conflate the two axes in one's mind and to perceive them as a cross, dividing the celestial dome into four quarters. This is precisely what the inhabitants of Misminay do, and this perception is reflected on the ground in a very direct way.

The central gathering place in the village is a chapel, standing at a crossroads from which four paths head off in the intercardinal directions. Apart from anything else, this alignment reflects a widespread tendency to organize space around lines radiating out from a central place, a tendency that stretches back to Inca and even pre-Inca times. The paths also mark out social divisions, again reflecting broader practices long established in the region. But the specifics of the layout of Misminay also bear important relationships to the sky. The center of the village is the crossing point of two irrigation canals. Their directions, as well as those of the paths they follow, reflect the two axes of the Milky Way in the sky. Terrestrial water flows downhill, southward to northward, and the Milky Way is seen as a celestial river that carries water back in the opposite direction, southward and up into the sky, from which it falls as rain. By this means, the village is appropriately placed within an integrated system that serves to circulate water through the cosmos as a whole, and by so doing ensures the continued success of the crops and the survival of the community.

In fact, these are just some of the principal elements of a much more complex conceptual scheme that links together, both in time and space, the layout of things on the ground with different objects and events in the sky. The directions of the four paths also correspond quite closely to the rising and setting positions of the sun at the solstices. The June solstice, when the sun rises in the northeast, corresponds to the time of year when the Milky Way appears after sunset oriented in this direction. The same is true with regard to the southeast and the December solstice. Then again, the rising direction of the Pleiades, and the opposite setting direction of the tail of Scorpius, also coincide quite closely with the directions of the paths, the Milky Way, and the solstitial directions. The heliacal rise of the Pleiades coincides with the time when the sun rises in the same direction at the June solstice. In short, the quartering of the earth by the cross-shaped paths and ditches reflects many significant aspects of the quartering of the sky. (The quarters are actually far from equal in size—the solstitial azimuths at Misminay are roughly 63 degrees, 117 degrees, 243 degrees and 297 degrees—but this is a detail that has no significance to the villagers.) The chapel at the center of the village is appropriately named: its name is *Crucero* (cross).

Another interesting aspect of the Misminay villagers' conception of the sky is the "dark cloud constellations." They perceive the shapes of dark patches in the Milky Way to be creatures such as a llama, toad, fox, and snake. The part of the Milky Way that appears in southern skies is especially bright and impressive, and it is scarcely surprising to see these dark shapes, hugely imposing against the foreground of bright stars, singled out as significant in southern hemisphere cosmologies. The celestial llama is one of the largest and most imposing shapes of all, and broadly corresponds to what is seen as an emu across Aboriginal Australia. It also reminds us that the practice of imagining stars as bright points on invisible objects and creatures, which underlies the Western concept of constellations, is not the only way of perceiving entities in the sky.

Finally, according to the villagers at Misminay, there is a correspondence between the gradual rise of the celestial llama in the sky each morning and the breeding period of terrestrial llamas. This is one of many examples from around the world of people perceiving a direct correspondence between linked events on the earth and in the sky, another being the Barasana "Caterpillar Jaguar" constellation.

See also:
Cosmology; Solstitial Directions.
Barasana "Caterpillar Jaguar" Constellation; *Ceque* System; Emu in the Sky; Nasca Lines and Figures.
Azimuth; Heliacal Rise.

**References and further reading**
Urton, Gary. *At the Crossroads of the Earth and Sky: An Andean Cosmology.* Austin: University of Texas Press, 1981.

# Mithraism

The Mithraic religion was one of the most powerful sun cults in human history. It took shape in Persia around the time of the conquests of Alexander the Great in the fourth century B.C.E., spread into Babylonia and Hellenistic Greece, and subsequently into ancient Rome in the second century C.E. All this time it was developing and changing as it came into contact with other religions and new influences. Mithra, a solar deity, became identified in Babylonia with the Babylonians' own ancient sun god, Shamash; in Greece with Helios, the sun, and the god Apollo; and in the Roman empire with Sol Invictus, the unconquered, or unconquerable, sun.

In the Roman world, Mithraism had only modest support at first, as one of a number of "mystery" cults from the east. In the second century C.E., however, it became popular among soldiers and spread rapidly with the legions to the far corners of the empire. Numerous temples, monuments, and inscriptions are found from Asia Minor in the east to Britain in the west. Mithraic temples are characterized by altars containing a scene depicting Mithras ritually sacrificing a bull. From this primordial act, according to the Mithraic belief, sprang life on earth.

By the late second century, Mithraism had attracted the support not only of a range of government officials but even of the emperor himself. Commodus, emperor from C.E. 180 to 192, was initiated into the sect, and a frieze at Ephesus shows the sun taking his father, Marcus Aurelius, up to heaven in a chariot escorted by the moon and stars. Commodus was the first of a succession of Roman emperors for whom popular worship of the sun as the supreme body in the heavens affirmed their own earthly powers. (An interesting comparison can be made here with the role of sun worship in Inca society.) Caracalla, emperor from C.E. 212 to 217, seems to have portrayed himself as the son of the sun god; and the mother of Aurelian (C.E. 270–275) was believed, at least by some, to be a priestess of the sun. The culmination came in C.E. 274, when, under Aurelian, the cult of Sol Invictus became the official state religion of the empire.

Following Constantine's conversion to Christianity early in the fourth century, Christianity began its inexorable rise and pagan alternatives declined. Mithraism certainly presented the most serious pagan challenge to early Christianity, and the bitter rivalries that ensued were sometimes overcome by a process of acculturation—subsuming pagan rituals within Christian ones. This process is most obvious in its lasting effect on the Christian calendar. It is no coincidence, for example, that the date of Christ's nativity was fixed as December 25, the date the Romans believed to be the winter solstice, and which had been established by Aurelian in C.E. 274 as the feast of the birth of the unconquered sun (Natalis Solis Invicti). Neither is it any coincidence that the Christian holy day is named Sunday.

The conflict between early Christianity and astral religions such as Mithraism also hardened Christian attitudes toward astrology.

See also:
Astrology; Christianization of "Pagan" Festivals.
Island of the Sun; Roman Astronomy and Astrology.
Solstices.

References and further reading

Cumont, Franz. *The Mysteries of Mithra*, trans. by Thomas J. McCormack. New York: Dover, 1956.

Hawkes, Jacquetta. *Man and the Sun*, 181–204. London: Cresset Press, 1962.

McCluskey, Stephen C. *Astronomies and Cultures in Early Medieval Europe*, 41–43. Cambridge: Cambridge University Press, 1998.

Ulansey, David. *The Origins of the Mithraic Mysteries: Cosmology and Salvation in the Ancient World*. New York and Oxford: Oxford University Press, 1989.

# Monuments and Cosmology

The architecture of public monuments, great and small, can give the archaeologist important clues about the worldviews of those who built them. One reason is that their designs may reflect the perceived cosmos, revealing associations that existed in the minds of the builders, just as the solar alignment at the passage tomb of Newgrange in Ireland reveals a perceived connection between the sun and the ancestors. Monumental architecture may also reflect a ritually defined order of things that was relatively widespread and relatively stable, even where the nature of the society at the time was constantly changing in other ways. It is easy to see how existing monuments may have helped reinforce the perceived order of things, since they created an indelible mark on the landscape that was bound to have influenced future worldviews—if this is how the ancestors did things, people would surely have thought, then this is how we must do things too. Conversely, fundamental changes in ideology may indicate major social disruption.

This inherent stability of the perceived world order means that associations of ideological and cosmological significance have a greater chance of leaving their mark on the material record, and thus have a greater chance of being detectable by modern archaeologists, than many other, more volatile aspects of a past society. The reason is that we can expect them to be repeated over and over again. A single association, such as the solstitial alignment at Newgrange, may have meant nothing at all to the builders; it may have arisen entirely fortuitously. But if we see other, similar solstitial alignments repeated again and again, then the likelihood that they could have arisen fortuitously is rapidly diminished. Where we find significant numbers of monuments with similar designs, similar orientation, and even consistent patterns of astronomical alignment, we can be confident that common practices prevailed over a considerable area and time.

Throughout Neolithic and Bronze Age Europe we find discernible traditions of monument construction operating at various levels and scales. Some of these are remarkably widespread in space and time. In Britain, Ireland, and Brittany, for example, a propensity to build circular and linear ceremonial monuments resulted in the construction of many hundreds of stone circles and stone rows over as long as two millennia. Patterns of orientation tend to be more localized, and patterns of astronomical orientation more so still. For example, among the recumbent stone circles, a group of stone circles of a particular form found in northeastern Scotland, and among the short stone rows of southwestern Ireland, there appear to be more specific traditions of orientation upon prominent features in the landscape or rising or setting positions of the sun or moon than in the larger area.

**See also:**
Cosmology; Landscape.
Megalithic Monuments of Britain and Ireland; Newgrange; Prehistoric Tombs and Temples in Europe; Recumbent Stone Circles; Short Stone Rows; Stone Circles.

**References and further reading**
Barrett, John, Richard Bradley, and Martin Green. *Landscape, Monuments and Society.* Cambridge: Cambridge University Press, 1991.
Bradley, Richard. *Altering the Earth.* Edinburgh: Society of Antiquaries of Scotland, 1993.
———. *The Significance of Monuments.* London: Routledge, 1998.
Scarre, Chris, ed. *Monuments and Landscape in Atlantic Europe.* London: Routledge, 2002.

## Moon, Motions of

Just as the rising and setting positions of the sun move north and south along the eastern and western horizons over an annual cycle, those of the moon do the same over the course of a month. This means that, unlike the sun, the day-to-day change in the rising or setting position of the moon is considerable, since the passage from one "lunistice" to the other is completed in a period of just two weeks. Observing these lunar changes is not an easy matter, even from a place with perfect weather, because of the lunar phase cycle. The moon rises and sets approximately an hour later each day, meaning that over the course of the month, the aspiring observer will have to rise at all hours of the night. Furthermore, for half of the month the moon will be rising (or setting) in daylight, rendering the event almost certainly invisible.

The length of time it takes for the moon to complete a whole cycle from north to south and back again, known technically as the *tropical month*, is significantly shorter than the synodic (phase-cycle) month: 27.3 days as opposed to 29.5. This means that the phase cycle and the cycle to and fro along the horizon get significantly out of step. However, a pattern emerges over the course of a complete year. At the time of the June solstice, the moon is full when it is farthest south, and new when it is farthest north. At the time of the December solstice, on the other hand, the opposite is the case. This means that in the northern hemisphere, the midsummer full moon—or more precisely the full moon nearest to the summer solstice, which can occur up to two weeks before or after the solstice itself—will rise and set at, or very close to, the southern limit of its monthly motions. Conversely, the midwinter full moon will be at or close to the northern limit. Since the full moon is the most prominent phase, and also the phase at which the moon lights up the entire night, it may well be that it was the annual motions of the *full* moon northward and southward, rather than the monthly cycle to and fro, that was most important to many human cultures in the past.

From what has been said so far, the northward and southward swing of the rising or setting moon along the horizon would seem to be similar to that of the sun, but on a shorter time scale. However, there is another critical complication. The directions of sunrise and sunset at the solstices are effectively constant from year to year and only change very slowly over the centuries, owing to the gradual change in the obliquity of the ecliptic. The directions of moonrise and moonset at the monthly "lunistices," in contrast, vary noticeably from year to year. They themselves swing back and forth over a cycle lasting 18.6 years, which is known as the *lunar node cycle*. At one point in this cycle, the monthly swing of the moon from north to south is at its maximum and considerably wider than the annual swing of the sun. At such times, the most northerly monthly rising or setting path of the moon will reach beyond the most northerly annual rising or setting path of the sun by more than five degrees, or ten solar or lunar diameters; the most southerly path will do the same in the other direction. Some nine years later, in contrast, the monthly swing of the moon will be at its minimum, falling short of the annual swing of the sun at both ends by a similar amount. Alexander Thom coined the terms *major standstill* and *minor standstill*, respectively, to signify these times. Although the terms have been criticized as misleading, since nothing ever physically stands still, they are convenient and have become quite commonly used.

Again, it is likely to have been the changes in the position of the *full* moon that stood out most clearly to many human communities in the past. If they noticed the lunar node cycle at all, it may have been because of the changing path through the sky of the midwinter or midsummer full moon, rising and setting unusually far north or south for only one or two short periods during a lifetime. The effect would be most striking at latitudes of about sixty degrees, which is reached in the northern hemisphere, for example, in northern parts of Scotland. Here, in years close to a major standstill, the midsummer full moon would appear only briefly in the south, passing very low over the southern horizon, where its proximity to terrestrial landmarks would make it appear larger and closer.

See also:
Solstitial Directions; Thom, Alexander (1894–1985).
Lunar Phase Cycle; Obliquity of the Ecliptic; Solstices; Sun, Motions of.

**References and further reading**

Aveni, Anthony F. *Stairways to the Stars: Skywatching in Three Great Ancient Cultures*, 27–33. New York: Wiley, 1997.

Ruggles, Clive. *Astronomy in Prehistoric Britain and Ireland*, 36–37. New Haven: Yale University Press, 1999.

## Mursi Calendar

The Mursi live in the northern part of the basin of the river Omo in southwest Ethiopia. They depend for their subsistence upon both cattle and cultivation, mainly of sorghum but including some maize and cowpeas. Their subsistence activities span two ecological zones: the Omo and its flanking belt of bushland thicket, which is suitable for cultivation following annual floods, and the higher, wooded grassland, where the cattle can be herded safe from infestations of tsetse fly. The two most significant seasonal events in Mursi country are the onset of the "big rains" in March or April, when the entire Mursi population comes together at the border of the two zones and rain cultivation is possible, and the flooding of the Omo six months later, when some of the population, mainly women and girls, move to the Omo and plant a flood cultivation along its banks, while the remainder, mainly men and boys, move in the opposite direction to cattle camps where they remain during the driest part of the year.

If one were trying to piece together an account of Mursi calendrical practice from informants' accounts, the first step in the process would be easy. The Mursi have a seasonal calendar based upon the phase cycles of the moon, and every member of the group, including children, can recount the seasonal events associated with each month. Thus, for example, one would learn immediately that in month eight the "big rains" fall. One would begin to encounter difficulties, however, in trying to match up the Mursi months with the seasonal year. There are between twelve and thirteen synodic (lunar-phase-cycle) months in the seasonal year, as we know. This means that in order to keep a lunar calendar in step with the seasons, some years must contain twelve months while others have thirteen. And yet, each Mursi year contains twelve numbered months together with a thirteenth, which has a name: *gamwe*. Mursi informants insist that no month is ever missed.

Only by living with the Mursi for a period of time could one discover the solution to this problem. It would then become clear that, in practice, no Mursi is ever quite sure what the current month is. This is despite the fact that everyone is confident that there are experts who are. In actuality, the regulation of the calendar is a matter of public consensus. When certain seasonal events occur, their effect is to sway the balance of opinion. Even major seasonal events such as the coming of the big rains do not serve to fix the calendar, as might be expected from the rule of thumb that everyone recounts: this is because everyone acknowledges that seasonal events can sometimes occur early or late. To an outsider, the Mursi calendar seems haphazard and imprecise; yet for the Mursi it is self-consistent and works perfectly well.

Among the Mursi are people who have a particular interest in watching the sun rising behind the mountains on the eastern horizon. They recognize the sun's two "houses" in which it rises for a period of time at the extremes of its sojourn along the horizon, around the time of the solstices. A visitor might be told, for example, that the sun enters its southern house in the first half of month five and leaves it again in the first half of month six. A future archaeologist, on discovering evidence of the existence of such observations, might well conclude that the Mursi had developed a horizon-based solar calendar. However, as we know, the sun cannot, for example, always enter its house in the first half of a month, since the lunar phases fall differently in each solar year. When pressed on the issue, the Mursi specialists admit that the sun can sometimes be early or late in its movements along the horizon. The behavior of the sun, in other words, is viewed as no more reliable a seasonal indicator than that of animals, plants, and the weather. The lunar calendar remains intact.

Despite the haphazard nature (to an outsider) of the Mursi calendar, it works well enough for nearly all of the year. There is, however, one event that must be timed critically. This is the time of migration to the banks of the Omo. The river floods its banks several times, and the crop of cowpeas must be planted within a few days of the final flood; otherwise, if planted too late, the seeds will be too dry to germinate or, if planted too early, they will be washed away by the subsequent flood. How do people know the right time to move away from their bushland clearings down to the river? The answer is: by observing the successive disappearance of four stars in the evening twilight. The stars in question are the middle two stars of the Southern Cross ($\delta$ and $\beta$ Crucis) and the Pointers ($\beta$ and $\alpha$ Centauri), which are more or less in a straight line falling vertically to the horizon at the latitude (close to the equator) where the Mursi live.

The successive disappearance (heliacal setting) of the four stars coincides with the time of year of the Omo floods, which are very regular. When *imai* ($\delta$ Crucis) ceases to appear in the evening sky at dusk (around the end of August), it is said that the Omo rises high enough to flatten the *imai* grass that grows along its banks, and then subsides. This is followed a week or two later by the disappearance of *thaadoi* ($\beta$ Crucis) when the Omo rises and falls but does not yet reach its full height. By the time that *waar* ($\beta$ Centauri) disappears, the Omo (for which the Mursi name is *waar*) has risen to its full height and has flooded the levee forest along its banks. Finally, as *sholbi* ($\alpha$ Centauri) disappears from view (around the beginning of October), the flood waters finally recede, carrying away the fallen petals of the *sholbi* (acacia) tree. Planting can now proceed down to the water's edge.

To us it might seem that the successive disappearance of these four stars simply provides a practical rule of thumb that helps to determine when the move to the river bank should take place. But the use of common terms for stars and terrestrial objects indicates deeper associations in the minds of the Mursi, direct connections that tell us something of their worldview. For them, the behavior of the Omo is *understood* in relation to other factors in the natural world, both terrestrial and celestial. It is this knowledge of the world that is being applied in order to determine when to move down to the banks of the Omo and begin planting.

The Mursi example serves to undermine virtually every generalization we might try to make about calendrical development. If people have a basic lunar calendar, so the argument goes, in which they count off months according to the changing phases of the moon, then they must inevitably notice that it soon gets out of step with the seasons. Whether their annual cycle consists of twelve or thirteen months, they will find themselves adding or subtracting intercalary months to keep the lunar calendar in phase with the seasons, at first on an ad hoc basis, and then, if they progress beyond this point, more systematically. The Mursi example shows how a basic lunar calendar, without intercalation, can work perfectly well through the mechanism of institutionalized disagreement. Another common assumption is that if people observe the changing position of sunrise or sunset on the horizon, then they have reached a stage of development at which they recognize that it is the sun, rather than the moon, that is actually tied to the seasons, and they are likely to shift to a horizon-based solar calendar. Again, the Mursi example repudiates this notion completely.

Finally, the direct observation of the four stars to determine the correct time of arrival at the river might also be seen as a developmental step. It would be easy to conclude that, while the Mursi are haphazard the rest of the time, in this instance, when it really matters, they are capable of being precise. This would be a misleading conclusion. As far as the Mursi are concerned, there is a direct connection between the successive disappearance of each star in the evening twilight and the corresponding terrestrial events. To say that the Mursi are precise when they need to be is missing the point. It is more correct to say that their understanding of how the world works is completely different from ours, and yet it is both self-consistent and has a predictive capability. It is the application of this knowledge that successfully prevents the loss of the crop that provides their main food source for half of the year.

In short, as the Mursi example shows, it is misguided to attempt to measure the sophistication of an indigenous calendar against any sort of developmental yardstick. Instead, we need to try to understand such calendars in the context of a framework of indigenous knowledge. By doing so, we better appreciate that knowledge for what it is, and for what it can add to our appreciation of the richness and diversity of indigenous knowledge in general. On the last point, a comparison of the Mursi with the nearby Borana, who have a much more precise but highly unusual luni-stellar calendar, demonstrates very clearly how two communities in similar circumstances, with similar skies, can nonetheless think and act in completely different ways.

**See also:**
Lunar and Luni-Solar Calendars.
Borana Calendar.
Heliacal Rise; Solstices.

**References and further reading**
Aveni, Anthony. *Ancient Astronomers*, 93–95. Washington, DC: Smithsonian Books, 1993.
Chamberlain, Von Del, John Carlson, and Jane Young, eds. *Songs from the Sky: Indigenous Astronomical and Cosmological Traditions of the World*, 298–309. Bognor Regis, UK: Ocarina Books, and College Park, MD: Center for Archaeoastronomy, 2005.
Turton, David, and Clive Ruggles. "Agreeing to Disagree: The Measurement of Duration in a Southwestern Ethiopian Community." *Current Anthropology* 19 (1978), 585–600.

# Nā Pali Chant

In the Hawaiian islands prior to European contact, people's awareness and use of the skies took place within a cognitive framework that combined an extensive practical knowledge used in long-distance navigation with strongly developed elements of ritual and ceremony. In this resource-rich corner of ancient Polynesia, where powerful social hierarchies developed, religious practices and cosmological principles are reflected archaeologically in sacred landscapes containing a variety of temples and shrines (*heiau*) as well as prominent natural places named after celestial objects. There is also an abundant heritage of oral literature—creation myths and formal chants recorded after European contact, some of which contain detailed astronomical information.

The *hula*—despite its modern image as public entertainment and tourist attraction—actually constitutes a very significant aspect of the Hawaiian oral tradition dating back to pre-contact times. In those days it took the form of a formal performance involving the recital of a chant (*mele*) together with associated bodily movement. Certain types of *hula* were deeply sacred. The repertoire in these cases was fixed, with no improvisation allowed, and the nature and circumstances of their performance were bound by strict protocols. It is likely that chants of this nature that were recorded after European contact could have been passed down from one generation to another, virtually unscathed, for some considerable time.

The following chant, recorded by the ethnographer Nathaniel Emerson in 1909 (in the days before diacritical marks were added to Hawaiian words to aid comprehension), may well be in this category:

> Hiki mai, hiki mai ka La, e.
> Aloha wale ka La e kau nei,
> Aia malalo o Ka-wai-hoa,
> A ka lalo o Kauai, o Lehua.
> A Kauai au, ike i ka pali;
> A Milo-lii pale ka pali loloa.

> E kolo ana ka pali o Makua-iki;
> Kolo o Pu-a, he keiki,
> He keiki makua-ole ke uwe nei. (Emerson 1997, p. 114)

"Literal" translations vary, but the following represents a broad consensus:

> It has come, it has come; lo the Sun!
> How I love the Sun that's on high;
> Below it swims Ka-wai-hoa,
> On the slope inclined from Kaua'i to Lehua.
> On Kauai met I a pali,
> A beetling cliff that bounds Milo-lii,
> And climbing up Makua-iki,
> Crawling up was Pua, the child,
> An orphan that weeps out its tale.
> (Warther and Meech, in Koleva and Kolev 1996, p. 28,
>  following Emerson 1997, p. 114)

According to a legend recorded in 1899, the chant was first performed on the island of Ni'ihau by a chief who was possessed by the spirit of Kapo, an alternative manifestation of the goddess Laka, principal deity of the *hula*. It refers to a number of landscape features both on the island of Ni'ihau and on the rugged northwestern coastline of Kaua'i some thirty kilometers (twenty miles) to the east, which, viewed from the northern part of Ni'ihau, stretches for some twenty kilometers (twelve miles) directly away from the observer. This remote part of Kaua'i, impassable even by modern roads, is known as Nā Pali ("the cliffs"), although its deep valleys, deserted today, are accessible from the sea and had thriving populations in pre-contact times. Of the places named in the chant, Kawaihoa is a volcanic cone on Ni'ihau; Lehua is another volcanic cone forming a separate islet; and Makua-iki is a point about halfway along the Nā Pali cliffs, with the Miloli'i valley close by.

But could the chant be doing more than just describing the landscape? It is certainly true that Hawaiian chants contained copious quantities of riddles and word play, something the whole language lends itself to very naturally. Often they also contained a hidden level of meaning, known as *kaona*. In 1996, a local resident, Francis Warther, together with astronomer Karen Meech, suggested the following interpretation of the hidden (*kaona*) meaning:

> The sun is rising, east.
> The sun rises to its zenith.
> We stand on Ni'ihau island,
> Sun reflected down from Kaua'i island.
> Look towards Kaua'i; see the cliff.

> A big cliff that bounds Milo-liʻi valley.
> The [baby] sun climbing up the cliff named Makua-iki.
> Alone, it takes five days to return, or eventually climb the
>   cliff and return on its six-month trip to the south.
> (Warther and Meech, in Koleva and Kolev 1996, p. 28)

According to Warther and Meech, the whole chant was in fact describing a solstitial alignment along the Nā Pali coastline. This could actually be viewed both ways, and it so happens that at the far (eastern) end of the cliffs is a large *heiau* sitting at the foot of a number of terraces leading up to a sacred altar dedicated to Laka herself, Ke-Ahu-a-Laka. According to traditions recorded early in the twentieth century, this was a place where priests or priestesses of the *hula* underwent periods of instruction and came to "graduate." The spot is very sheltered, and only from the very lowest corner of the heiau can one see along the coast. From this point, the cliffs of Makua-iki can be seen jutting out into the sea beyond the immediate cliffs, and Lehua appears in the distance a short way along the skyline. The flat northern part of Niʻihau is in the line of sight but lost beneath the horizon.

The chant and its double connection with Laka/Kapo strongly hints that this alignment in the landscape had a deep resonance for ancient Hawaiians, notwithstanding the astronomical connection. But what was the nature of that connection? Archaeoastronomical work has shown that the alignment may have had as much, if not more, to do with sunrise and sunset on the days of zenith/antizenith passage of the sun than the solstices. The moment when the sun passed directly overhead was certainly a moment of great sacred power (*mana*) for ancient Hawaiians.

The alignment along the Nā Pali cliffs is, of course, naturally occurring rather than artificially constructed. This does not mean, though, that astronomical significance could not have been attributed to it: on the contrary, in fact. By an accident of nature (as a Western investigator would see it) resulting from the process of their geological formation, the highest points of various islands along the whole Hawaiian chain are quite closely aligned along an axis pointing toward June solstice sunset in one direction and toward December solstice sunrise in the other. Such fortuitous alignments preexisting in the landscape might not impress a modern astronomer, but from an indigenous point of view it is difficult to imagine a more direct affirmation of the natural harmony between the land and the sky.

It is possible that this so-called Nā Pali chant will stand the test of time as a rare example of a "documented" ancient alignment. Unfortunately, the evidence is very inconclusive. In particular, the freedom with which double meanings can be expressed in Hawaiian, and with which *kaona* meanings can be imagined, cuts both ways, since it gives the modern investigator a great deal of flexibility to postulate much that may not have been in the ancient Hawaiians' minds at all. For now, no firm verdict has been reached.

**See also:**
Antizenith Passage of the Sun; Cosmology; Methodology; Solstitial Directions; Zenith Passage of the Sun.
Navigation in Ancient Oceania; Polynesian Temple Platforms and Enclosures.
Solstices.

**References and further reading**
Barrère, Dorothy B., Mary K. Pukui, and Marion Kelly. *Hula: Historical Perspectives,* 17–67, 106–115, 123–126. Honolulu: Bishop Museum Press, 1980.
Emerson, Nathaniel B. *Unwritten Literature of Hawaii: The Sacred Songs of the Hula,* 114. Honolulu: 'Ai Pōhaku Press, 1997. Originally published in 1909 by Government Printing Office, Washington, DC.
Kirch, Patrick V. *Legacy of the Landscape: An Illustrated Guide to Hawaiian Archaeological Sites,* 20–21. Honolulu: University of Hawai'i Press, 1996.
Koleva, Vesselina, and Dimiter Kolev, eds. *Astronomical Traditions in Past Cultures,* 25–33. Sofia: Bulgarian Academy of Sciences, 1996.
Ruggles, Clive. "Astronomy, Oral Literature, and Landscape in Ancient Hawai'i." *Archaeoastronomy: The Journal of Astronomy in Culture* 14 (2) (2000), 33–86.
Selin, Helaine, ed. *Astronomy across Cultures,* 122–123. Dordrecht, Neth.: Kluwer, 2000.
Tava, Rerioteria, and Moses K. Keale. *Ni'ihau: The Traditions of an Hawaiian Island,* 18. Honolulu: Mutual Publishing, 1989.

# Nabta Playa

Between about 9000 B.C.E. and 3000 B.C.E., parts of what is now the Sahara Desert became habitable for periods lasting many centuries, when summer monsoons reached farther northward than at present. Nabta Playa is a natural basin in southern Egypt, roughly fifty square kilometers (twenty square miles) in area, where water accumulated in the summer rainy season. It was a focus for seasonal occupation and settlement over several millennia, including ritual activities such as the burial of slaughtered cattle in roofed chambers covered in sandstone blocks. At some time between 7000 B.C.E. and the final abandonment of the playa in c. 3000 B.C.E., a number of stone slabs up to three meters (ten feet) in length seem to have been erected over a area of about three square kilometers (1.5 square miles), including three alignments several hundred meters long. There is also an oval arrangement of small standing and recumbent stones about three meters (ten feet) across. In many respects these stones bear an uncanny resemblance to European megaliths but predate them, possibly by several millennia.

Surveys of the Nabta "megaliths" by the American archaeoastronomer Kim Malville in the late 1990s revealed that one of the alignments faced due east-west, while pairs of small standing stones sighted on opposite sides of

*Some of the the stone slabs at Nabta Playa, in the Egyptian Sahara. (Courtesy of John McKim Malville)*

the circle defined two sightlines across it, one oriented north-south and the other aligned upon sunrise on the June solstice. The circle has accordingly been dubbed a *calendar circle* and seized upon by those keen to challenge the idea that the oldest astronomical alignments must have developed in Europe. It is certainly true that there are configurations of standing stones in many parts of the world that have not received the attention lavished upon their European counterparts. However, the history of the investigations of the European megaliths readily demonstrates the many dangers that await those overeager to find astronomical alignments and to overinterpret them in the absence of supporting archaeological evidence.

Nonetheless, it is possible that the alignments at Nabta provide a modest but valuable indication of early cosmological beliefs and calendrical practices among nomadic peoples in northeast Africa before the rise of the Egyptian kingdom. One suggestion is that climatic change in the fourth millennium B.C.E. forced these peoples to move north, not only forcing social change that precipitated the development of the kingdom, but introducing beliefs and practices that, in a later form, became enshrined in the architecture of large temples and pyramids.

See also:
Solstitial Directions.
Egyptian Temples and Tombs; Namoratung'a.

**References and further reading**
Malville, J. McKim, et al. "Megaliths and Neolithic Astronomy in Southern Egypt." *Nature* 392 (1998), 488–491.
Selin, Helaine, ed. *Astronomy across Cultures,* 43, 457–459. Dordrecht, Neth.: Kluwer, 2000.

# Nadir Passage of the Sun
*See* Antizenith Passage of the Sun.

# Namoratung'a

Namoratung'a II (or Kalokol) is an archaeological site close to Lake Turkana in northwest Kenya. It consists of a collection of nineteen small standing stones—basalt pillars—associated with two burial cairns. It achieved notoriety in the late 1970s as an ancient calendrical site marking the horizon rising positions in 300 B.C.E. of seven stars and star groups still used in the present-day calendar of the Borana, who live about three hundred kilometers (two hundred miles) to the east.

This interpretation serves as a warning against the dangers of heaping supposition upon supposition without paying enough attention to the broader evidence. Before the Borana were revisited by Italian anthropologist

Marco Bassi in the mid-1980s, the only first-hand account of their calendar was that obtained by the Ethiopian anthropologist Asmerom Legesse around twenty years earlier. Legesse's informants had said that successive months were recognized by the newly sighted crescent moon rising "in conjunction with" seven successive stars or star groups: Triangulum, the Pleiades, Aldebaran, Bellatrix, Orion's belt and sword, Saiph, and Sirius. Astronomer Laurence Doyle from NASA examined Legesse's account but could not make astronomical sense of it, unless "rising in conjunction" was taken to mean rising at the same horizon position, that is, in astronomical terms, at the same declination. The only problem was that this didn't work correctly in the present era; however, it would have worked perfectly in around 300 B.C.E. The answer, suggested Doyle, was that the present-day Borana calendrical system derives from a more ancient calendar set up by Cushitic peoples around 300 B.C.E. A survey of alignments between the pillars at the Namoratung'a II site revealed alignments upon all seven of the Borana stars and asterisms at their rising points in around 300 B.C.E.

Unfortunately, subsequent reassessments challenged many aspects of this attractive picture: the actual dating and purpose of the Namoratung'a stones; their cultural connection with the later Borana; whether the alignment evidence was statistically significant (there exist a lot of pairwise alignments between nineteen stones); and whether the measured alignments were accurate in the first place. Finally, it is difficult to understand why, and how, the observational principles underlying the Borana calendar could have remained static for over two thousand years when they no longer fit the actual motions of the stars.

Taken together, these problems made the idea that the Borana calendar was some sort of frozen remnant of a much earlier calendrical system seem extremely shaky. The house of cards collapsed completely when Bassi's fieldwork confirmed that "in conjunction with" should be interpreted as "side by side with," in other words, in astronomical terms, "rising at the same right ascension as." When properly understood, the Borana system works perfectly well in the present day.

**See also:**
Calendars.
Borana Calendar.
Celestial Sphere; Declination.

**References and further reading**
Bassi, Marco. "On the Borana Calendrical System." *Current Anthropology* 29 (1988), 619–624.
Doyle, L. R. "The Borana Calendar Reinterpreted." *Current Anthropology* 27 (1986), 286–287.
Legesse, Asmerom. *Gada: Three Approaches to the Study of African Society.* New York: Macmillan, 1973.

Lynch, B. M., and L. H. Robbins, "Namoratunga: The First Archaeoastronomical Evidence in Sub-Saharan Africa." *Science* 200 (1978), 766–768.

Ruggles, Clive. "The Borana Calendar: Some Observations." *Archaeoastronomy* 11 (1987), S35–53.

Ruggles, Clive, ed. *Archaeoastronomy in the 1990s,* 117–122. Loughborough, UK: Group D Publications, 1993.

Selin, Helaine, ed. *Astronomy across Cultures,* 457–460. Dordrecht, Neth.: Kluwer, 2000.

Soper, Robert. "Archaeo-Astronomical Cushites: Some Comments." *Azania* 17 (1982), 145–162.

## Nasca Lines and Figures

Few archaeological enigmas have excited so much fanciful speculation as the lines and figures etched into the desert near Nasca (or Nazca) in southern Peru. Few of the theories are scientifically tenable, and many are pure fantasy. However, behind the speculation lies a unique cultural phenomenon that for almost a century has attracted the attention of scientists and archaeologists alike.

The coastal strip of southern Peru, which is in effect the northern extension of the Atacama desert in Chile, is one of the most arid and desolate regions of the world. The landscape here comprises a series of flat desert plains, or pampas, separated by oasis-like river valleys. Measurable rainfall occurs on average only once in several years. Rivers are dry for much of the year, and water is plentiful for a short period only, when the seasonal meltwater flows down from the snow-capped Andes to the east. Yet despite the severity of the climatic regime, prehistoric societies flourished in the area for several millennia. Human occupation was concentrated in the valleys, but cultural activity extended onto the adjacent pampas, where it left a distinctive concentration of prehistoric remains.

The small town of Nasca, some four hundred kilometers south of Lima, is situated in such a valley. A distinctive culture flourished here between the first and sixth centuries C.E., leaving an abundant archaeological record including a fine and distinctive style of pottery, brightly colorful and richly decorated. During the centuries before the arrival of the Europeans, the people who lived in this area also seem to have channeled considerable efforts into etching monumental drawings on the desert. The Nasca pampa, an arid plain to the north of the town, covers an area of some two hundred square kilometers (about eighty square miles) and is covered in a vast array of long, straight lines, rectangles and trapezoids, labyrinths and spirals. The greatest concentration of markings is in the northern corner, where a number of large stylized bird and animal figures, as well as less readily identifiable forms, are also found. The overall impression, as viewed from one of the many light aircraft that carry tourists over the plain, is one of a giant

"The lady of the lines": Maria Reiche walking a line in 1984. (Courtesy of Clive Ruggles)

*A replica line and spiral built by a team of volunteers in 1984. (Courtesy of Clive Ruggles)*

sketchpad, much scribbled upon.

The desert surface here is composed of black ferrous oxide pebbles darkened by oxidation over many centuries. By simply brushing them aside, a bright yellow sandy soil is revealed beneath. This means that the desert markings, often termed *geoglyphs,* are highly susceptible to modern damage. Merely walking on the pampa is often enough to leave conspicuous footprints, which, owing to the lack of precipitation, will endure almost indefinitely. Worse still, many of the ancient lines and figures are scarred by deep ruts created by cars and even large commercial vehicles, which drive across the open desert in order to avoid paying highway tolls. On the other hand, there is no great mystery about how the Nasca geoglyphs were created, at least in principle. Armed with nothing more than a piece of string and a few sighting poles, a group of six volunteers was able to produce a ten-meter- (thirty-foot-) long straight line ending in a spiral on a nearby pampa in less than ninety minutes (see photo).

Yet the Nasca markings were more than casual doodles. Some lines run for several kilometers, remaining dead straight even where they pass over small hills and dips. The figures, generally too large to be seen for what they are when standing close by, must have been constructed by scaling up from a template of manageable size. The enigma lies not in how the etchings were constructed per se, but in the reasons for their construction on such a vast scale. One suggested motivation for the anthropomorphic figures, which are clearly visible only from the air, is that they were for the benefit of shamans "flying" above the figures under the influence of the powerfully hallucinogenic San Pedro cactus. There is actually some evidence to support this speculation: the San Pedro cactus is depicted upon Nasca pottery; images of birds are very common amongst Nasca artifacts (including, of course, the desert drawings themselves); and traditional beliefs about the spirit world that survive today have much to do with flying. Be this as it may, it is clear that one does not need to resort to explanations totally divorced from the cultural evidence (such as the suggested existence of hot-air balloons), let alone extraterrestrial involvement, to provide plausible theories as to why the Nasca people were motivated to etch vast figures in the desert.

The lines are a more widespread phenomenon than the figures, both in time and space. Pottery fragments scattered in the vicinity of the geoglyphs indicate that the figures are attributable to the "Classic" Nasca period, during the first six centuries C.E., while the lines were created over a longer period of time. Not only do the lines (as opposed to the figures) cover the whole Nasca pampa; they also spread off into the valleys, where they become obliterated by modern cultivation. Similar lines are known on adjacent pampas as well as much farther afield. It just happens that on the Nasca pampa they are exceptionally well preserved, owing to the nature of the surface geology.

Survey work has revealed a network of line centers, mostly located around the edge of the Nasca pampa, with lines of various types radiating out from them. Some of the lines stop dead, others turn corners and continue into labyrinths, but some run on, absolutely straight, for up to several kilometers, connecting different line centers. They are not pathways, at least not casual pathways; they are too perfectly straight. So what was their purpose?

The Peruvian archaeologist Toribio Mejía Xesspe, working in the 1920s, was one of the first to suggest a possible interpretation, dismissing the possibility that the lines were part of an irrigation system in favor of the idea that they were associated with some form of religious ritual. A decade or so later the North American geographer Paul Kosok chanced to observe the sun setting along one of the lines on June 22, the winter solstice in the southern hemisphere. This single fortuitous observation apparently led him to the conviction that the Nasca lines had a calendrical function (he later described them as the "largest astronomy book in the world"), thereby setting the seal on an astronomical interpretation that dominated Nasca studies for many years.

It was this interpretation that inspired a young German mathematics teacher living in Lima, Maria Reiche, to devote a lifetime to studying the

"Walking the lines": a team of volunteers maps some of the Nasca lines in 1984. (Courtesy of Clive Ruggles)

lines. A few visits to the pampa convinced Maria that the lines were directed toward horizon directions where the sun, moon, or stars appeared and disappeared; solving the riddle of the mathematical and astronomical meaning of the lines and figures subsequently became her mission in life. She began to visit the pampa regularly, living the life of a recluse, spending hours, days, and weeks walking on the desert and making measurements. Despite Reiche's unremitting devotion to the investigation of the lines, which lasted for the rest of her life (she died in 1998, aged ninety-five), it produced precious little published hard data. Reiche's book *Mystery on the Desert,* which has run to several editions, concentrates mainly on descriptive material. In 1968, the astronomer Gerald Hawkins, who had proposed that Stonehenge in England was an astronomical observatory or computer, visited the pampa and carried out a statistical examination of the line orientations. His conclusion, which came as a surprise for many, was that they had no astronomical significance whatsoever, beyond what might be expected by chance. Although very different, both these approaches failed in one fundamental respect: they were divorced from the cultural context. Each, in its different way, was an intellectual exercise dictated by Western concepts of science and mathematics but unrelated to the rich cultural traditions of pre-Columbian America.

Any meaningful explanation for the Nasca lines needs to be couched in a wider, pan-Andean framework. There is, for example, an evident similarity between the network of line centers and radial lines at Nasca and the system of *ceques* that arose later at the imperial Inca capital of Cusco. A radial structure is also evident in the *quipu*—Andean recording devices consisting of knotted strings—which it seems natural to lay out so that the strings radiate from the single, "primary," string to which each is attached. Added to this, ethnographic work has revealed the existence of a number of modern, traditional Andean villages, some in the vicinity of Cusco but others as far afield as northern Chile and Bolivia, organized around lines radiating from a central place such as a plaza or church. Attaching importance to straight lines radiating from central places is something that emerges in a variety of cultural contexts in pre-Columbian, historical, and modern South America. The Nasca lines seem to represent an early manifestation of conceptual principles attaching significance to radial lines that may have been remarkably widespread in Andean thought even in pre-Inca times, before becoming integrated into Incaic worldview. Fragments of this principle survive in the present day in a few modern, remote Andean villages.

What does this suggest about the purpose of the Nasca lines? The Incaic *ceque* system, as we know from historical evidence, was a basic organizing principle of empire, a mechanism of social and political control. Something similar may have been true at Nasca, at least in later, Incaic times. Post-conquest documents relating to land and water use in the vicinity of the pampa shortly after the Spanish conquest suggest that a scheme of social organization not unlike the one that prevailed in Cusco also existed in that region.

The Nasca lines find a counterpart at an elevation of over four thousand meters (thirteen thousand feet) in a series of lines high in the Bolivian Andes. Some of these connect villages to the summits of nearby hills and appear to have been maintained until very recently. In a village southwest of La Paz, anthropologist Johan Reinhard chanced to witness an annual procession along one such line to a hilltop shrine. Here is a community still maintaining and using a straight line during an important springtime festival. After dark, the devotees make their way along the line, which is over one kilometer long. At the summit of a nearby hill, they worship local mountain gods, petitioning for fertility for animals (which are believed to be owned by the mountain) and rain for crops. They descend again to the village in the morning. At similar rituals in other villages there is a sacrificial element, where offerings are made to mountain gods that involve a llama sacrifice. Comparable practices took place during the Inca period.

Some of the Nasca line centers are located at higher points on the predominantly flat plain. Others are on the edges of the pampa, and in some cases lines running up to them from the valleys are still visible. It may well be, then, that the Nasca lines were meant to be walked upon after all, but in a formal sense—according to strict protocols, at the correct times, and in the context of the appropriate rituals. In such an arid environment, it is scarcely surprising that some of these ceremonials may have related to rain and fertility. Nor it is surprising, then, that statistical analyses of the line orientations demonstrate a degree of correlation with local directions of water flow. On the other hand, we should not conclude that there was no connection with astronomy at all. A few evident alignments, for example on solstitial sunrise or sunset, may well have been formed as a deliberate part of the general scheme of things, just as a few of the Cusco *ceques* seem to have been astronomically aligned.

In short, there is no simple explanation for the lines and figures on the pampa. Looking for simple solutions to a mystery and ignoring the wider cultural context is not the right approach. The Nasca geoglyphs can only ever be understood by trying to understand more about the totality of that cultural context viewed, at best, obscurely through a historical haze.

**See also:**
Andean Mountain Shrines; *Ceque* System; Misminay; Quipu.

**References and further reading**
Aveni, Anthony F. *Between the Lines: The Mystery of the Giant Ground Drawings of Ancient Nasca, Peru.* Austin: University of Texas Press, 2000. (Published in the UK as *Nasca: Eighth Wonder of the World?* London: British Museum Press, 2000.)
Aveni, Anthony F., ed. *The Lines of Nazca.* Philadelphia: American Philosophical Society, 1990.
Hadingham, Evan. *Lines to the Mountain Gods: Nazca and the Mysteries of Peru.* London: Harrap, 1987.
Morrison, Tony. *Pathways to the Gods: The Mystery of the Andes Lines.* London: Paladin/Granada, 1980.
———. *The Mystery of the Nasca Lines.* Woodbridge, UK: Nonesuch Expeditions Ltd, 1987.
Reiche, Maria. *Mystery on the Desert* (4th ed.). Stuttgart: Heinrich Fink, 1982. [In German, English, and Spanish.]
Reinhard, Johan. *The Nazca Lines: A New Perspective on their Origin and Meaning* (4th ed.). Lima: Editorial Los Pinos, 1988.
Silverman, Helaine, and Donald A. Proulx. *The Nasca.* Oxford: Blackwell, 2002.

# Nationalism

The idea that certain peoples in the past had sophisticated astronomical knowledge has sometimes been conflated with issues of national identity and invoked in support of nationalist agendas. This happened most notoriously when the Nazis used ancient astronomical achievements as part of their demonstration of supposed Aryan superiority. Apart from the fact that political motivations as strong as this are almost certain to compromise scientific objectivity in assessing the actual evidence, the whole agenda stems from a misguided belief that the achievements of past communities can be rightly judged against our own—a clear manifestation of ethnocentrism. It is quite different from the proper agenda of archaeoastronomy, which, in common with many aspects of anthropology and other social sciences, strives to highlight human diversity in order to promote wider understanding of the breadth of human achievements and tolerance of humanity in general.

See also:
Ethnocentrism; Methodology.

**References and further reading**
Michell, John. *A Little History of Astro-Archaeology*, 58–65. London: Thames and Hudson, 1989.

# Navajo Cosmology

Navajo country stretches across the four-corners region of the U.S. Southwest. The bulk of it lies in northern New Mexico and northern Arizona, with a small part spilling over into southern Utah and the southwestern tip of Colorado. It is bounded by four sacred mountains: Turquoise Mountain in the south (Mount Taylor, New Mexico, between Gallup and Albuquerque); Obsidian Mountain in the north (Hesperus Peak, Colorado, in the La Plata Mountains); in the west Mount Humphrey, Arizona, one of the San Francisco Peaks near Flagstaff; and in the east by a mountain called *Sis Naajíní*, which is probably Blanca Peak, Colorado, in the Sangre de Cristo range.

Navajo territory exists at the center of things, integrated into the cosmos through a fourfold harmony: four directions, four seasons, four times of day, four sacred colors, four precious stones, and so on. Each direction is associated with a different one of the four things in a particular category: thus, for example, south is associated with the color blue and the middle of the day. In this four-ness, Navajo cosmology has much in common with the worldview of other native American groups, but it is fixed geographically by a particular sacred topography. The four-directional awareness perpetrates all aspects of life. West, for instance, portrays family life and is associated with being at home in the evening with the family around the fire. In learning, east is associated with the arts, west with the social sciences, north with the hard sciences, and south with technical skills. These associations are reflected in the layout of Diné College (formerly Navajo Community College) at Tsaile, Arizona.

The relationship between earth and sky is central to Navajo thought. Father Sky stretches out over Mother Earth: the space in between them is inhabited by living beings that live on top of the earth, and the stars which adorn Father Sky. The life force of all living and growing things derives from Mother Earth and Father Sky (or Father Sun).

Various supernatural beings helped create the world, and they inhabit it alongside ordinary mortal humans. These Holy People can be dangerous and harmful, but are generally concerned for the welfare of the Earth Surface People. The Holy People include First Man and First Woman, who showed, by building the first dwelling, or hogan, how all hogans should be constructed, modeling the cosmos in various ways in order to ensure the well-being of their inhabitants. Black God was responsible for placing the stars in the sky—by some accounts helped by First Man and First Woman. *Dilyéhé* (the Pleiades) were his own stars. Each time he stamped his foot, they moved up his body until they finally settled on his forehead. The crystals that would form the other stars were carried in his pouch, or in some versions of the story, in a blanket. Black God carefully placed various stars in the sky: Fire Star (Polaris), Revolving Male (Ursa Major), Revolving Female (Cassiopeia), First Slim One (Orion), Man With Legs Ajar (Corvus), First Big One (head of Scorpius), and Rabbit Tracks (tail of Scorpius).

These principal constellations not only lit up the night in the absence of the moon, but also invoked stories and reinforced social laws, confirming principles of orderly behavior needed to ensure that the community as a whole would survive and flourish. Thus the sight of Revolving Male and Revolving Female turning continuously around Fire Star set an example to be followed: only one couple should live in a single hogan and cook over the same fire. This served to emphasize the importance of family responsibilities and to reinforce the social convention that sons-in-law should avoid looking their mothers-in-law in the face. Rabbit Tracks directed social behavior in a different way, by means of its seasonal patterns of appearance and disappearance, which defined the periods when hunting was permitted.

According to the legend, Black God placed *Dilyéhé* in the sky close to First Slim One, then rested. At this time, Coyote was allowed to place one star in the south. But Coyote was a trickster. He seized the pouch or blanket containing the rest of the stars and flung them across the sky. This is why most of the stars are chaotic. In this way, the Navajo creation story neatly explains both order and disorder in the heavens.

A distinctive artifact of Navajo culture is the sandpainting or dry painting. These paintings have a life and power of their own. Creating one is a formal affair carried out according to strict protocols. When completed, the painting functions (in the context of the appropriate ceremonial) to call supernatural beings and focus their healing powers, for example, in curing sick or injured people. Navajo sandpaintings depict a variety of living beings and Holy People. A number contain recognizable star groups within representations of the sky, for example in the body of Father Sky.

In sum, the cosmos in general, and the sky in particular, pervades everyday life for traditional Navajo people. The night sky with its key constellations serves as a text, underlying stories that chanters elaborate during ceremonials (each in their own way, resulting in many variants). Elements of Najavo worldview are also reflected in a variety of ways in material culture, for example in the design of the hogan and in the content and use of sandpaintings.

**See also:**
Cosmology; Ethnoastronomy.
Navajo Hogan; Navajo Star Ceilings.

**References and further reading**
Chamberlain, Von Del, John Carlson, and Jane Young, eds. *Songs from the Sky: Indigenous Astronomical and Cosmological Traditions of the World,* 49–79. Bognor Regis, UK: Ocarina Books, and College Park, MD: Center for Archaeoastronomy, 2005.
Griffin-Pierce, Trudy. *Earth Is My Mother, Sky Is My Father: Space, Time and Astronomy in Navajo Sandpainting.* Albuquerque: University of New Mexico Press, 1992.
Selin, Helaine, ed. *Astronomy across Cultures,* 269–301. Dordrecht, Neth.: Kluwer, 2000.
Williamson, Ray A., and Claire R. Farrer, eds. *Earth and Sky: Visions of the Cosmos in Native American Folklore,* 101–109. Albuquerque: University of New Mexico Press, 1992.

# Navajo Hogan

Traditionally, Navajo families do not live in concentrated villages but in homesteads scattered about in the landscape. The traditional family dwelling is known as a *hogan* (or *hooghan*). Today Navajo families may live in separate houses or caravans, but the hogan remains the sacred dwelling, a place of instruction, healing, or simply a quiet refuge.

Round or octagonal in shape, the hogan is built in accordance with strict cosmological principles. The doorway, for example, always faces east, toward the rising sun. The roof reflects the sky, the walls reflect the surrounding trees and mountains, and the earth floor reflects the earth as a whole. Four poles are placed in the four cardinal directions, reflecting the overall structure of the cosmos. Different quarters of the hogan are used for different activities: everyday crafts and handiwork in the south, telling stories and entertaining visitors take place in the west, sacred activities (such as preparing masks for ceremonials) in the north. Movement within the hogan, between these different areas, is sunwise.

In short, the hogan is a reflection of the world as a whole—a microcosm. As a result, its occupants, while occupying their home, are also occupying the whole cosmos, at harmony with all things, and hence assured long life and happiness.

There is a lesson here for archaeologists who might seek to interpret orientation trends among older dwellings for which we have no historical or modern informants to interpret their meaning. If, as archaeologists of the future, we measured the orientations of a group of Navajo hogans, we might find their entrances facing a spread of directions within the range of horizon where the sun rises at some time during the year. If we then became preoccupied with questions such as "which day of the year did the sun rise directly in line with the entrance of a particular hogan?" and "was a hogan built to face sunrise on the day of its construction?," we might lose sight of the broader questions and meanings that really mattered to the Navajo people. "East" is the quarter of the world—that part of the horizon—where the sun shines in at some time of the year. Hogans face "east," and that may be all that matters. Even if there is more, Navajo elders may never wish to tell us, and the archaeological record may never reveal it.

See also:
Orientation.
Navajo Cosmology; Navajo Star Ceilings; Pawnee Earth Lodge.

**References and further reading**
Griffin-Pierce, Trudy. *Earth Is My Mother, Sky Is My Father: Space, Time and Astronomy in Navajo Sandpainting*, 92–96. Albuquerque: University of New Mexico Press, 1992.
Williamson, Ray A., and Claire R. Farrer, eds. *Earth and Sky: Visions of the Cosmos in Native American Folklore*, 110–130. Albuquerque: University of New Mexico Press, 1992.

## Navajo Star Ceilings

Certain stars and constellations that were of particular cultural significance can be recognized in Navajo sandpaintings, rock art, and certain portable artifacts such as gourd rattles used in sacred ceremonies. Small crosses representing stars are also found painted on the ceilings of rock shelters and on horizontal, downward-facing surfaces of rock overhangs. Over seventy-five examples are known, concentrated mainly in the Canyon de Chelly area of Arizona. They contain anything from a single star to more than two hundred, sometimes mixed with other elements such as stylized birds or dragonflies.

According to Navajo myth, Black God, who put the stars in the sky, only had the chance to place a few carefully before the rest were flung across the sky haphazardly. As a result, only a few bright stars and constellations are important to the Navajo; the rest are regarded as random. Some of these constellations are depicted on certain Navajo artifacts, but the important thing was to identify them rather than to represent their shape literally (as we would see it). Thus where *Dilyéhé* (the Pleiades) are represented on masks of Black God, the seven dots do not reflect the actual form of the constellation in the sky.

*Pueblo rock art at Comanche Gap petroglyph field, Lamy, New Mexico, featuring representations of stars. (Courtesy of Clive Ruggles)*

Star ceilings, it seems, do not even depict actual stars. Though there are inviting scatters of stars on many ceilings, attempts to identify particular constellations are unconvincing. A number of ceilings have the stars arranged in regular rows; others are as much as twenty meters (sixty-five feet) above the ground, totally out of reach. The stars were likely placed there by using bows to fire "arrows" carrying paint-covered stamps—a process that could not have resulted in great precision.

What were star ceilings for? In Navajo thought, stars are regarded as beneficent supernatural beings (Holy People) who can restore well-being in times of misfortune. A likely answer, then, is that the stars had a restorative or protective role. Perhaps the act of painting of a ceiling created a place—a shrine—where the healing power of the stars could be harnessed through the performance of the appropriate rites. Another suggestion is that they created a supernatural presence within a dangerous place to protect people nearby against disasters such as rock-falls. These and other possibilities are not mutually exclusive.

In short, it seems that stars on Navajo star ceilings were symbols conveying the stars' protective power rather than literal depictions of stars. This example cautions against jumping to "literalist" conclusions, especially where we do not have access to the cultural evidence that could counter them.

See also:
Navajo Cosmology; Pawnee Star Chart.

**References and further reading**
Aveni, Anthony F., ed. *World Archaeoastronomy,* 331–340. Cambridge: Cambridge University Press, 1989.
Chamberlain, Von Del. "Navajo Constellations in Literature, Art, Artifact and a New Mexico Rock Art Site." *Archaeoastronomy* (Center for Archaeoastronomy) 6 (1983), 48–58.
Chamberlain, Von Del, John Carlson, and Jane Young, eds. *Songs from the Sky: Indigenous Astronomical and Cosmological Traditions of the World,* 80–98. Bognor Regis, UK: Ocarina Books, and College Park, MD: Center for Archaeoastronomy, 2005.
Griffin-Pierce, Trudy. *Earth Is My Mother, Sky Is My Father: Space, Time and Astronomy in Navajo Sandpainting,* 120–122. Albuquerque: University of New Mexico Press, 1992.
Ruggles, Clive, ed. *Archaeoastronomy in the 1990s,* 227–241. Loughborough, UK: Group D Publications, 1993.

# Navigation

One of the obvious ways in which ancient peoples may have made practical use of astronomy was in finding their way through unknown territory, where they could not rely on familiar landmarks. Of course travelers in the past could always attempt to identify a succession of distinctive places and landscape features, preferably ones that were clearly visible from some distance along the route. However, in some circumstances—in great plains, deserts and tundra, and especially at sea—these were scarce or nonexistent. In such situations, the sky can be critical in maintaining a course and, for longer journeys, indicating the current location.

There are obvious broad indicators of direction: the easterly sunrise, the westerly sunset, or (in locations north of the tropics) the northerly-facing shadows cast by upright objects in the midday sun (in locations south of the tropics, the shadows are southerly-facing). However, these only provide crude indications, since the rising and setting points of the sun depend upon the time of year, and the midday shadow technique presupposes the means to accurately determine the time of local noon.

Stars provide a more direct and more accurate means of determining directions. Any given star, viewed from a particular place, always rises and sets at the same point and follows the same path through the sky. By acquiring a knowledge of guiding stars whose rising or setting positions mark different directions, it is possible to build up a mental *star compass*. In practice, adverse weather conditions can create difficulties, so the more stars a traveler knows, the better. Another problem is *extinction*, the dimming of a star close to a low horizon as its light is absorbed by the earth's atmosphere. Determining direction by using stars that are already high in the sky, on the other hand, is complicated by the fact that the direction depends on the altitude above the horizon, something that is itself generally difficult to determine accurately. This only ceases to be a problem close to the equator, where the celestial bodies rise and set almost vertically.

From anywhere on earth, there is one point in the heavens whose direction is unchanging: the celestial pole. The north celestial pole, visible to observers north of the equator, is marked by the Pole Star, Polaris (although, owing to the steady drift in the position of the pole due to precession, this was not the case before modern times). Observing this star, whatever the time of night and the time of year, gives a direct indication of due north to within about a degree (since Polaris is not quite at the celestial pole). Observers in the southern hemisphere have no such bright star to mark the south celestial pole, but a good knowledge of the star formations that surround it, such as the Southern Cross, can be used to obtain quite a good estimate of its position and hence of the direction of due south.

Voyagers who traveled significant distances to the north or south would have begun to notice the changing appearance of the sky with latitude. The altitude of Polaris, for example, is a direct indication of the latitude for an observer north of the equator. Even without instruments, travelers in tropical latitudes can estimate their latitude north of the equator by such techniques as holding a finger or hand at arm's length to estimate the angle of Polaris above a level horizon. Another way of keeping track of one's latitude is to identify the stars that pass across the zenith, the point directly overhead.

The value of navigation by the stars is most obvious in the open sea, where there are no terrestrial points of reference and distances traveled can be great indeed. The most impressive feats of navigation in the past were undoubtedly those of the ancient Polynesians, who successfully discovered and colonized innumerable islands, most of which were mere dots in the vast expanse of the Pacific Ocean. The Polynesian navigators used a combination of astronomical observations and other techniques. Estimating one's longitude is much trickier than estimating latitude, since it involves relating the appearance of the sky to the time of night and the time of year. It is generally easier to use other means, such as *dead reckoning:* estimating the distance traveled in order to compute the current position. A method of locating known islands was to reach the appropriate latitude and then sail due east or west, using means such as ocean currents or the appearance of certain birds to indicate that land was close.

Navigation by land using the stars is much less well documented. For example, despite the vast and open nature of much of the landscape of central Australia, there is little or no evidence of Aboriginal groups—despite their detailed knowledge of the skies in other ways—using the appearance of the skies for navigation purposes at night. Inuit groups, on the other hand, made extensive use of both the sun and stars for wayfinding in the Canadian Arctic.

**See also:**
Aboriginal Astronomy; Inuit Cosmology; Navigation in Ancient Oceania.
Altitude; Celestial Sphere; Extinction; Precession; Solstices; Star Rising and Setting Positions.

**References and further reading**
Finney, Ben. *Voyage of Rediscovery,* 51–65. Berkeley, University of California Press, 1994.
Lewis, David. *We the Navigators* (2nd ed.). Honolulu: University of Hawai'i Press, 1994.
MacDonald, John. *The Arctic Sky,* 160–191. Toronto: Royal Ontario Museum and Iqaluit: Nunavut Research Institute, 1998.
Selin, Helaine, ed. *Astronomy across Cultures,* 32–33, 62–63, 106–107. Dordrecht, Neth.: Kluwer, 2000.

## Navigation in Ancient Oceania

The Pacific Ocean is so huge that it covers a third of the earth's surface and is larger in area than all of the world's continents put together. When James Cook led the first scientific expeditions to the Pacific in the mid-eighteenth century, he was staggered to discover that small and widely dispersed islands spread over thousands of miles were not only inhabited, but by peoples who spoke essentially the same language. In this way, Polynesia became recognized as the most widespread nation in the world, dispersed over an area that—if shifted to the other side of the world—would simultaneously encompass London, Vladivostok, and the southern tip of India. The islands of Micronesia and Melanesia in the western Pacific, generally closer to the Asian continent, were also inhabited prior to European contact, but by a mixture of different ethnic and linguistic groups.

Ever since 1520, when Ferdinand Magellan sailed west into the unknown ocean from the tip of South America, arriving in the Philippines with a starving crew nearly four months later, Europeans had known of the existence of (as one of Magellan's chroniclers described it) "a sea so vast the human mind can scarcely grasp it" (Beaglehole 1974, p. 109). For another two and a half centuries before Cook's expedition, long-distance trading ships occasionally stumbled upon Pacific islands and to their complete bewilderment found them occupied by thriving communities. Having themselves only recently developed the technology to build ocean-spanning ships, the Europeans were simply unable to comprehend how Stone Age people, traveling in canoes built using stone tools and with no navigational devices, could possibly have got there first.

So how did Polynesian and other Oceanic peoples come to colonize such tiny pieces of land in a vast ocean? The navigators must have explored widely in order to locate them in the first place. They must have been confident of finding them again, in order to bring boatloads of people, tubers, and animals. And they must have been sure of returning repeatedly in order to maintain communications. Ocean-going canoes were large and generally double-hulled or with outriggers. Their sails were woven from pandanus leaves (rather like bulrushes) and their ropes manufactured from coconut husks. Polynesian canoes typically used a distinctive crab-claw-shaped sail, wider at the top than the bottom, which enabled them to catch the wind continuously, even in quite heavy swells. Nonetheless, longer journeys would still have taken several weeks. During the 1960s, it was seriously proposed that the Pacific had been colonized by luck rather than by skill. However, this always begged many questions, such as why people left safe islands and sailed off into the unknown, taking all their possessions with them, unless they were forced; and what proportion of the voyagers (surely large) perished at sea, making no landfall. The idea of accidental colonization was firmly laid to rest by the construction in the 1970s of *Hōkūle'a,* a traditional-style Polynesian canoe, reconstructed as authentically as possible and sailed without a compass or any other modern instruments. Between 1975 and 2000, *Hōkūle'a* made several successful voyages between the farthest corners of Polynesia.

Polynesia stretches from the Hawaiian islands in the north to Aotearoa (New Zealand) in the west and tiny, isolated Rapa Nui (Easter Island) in the east. The idea that Polynesia was colonized by people drifting westwards from South America, famously championed by the Norwegian explorer Thor Heyerdahl in the 1950s following the successful voyage of the balsa-wood raft *Kon Tiki* from Peru to the Tuamotus, is completely refuted by a range of archaeological and linguistic evidence. This shows conclusively that Polynesians ultimately derived from south-east Asia. As early as the second milennium B.C.E., peoples characterized by a distinctive type of pottery (Lapita ware) had spread out along the islands of Melanesia to reach the central Pacific islands of Tonga, Samoa, and Fiji. Distinctively Polynesian characteristics developed here in the central Pacific, and these Polynesians subsequently moved eastward to the Society Islands, Marquesas, and Tuamotus, and finally dispersed to the farther islands. The chronology is far from certain, but the Society Islands and Marquesas had probably been settled by about the year 0, Hawai'i and Rapa Nui by C.E. 400, and Aotearoa by C.E. 900.

One problem remains. The assortment of domestic animals and staple crops dispersed by the Polynesians to their many islands included the sweet potato, which comes from the Andes. It seems almost ludicrous to suppose that—as a one-off occurrence at a remarkably early date, before the main

Polynesian expansion—a canoe managed somehow to travel halfway across the Pacific from central Polynesia and happen across South America, to collect the vital tubers, and to successfully return with them. Surely, some have argued, there must have been more prolonged contact, a period when regular voyaging took place across the Pacific, and when regular contact may even have been made with other parts of the Pacific rim. Others take the view that, since the island societies went on to develop in such different ways, any regular contact between distant groups of islands was relatively short-lived.

Be this as it may, regular long-distance voyages of anything over a few hundred kilometers had essentially vanished by the time of European contact. Nonetheless, we know a remarkable amount about Polynesian navigational astronomy from the survival of various oral traditions, many recorded in the early years after European contact. Polynesian navigators enjoyed special status. One of the most famous contacts made by Cook on his first voyage in the *Endeavour* was with a Tahitian navigator, Tupaia. Tupaia could name scores of islands in central Polynesia, specifying (with remarkable accuracy) the direction and distance of each. While the practice of traveling these distances had apparently died out, the knowledge of how to do it had remained, as was demonstrated when Tupaia subsequently sailed away with Cook in the *Endeavour*.

Our fragmentary knowledge of Polynesian and Micronesian navigation techniques owes much to a succession of ethnographers and cultural informants as well as to the experimental voyagers who have researched, and attempted to use, traditional navigation techniques. Their achievement in acquiring and preserving much of what we do know is all the more impressive given that navigational knowledge was often sacred and known only to a very few privileged experts, who were reluctant to reveal it to outsiders. Throughout Oceania, navigators were highly respected people and generally had high social status. Tupaia was a priest, formerly a chief. In Tonga, knowledge of navigation was the preserve of specialist families who had chiefly rank.

Astronomy was crucial in navigation. By day, direction-finding had to depend upon the sun, the winds, currents, and swells. The presence of an island not far beyond the horizon could sometimes be detected from cloud formations or the appearance of certain birds. By night, there were the stars. For relatively short journeys to familiar destinations, one could learn the succession of rising or setting *directional stars* that marked a desired course. More effectively for general use, the navigator could commit to memory a so-called *star compass,* a mental image of the rising and setting positions around the horizon of key stars and star groups, some of which would be visible not too far above the horizon at any particular time. One of the best-documented star compasses derives from the Caroline Islands in Micronesia.

Techniques were also needed to determine the current position. From a

Western perspective it seems natural to determine the latitude by estimating the altitude of the relevant celestial pole above the horizon. This is facilitated in the northern hemisphere by the presence of the bright star Polaris close to the pole. Its height above the horizon can be estimated, for example, by holding out a hand at arm's length. Another method is to observe which stars pass across the zenith. Determining one's longitude, on the other hand, is extremely tricky without instruments. But Oceanic navigators did not think in terms of latitude and longitude. Instead, it seems that they employed (what seem to us) highly ingenious and (to our way of thinking) very strange techniques for *dead reckoning*—determining the absolute distance traveled. This, in combination with a knowledge of the direction of travel, is sufficient to specify the present position.

The colonization of the Pacific was achieved, as far as we know, without any tools apart from the navigator's own body and brain. The only suggestion of the use of what one might call "navigational instruments" comes from Hawai'i, where there is some evidence for the use of hollowed-out gourds as star compasses and sextants, but this is a controversial topic.

Why was the Pacific settled? To European sailors of the sixteenth to eighteenth centuries, this vast ocean was a terrifying void that had to be crossed out of commercial necessity. Whatever motivation prehistoric peoples had had for setting out across it in no more than dugout canoes was simply unimaginable. The ancient Polynesians and other Oceanic peoples, however, thought very differently. To them, this great sea *was* the world: "a familiar, life-giving world . . . strewn with fertile islands on which they could settle, plant their taro, bananas, and other crops, and raise their children" (Finney 1994, p. 3). It is only by starting to appreciate the second of these contrasting worldviews that we can even begin to understand their motivations.

**See also:**
Navigation.
Easter Island; Star Compasses of the Pacific; Zenith Stars in Polynesia.
Celestial Sphere.

**References and further reading**
Åkerblom, Kjell. *Astronomy and Navigation in Polynesia and Micronesia.* Stockholm: Stockholm Ethnographical Museum, 1968.
Beaglehole, J. C. *The Life of Captain James Cook.* Stanford, CA: Stanford University Press, 1974.
Chamberlain, Von Del, John Carlson, and Jane Young, eds. *Songs from the Sky: Indigenous Astronomical and Cosmological Traditions of the World,* 336–347. Bognor Regis, UK: Ocarina Books, and College Park, MD: Center for Archaeoastronomy, 2005.
Finney, Ben. *Voyage of Rediscovery.* Berkeley: University of California Press, 1994.
Flenley, John, and Paul Bahn. *The Enigmas of Easter Island,* 27–40. Oxford: Oxford University Press, 2002.

Gladwin, Thomas. *East Is a Big Bird: Navigation and Logic on Puluwat Atoll.* Cambridge, MA: Harvard University Press, 1970.

Hodson, F. R., ed. *The Place of Astronomy in the Ancient World,* 133–148. London: Royal Society, 1976.

Johnson, Rubellite K., and John K. Mahelona. *Nā Inoa Hōkū: A Catalogue of Hawaiian and Pacific Star Names.* Honolulu: Topgallant Press, 1975.

Kirch, Patrick V. *On the Road of the Winds: An Archaeological History of the Polynesian Islands before European Contact,* 238–243. Berkeley: University of California Press, 2000.

Lewis, David. *We the Navigators* (2nd ed.). Honolulu: University of Hawai'i Press, 1994.

Makemson, Maud. *The Morning Star Rises: An Account of Polynesian Astronomy.* New Haven: Yale University Press, 1941.

Selin, Helaine, ed. *Astronomy across Cultures,* 100–113. Dordrecht, Neth.: Kluwer, 2000.

Woodward, David, and G. Malcolm Lewis, eds. *The History of Cartography, Volume Two, Book Three: Cartography in the Traditional African, American, Arctic, Australian, and Pacific Societies,* 443–492. Chicago: University of Chicago Press, 1998.

## Nebra Disc

In 2002, a spectacular discovery came to the world's attention. It had been made three years earlier in the region of Sachsen-Anhalt in central Germany, close to the small town of Nebra, thirty-five kilometers (twenty-two miles) southwest of Halle, at the summit of a 250-meter (800-foot) hill known as the Mittleberg. Here, inside a pit within a palisaded enclosure dating to the Bronze Age, treasure hunters unearthed a disc thirty-two centimeters (thirteen inches) across, about the size of a dinner plate, made of arsenic-rich bronze. On one side was a decorative pattern of inlaid shapes in thin plate gold. Although corroded and slightly damaged, the original appearance of the pattern is clear enough. To one side of the center was a circle of gold, occupying about a tenth of the whole area. There was also a crescent of comparable size, and three bands in the shape of curved arcs, two of which were placed along the rim on opposite sides, while the other follows a tighter curve, with only its central part touching the rim. The remaining space contained thirty-two smaller dots, scattered unsystematically but generally filling it reasonably evenly, apart from one group of seven dots placed noticeably closer together.

Few could fail to imagine, on first sight of the disc, that these were meant as representations of celestial objects. The crescent moon is obvious; the complete circle could be interpreted either as the sun or the full moon. The fact that there are seven stars in the cluster invites suggestions that it portrayed the Pleiades, one of the most conspicuous groups of stars in the sky. However, the pattern on the disc does not bear any close resemblance to the actual pattern of those stars in the sky—it was not a literal depiction.

*The Nebra disc. (Kenneth Garrett/National Geographic Image Collection)*

Likewise, the disc as a whole was clearly not a literal depiction of the sky. The sun (or full moon) and crescent moon are vastly out of proportion, and the majority of the stars are clearly space fillers.

Could the disc have functioned as some sort of star map or astronomical aid? A number of detailed and highly speculative theories have been put forward, but the strongest evidence in support of this proposition comes from the two golden arcs along opposite rims. Their lengths are almost identical, and each covers almost exactly eighty-three degrees, which at the latitude of Sachsen-Anhalt matches the extent of the eastern and western horizon over which the sun rises and sets, respectively, during the year. Furthermore, the location where the disc was found may well have had calendrical significance. From this location, the Brocken, the highest mountain in northern Germany, is visible on clear days eighty-five kilometers (fifty-three miles) away on the northwestern horizon and is in the direction of sunset on the summer solstice. This has led some to suggest that the disc had a practical use as a device for fixing the time of year by observing the position of sunrise or sunset along the horizon. However, this makes no sense. There are no reference points on the disc, whereas the distant horizon, like any horizon, was full of distinguishing features that could have been used directly to identify different days in an agricultural or ceremonial calendar.

Even those who have sought hardest to find a pragmatic function for the disc tend to acknowledge that the third curved arc, touching the "southern" part of the rim (the left side in the photo), presents a problem. Some say that it resembles a ship sailing through the night: tiny lines on the upper and lower edges might even represent oars. If this interpretation is correct, then the whole disc may depict a legend relating to a ship sailing under the stars, or perhaps sailing through the sky. More speculatively, it could represent a myth in which the sun is transported from the western to the eastern horizon during the night in a ship. On the other hand, if the disc is held the other way up, the mysterious curve may simply be a rainbow.

How can we resolve the apparent contradiction between the interpretation of the disc as a portrayal of a cosmic myth and the apparent exactitude of the sunrise and sunset horizon arcs? If we interpret the disc as a representation of the cosmos, there is not necessarily any contradiction. Indigenous worldviews commonly incorporate (on an equal footing) aspects that we would see as purely mythical and aspects that we would see as reflecting scientific actuality, sometimes observed with impressive exactitude. Many indigenous cosmologies assign distinct sets of meanings and attributes to the "quarters of the world" represented by a fourfold partitioning of the horizon around the solstitial directions. (Two of the dots, or stars, on the Nebra disc were actually found underneath one of the rim bands, which implies that these were added as an afterthought or later elaboration.)

Unlike the Bush Barrow gold lozenge, another Bronze Age artifact for which detailed but highly speculative astronomical interpretations have been put forward, the Nebra disc indisputably contains astronomical/astrological symbols. On the other hand, to think of the disc as a portable star map is probably wide of the mark. It was more likely a remarkable status object of aesthetic rather than practical value, owned by someone with considerable prestige. As a representation of the cosmos it may well have been an object of considerable sacred power.

**See also:**
Science or Symbolism?
Bush Barrow Gold Lozenge; Pawnee Star Chart.
Solstices.

**References and further reading**
Nebra disc "Official website" http://www.archlsa.de/sterne/ [in German, partly translated into English].

# Necker Island

Necker Island is one of the farthest flung of the Hawaiian islands, an isolated rocky outcrop protruding above sea level more than five hundred kilometers (three hundred miles) to the northwest of the main islands in the chain. It is certainly not habitable. It is totally exposed, without fresh water or a natural harbor. It is only some twelve hundred meters (four thousand feet) from end to end and comprises a single narrow ridge, little more than a hundred meters (three hundred feet) at its widest, falling (often precipitously) down to the sea on both sides. And yet, staggeringly, it is covered in no fewer than thirty-four temple platforms. It has been suggested that a shipwrecked party may have been stranded here, building a succession of shrines to appease the gods while surviving by catching trickles of rainwater and snaring birds. A more likely, although still extraordinary, explanation is that Polynesians came here deliberately and repeatedly, evidently in significant numbers, risking life and limb to construct one temple after another in this remotest of places. Why?

One reason might have been that the island was regarded as a highly sacred place. Its traditional name in Hawaiian is Moku-manamana, which has been taken by some to mean "the island with a great deal of sacred power." Unfortunately, more than one translation is possible: the name may simply have meant "branching island." In any case, this does not in itself tell us why Necker might have been regarded as so sacred.

Archaeology does not provide clear answers; it poses more questions. The Hawaiian archaeologist Kenneth Emory, who visited the island in the 1920s, noted that the design of the rectangular temples on Necker bore a striking resemblance to the *marae* of central Polynesia, even though those islands were thousands of kilometers more distant than the other Hawaiian islands. In addition, a number of images of male faces carved out of stone, found on one of the platforms, are similar to stone idols found in the Marquesas, but are quite unlike anything found elsewhere in Hawai'i. It is possible, then, that the activities on Necker have a very early origin, dating to an erstwhile phase of long-distance Pacific exploration preceding many centuries of cultural developments in Hawai'i prior to European contact. In the absence of firm archaeological dating evidence, however, the question remains an open one.

Studies of the orientations of the temple platforms on Necker by the astronomer William Liller have shown that as many as nine are broadly aligned with the December solstice sunrise and June solstice sunset. However, as this is generally in line with the orientation of the ridge, this does not in itself provide conclusive evidence that these alignments were intentional.

A more controversial but also intriguing astronomical explanation has to do with the latitude of Necker Island—23°35' N. There is reasonable evidence to suggest that Polynesians—and Hawaiians in particular—had a keen interest in the passage of the sun through the zenith—the highest point in the heavens. At any location in the tropics, the sun passes through the zenith on two different days in the year, the actual dates depending on the latitude. The Tropics of Cancer and Capricorn represent the limits of this region. At a location on the Tropic of Cancer, the sun passes directly overhead on only a single day in the year, the June solstice. Necker Island lies almost exactly on the Tropic of Cancer, whereas the larger Hawaiian islands lie farther southwards, well inside the tropics. Could Necker's sacred significance have been because it lay on this limit—the symbolic limit of Polynesia?

The two Tropics are, in fact, drifting gradually closer together owing to the slow decrease in the obliquity of the ecliptic. Necker was just within the tropics (located just south of the Tropic of Cancer) before about C.E. 900. Then at around this time the Tropic passed by on its course southwards. Thereafter the zenith sun could be seen from the island no more. Could the sacred power of the island have been at its greatest when the island was precisely on the tropic, and quickly waned thereafter? Such an idea is not inconsistent with the broader chronological evidence: in around C.E. 900 it is possible that Polynesians were regularly voyaging over long distances in the central Pacific. However, all this assumes that people were interested in determining the zenith passage of the sun very precisely, and for this there is no direct supporting evidence.

If achieving such a high level of precision were not an issue, then the idea that Necker was sacred because it marked the limit of the zenith sun would fit with a wider range of dates. In this case, it may be relevant that unusually large numbers of temple platforms are also found on some islands close to the Tropic of Capricorn south of the equator, although the concentrations are nowhere near as dramatic as on Necker. However, it may not be necessary to imagine such a specifically astronomical motivation at all. It may be that Necker's sacredness derived simply from its extreme location—an end place beyond which there were no more habitable islands.

> See also:
> Solstitial Directions; Zenith Passage of the Sun.
> Navigation in Ancient Oceania; Polynesian Temple Platforms and Enclosures; Zenith Stars in Polynesia.
> Obliquity of the Ecliptic; Solstices.
>
> **References and further reading**
> Cleghorn, Paul L. *Prehistoric Cultural Resources and Management Plan for Nihoa and Necker Islands, Hawai'i*. Honolulu: Bishop Museum Press, 1987.
> Emory, Kenneth P. *Archaeology of Nihoa and Necker Islands* (repr.), 51–122. Honolulu: Bishop Museum Press, 2002.
> Kirch, Patrick V. *Feathered Gods and Fishhooks: An Introduction to Hawaiian Archaeology and Prehistory*, 94–98. Honolulu: Honolulu: University of Hawai'i Press, 1985.
> Liller, William. "Necker Island, Hawai'i: Astronomical Implications of an Island Located on the Tropic of Cancer." *Rapa Nui Journal* 14 (4) (2000), 103–105.

# Newgrange

Newgrange is one of three large Neolithic passage tombs that are found within three kilometers of one another on the northern side of the river Boyne west of Drogheda, County Meath, in central Ireland. It incorporates one of the most famous astronomical alignments in the world. Around the shortest day every year, just after sunrise, a shaft of sunlight suddenly enters the tomb and shines directly down the long passage, penetrating right through to the central chamber for a precious few minutes before, equally swiftly, it plunges once again into almost interminable darkness.

Newgrange is a supreme example of the passage tomb genre, of which there are more than three hundred examples to be found across Ireland, with many more in northern Scotland. Built in the later part of the fourth millennium B.C.E., it is an enormous mound some eighty meters (260 feet) across, entered by a passage leading in from an entrance on the southeastern side. Almost twenty meters (sixty-five feet) long, with a roof formed of enormous, flat slabs of rock held up by large, upright side-stones (*orthostats*), the passage leads to a central chamber deep within the tomb. This chamber has an impressive corbelled roof six meters (twenty feet) high and is so large that several people can stand together inside comfortably. Three smaller chambers open off the main central chamber, one opposite the en-

*The entrance to the passage tomb at Newgrange, Ireland, showing the roof-box above it. (Courtesy of Clive Ruggles)*

trance and one on each side. Fragments of human bone, both burned and unburned, together with animal bones, pendants, beads, and other grave goods were found within the tomb.

Before Michael O'Kelly began excavations at the site in the early 1960s, visitors were perplexed by what appeared to be an additional roof slab above the entrance, apparently a false lintel about one meter (three feet) higher than the level of the main roof of the passage. This turned out to be the top of a *roof-box* above the passage entrance. It is through this roof-box that the sunlight enters at midwinter. This is not a precise phenomenon: it is repeated each day for a few days after the solstice, though for a progressively shorter period of time each day; and likewise (increasing up to the maximum duration) beforehand. Owing to the changing position of the solstitial sun over the millennia (due to the changing obliquity of the ecliptic), the shaft of sunlight would have appeared for somewhat longer, and for a few more days, at the time when the tomb was built.

There are many beautifully decorated stones at Newgrange, and it has been suggested that the interplay of sunlight and shadow across various spirals, lozenges, and other carvings at certain times of year might also have been significant to the builders. However, since eye-catching associations of this nature could easily have arisen fortuitously, such ideas are more controversial. Similarly, claims of alignments involving the circle of standing stones that surround the tomb also remain speculative.

One thing that can be stated with relative certainty, however, is that Newgrange was not an ancient observatory. It is ludicrous to suggest that people entered the tomb and sat among the bones of the dead, merely to determine whether or not the shortest day of the year had arrived. Frivolous as this point sounds, it is important in that for many years a number of people—most famously the Scottish engineer Alexander Thom—claimed that free-standing megalithic sites that had no other obvious primary function had been constructed as astronomical observing instruments. In fact, this theory runs counter to common sense: people in prehistory would clearly not have gone to all the bother of constructing monumental architecture, even just one or two standing stones, to do what could be achieved perfectly effectively using non-permanent markers and the natural horizon. Where such monuments did incorporate intentional astronomical alignments, they expressed what was already known, and there was surely some greater purpose. At Newgrange the greater purpose is self-evident: the site was first and foremost a tomb, or (to use a form of words that better reflects its continual use over a considerable period) a shrine for offerings relating to the ancestors. Either way, the solstitial alignment surely expresses some link that existed in people's minds between the sun, the seasonal cycle, and the ancestors, although in itself it tells us nothing about the nature of that association.

At some date, possibly many generations after its initial construction, the entrance was blocked off using a *closing stone* weighing about a ton. However, because the critical shaft of sunlight entered through the roof-box rather than the entrance itself, the solstice hierophany was unaffected. Some have argued that this was intentional: by using the roof-box device, the builders ensured that the sun would continue to enter the tomb even after it had been sealed off. And yet this interpretation brings its own problems: is it not somewhat surprising that the original builders had the intention—or at least planned for the possibility—that what was then a dominant practice would some day come to a sudden end?

In Irish legend, Newgrange tends to be identified as Brug na Bóinne ("the mansion of the Boyne"), and thereby as the abode of some of the old gods of Ireland, including one of their principal figures, the Dagda (or Daghdha). His cauldron, it is said, was the vault of the sky, and this has led some to propose that the solstitial orientation of the tomb could be a manifestation of some long forgotten version of this legend. But the connections are too few and tenuous, and the timespan far too long between Neolithic times and even the earliest recorded legends in early Christian times, for much credulity to be attached to any suggestion of a direct continuity of tradition.

What, then, might have been the significance of the alignment? Although we cannot draw direct parallels, modern indigenous practices can sometimes give us useful insights in our attempts to interpret practices in the past. Thus, for example, it was important to the Skidi Pawnee of the U.S. Midwest that the first rays of the morning sun should enter their traditional dwelling ("earth lodge") to bring life-giving strength and power. Could a similar conviction explain why people in the past appear to have been so interested in sunlight entering dark places at special times? This seems evident at a variety of prehistoric tombs, of which the solstitial alignment at Newgrange is merely the most famous. Many of these tombs reflect the structure of houses for the living in various ways, and it is not unreasonable to suppose that they were seen as houses for the dead, or for ancestral spirits. Did people believe that the light of the sun imparted strength and power to the ancestors, as it did to them?

The example of Newgrange shows how even a simple, spectacular, and relatively uncontroversial astronomical alignment can give rise to many subtleties of interpretation when we try to explore the nature of its significance to people in the past.

**See also:**
Science or Symbolism?; Solstitial Directions; Thom, Alexander (1894–1985). Boyne Valley Tombs; Pawnee Earth Lodge. Obliquity of the Ecliptic; Solstices.

**References and further reading**
Jestice, Phyllis J. *Encyclopedia of Irish Spirituality,* 42, 101. Santa Barbara: ABC-CLIO, 2000.
MacCana, Proinsias. *Celtic Mythology,* 66–67. London: Hamlyn, 1970.
O'Kelly, Claire. *Illustrated Guide to Newgrange* (3rd ed.). Wexford, UK: John English, 1978.
O'Kelly, Michael. *Newgrange: Archaeology, Art and Legend.* London: Thames and Hudson, 1982.
Ruggles, Clive. *Astronomy in Prehistoric Britain and Ireland,* 12–19. New Haven: Yale University Press, 1999.
Stout, Geraldine. *Newgrange and the Bend of the Boyne,* 40–47, 62, 65–67. Cork: Cork University Press, 2002.
Waddell, John. *The Prehistoric Archaeology of Ireland,* 59–62. Galway: Galway University Press, 1998.
Walker, Christopher, ed. *Astronomy before the Telescope,* 21–23. London: British Museum Press, 1996.
Whittle, Alasdair. *Europe in the Neolithic,* 244–248. Cambridge: Cambridge University Press, 1996.

# Nissen, Heinrich (1839–1912)

Heinrich Nissen, professor of history at the University of Bonn, deserves more than anyone else to be recognized as the earliest pioneer of modern archaeoastronomy. His interests ranged from the orientations of Egyptian and Greek temples—on which he published his first paper in 1885—to the orientations of churches, where he was responsible for demolishing the popular myth that all churches faced (precisely) east. He was an almost exact contemporary of Sir Norman Lockyer in England, but he anticipated Lockyer's own measurements of Egyptian temples. Indeed, Lockyer was dismayed to discover the existence of Nissen's published work after returning from his own first season in Egypt and, when putting forward his own ideas that the temples were aligned upon the rising and setting of the sun and stars, fully acknowledged that Nissen had anticipated him in suggesting the possibility.

The fact that Nissen is rarely mentioned in accounts of the early development of ideas and methods in what subsequently became known as archaeoastronomy, whereas Sir Norman Lockyer is generally considered the earliest serious pioneer in the subject, is explained by the fact that initial developments in modern archaeoastronomy in the 1960s and 1970s primarily took place amongst English-speaking scholars. Nissen's great work on the subject, the three-volume *Orientation: Studien zur Geschichte der Religion*, was published between 1906 and 1910 in German. The first volume appeared in the very same year as Lockyer's seminal *Stonehenge and Other British Stone Monuments Astronomically Considered*.

See also:
Archaeoastronomy; Lockyer, Sir Norman (1836–1920).
Church Orientations.

**References and further reading**
Lockyer, Norman. *Stonehenge and Other British Stone Monuments Astronomically Considered*, 459. London: MacMillan, 1906.
Michell, John. *A Little History of Astro-Archaeology*, 29. London: Thames and Hudson, 1989.
Nissen, Heinrich. *Orientation. Studien zur Geschichte der Religion* (3 vols.). Berlin: Weidmannsche Buchhandlung, 1906–1910. [In German.]

# Novae
*See* Comets, Novae, and Meteors.

# Nuraghi

Nuraghi are tall towers, remarkable because of their scale and the fact that they are built entirely of dry stone. Found on the Mediterranean island of Sardinia, they are also remarkable for having been built in extraordinary numbers. No fewer than seven thousand are known to have been constructed, during the second millennium B.C.E., on an island measuring no more than about four hundred kilometers (250 miles) by two hundred kilometers (120 miles). They are characterized by a central chamber with a corbelled roof, often of magnificent proportions. Several have a second corbelled chamber built directly above the first, and a few have yet a third on top of that. An example of this last category is Santu Antine, situated some fifty kilometers from Alghero in the northwest of the island. Its lowest chamber rises to a height of just under eight meters (twenty-six feet), the next to over five meters (seventeen feet), and the highest was of comparable size but its roof has now collapsed. The remaining tower still stands to a height of over 17.5 meters (fifty-seven feet). The majority of the nuraghi were simple towers, but some were elaborated by the addition of a surrounding complex of walls, corridors, chambers, and smaller towers. Commonly, these *complex nuraghi*—like Santu Antine itself—have a three-fold symmetry, the compound stretching out to rounded lobes at three corners of a roughly equilateral triangle; others have four or even five corners.

The function of the nuraghi remains an archaeological puzzle. Astronomical associations have been suggested at a number of them, including solstitial alignments in both directions along two of the three outer walls at Santu Antine. Alignments between nuraghi may also have been important: local archaeoastronomer Mauro Zedda has identified a number of solstitial and also lunar alignments between nuraghi in the vicinity of the town of Isili in the center of the island. For example, from the now-ruinous nuraghe Nueddas, the tower of nuraghe Is Paras is prominent on the skyline in the direction of midsummer sunset, while that of nuraghe Longu marks midsummer sunrise in a similar way.

However, such alignments can arise by chance—especially considering the large numbers of nuraghi in the Sardinian landscape—and systematic studies are clearly needed. Recently such a study has been attempted. It concentrated upon the orientation along the main axis toward the entrance and measured this direction at 450 nuraghi from all over the island. The study concluded that nuraghi are predominantly oriented southeastward and southward, a pattern that fits an "orientation signature" common among tombs and temples in western Europe, and in particular applies also to the *tombi di giganti* ("tombs of giants"), a type of megalithic tomb attributed to the same culture that built the nuraghi. Although this pattern corresponds to a general practice of orientations toward where the sun climbs in the sky, a strong preference for orientations close to azimuth 150 degrees may be better explained in relation to the rising of the Southern Cross. This suggestion appears to be supported by evidence of a gradual orientation change with time, which would have resulted if monument alignments had followed the inexorable southward movement of this asterism over the centuries that occurred as a result of precession.

See also:
Solstitial Directions.
Is Paras; Prehistoric Tombs and Temples in Europe.
Azimuth; Precession.

**References and further reading**

Contu, Ercole. *Il Nuraghe Santu Antine.* Sassari: Carlo Delfine Editore, 1988. [In Italian.]

Hoskin, Michael. *Tombs, Temples and Their Orientations,* 183–185. Bognor Regis, UK: Ocarina Books, 2001.

Littarru, Paolo, and Mauro Peppino Zedda. *Santu Antine: Guida Archeoastronomica al Nuraghe Santu Antine di Torralba.* Cagliari, Italy: Associazione Culturale Agora' Nuragica, 2003. [In Italian.]

Zedda, Mauro Peppino. *I Nuraghi: Il Sole La Luna.* Cagliari, Italy: Ettore Gasperini Editore, 1991. [In Italian.]

Zedda, Mauro Peppino. *I Nuraghi tra Archeologia e Astronomia.* Cagliari, Italy: Agorà Nuragica, 2004. [In Italian.]

Zedda, Mauro Peppino, and Juan Antonio Belmonte. "On the Orientations of Sardinian Nuraghes: Some Clues to their Interpretation." *Journal for the History of Astronomy* 35 (2004), 85–107.

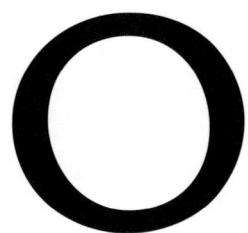

## Obliquity of the Ecliptic

*Obliquity of the ecliptic* is not a term that trips easily off the tongue, nor is it an easy concept to grasp technically. Yet it is vitally important for anyone investigating possible architectural alignments upon the sun, moon, or planets to take account of the way the obliquity of the ecliptic changes over the centuries, just as it is necessary for investigations of possible stellar alignments to take account of precession.

Each day, as the stars and sun move together around the rotating celestial sphere, the sun moves by a small amount against the stars. Only after a full year does it return to the same position, having traced out a path through the stars known as the *ecliptic*. The ecliptic is a circle, but it is inclined at an angle to the daily paths of the stars, which move around on lines of constant "latitude," or declination. In particular, if we focus on the celestial equator because it is the one line of constant declination that is also a *great circle* (a circle of the largest possible size that can be drawn on a sphere, its center coinciding with the center of the sphere), then we find that the angle of inclination between the ecliptic (also a great circle) and the celestial equator is between twenty-three and twenty-four degrees. It is this angle that is known as the obliquity of the ecliptic.

At a certain time of year, the sun reaches the northernmost point on the ecliptic. This is the June solstice, and its declination at this time is approximately 23.5 degrees. Six months later, it reaches its southernmost point: this is the December solstice, and its declination now is approximately –23.5 degrees. When the sun crosses the celestial equator, in the spring and autumn, its declination is 0. These times are known as the equinoxes.

To understand what the obliquity is and hence to understand how it changes, it is best to switch to the astronomer's perspective, looking in on the solar system. The earth, as it orbits annually around the sun, also spins daily about its own axis. If this axis were perpendicular to the plane of the earth's orbit, then the sun would simply appear to us to move around the celestial equator. But it isn't. The axis of spin (the earth's polar axis) actu-

317

ally leans over by about 23.5 degrees, and this is the reason that the ecliptic is inclined to the celestial equator by this amount; in other words, it is this leaning over that gives rise to the obliquity.

Putting to the back of our minds for a moment that it is also orbiting around the sun, the earth in fact resembles a spinning top, in that it not only spins daily about its axis, but the axis itself gradually turns, going once round approximately every twenty-six thousand years. It is this turning that is known as precession. But a spinning top also wobbles; and so does the earth. The difference is that a top gradually spins down, so its "obliquity" gets greater and greater until the top falls over. The tip of the earth's axis, on the other hand, oscillates slightly up and down. The obliquity in fact swings between about twenty-five degrees and about twenty-two degrees over a timescale of some forty-one thousand years.

The last five thousand years represent a small segment of this cycle. During this time the obliquity has been gradually decreasing, from approximately 24.0 degrees in 2500 B.C.E. to about 23.5 degrees in the present day. One practical effect of this is that the ranges over which the rising and setting position of the sun swings during the year were somewhat wider in, say, Neolithic times. In particular, the rising and setting positions of the sun at the solstices were about one degree (two solar diameters) further away from each other then than they are now. (The shift is actually in declination: at higher latitudes, where the rising and setting path is shallower, the shift in azimuth is greater than this.) This is important in assessing the astronomical significance of later prehistoric monuments such as Newgrange (Ireland) or Brainport Bay (Scotland). The rising and setting ranges of the moon and planets have contracted correspondingly over recent millennia, and it is important for archaeoastronomers to take these changes into account. But for the moon and planets the effects of the changing obliquity are superimposed upon much more complex shorter-term variations.

The longer-term cycle of change in the obliquity of the ecliptic becomes important if it is proposed that much older (Palaeolithic) sites might have been solstitially aligned. An example of this is Parpalló, a cave in eastern Spain that was occupied from about 19,000 B.C.E. to 8000 B.C.E. and into which the light of the sun would have penetrated after sunrise at the winter solstice.

Another practical effect of the changing obliquity is that the Tropics of Cancer and Capricorn, and the Arctic and Antarctic circles, move gradually up and down on the earth. Apart from affecting the size of the tropics and polar regions, this movement might also have had a direct effect upon myths associated with, and ritual significance accorded to, places that were or had been on the tropic. A classic example of this is Necker Island in the Hawaiian chain.

In summary, precession causes the rising and setting positions of the stars to change over the centuries, but it is the change in the obliquity of the ecliptic that affects the rising and setting positions of the sun, moon, and planets. Compared with the changes to stellar rising and setting positions due to precession, the changes in the rising and setting positions of the sun, moon, and planets due to the changing obliquity are not great, but they do become significant over a time scale of millennia.

**See also:**
Equinoxes.
Brainport Bay; Necker Island; Newgrange.
Azimuth; Celestial Sphere; Declination; Ecliptic; Precession; Solstices; Star Rising and Setting Positions; Sun, Motions of.

**References and further reading**
Aveni, Anthony F. *Skywatchers*, 102–103. Austin: University of Texas Press, 2001.
Ruggles, Clive. *Astronomy in Prehistoric Britain and Ireland*, 57. New Haven: Yale University Press, 1999.
Ruggles, Clive, Frank Prendergast, and Tom Ray, eds. *Astronomy, Cosmology and Landscape*, 8–14. Bognor Regis, UK: Ocarina Books, 2001.

# Orientation

Many different factors might have influenced the orientation of an ancient dwelling house, sacred building, or tomb. Some would have had clear practical benefits. Thus anyone building a house in temperate regions of the northern hemisphere might well consider orienting its entrance southeastward so that the early rays of the sun enter the house and help warm it up in the mornings. Similarly, it would be beneficial to avoid having entrances facing into the prevailing wind. Tomb entrances might face downhill so that water would not run down into them and the interior stayed dry. The possibilities seem almost endless.

But studies of historical and indigenous peoples all over the world reveal a huge variety of less tangible considerations, such as orienting a temple upon a sacred mountain seen as the dwelling place of a god, or a tomb in the direction whence ancestors are believed to have come. This "intangible" category includes a range of astronomical considerations, such as aligning temples and tombs upon sunrise at one of the solstices or the rising point of a significant star or constellation.

A common reason for choosing particular orientations has to do with cosmology—people's understandings and beliefs concerning the nature of the world—and is motivated by a desire to keep human activity in harmony with the cosmos to ensure the continuity of life, good health, and so on. Thus one reason that the traditional earth lodges of the Pawnee face east is so that every time people enter their home, they do so in the same manner—

the same direction—as the stars enter the sky. To a Western mind this might amount to little more than superstition, but such considerations are often bound up inextricably with others that we might consider more pragmatic. Thus, while some archaeologists argue that the majority of Iron-Age roundhouses in Britain faced the rising sun for primarily cosmological reasons, there would also have been practical benefits in terms of warmth from the sun in the mornings and shelter from the prevailing wind. Indeed, to the minds of the time, the distinction we might make between pragmatism and religious or superstitious belief would not have been a meaningful one. All these different considerations served the common purpose of sustaining life.

Interpreting the orientations of archaeological structures can be highly problematic, especially if we only have the material evidence to go on. Quite simply, everything must point somewhere. The mere fact that something points at what we might consider a prominent topographic feature, or an important astronomical event, or anything else of apparent significance, does not mean that this was deliberate. One solution, where possible, is to seek groups of monuments or buildings that are culturally related and similar in form and structure (and hence, probably, purpose) and to look for common trends in their orientation. If a certain pattern emerges strongly enough, especially if we can obtain a suitable statistical verification, we can be reasonably confident that a certain set of orientations was deliberate. But even here there are a number of problems. For a start, we have to be careful about data selection. We delude ourselves, and invalidate any statistical results, if we include a sample of orientations that fit a desired pattern but exclude others that do not. Secondly, it is quite possible that more than one factor was important in determining an orientation, and that compromises may have been reached: consider a culture where the prevailing practice was to align a temple both upon a sacred mountain and upon the rising sun. Another problem is that only the most dominant and widespread common practices will be revealed in this way. People do not act like laws of physics: even when constrained by the strongest social traditions, people often do things in their own manner. According to a burial practice recorded among the Ashanti people of Ghana and the Ivory Coast, for example, the dead are buried facing away from the village. However, a small percentage of people believe that the dead turn themselves around in the grave, so they bury their dead in exactly the opposite direction, facing toward the village. This would create a tricky problem for an archaeologist of the future trying to make sense of the orientations of Ashanti graves.

Despite all these difficulties, the approach of looking at groups of similar monuments has achieved some startling successes. Perhaps the best example of this is the orientations of later prehistoric tombs and temples in western Europe, no fewer than three thousand of which have been measured and analyzed by the British historian of astronomy Michael Hoskin. There are a multitude of local groups—distinctive types of stone monument built in different localities during some two millennia of prehistory—and in nearly every case the orientations are bunched, clearly intentionally, into a restricted range of directions forming a characteristic *orientation signature*. A variety of such orientation signatures is encountered. This shows that orientation was of near universal importance, although traditions and practices—just like other aspects of the design and construction of the monuments—varied considerably from place to place and time to time.

Even where we do have demonstrably strong and consistent trends such as these, we cannot always be confident about the motivation. For example, three quarters of the Early Neolithic long barrows on Cranborne Chase in southern England face southeast. These might be interpreted as alignments upon midwinter sunrise until it is realized that their orientation is constrained by the hill ridges along which the barrows are placed; these tend to run northwest-southeast as a result of the local geology. And even if we have few doubts about the intended target, we need to look beyond mere statistics—to the social context—in order to infer anything meaningful about their purpose and significance.

See also:
Cardinal Directions; Cosmology; Methodology; Solstitial Alignments; Statistical Analysis.
Church Orientations; Iron Age Roundhouses; Navajo Hogan; Pawnee Cosmology; Prehistoric Tombs and Temples in Europe.

**References and further reading**
Aveni, Anthony F. *Skywatchers,* 217–222. Austin: University of Texas Press, 2001.
Hoskin, Michael. *Tombs, Temples and Their Orientations,* 7–20. Bognor Regis, UK: Ocarina Books, 2001.
Ruggles, Clive. "Megalithic Astronomy: The Last Five Years." *Vistas in Astronomy* 27 (1984), 231–289, pp. 271–283.
———. *Astronomy in Prehistoric Britain and Ireland,* 89–90. New Haven: Yale University Press, 1999.
Ruggles, Clive, and Alasdair Whittle, eds. *Astronomy and Society in Britain during the Period 4000–1500 B.C.,* 243–274. Oxford: British Archaeological Reports (British Series 88), 1981.

# Orion

Except insofar as their visibility and position in the sky is affected by latitude and local conditions, the patterns of bright stars are the same for all human cultures. However, what different people "see" in the sky can vary enormously from one cultural context to another. First there is the question of what people notice at all: it is not always the brightest objects. There are also significant differences in which stars people perceive as grouped to-

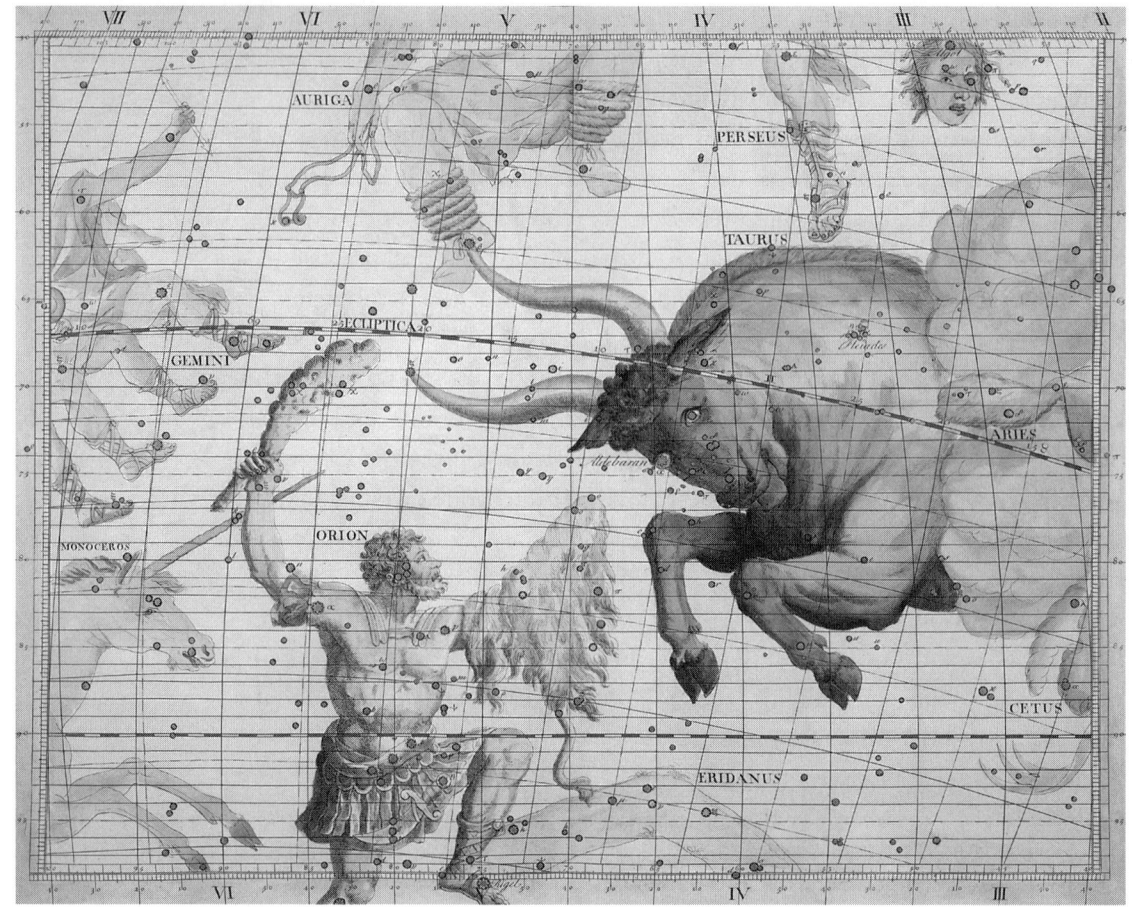

*The constellations of Orion and Taurus, by James Thornhill (1725). (Stapleton Collection/Corbis)*

gether to form constellations, and how people interpret what they see varies widely. This said, certain configurations of bright stars are so eye-catching that they have assumed importance in a great range of human cultures. One of the most obvious is the trapezium of four bright stars with a short line of three at its center that we know as Orion.

The shape of Orion lends itself very easily to being interpreted as a person wearing a tunic and belt, especially in temperate latitudes where the figure is seen upright. Its nightly motion across the sky encourages the idea the Orion figure is chasing, or being chased by, other characters. Our own image of the constellation as Orion the hunter derives from Greek tradition, but the idea of the Orion figure as a male in pursuit of a group of young women (the Pleiades) can also be found in legends as far afield as Aboriginal Australia. A variant is found in Vedic myth, where Orion is the creator figure Prajapati in pursuit of his daughter Rohini (Aldebaran). Orion's belt is an arrow, shot at him from behind by an angered hunter (Sirius). In ancient China the stars of Orion and Scorpius were two quarreling brothers who ended up on opposite sides of the sky.

The calendrical importance of a constellation is often inseparable from the mythology and meaning that attaches to it. In Navajo tradition Orion was First Slim One, the keeper of the months, whose heliacal phenomena (along with those of certain other star groups) were used to keep track of the seasons.

Not all cultures see Orion as a person. Close to the equator, where the constellation appears on its side, other interpretations are more common. On Kapingamarangi in the Caroline Islands (Federated States of Micronesia), for example, it was seen as a canoe house with three men inside. In rural Java, Indonesia, the belt plus three of the four outer stars—Rigel ($\beta$ Ori), Bellatrix ($\gamma$ Ori), and Saiph ($\kappa$ Ori), but not Betelgeuse ($\alpha$ Ori)—form the constellation of the plough, whose annual cycles of appearance and disappearance traditionally helped regulate the agricultural year.

Not everyone groups the same set of stars that we see as Orion into a single constellation. The three stars of Orion's belt—Mintaka ($\delta$ Ori), Alnilam ($\varepsilon$ Ori), and Alnitak ($\zeta$ Ori)—form a prominent asterism in their own right: to the Tswana of South Africa, for example, they are three pigs, being chased by three dogs (a line of three fainter stars that we know as Orion's sword). To the Lakota of the United States, the stars of Orion merely form the central part of a larger constellation, which includes the Hyades and Sirius: a large buffalo being born out of an even larger circle of stars.

It is problematic to postulate stellar alignments at prehistoric sites where the date is uncertain, but plausible alignments upon the three stars of Orion's belt have been recognized at a few. One of these is the Thornborough group of henges in Yorkshire, England. It has also been suggested, both at Thornborough and more notoriously in the case of the three Pyramids of Giza, that sets of monuments on the ground actually form a spatial representation of these stars. Such arguments are a lot more problematic because of the total lack of any convincing evidence for any human society making *constellation maps* of this sort. At Giza, for example, the historical context and the local topography provide a far simpler explanation for the layout of the three pyramids.

Orion is close to the celestial equator and hence visible at most times of the year from nearly everywhere on Earth. Yet it was not always so. On a time scale of centuries and millennia, the position of every constellation shifts as a result of precession. Orion may be close to the celestial equator now, but this is just about as far north on the celestial sphere as it ever reaches. For most of the next twenty-six thousand years, as for most of the last, it will be situated well to the south, prominently visible in the southern hemisphere but much less conspicuous in northern latitudes.

**See also:**
Constellation Maps on the Ground.
Aboriginal Astronomy; Javanese Calendar; Lakota Sacred Geography; Navajo Cosmology; Pyramids of Giza; Thornborough.
Celestial Sphere; Precession; Star Names.

**References and further reading**
Allen, Richard H. *Star Names: Their Lore and Meaning,* 303–320. New York: Dover, 1963.
Brown, Dayle L. *Skylore from Planet Earth: Stories from Around the World . . . Orion.* Bloomington, IN: AuthorHouse, 2004.
Chamberlain, Von Del, John Carlson, and Jane Young, eds. *Songs from the Sky: Indigenous Astronomical and Cosmological Traditions of the World,* 320–335. Bognor Regis, UK: Ocarina Books, and College Park, MD: Center for Archaeoastronomy, 2005.
Ridpath, Ian, ed. *Norton's Star Atlas and Reference Handbook* (20th ed.), 152–153. New York: Pi Press, 2004.
Ruggles, Clive, and Nicholas Saunders, eds. *Astronomies and Cultures,* 1–8. Niwot: University Press of Colorado, 1993.
Selin, Helaine, ed. *Astronomy across Cultures,* 9–13, 75–79, 295. Dordrecht, Neth.: Kluwer, 2000.

# Orion's Belt
*See* Orion.

# Palaeoscience

Where might we hope to find the earliest signs of human enquiry into the causes of what we understand as natural phenomena, and what was their nature? Many different answers may be found in different parts of the world, and not only in the prehistoric past. But they emerged in the context of conceptual frameworks that bear little resemblance to the way that we classify things we perceive in the natural world according to the Western scientific tradition. Within these other frameworks—as in the development of our own—understanding was not sought as an end in itself but was inextricably bound up with ideology and religion, myth and ritual, politics and power, and what to us seem the more practical, everyday aspects of life—in short, with all aspects of sacred and mundane knowledge and practice. Our own ideas about how people conceived of the world in times and places where no form of writing had developed, and where no historical account is available, can only be formulated in the context of our knowledge and theories about human development and behavior in general, and can only be directly enlightened by what is left to us in the form of material remains. Our own interpretations can only be achieved through the filter of our own cultural background.

The very earliest stages in the development of human thought remain obscure. Only a few have ventured to explore the development of cognition on an evolutionary time scale. Physically, the human mind has changed little since the emergence of *Homo sapiens sapiens* between about 100,000 and 50,000 B.C.E., although some scholars suggest that certain key developments, such as the ability to communicate using language, had already taken place some two million years before. By combining ideas from evolutionary psychology and evidence from palaeolithic archaeology, archaeologist Steven Mithen has suggested that in the earlier stages of human evolution, human intelligence was compartmentalized into four separate specialist domains—technical, social, natural historical, and linguistic—and these only came together at the time of transition to the Upper Palaeolithic period, around 30,000 B.C.E. This is the time when fine cave-art depictions of animals appeared in western Europe and when the first clear evidence emerges of decorative items being buried with the dead, although deliberate burial itself may date back at least to the Neanderthals of the Middle Palaeolithic, up to 100,000 years earlier. Though controversial, Mithen's ideas draw attention to what is perhaps a key notion in the development of human thought: the way people come to terms with, and make sense of, themselves and their environment has much to do with drawing associations between different things that they perceive in the world around them.

One challenge is to try to trace evidence for the development of certain fundamental concepts. This is not a trivial matter. No concept seems more abstract and inviolable than that of number, and yet the anthropologist Thomas Crump has identified three types of symbolic number, which he takes to represent three levels of conceptual evolution, and examined their use in different cultural contexts. This study begins to suggest orders of relationship between levels of numerical notation and other aspects of cognition. Nevertheless, it may be inappropriate to try to trace the evolution of most human concepts in this way; they may simply be different in different worldviews. For example, Westerners have a concept of direction as something precisely defined: two buildings aligned, respectively, due east and two degrees south of east would be said to point in slightly different directions. There is some evidence, on the other hand, that the four directions that are a common feature of indigenous worldviews in North and Central America were conceived rather more as (what we would see as) ranges of directions centered upon (what we would see as) the cardinal points. Wherever possible, we must try to avoid assuming that concepts that seem fundamental to us are in fact universal and inevitable.

Concepts of measurement have a special place in the development of thought. These can be recognized in various ways in the archaeological record. A classic example is a set of carefully manufactured cubical stones from the city of Mohenjodaro in the Indus valley civilization of c. 2000 B.C.E. The masses of the stones are very close to exact multiples of a unit of 0.836 grams. From this it can be deduced, among other things, that the people who made and used these stone weights possessed a concept equivalent to our notion of weight or mass, a concept of modular measure, a hierarchical system of numeration based on units of four or sixteen, and a notion of equivalence. Measuring things is one aspect of trying to make sense of the world. But even this does not in itself tell us anything of how people *explained* natural phenomena. Such explanations stem from correspondences that people perceive in the world. The mental process of scientific discovery involves bringing together phenomena and concepts that were not previously seen as related.

The Scottish engineer Alexander Thom suggested that a sophisticated "megalithic science" (although he never himself used this term) existed in later prehistoric Britain. Thom reasoned that "megalithic man," as he styled him, had

> a solid background of technological knowledge. Here I am thinking not only of his knowledge of ceramics, textiles, tanning, carpentry, husbandry, metallurgy, and the like, but of his knowledge of levers, fulcrums, foundations, sheerlegs, slings, and ropes. . . . There was also his ability to use boats: he travelled freely as far as Shetland, crossing the wide stretch of open water north of Orkney, as well as the exceedingly dangerous Pentland Firth and the North Channel between Kintyre and Ireland. This involved a knowledge of the tides and tidal currents that rule those waters. (Thom 1971, pp. 9–10)

With this in mind, Thom measured and analyzed the geometrical groundplans and potential astronomical alignments at several hundred megalithic monuments (mostly settings of free-standing megaliths such as stone circles and stone rows) in northern and western Britain. One of his main conclusions was that high-precision observations were made of the sun, moon, and stars using features such as notches and hilltops on distant horizons as natural foresights against which minute changes in their rising and setting positions could be measured. There was a "megalithic calendar," which divided the year into eight, or possibly sixteen, equal periods, epoch dates being marked by precise solar alignments at various megalithic sites around the country. He also concluded that sites all over Britain were laid out using a small number of precise geometrical constructions using a standard unit of length (the "megalithic yard") accurate to about one millimeter.

Thom's theories have not stood the test of time. On the archaeological side, detailed reassessments have revealed a variety of subtle biases in the selection of data that have destroyed the statistical conclusions. Astronomical critiques have drawn attention to day-to-day changes in atmospheric refraction and to the effects of extinction near the horizon. These have shown that programs of naked-eye observations of the precision envisaged by Thom could never have been successfully carried out. Similarly, statistical reappraisals have shown that Thom's ideas about precise megalithic mensuration and geometry cannot be supported on the evidence available.

Observations and perceptions of astronomical phenomena do, nonetheless, frequently feature in symbolism associated with places of significance in the landscape or the design of monumental architecture. This is because human activities tend to reflect the perceived cosmos, of which the sky is viewed as an integral part. The alignment of numerous European prehistoric temples and tombs upon horizon astronomical events is probably best understood in this way. But such alignments are not the only manner in which ancient obsessions with the skies may show up in the material record. There is widespread evidence, for example, of the deliberate positioning and design of art and architecture to permit the interaction of sunlight and shadow at certain times, a famous example being the "sun dagger" at Fajada Butte in Chaco Canyon, New Mexico.

It is tempting to speak of astronomy as the oldest science, but in doing so we may fail to recognize that attempting to make sense of what was perceived in the sky merely formed an integral part of human understanding of the natural world in an integrated way. The solstitial orientation of the passage tomb at Newgrange, Ireland, expressed something about the perceived association between ancestors and the sun; it was not set up to measure or track the solar motions. Likewise, in designing their *earth lodges* the Skidi Pawnee of North America paid attention to details of light and shadow and used viewing stations for astronomical observations. However, we would be wrong to view the earth lodge as an observatory in the Western sense. It was, according to ethnoastronomer Von Del Chamberlain, "primarily a shelter to protect its builders from the intense solar heat, cold winter air, wind, rain, snow, and creatures that might molest them" (Chamberlain 1982, p. 166).

See also:
Cosmology; "Megalithic" Calendar; Megalithic "Observatories"; Science or Symbolism?; Thom, Alexander (1894–1985).
Fajada Butte Sun Dagger; Newgrange; Pawnee Earth Lodge; Prehistoric Tombs and Temples in Europe; Stone Circles.

**References and further reading**
Chamberlain, Von Del. *When Stars Came Down to Earth: Cosmology of the Skidi Pawnee Indians of North America,* 166, 178. Los Altos, CA, and College Park, MD: Ballena Press/Center for Archaeoastronomy, 1982.
Crump, Thomas. *The Anthropology of Numbers.* Cambridge: Cambridge University Press, 1990.
Donald, Merlin. *Origins of the Modern Mind: Three Stages in the Evolution of Culture and Cognition.* Cambridge, MA: Harvard University Press, 1991.
Heggie, Douglas. *Megalithic Science: Ancient Mathematics and Astronomy in Northwest Europe.* London: Thames and Hudson, 1981.
Mithen, Steven. *The Prehistory of the Mind: The Cognitive Origins of Art, Religion and Science.* London: Thames and Hudson, 1999.
Renfrew, Colin. *Towards an Archaeology of Mind.* Cambridge: Cambridge University Press, 1982.
Thom, Alexander. *Megalithic Lunar Observatories.* Oxford: Oxford University Press, 1971.

# Pantheon

This extraordinary temple, one of the main tourist sites of Rome, was completed in the first half of the second century C.E. The building itself survives almost intact and is undoubtedly one of the architectural wonders of the world. Its enormous dome, more than forty-three meters (142 feet) in diameter, remained the largest in the world for thirteen centuries. It is made of solid concrete and is near-perfectly hemispherical on the inside (although, for practical reasons, it is thinner and less dense toward the top), and supported upon masonry walls more than six meters (twenty feet) thick. At its apex is a circular hole nine meters (twenty-nine feet) across, so that the visitor feels at once enclosed and yet exposed to the elements. Though now largely empty, the building was once richly and colorfully decorated. The ceiling of the dome was stuccoed, and stones of different colors (porphyry, marble, and granite) were used in the walls and floor. We can imagine the numerous recesses filled with statues of the gods.

Although more recently consecrated as a Catholic church, the temple was originally dedicated to the sun and stars. An intriguing question is the extent to which religious and astronomical symbolism was reflected in the architecture. Proportionality evidently mattered, in that the height is precisely the same as the width; in other words, if one were to imagine the dome being extended into a complete sphere it would just touch the ground in the middle. A more controversial matter is whether sun-and-shadow effects were deliberately incorporated. It is certainly true that as the sun moves through the sky at different times of the year and day, natural light entering through the apex creates eye-catching patterns on the walls and floor. To what extent particular effects were intentional and orchestrated to occur at significant times remains largely a matter of speculation.

See also:
Roman Astronomy and Astrology.

**References and further reading**
MacDonald, William L. *The Pantheon: Design, Meaning and Progeny*. Cambridge, MA: Harvard University Press, 1976.
Masi, Fausto. *The Pantheon as an Astronomical Instrument*. Rome: Edizioni Internazionali di Letteratura e Scienze, 1996.

# Parallax
*See* Lunar Parallax.

# Pawnee Cosmology
The Pawnee are one of the best-known native American groups in the U.S. Midwest. Before the spread of Europeans into the area, they inhabited what was to become northern Kansas and southern Nebraska. The Skidi (or Skiri), one of four Pawnee bands, received particular attention from ethnographers in the early twentieth century, which provided a strong basis for

more recent studies of many different aspects of traditional Pawnee lifestyle, including myths and rituals relating to the sky.

The Skidi Pawnee creation myth tells of how Tirawahat (or Tirawa), creator of all things, instructed various sky gods, including the Sun, the Moon, Bright Star (Evening Star), Great Star (Morning Star), and Star-that-does-not-move (Polaris). The four World Quarter Stars were charged with holding up the heavens and were given the power to create people. Sun pursued Moon to produce the first boy; Great Star pursued Bright Star to produce the first girl. The gods in the heavens provided people with materials for tools, weapons, and clothing, and also with precious objects that they would need to create sacred bundles. Different star gods created different groups of people with complementary knowledge and different sacred bundles. By coming together and keeping their lives in tune with the patterns established by the gods in the heavens, and in particular by undertaking the appropriate ceremonies associated with each of the bundles at the right times, the Skidi people as a whole were able to live in harmony with themselves and with the cosmos for many generations.

The traditional *cosmology*, or worldview, of the Skidi Pawnee is based upon a rich set of associations that are perceived to exist between people, the earth, and the sky, intricately woven together to form a complex framework of understanding. Thus each of the four World Quarter Stars occupied one of the intercardinal directions and was associated with a color, one of the four seasons of nature, one of the four seasons of life, a meteorological phenomenon, an animal, a type of wood, and a type of corn. The various ethnographic accounts differ slightly in some of the finer details, but we can say with reasonable confidence that Yellow Star occupied the northwest, controlled the setting sun, and was associated with the spring, childhood, lightning, the mountain lion, willow, and yellow corn. Likewise Red Star occupied the southeast and was associated with summer, youth, cloud, the wolf, box elder, and red corn. Big Black Star occupied the northeast and was associated with autumn, adulthood, thunder, the bear, elm, and black corn. Finally, White Star occupied the southwest and was associated with winter, old age, wind, the wildcat, cottonwood, and white corn. Black Star (male) and White Star (female) were mates, as were Red Star (male) and Yellow Star (female). Red Star controlled the coming of day, while Black Star controlled the coming of night. White Star stood by the Moon. Black Star was the patron of knowledge. The associations seem endless. (The question of whether the World Quarter Stars represented actual stars, and if so, which, has been investigated extensively but remains unresolved.)

The Skidi view of the world was reinforced by story and by observations of nature itself. For example, the Sun continually chased the Moon across the sky, regularly (once every month) catching up and causing her to disappear. It was also continually enacted in activities and ceremonies that ensured harmony with the cosmos. As an outsider might see it, the Skidi Pawnee based their lives around observations of the heavens. However, it would be a more accurate representation of the complete picture to say that the pattern of their lives reflected the structure of the cosmos in many different ways, both in space and time.

This is particularly evident in the layout of individual dwellings, often known as *earth lodges*. They are round because the world is round; the floor represents the earth and the ceiling the sky. The lodge entrance faces east, toward Morning Star, so that people can enter the lodge in the same way as the stars enter the sky; the altar is located in the western part, the domain of Evening Star, to symbolize creation and renewal. The roof is held up by four main posts that represent the World Quarter Stars. They are placed in the intercardinal directions and each is painted with the appropriate color. These reflect the four quarters of the world, and symbolize the connection between the earth and the sky.

The structure of the cosmos was not only reflected in individual dwellings; the whole landscape reflected the arrangement of the sky gods. Four villages were particularly associated with the World Quarter Stars and were situated in appropriate intercardinal directions in the landscape. Each possessed sacred bundles bestowed by the World Quarter Star in question and led ceremonials at certain times of year. These ceremonies, in turn, were related to seasonal activities such as planting, harvesting, and buffalo hunting.

The Skidi Pawnee provide an excellent example of why, and how, human communities strive to keep their lives in tune with the cosmos as they perceive it.

See also:
Cosmology; Ethnoastronomy.
Pawnee Earth Lodge.

**References and further reading**
Chamberlain, Von Del. *When Stars Came Down to Earth: Cosmology of the Skidi Pawnee Indians of North America.* Los Altos, CA, and College Park, MD: Ballena Press/Center for Archaeoastronomy, 1982.
Selin, Helaine, ed. *Astronomy across Cultures,* 269–301. Dordrecht, Neth.: Kluwer, 2000.

# Pawnee Earth Lodge

The traditional dwelling of the Pawnee, often referred to as an *earth lodge,* symbolized the cosmos in many ways: in the materials used, the techniques of construction, in its orientation, and in the spatial configuration of activities taking place there. The lodges of one of the four Pawnee bands, the Skidi Pawnee, have been studied in particular detail. They faced eastwards, with altars in the west. Their four main supporting posts represented the World Quarter Stars, each placed in the appropriate intercardinal direction, thus reflecting the four quarters of the world and symbolizing the connection between the earth and the sky. At the top of the lodge, directly over the hearth, was a smoke-hole, which was said to reflect the shape of the Council of Chiefs, a group of stars corresponding to our constellation Corona Borealis. In short, the earth lodge was a model of the world—a *microcosm*—in which people, in carrying out their daily lives within their homes, also lived at one with, and within, the cosmos as a whole.

The lodge also functioned as a place for observing the heavens. The first rays of the morning sun, passing in through the entrance, imparted life-giving strength and power. During the day, and depending upon the season of the year, sunlight slanted down through the roof hole and passed around the lodge, touching objects and people, bringing light and warmth. It is also recorded that Skidi priests used the smoke hole to observe stars passing overhead, in order to schedule ceremonies. In this sense, the lodge had a pragmatic function for calendrical observations: it truly was an observatory.

This juxtaposition of the practical and the symbolic may come as a surprise. It also has implications for some of the questions that have been asked in the past about the significance and meaning of prehistoric remains, and about whether astronomical alignments at prehistoric monuments in Britain represented science or symbolism. Earth lodges clearly combined elements of both.

See also:
Science or Symbolism?
Pawnee Cosmology.

**References and further reading**
Aveni, Anthony F., ed. *Archaeoastronomy in the New World,* 183–194. Cambridge: Cambridge University Press, 1982.
Chamberlain, Von Del. *When Stars Came Down to Earth: Cosmology of the Skidi Pawnee Indians of North America.* Los Altos, CA, and College Park, MD: Ballena Press/Center for Archaeoastronomy, 1982.
Selin, Helaine, ed. *Astronomy across Cultures,* 269–301. Dordrecht, Neth.: Kluwer, 2000.

## Pawnee Star Chart

In 1906, the Field Museum of Natural History in Chicago acquired an oval-shaped piece of thin leather measuring fifty-six by thirty-eight centimeters (twenty-two by fifteen inches) with lace holes around the edge. One side is covered with several hundred four-pointed stars of various sizes, painted in irregular patterns, with a band of smaller dots running across the middle (on the narrower axis). Several configurations of larger crosses stand out as possible bright stars or constellations, and the band of dots could be interpreted as the Milky Way. That this was a representation of the sky seems beyond question.

Early attempts to identify particular stars on the *star chart* tended to start from the assumption that it was a literal map of the night sky. Larger crosses were thought to represent brighter stars and positions were assumed to be marked with considerable accuracy. Yet even though some groups of stars such as the constellations Ursa Major and Corona Borealis, and the Pleiades, are certainly recognizable, it quickly becomes clear that their shapes, positions, and relative sizes are approximate at best. Most of the smaller stars seem to be just a backdrop, not representing any stars or groups of stars in particular.

It is a mistake to take an indigenous artifact out of context and to judge it by our own standards of precision. Before anything else we should ask what its purpose was. Fortunately, cultural data do exist in this case: ethnographic notes associated with the chart indicate it was part of a medicine bundle used in various ceremonies that ensured well-being. It was used, in other words, to harness the sacred power of the stars.

Examples such as this expose the dangers of imposing our own criteria when trying to interpret an astronomical artifact from another culture. We might seek a literal representation of the configurations of the stars and be impressed if it achieved a high degree of accuracy, but in doing so we are being ethnocentric: imposing our own predilections on another culture. Accuracy in this literal sense might matter to *us,* but we cannot expect that it mattered to *them*. These dangers are especially great where we do not have a cultural context, as in the case of the prehistoric Nebra disc. A similar lesson can also be extended to those who try to find literal representations of constellations in the positions of monuments on the ground.

**See also:**
Constellation Maps on the Ground; Ethnocentrism.
Angkor; Nebra Disc; Pyramids of Giza.

**References and further reading**
Aveni, Anthony. *Ancient Astronomers*, 132–133. Washington, DC: Smithsonian Books, 1993.
Chamberlain, Von Del. *When Stars Came Down to Earth: Cosmology of the Skidi Pawnee Indians of North America*, 185–205. Los Altos, CA, and College Park, MD: Ballena Press/Center for Archaeoastronomy, 1982.

# Pecked Crosses
*See* Mesoamerican Cross-Circle Designs.

# Pilgrimage
The term *pilgrimage* tends to be applied to journeys in search of spiritual fulfillment made in historical and modern times, particularly in the context

of global faiths such as Christianity and Islam. However, the idea of mystical journeys that are undertaken exceptionally—representing a complete break from familiar places and/or practices—may well be more widely applicable to peoples in the past, and even in prehistory. Of course, such an experience can be totally individual and personal, but it is very often motivated by the perceived need to be in a particular place to undertake one or more specific acts of idolatry, and very often at a particular time. If similar ideological perceptions are widely shared, then *mass pilgrimages* can easily result. Small-scale, and even individual, acts of reverence in the past could have resulted in the placing of votive offerings that are discoverable by archaeologists—a possible example is the "formal" deposits found in the ditch and postholes of the early earth and timber circle at Stonehenge in England; however, mass pilgrimages seem most likely to be archaeologically visible. How do we recognize potential places of pilgrimage? And how do we proceed to address the questions of who came to them, from where, when, and why?

Answering even the first of these questions is a task that is far from straightforward. It seems clear that for a place to become the focus for pilgrimage, it must be exceptional in people's minds. This idea is fine as far as it goes, but a place could have been seen as exceptional for a whole variety of reasons that could apply equally well to distinctive natural places or to human constructions on any scale. Just because a place was large and important enough to have attracted mass gatherings does not mean that we would necessarily wish to describe it as a pilgrimage center—consider trading centers on market day; nor does a place need to be remote or to attract participants from great distances in order to qualify (a good example is the peak sanctuaries of Minoan Crete). Nonetheless, evidence of religious or ritualistic activity and participation from afar are key indicators for the archaeologist, and a place of pilgrimage may be more obvious archaeologically if these pilgrims from afar deposited "foreign" artifacts that can be convincingly interpreted as cult objects or votive offerings.

Another characteristic aspect of pilgrimage is the journey itself. The act of making a pilgrimage can be as important as the place finally reached, and the idea of pilgrimages in the distant past may be supported by the existence of "formalized" routes of approach or by sets of shrines strung out along routes through the landscape. A possible example of the former is the enigmatic Cursus Monuments of Neolithic Britain, whose low earthen ditches and banks could have prescribed movement at special times or for special purposes without constraining it at other times, given (as seems likely) that they were easily walked across. Elsewhere, and later, stone and timber avenues may have served a similar purpose. Such arguments are more convincing when the formal route in question leads directly to an obvious candidate for a pilgrimage center, as did the three-kilometer-long avenue that led up to Stonehenge in the late second millennium B.C.E. Far away in space and time, the impracticality for normal purposes of many of the roads and trackways converging upon Chaco Canyon, New Mexico, around the eleventh century C.E. implies that their function was largely symbolic and adds strength to arguments that Chaco was a huge pilgrimage center, its Great Houses full to capacity only at certain times. Likewise, the ceremonial complex of Cahuachi in the valley south of the Nasca pampa in Peru has been interpreted as a focus for pilgrimage, and it is possible that some of the famous *geoglyphs*—long straight lines running across the arid pampas surrounding the site—could have been ritualistic paths of approach.

One of the most important aspects of many pilgrimages is timing. Many places acquire particular significance and are seen as charged with particular sacred power at certain times. Astronomical phenomena such as sunrise alignments and shadow hierophanies are often of key importance in this context—witness the modern crowds eager to view the equinox hierophany at Kukulcan in Mexico, midsummer sunrise at Stonehenge, or midwinter sunrise at Newgrange in Ireland. (These examples also demonstrate that such practices do not necessarily bear any relation to the actual original intention of the builders—as opposed to the perceived or assumed intention—something that may well have also been true in the past when people ascribed sacred significance to even older monuments.) If such phenomena inspired mass pilgrimages, then an interesting question arises: how did people from afar know when to set out?

A possible answer is that they observed the stars. The heliacal rising or setting of a prominent star or asterism can, for example, provide a sure way of determining the time of year to within a few days, even if there are periods of bad weather. If the astronomical event in question occurred sufficiently far ahead of the sacred ceremony (which might itself have been triggered by an astronomical event), then each pilgrim would have sufficient time to prepare for and make the journey. For example, the Delphic oracle in ancient Greece was originally only available for consultation on one day in the year. It has been suggested that the trigger, both for the oracle itself and for pilgrims to set out, was the first pre-dawn appearance of the constellation Delphinus. At Delphi itself, however, the heliacal rise of Delphinus was considerably delayed owing to the high eastern horizon, giving pilgrims time to arrive there.

The Thornborough henges in northern England demonstrate how various strands of evidence can be brought together to make a plausible argument for a prehistoric site being a pilgrimage center. These include a concentration of non-local flints that suggest a mass influx of people from a considerable distance, and repeated alignments upon the rising position of

Orion's belt and Sirius (which reflect their changing position over many centuries), suggesting that ceremonial activity here might have been directly related to these two prominent objects in the sky. It makes sense that a major ceremony would have been held in the early autumn, when the harvest was completed but before the onset of cold weather. The heliacal rising of Orion's belt and Sirius took place some three weeks apart in the late summer and early autumn. The first appearance of Orion's belt could have provided the sign for pilgrims to set out on their journey, while the subsequent appearance of Sirius provided the signal for the festivities to begin.

Although the term *pilgrimage* is broad and in some ways ill-defined, the past is undoubtedly full of instances where large numbers of people—often coming from large distances—converged on sacred places at particular times. Where this happened, watching the skies would have provided an easy and reliable way for pilgrims to judge when to commence their journeys. Pilgrimages of this type, even in prehistory, may be detectable by carefully combining evidence from archaeology and archaeoastronomy.

See also:
Equinoxes.
Chaco Meridian; Cursus Monuments; Kukulcan; Minoan Temples and Tombs; Nasca Lines and Figures; Newgrange; Stonehenge; Thornborough.
Heliacal Rise; Solstices.

**References and further reading**
Aveni, Anthony F., ed. *The Lines of Nazca,* 207–244. Philadelphia: American Philosophical Society, 1990.
Barba de Piña Chán, Beatriz, ed. *Caminos Terrestres al Cielo.* Mexico City: INAH, 1998. [In Spanish.]
Carlson, John, ed. *Pilgrimage and the Ritual Landscape in Pre-Columbian America.* Washington, DC: Dumbarton Oaks, in press.
Gibson, Alex, and Derek Simpson, eds. *Prehistoric Ritual and Religion,* 14–31. Stroud: Sutton, 1998.
Malville, J. McKim, and Nancy J. Malville. "Pilgrimage and Astronomy in Chaco Canyon, New Mexico." In D. P. Dubey, ed., *Pilgrimage Studies: The Power of Sacred Places,* 206–241. Allahabad, India: Society of Pilgrimage Studies, 2000.
Renfrew, Colin, ed. *The Prehistory of Orkney,* 243–261. Edinburgh: Edinburgh University Press, 1985.
Ritchie, Anna, ed. *Neolithic Orkney in its European Context,* 31–46. Cambridge: McDonald Institute for Archaeological Research, 2000.
Smith, A.T., and A. Brookes, eds. *Holy Ground: Theoretical Issues Relating to the Landscape and Material Culture of Ritual Space,* 9–20, 91–97. Oxford: Archaeopress (BAR International Series 956), 2001.
Stopford, Jennifer, ed. *Pilgrimage Explored,* 1–23. Rochester, NY: Boydell and Brewer, 1999.

# Polynesian and Micronesian Astronomy

The ancient Oceanic peoples populated a vast area of the earth, successfully colonizing a diverse range of island environments prior to European contact. Astronomy was a major factor in their ocean navigation. Directional stars were used to guide navigators to familiar and not-too-distant islands. On land, directional stones or even *stone canoes* were set up so that would-be voyagers could sight along them and learn the appearance of the stars in the required direction of travel. For longer journeys, one of the most important prerequisites was acquiring a knowledge of the rising and setting positions of various stars around the horizon, the so-called star compass. This, combined with various techniques for *dead reckoning*, allowed navigators to estimate their current position in relation to the island they had left or to the one they were trying to reach. There is also patchy evidence for the use of a variety of other techniques, such as the use of zenith stars passing directly overhead to estimate the current latitude, and even (in the Hawaiian islands) the use of gourds as instructional aides or navigational devices.

But objects and events in the skies were also important to ancient Oceanic peoples in a variety of other ways. They certainly had an extensive knowledge of astronomy: ethnographers in the nineteenth and early twentieth centuries recorded a great many names for stars, planets, nebulae (such as the Magellanic Clouds), areas of the Milky Way, and so on—things actually visible in the sky—as well as for purely conceptual constructs related to the motions of the heavenly bodies. As an example of the latter, the Hawaiians had names for what we might call the *celestial tropics*—the most northerly and southerly paths followed by the sun around the sky at the times of the June and December solstices, respectively. The northern tropic they termed "the black shining road of Kane" and the southern one "the black shining road of Kanaloa," Kane and Kanaloa being two principal creator-gods. The same or similar names for certain celestial objects (with dialectic variants) can often be found right across the linguistically homogenous area of Polynesia and even farther afield, which indicates considerable antiquity. For example, the Pleiades were known in Hawai'i as *Makali'i*, in Samoa as *Li'i*, in Tonga as *Mataliki*, in Tahiti as *Matari'i*, and by the Maoris of New Zealand as *Matariki*. To the west of Polynesia they were known, for example, within Vanuatu (Melanesia) as *Matalike* and in Pohnpei (Micronesia) as *Makeriker*.

Stars and constellations were frequently associated with gods, culture heroes, or living chiefs, as well as featuring in stories of ocean voyaging and of ancient homelands. A form of genealogical prayer chant common in Polynesia served to place those of the highest rank in a cosmic scheme of things that includes everything in the sky as well as on earth. A famous example of this is the Hawaiian *Kumulipo*.

Another way in which astronomy permeated the lives of Oceanic peoples was through the calendar, which had a common form all over Polynesia and Micronesia. It consisted of an annual cycle of months based on the lunar phases, with a named sequence of days (or rather, nights) in each month. Many of the same or similar month or day names occur over a wide area, indicating considerable antiquity, and in certain places there is clear evidence that months shared their names with stars or constellations used to identify them. Yet some of the variants from region to region were major, such as when in the solar year a given month actually occurred. There were also significant, though generally more minor, local variations (for example from one Hawaiian island to another, or between different Maori groups in New Zealand). Systems of prognostications and observances associated with the calendar seem to have grown to their most complex on those groups of islands where powerful social hierarchies developed. The Hawaiian calendar at the time of European contact, for example, involved an elaborate system of cycles of ritual observances and taboos tied both to the time of the year and the time of the month.

Traditional astronomical practices in Oceania have left their mark in the material record in a number of ways. The purposes of some of the more recent artifacts, such as sighting stones, can be verified using recorded oral history or ethnography. Clues to older, pre-contact practices within Polynesia may be encapsulated in the temple enclosures and platforms that were ubiquitous throughout this region. These do not manifest any systematic astronomical alignments on a large scale, but a few are known from oral history to have been used for sighting the sun or stars. Occasionally we find firmer archaeological evidence of deliberate orientation, such as a group of temple enclosures (*heiau*) on the Hawaiian island of Maui that are aligned upon the rising position of the Pleiades, apparently for reasons associated with the calendar and worship of Lono, a prominent god of agriculture.

Although a number of claims have been made that Oceanic peoples determined the solstices and equinoxes to high precision, most of these stem from a desire (witting or otherwise) to impose Western aspirations and do not fit the evidence. However, ethnography confirms that the inhabitants of one Polynesian island, Mangareva, did indeed mark the solstices to considerable precision by erecting pairs of stones as horizon foresights.

In the Hawaiian islands there is a particularly rich heritage of *oral literature* in the form of creation myths, stories, and chants. Some were recorded by early missionaries or published in the Hawaiian language in local newspapers in the nineteenth and early twentieth centuries. The small proportion that have been translated and republished contain numerous references to the sun, moon, stars, and planets. Some chants clearly describe places in the landscape and appear to provide explicit descriptions of astronomical alignments. An example is the so-called "Nā Pali chant" from the islands of Niʻihau and Kauaʻi.

Throughout Oceania, astronomical knowledge was sacred knowledge, much of it the exclusive preserve of navigators and priests. This fact, added to all the other processes of change that have occurred since European contact, means that only a few fragments have been preserved of what was without doubt an enormously rich and far-reaching knowledge of astronomy. Sky knowledge was nonetheless framed within worldviews that governed all aspects of existence and were woven into everyday life in a variety of ways. In the Gilbert Islands, for example, conceptual divisions of the sky (the "roof of voyaging") were based upon the structure of actual house roofs. If we are trying to understand more about that worldview, then it is misleading to divide Oceanic astronomical knowledge using such categories as pragmatic and ritualistic, or useful and conceptual. These are divisions that only have meaning from a Western perspective. To ancient Polynesian and Micronesian peoples, all their sky knowledge was useful, all vital within their framework of understanding the cosmos and how it functioned.

See also:
Lunar and Luni-Solar Calendars; Magellanic Clouds.
Hawaiian Calendar; Kumulipo; Mangareva; Nā Pali Chant; Navigation in Ancient Oceania; Polynesian Temple Platforms and Enclosures; Star Compasses of the Pacific; Zenith Stars in Polynesia.
Celestial Sphere.

**References and further reading**
Åkerblom, Kjell. *Astronomy and Navigation in Polynesia and Micronesia.* Stockholm: Stockholm Ethnographical Museum, 1968.
Aveni, Anthony F., and Gary Urton, eds. *Ethnoastronomy and Archaeoastronomy in the American Tropics*, 313–331. New York: New York Academy of Sciences, 1982.
Beckwith, Martha W. *Hawaiian Mythology* (repr.), 60–66. Honolulu: University of Hawaiʻi Press, 1970.
Best, Elsdon. *The Astronomical Knowledge of the Maori.* Wellington, NZ: Dominion Museum, 1922.
Buck, Peter H. [Te Rangi Hiroa]. *Ethnology of Mangareva*, 403–416. Honolulu: Bishop Museum Press, 1938.
Finney, Ben. *Voyage of Rediscovery.* Berkeley: University of California Press, 1994.
Grimble, Arthur. "Gilbertese Astronomy and Astronomical Observances." *Journal of the Polynesian Society* 40 (1931), 197–224.
Hodson, F. R., ed. *The Place of Astronomy in the Ancient World,* 133–148. London: Royal Society, 1976.
Johnson, Rubellite K., and John K. Mahelona. *Nā Inoa Hōkū: A Catalogue of Hawaiian and Pacific Star Names.* Honolulu: Topgallant Press, 1975.
Kamakau, Samuel M. *The Works of the People of Old,* 13–17. Honolulu: Bishop Museum Press, 1976.
Kirch, Patrick V., and Roger C. Green. *Hawaiki, Ancestral Polynesia: An Es-*

say in Historical Anthropology, 260–273. Cambridge: Cambridge University Press, 2001.

Makemson, Maud. *The Morning Star Rises: An Account of Polynesian Astronomy.* New Haven: Yale University Press, 1941.

Ruggles, Clive, ed. *Archaeoastronomy in the 1990s,* 128–135. Loughborough, UK: Group D Publications, 1993.

Selin, Helaine, ed. *Astronomy across Cultures,* 91–196. Dordrecht, Neth.: Kluwer, 2000.

## Polynesian Temple Platforms and Enclosures

Undoubtedly the best known and most conspicuous material remains of Polynesian culture are the temple platforms and enclosures, variously known as *marae* in much of central Polynesia (the Marquesas, the Society Islands) and Aotearoa (New Zealand), *ahu* in the Tuamotus and remote Rapa Nui (Easter Island), and *heiau* in the Hawaiian Islands (where the term *ahu* refers just to the altar). The *ahu* of Easter Island, with their famous statues, or *moai,* are undoubtedly the best known, but many hundreds of these temple platforms and enclosures are found throughout Polynesia, and very many more existed in the past.

The temple platform or enclosure was a sacred place constructed, used, and maintained by a particular family or social group, or for people of a certain social rank. In the Society Islands, for example, there existed *marae*s serving individual households; the inhabitants of a district; people in the same kinship group; those in particular occupations such as healers, canoe builders, and fishermen; and the high chiefs or *ari'i nui*. Temple sites were used for a variety of purposes. Thus in Hawai'i, the large and elaborate *luakini heiau,* which were the exclusive preserve of high chiefs, might be used for investitures or religious rites including sacrifices, while at the other end of the scale, modest *ko'a* were built by commoners as simple fishing shrines. There existed war *heiau,* medicinal *heiau,* navigational *heiau,* agricultural *heiau,* and a host more, all dedicated to the appropriate gods.

Why a given temple site was located in a particular place or constructed in a particular way is generally lost to us, but there is every reason to suppose that a strict set of sacred protocols had to be followed, as can still remain true today. It is possible, for example, that an extraordinary concentration of temple platforms on the tiny and remote Hawaiian outlier of Necker Island may be due to the location of this isolated rocky outcrop precisely on the Tropic of Cancer. It has been suggested that many of the temple sites were astronomically oriented, as was, for exmple, the inland *ahu* Huri a Urenga on Easter Island. However, attempts to identify overall patterns have met with little success. On many islands, Rapa Nui being a particularly clear example, it is evident that the predominant trend is simply to orient temple sites with respect to the coast, either reflecting a general rela-

*The ruins of the highest* heiau *in Hawai'i, Ahu a 'Umi, at an elevation of 1600 meters (5250 feet) in the saddle of the Big Island, viewed from the first of two different angles. In the background of this view is Mauna Kea with the modern observatory at its summit. (Courtesy of Clive Ruggles)*

tionship with the sea or the constraints of the local topography. This has caused some archaeologists to conclude that astronomy played no part at all in determining temple orientations.

The truth undoubtedly lies between these two extremes. The fact that there are no obvious systematic astronomical associations is hardly surprising, given the widely ranging functions and purposes of the different types of temple site. Even where they were important, astronomical considerations are likely to have been just one of a range of factors affecting temple location and design, and were not necessarily reflected in the orientation. Some surviving snippets of oral history do indicate an interest in the annual progress of the rising sun up and down the eastern horizon (one of the few observation points mentioned explicitly is Cape Kumukahi in the Big Island of Hawai'i), but this account does not point to a direct connection with any *heiau*.

Accounts do exist of temple platforms being used for sun observations in Hawai'i, and of the use of artificial horizon foresights in Mangareva, an island in French Polynesia. But no accounts mention platforms being actually aligned upon the sun. In one case, a surviving sacred *hula* chant appears to describe a solstitial alignment from a *heiau* on the Hawaiian island of Kaua'i stretching along the high cliffs of the Nā Pali coastline of the island. But even here, there is no suggestion that the alignment is directly reflected

The ruins of the highest heiau in Hawai'i, Ahu a 'Umi, at an elevation of 1600 meters (5250 feet) in the saddle of the Big Island. In the background of this view is the volcano Mauna Loa. (Courtesy of Clive Ruggles)

in the orientation of the platform itself. Similarly, certain navigational *heiau* are said to have been associated with voyaging stars, without suggesting that they were oriented upon them. All this complicates efforts to tie together archaeoastronomical field data with relevant fragments of oral tradition, but does not make it impossible, as the following example shows.

Kahikinui is a remote area on the leeward side of the Hawaiian island of Maui, on the southern slopes of the volcano Haleakalā. Recent survey work here has revealed an unusually well preserved pre-conquest landscape including households, cultivation areas, and almost thirty *heiau*, mostly of a distinctive *notched enclosure* design, which all date to a similar period around C.E. 1600. It is very unusual to find so many *heiau* in such a restricted area and within such a well-defined archaeological context. This enables us both to spot consistent orientation patterns that are likely to have been intentional and meaningful and to suggest plausible interpretations related to our broader knowledge of the cultural context. The structure, location, and orientation of the Kahikinui *heiau* suggest that they may be divisible into four categories, each of which relates to one of the four main gods in the Hawaiian pantheon. One set face north, one east, and one relate to the sea. The last tend to overlook plantation areas and may relate to Lono, the god associated with dryland agriculture. Their orientations are consistently around twenty degrees north of east, which strongly suggests a particular celestial target—the horizon rising position of the Pleiades. This asterism is known to have been of particular significance in the Hawaiian agricultural and ritual calendar, because its heliacal and acronical rise marked the beginning of the two halves of the Hawaiian year. On the northern side of Maui, on the isolated peninsula of Kalaupapa, is another *heiau* incorporating a similar alignment. It is found on a promontory whose traditional name—Makaliʻi—means Pleiades in Hawaiian.

The example of the Kahikinui *heiau* serves to illustrate how little we would know (and generally do know in other parts of Polynesia) about the meaning of temple orientations but for being able to combine alignment evidence from a well-defined group of sites with oral history.

See also:
Methodology.
Easter Island; Hawaiian Calendar; Kumukahi; Mangareva; Nā Pali Chant; Necker Island; Polynesian and Micronesian Astronomy.
Heliacal Rise.

**References and further reading**
Hodson, F. R., ed. *The Place of Astronomy in the Ancient World*, 137. London: Royal Society, 1976.
Kirch, Patrick V. *On the Road of the Winds: An Archaeological History of the Polynesian Islands before European Contact*, 207–301. Berkeley: University of California Press, 2000.
———. "Temple Sites in Kahikinui, Maui, Hawaiian Islands: Their Orientations Decoded." *Antiquity* 78 (2004), 102–114.
Ruggles, Clive, ed. *Archaeoastronomy in the 1990s*, 128–135. Loughborough, UK: Group D Publications, 1993.
Selin, Helaine, ed. *Astronomy across Cultures*, 127–160. Dordrecht, Neth.: Kluwer, 2000.
Valeri, Valerio. *Kingship and Sacrifice: Ritual and Society in Ancient Hawaii*, 172–188. Chicago: University of Chicago Press, 1985.

# Power

In many human societies of the past, knowledge and beliefs relating to the sky served to reinforce the political status quo. In some cases, the power of the ruler was directly linked to the sky or to sky gods in the form of celestial objects. Thus Old Kingdom Egyptian pharaohs' power depended upon sun worship, as did that of the ruling elite of the Inca empire and that of Roman emperors at the time when Mithraism held sway. Chinese emperors, on the other hand, established their place at the head of the political order by analogy to the north celestial pole—the hub of the turning sky—and underpinned this by aligning the buildings of the Imperial Palace (Forbidden City) along the meridian. The layout of the greatest medieval Hindu city, Vijayanagara, together with its position in the landscape in relation to sacred hills, helped to establish the place of its kings within the cosmic order. Similarly, the sarsen monument at Stonehenge, undoubtedly a potent symbol of the power of the chiefs who controlled it, incorporated a solstitial alignment that, by linking the monument to the workings of the cosmos, may have contributed to its symbolic power every bit as much as the exotic stones and the scale of its construction.

Considerable social prestige and political power could accrue to anyone who was seen as able to control or influence phenomena that could not be controlled by anyone else. For this reason, an astronomer-priest might hold the solstice ceremony around the time that the sun set in a particular horizon notch so that he could demonstrate the power of "turning it around." Alignments that "worked" only at special times—the appearance of the midsummer full moon low over the recumbent stone of a Scottish recumbent stone circle, for example, or the sudden entry of sunlight into places that were normally dark, such as the cave below the zenith tube at Xochicalco in Mexico—would have been highly impressive and could have been used to help justify the power and prestige of the priest in command. Such apparent control ultimately relies, of course, on the regularity of astronomical phenomena—reliable cycles that are tied to other, observable events in the natural world.

Where a priest or shaman professed the ability to predict phenomena such as eclipses or the appearance of comets, the results (as *we* know) would be a lot more unreliable. Indeed, unexpected events would undermine their authority and could even cause social and political instability. Certainly they were commonly taken as bad omens. In C.E. 549, an unexpected lunar eclipse led two Teutonic armies to flee the scene of battle and thus postponed the Roman occupation of the Carpathian basin by two years. The appearance of Halley's Comet in the spring of 1066 portended disaster for King Harold at the battle of Hastings. And a spectacular meteor seen in the Aztec capital of Tenochtitlan around 1515 caused great fear among the people. The Spanish conquest followed just a few years later.

Observations of the regular cycles of the skies had a clear practical value within a wide variety of human societies in the past by helping to keep subsistence activities in step with the seasonal cycles of nature. However, as often as not—and especially within larger political units with strong and powerful social hierarchies—they also played a significant role in perpetuating the power of the ruling elites.

**See also:**
Comets, Novae, and Meteors; Lunar Eclipses; Solar Eclipses.
Chinese Astronomy; Egyptian Temples and Tombs; Island of the Sun; Mithraism; Recumbent Stone Circles; Stonehenge; Zenith Tubes.

**References and further reading**
Aveni, Anthony F., ed. *World Archaeoastronomy*, 289–299. Cambridge: Cambridge University Press, 1989.

Krupp, Edwin C. *Echoes of the Ancient Skies*, 259–285. Oxford: Oxford University Press, 1983.

———. *Skywatchers, Shamans and Kings*. New York: Wiley, 1997.

Ruggles, Clive, ed. *Archaeoastronomy in the 1990s*, 98–106. Loughborough, UK: Group D Publications, 1993.

Ruggles, Clive, and Nicholas Saunders, eds. *Astronomies and Cultures*, 1–31, 139–162. Niwot: University Press of Colorado, 1993.

# Precession

Technically, precession is quite a difficult concept, but it is an important one for anyone wishing to study perceptions of the skies in the past. Precession has the effect of shifting the overall positions of the stars in the sky over a time scale of centuries, and (from any given place) can make some stars disappear completely from view while others become visible for the first time (in thousands of years). For example, when prehistoric farmers first colonized the area of southern England where they would eventually build Stonehenge, the Southern Cross was prominently visible in the night sky at certain times of year. However, it has not been visible from this latitude for over 5,000 years. Taking account of precession is critically important for anyone investigating possible architectural alignments upon stars. But it also enters into archaeoastronomy in a different way, in that some have claimed that the effects of precession are described and documented in myths that have been perpetuated over many centuries.

In order to explain precession, it is best to adopt the astronomer's perspective, looking in on the solar system.

As it progresses on its annual orbit around the sun, the earth also spins once daily around its own axis. This axis maintains the same orientation in space, so there is one point in the earth's orbit where its north pole leans directly toward the sun while the south pole leans directly away from it. When the earth is at this point in its orbit, for people in the northern hemisphere, the sun is seen following its highest daily path through the sky, while for people in the southern hemisphere its path is the lowest. This is one of the solstices: the summer solstice in the northern hemisphere and the winter solstice in the southern. In the modern (Gregorian) calendar this solstice falls on or very close to June 21. The opposite point in the earth's orbit, when the north pole is leaning directly away from the sun and the south pole toward it, is the December solstice. A quarter of the way around the earth's orbit, halfway between the solstices, are the two equinoxes, where the axis joining the two poles leans neither toward nor away from the sun.

The earth's axis maintains a fixed direction with respect to the stars. This is because the stars are so staggeringly far away compared with the dimensions of the solar system that the earth's motion around the sun makes no difference. If the sun were scaled down to the size of a beach ball, then the earth would be the size of a ball bearing about thirty meters (a hundred feet) away from it, and the farthest planet, Pluto, would orbit the sun at a distance of a little over one kilometer (a little under a mile). On this scale, the distance of the nearest star from the sun would be about the distance of London from New York, with some of the other visible stars considerably farther away still. Having established these vast distance scales, it is conceptually far easier to pretend that the solar system is surrounded by an enormous sphere with all the stars attached to it. This means that the earth's axis, extended out through the north pole, is always pointing at a fixed position on the celestial sphere: this is called the north celestial pole. Similarly, the axis extended out through the south pole is always pointing to the south celestial pole.

Over a time scale of centuries, the direction of the earth's axis in space (and with respect to the distant stars) does change. This is because the earth not only spins about its axis, but the axis itself gradually turns in a way that resembles a spinning top. One effect of this is that the position in the earth's orbit where we find (say) the north pole tilted toward the sun is now different: in fact, it has "precessed" around the earth's orbit. So have the positions corresponding to the other solstice and to both the equinoxes. This is the reason that the term *precession* is used to describe this phenomenon: it is in fact short for *precession of the equinoxes*.

From the point of view of someone watching the skies from the surface of the earth, precession makes it seem that the celestial poles gradually change position. In fact, each of them traces out a wide circle on the celestial sphere, completing one circuit every 25,800 years. Occasionally during this period, one of the poles will come close to a bright star, making it easy to identify the pole; this is the case with Polaris at the present time, but it is more often not so.

The principal effect of precession is that the declination of every star in the sky changes over the centuries. One consequence of this is that the horizon rising and setting positions of any particular star as viewed from any particular place on earth will generally change significantly over a period of a few centuries (although how significant this effect is depends upon the precision of the alignment being considered). This means that there is a nasty trap waiting to ensnare the modern investigator looking for possible astronomical alignments—at least, stellar alignments—at prehistoric monuments such as stone circles. The danger is this: these monuments are rarely datable to within a few centuries (and very often the range of possible dates is far wider than that). If we notice that a particular structure is aligned upon a horizon point fairly close to the rising or setting position of some bright star in the past, then we may well be able to find a date within the permissible range that fits the alignment perfectly. While there may be exceptions (such as the prehistoric sanctuary at Son Mas in Mallorca), this sort of astronomical dating generally proves nothing because of the ease by which such alignments can arise fortuitously. To give an idea, suppose we take the fifteen brightest stars in the sky and a date range of five hundred years. Then by choosing an appropriate combination of star and date we can cover about a third of the horizon. In other words, if an alien giant were to come along and scatter fake prehistoric aligned structures randomly over the landscape, it would be possible to choose a star and a date within the permitted range that would fit one in three of them. The problem is further confounded by atmospheric extinction. What all this means is that the aspiring archaeoastronomer must pay particular attention to problems of methodology. Two case studies that illustrate the dangers particularly well are the stone row at Ballochroy in Scotland and the stone configuration at Namoratung'a in Kenya.

**See also:**
Astronomical Dating; Equinoxes; Methodology.
Ballochroy; Gregorian Calendar; Namoratung'a; Son Mas; Stone Circles.
Celestial Sphere; Declination; Extinction; Obliquity of the Ecliptic; Solstices.

**References and further reading**
Aveni, Anthony F. *Skywatchers*, 100–103. Austin: University of Texas Press, 2001.
Krupp, Edwin C. *Echoes of the Ancient Skies*, 10–11. Oxford: Oxford University Press, 1983.

# Precision and Accuracy

These two terms, which mean similar things in common usage, have distinct meanings to archaeologists and others who undertake surveys as part of their fieldwork techniques. Suppose that we use a magnetic compass to measure the azimuth of a prominent mountain peak on the horizon. The compass, we find, is graduated in degrees but it is possible to estimate the angle to the nearest half-degree. When we take the same reading three or four times we get readings consistent to the nearest half-degree. Half a degree, then, is the *precision* of our compass reading.

Accuracy, on the other hand, refers to how close that reading is to the truth. We may have forgotten to take into account the fact that compasses point to magnetic north rather than true north. There may also be a local magnetic anomaly. As a result, it may be that the reading we obtain is actually four degrees higher than the true value. The *accuracy* of our reading, then, is only four degrees. Great precision is no guarantee of great accuracy.

Overprecision can be very misleading. If we use a theodolite with an electronic distance measurement (EDM) capability to measure the distance between, say, the center of a stone circle and an outlier, then we will be able to read off the result to the nearest millimeter. However, standing stones are generally large and irregular in shape and may have shifted somewhat during the millennia since their original construction. The circle may not be exactly circular and its center may be ill-defined. To measure, precise to one millimeter, the distance between two points which are themselves arbitrary to within several centimeters or more is completely pointless. Worse, it can be misleading: there is a danger that such "falsely accurate" measurements will be used subsequently, say, in calculations that explore the possible existence of a precise unit of measurement.

Finally, we may also speak of the *precision* with which prehistoric people set up an astronomical alignment. Thus it could be that a group of people in Bronze Age Britain constructed a row of standing stones aligned upon the solstitial sunrise to a precision of about half a degree (thirty arc minutes); that a team of surveyors in 1999 measured the alignment to an accuracy of a few arc minutes; and that the results were published in 2001 quoting declinations to a precision of one arc minute.

**See also:**
Compass and Clinometer Surveys; Field Survey; Theodolite Surveys.
Stone Circles.
Azimuth; Declination.

**References and further reading**
Ruggles, Clive. *Astronomy in Prehistoric Britain and Ireland,* ix. New Haven: Yale University Press, 1999.

## Prehistoric Tombs and Temples in Europe

Scattered through western Europe from Scandinavia and Scotland down to the Mediterranean (and extending into North Africa) are a staggering variety of later prehistoric monuments that still remain conspicuous in today's landscape. Built in some areas as far back as the fourth millennium B.C.E. and even earlier, and in some places only a few centuries before the coming of the Roman empire, many hundreds of great constructions in stone have stood the test of time remarkably well. More modest examples, however, tend to be little known, and many remain under serious threat. Some were tombs, some undoubtedly functioned as temples where sacred rites took place, and some—like the great alignments of hundreds of standing stones at Carnac in Brittany, northwest France—seem to defy explanation. Within this broad custom of constructing monumental architecture were a wealth of local traditions, resulting in groups of monuments as diverse as the recumbent stone circles of eastern Scotland, Irish wedge tombs, Causse-type dolmens of south-central France, Portuguese antas, Sardinian nuraghi, Menorcan taulas, and many more.

Orientation is a basic and relatively uncontroversial aspect of most of

*The Dolmen del Coll de Medàs I near to Cantallops in Cataluña, northern Spain, close to the French border. From its location in the Pyrenean foothills it commands a stunning view to the south. In this region, eastern and western practices of tomb orientation were commingled. (Courtesy of Clive Ruggles)*

these monuments, in that they have an axis of symmetry and face in some direction. The study of the orientations of these monuments, and in particular of their possible astronomical associations, has been extensive in Britain, Ireland, and Brittany. Indeed Britain was the birthplace of modern archaeoastronomy, which has resulted in a great deal of attention being lavished on sites such as Stonehenge. Yet, surprising as it seems, there was little or no interest in—let alone large-scale investigations of—monumental orientations in the remainder of western Europe until a systematic survey was undertaken by the British historian of astronomy Michael Hoskin during the 1990s. Hoskin measured and documented the orientations of over three thousand later prehistoric tombs and temples in Portugal, Spain, southern France, and the islands of the western Mediterranean. Within this considerable area and time span one can identify numerous local traditions resulting in characteristic types of monument broadly consistent in their design and methods of construction. Some occur in large numbers, like the Sardinian *tombi di giganti* ("tombs of giants"), of which over 250 remain; others consist of only a handful of extant examples, such as the *dolmenic hypogea* of the Fontvieille region of southern France, of which there are only four.

The remarkable fact that emerges is that nearly every local group has a discernible *orientation signature,* as characteristic as the architectural design of the tomb or temple itself. The vast majority of the three thousand monuments faces broadly toward the east, southeast, or south. Some local groups cover the whole range from northeast to south, while others fall exclusively within a much narrower span. A few groups of monuments, on the other hand, face predominantly westerly or southwesterly. South-facing groups are rare and north-facing monuments are virtually unheard-of, there being only two examples in the whole set. This shows beyond any doubt that, throughout a significant area of the world and through three or more millennia of prehistory, the orientation of a monument was of almost universal importance—as important, evidently, as the materials used or the techniques of construction. At the same time, it is clear that dominant practices with regard to orientation varied from place to place and time to time just as much as those relating to design and construction. The orientation data give us some clues about how certain practices may have propagated from one region to another. West-facing tombs, for example, are rare in the Iberian peninsula and western and central France, but predominate in the Mediterranean coastal region of France. It seems likely that this practice developed in Provence or Languedoc, where it prevailed, spreading both east toward Italy and southwest toward Spain, but then (in both cases) became increasingly mixed with rival traditions.

It is much trickier to identify the targets that determined a set of orientations, let alone their meaning and significance to the people who built and used the monuments. However, one thing is clear: even the most modest local groups of similar monuments are scattered through a sufficiently large area, and placed in a sufficient variety of landscape situations, to rule out any suggestion that the determinant was topographic, or even that the orientations were due to factors such as the prevailing wind. We are forced to conclude that reference must have been made to the north-south line defined by the diurnal motion of the heavens. This demonstrates, at least at the most basic level, that reference was made to the skies.

How reliably particular targets, astronomical or otherwise, can be identified for particular groups is debatable. One of the clearest cases is that of the seven-stone antas (dolmens) of central Portugal, where the orientations of all 177 examples measured fall within the sunrise arc on the eastern horizon. A common orientation signature stretches from northeast round to due south, and Hoskin suggests that the motivation here was to orient monuments either in the direction of sunrise or of the sun climbing in the sky. Similarly, some of the westward-oriented groups correspond to the range of directions of sunset and/or sun descending in the sky. An exceptional group of monuments are the taulas of Menorca, whose orientations fall in a range centered upon the south. It is dangerous to try to fit stellar alignments in such a case, since there are many bright stars, and their positions shift significantly over the centuries, owing to precession. However, there is quite strong independent evidence to support the argument that the taulas were associated with particular stars, most notably the Southern Cross and Pointers.

Later prehistoric tombs and temples farther north, in Britain and Ireland, have been much more intensively studied. In some cases, more specific astronomical associations have been identified: for example, the Scottish recumbent stone circles appear to be related to the midsummer full moon. However, taking a broader European view it is clear that their general orientation signatures are very similar to those found elsewhere in western Europe. This implies that more detailed investigation of the monuments in continental Europe may have a great deal more to tell us.

**See also:**
"Green" Archaeoastronomy; Orientation.
Antas; Megalithic Monuments of Britain and Ireland; Nuraghi; Recumbent Stone Circles; Stonehenge; Taulas; Wedge Tombs.
Diurnal Motion; Precession.

**References and further reading**
Belmonte, Juan Antonio, and Michael Hoskin. *Reflejo del Cosmos*. Madrid: Equipo Sirius, 2002. [In Spanish.]
Chapman, Robert. *Emerging Complexity: The Later Prehistory of South-East Spain, Iberia and the West Mediterranean*. Cambridge: Cambridge University Press, 1990.
Hoskin, Michael. *Tombs, Temples and Their Orientations*. Bognor Regis, UK: Ocarina Books, 2001.

# Presa de la Mula

In northeast Mexico, beyond the northern frontiers of the great Mesoamerican civilizations, hunter-gatherer groups still roamed at the time of the Spanish conquest, as they had done for more than ten thousand years. By the end of the eighteenth century they had been annihilated, but several hundred of their rock art sites are known in this region. The dating is generally uncertain within wide bounds, but many are clustered in certain locations, usually where there are wide views of the surrounding landscape, horizon, and sky. These petroglyphs contain numerous depictions of animals and weapons but also, intriguingly, abstract designs consisting of short lines organized in rectangular arrays, and groups of dots organized in sets of parallel rows or columns, crosses, concentric circles, and meandering lines. Anthropologist Breen Murray has examined the numerical characteristics of these designs in an attempt to identify patterns of numerical order that might give some insight into their meaning.

The vertical face of a rock at Presa de la Mula, near Monterrey, contains one of the most extensive arrays of lines. This petroglyph, which measures some three meters (ten feet) across by one meter (three feet) high, contains 207 short vertical lines organized in a grid pattern with six rows and four columns, which divides them into twenty-four cells. Reading horizontally from left to right, and from the top row downwards, it is possible to discern sequences of 29, 27, 29, 28, 27, 7, 30, and 28 days. Each of the larger sequences is subdivided into subsequences such as 11+11+4+3 and 7+8+7+6.

Murray suggests that the lines are "tally marks" that reckon the days in relation to the phases of the moon. The number 207 corresponds exactly to the number of days in seven synodic (lunar-phase-cycle) months, and there are seven subtotals of between twenty-seven and thirty days. The fact that these are generally shorter than the true length of the synodic month (29.5 days) could be attributed to the vagaries of observing the moon (especially the thin crescent before and after new moon), and the additional seven days after the fifth month could be a correction, required to bring the "calendar" back into line at this point. The fact that this cell of seven days is marked by two otherwise unexplained arrow-shaped lines backs up this speculation. The whole design, then, appears to be a record of actual day-by-day observations of the phases of the moon lasting over half a year. Perhaps they had a seasonal camp near here, and after this period they moved on.

This interpretation has been arrived at by post hoc reasoning. Reading the count from left to right gives the best fit to a lunar interpretation, so that is what is suggested. A skeptic might argue that the various cells could be conjoined in different ways, or read in a different sequence, and to some extent this is undeniable. However, the interpretation is inherently plausible and is backed up by the presence of circles that seem to mark the completion of the months. Furthermore, the number 207 occurs in at least one nearby petroglyph—an array of dots found forty kilometers away at a site called Boca de Potrerillos. Finally, the total numbers of dots in several smaller dot counts on other petroglyphs in the area seem to have possible lunar significance.

See also:
Abri Blanchard Bone.
Lunar Phase Cycle.

**References and further reading**
Aveni, Anthony F., ed. *Archaeoastronomy in the New World*, 195–204. Cambridge: Cambridge University Press, 1982.
Koleva, Vesselina, and Dimiter Kolev, eds. *Astronomical Traditions in Past Cultures*, 14–24. Sofia: Bulgarian Academy of Sciences, 1996.
Ruggles, Clive, ed. *Archaeoastronomy in the 1990s*, 264–269. Loughborough, UK: Group D Publications, 1993.

# Pyramids of Giza

The three Pyramids of Giza, to the west of Cairo, were built by three despotic pharaohs of the Fourth Dynasty, in the twenty-seventh to the twenty-fifth centuries B.C.E.: Khufu (Cheops), Khafre (Chephren), and Menkaure (Mykerinos). Khufu's pyramid, most likely constructed in c. 2528 B.C.E., was an extraordinary feat of engineering (and of labor organization), built using over two million blocks of limestone, each weighing some fifteen tons and rising to an incredible height of over 145 meters (480 feet). Khafre's was only slightly smaller, while Menkaure's, at sixty-seven meters (220 feet), was modest by comparison. Each was actually part of a funerary complex serving not only as the king's tomb but also as a place of continuing ritual. Khafre's complex included the famous Sphinx.

The sides of each of the Giza pyramids were carefully aligned upon the cardinal directions (north–south or east–west). This alignment followed established practice, but the accuracy with which it was achieved at Giza is truly impressive, particularly in the case of Khufu's pyramid. Each of its sides is cardinally aligned to within six arc minutes, or one tenth of a degree. This is equivalent to no more than one fifth of the apparent diameter of the sun or moon. The other pyramids are only slightly less well aligned: Khafre's to within about eight arc minutes and Menkaure's to within sixteen.

Why this careful alignment? The broader Old Kingdom practice of cardinal orientation of tombs and temples almost certainly had to do with preserving the perceived cosmic order. Thus the northerly direction, where the imperishable (circumpolar) stars resided, was the resting place of the immortal. The westerly part of the horizon, where the sun set each night, was the entrance to the *Duat*, the otherworld, which had to be traversed on the path to immortality. But why was it important to be so accurate? Here we are reduced to speculation. Perhaps the north celestial pole, the one fixed place in the heavens, was seen as the very resting place of the immortal soul, and the successful path of the pharaoh's soul to immortality could be ensured only if the pyramid faced this spot exactly.

What is certain is that great accuracy was achieved, and many different theories have been put forward to explain how. One of the most plausible answers is that this was done by direct reference to the celestial north pole. But there is a problem with this explanation. It is only by chance that we have, at the present time, a bright star close to the pole. In Old Kingdom times the closest star to the pole was Thuban in the constellation Draco. While it had been almost exactly at the pole just after 2800 B.C.E., by 2500 B.C.E. it was 1.7 degrees, or three and a half moon-diameters, away (Polaris is less than half that distance from the pole today). On the other hand, there were a number of bright stars circling the pole then, as now. By carefully observing these night after night, it would have become obvious that certain

*The Pyramids of Giza, near Cairo, Egypt. (Corel Corp)*

pairs were (more or less) on opposite sides of the pole, or in line on the same side. Suppose that the architect set up some form of vertical plumb line device to determine accurately when one of the chosen pair of stars was directly above the other. The direction of the stars when this happened would be close to due north. Even then, it is not possible to explain the extraordinary accuracy of the Khufu pyramid except as a result of good fortune, but if the same two stars were used in successive generations we can explain why the other two pyramids are not quite so accurate. After a few decades, precession would have moved them slightly farther out of line.

Celestial observations may well have had a greater part to play in the construction of the pyramids than merely determining their orientation. The Khufu pyramid contains various chambers and passages, but also two small shafts running directly from the walls of the King's Chamber deep in the interior up to openings on the outside of the pyramid, one on the north side and one on the south. Each shaft runs horizontally for a short distance at both ends, but their main parts rise perfectly straight and at a steep angle, the north one inclining at thirty-one degrees and the south one at forty-four degrees. To archaeologists, the obvious explanation for a long time was that they were for ventilation, although it seemed strange that they were not symmetrical. Then, in the 1960s, it was discovered that the northern shaft was aligned upon the star Thuban at the time of construction, and the southern upon one of the stars of Orion's belt. They were clearly not for observation, but we do not need to put these stellar alignments down to chance. The pharaoh needed to visit these stars in the afterlife; this is clear from the *pyramid texts* found in later Old Kingdom pyramids. It is likely that these alignments were quite deliberate and that the shafts' purpose was to symbolize, or perhaps to facilitate, his passage.

In the 1990s, the Pyramids of Giza attracted astronomical speculations of a different sort, following from an observation by writer Robert Bauval that the layout of the three pyramids on the ground appeared to mimic the formation of the three stars of Orion's belt in the sky. Bauval, and later Graham Hancock, elaborated on this in a number of bestselling books. The representation of Orion's belt, they argued, was part of a larger "sky map" on the ground, which included the Milky Way, represented by the river Nile itself. Unfortunately, the map was wrongly oriented in the third millennium B.C.E.: it would only fit correctly in the middle of the eleventh millennium B.C.E. This much older date, argued Bauval and Hancock, was the true date reflected in the design.

These ideas attracted a great deal of publicity in the 1990s, along with a torrent of criticism from Egyptologists, astronomers, and archaeoastronomers alike. The main problem is that there are more than seventy pyramids altogether in Egypt, and the selection of those claimed to be part of the star map is arbitrary. (The same problem is evident at the Holy City of Angkor in Cambodia, where a similar theory was proposed.) Another problem is that over a time scale of millennia, the course of the Nile is variable. But the main problem is the 10,500 B.C.E. date: it flies in the face of all the archaeological evidence—it is completely fanciful. One can only conclude that the resemblance to Orion's belt is fortuitous. The configuration of the three pyramids is perfectly well explained by the need of successive pharaohs to locate a pyramid with a clear view to the north in the same general area, given the constraints of the local topography.

The Giza pyramids illustrate two very different ways in which an attempt can be made to use astronomical alignments for archaeological dating. The first procedure presupposes a technique for fixing the pyramids' orientation that is plausible enough, given our knowledge of Egyptian construction techniques and instrumentation in later periods. Each candidate pair of circumpolar stars that could possibly have been used to fix the pyramids' orientation would have produced particular errors at particular dates. Examining the different possibilities may help archaeologists resolve outstanding questions concerning the precise dates of construction of these pyramids and of the reigns of the pharaohs concerned. This is in complete contrast to the second procedure, which rests upon a set of unsupported speculations and produces conclusions that are at complete odds with all the archaeological evidence.

See also:
Astronomical Dating; Cardinal Directions; Methodology.
Angkor; Egyptian Temples and Tombs; Pawnee Star Chart.
Circumpolar Stars; Meridian; Precession.

**References and further reading**

Bauval, Robert, and Adrian Gilbert. *The Orion Mystery: Unlocking the Secrets of the Pyramids.* London: Heinemann, 1994.

Belmonte, Juan. "On the Orientation of Old Kingdom Egyptian Pyramids." *Archaeoastronomy* 26 (supplement to the *Journal for the History for Astronomy* 32) (2001), S1–S20.

Edwards, Iorwerth E. S. *The Pyramids of Egypt* (rev. ed.). London: Penguin, 1991.

Hancock, Graham, and Santha Faiia. *Heaven's Mirror,* 94–99. London: Penguin, 1998.

Krupp, Edwin C. *Skywatchers, Shamans and Kings,* 285–295. New York: Wiley, 1997.

Krupp, Edwin C. "The Sphinx Blinks." *Sky and Telescope* 101 (3) (2001), 86–88.

Spence, Kate. "Ancient Egyptian Chronology and the Orientation of Pyramids." *Nature* 408 (2000), 320–324.

# Quipu

Unlike the city-states of Mesoamerica, the great Inca empire of South America produced no written records of which we are aware. However, information was encoded in mysterious knotted string devices known as *quipu,* or *khipu.* Many hundreds of examples are known, held in museums in Peru and all over the world, and new ones are still being uncovered. In 1996, thirty-two *quipu* were discovered in a remote region of northern Peru, in burial chambers built into a rock overhang surrounded by thick forest, close to a lake known as Lake of the Condors. At first the *quipu* appeared to have been mechanisms for recording fairly mundane numerical data, but there is reason to believe that at least some of them contained a more permanent record of sacred information relating, for example, to the calendar or astronomy. More intriguingly still, some of them were apparently read as texts by storytellers, so they seem to have conveyed histories and myths.

*Quipu* generally comprise a number (in some cases, as many as several hundred) of *pendant strings* attached to a single *primary string.* Sometimes some of the pendant strings have further strings (*subsidiaries*) hanging off them, and there are other minor variations, but overall the consistency in form is remarkable. *Quipu* can be bundled up, or held or laid out to display the various strings, which can be of many and various yarn types and colors. On the strings are knots. These normally contain up to nine turns, and it is generally accepted that they encode digits from one to nine. Furthermore, in the majority of *quipu* the knots on each string are carefully arranged in a way that suggests that they represent a number of two or more digits—in other words, a number expressed in a base-ten counting system the same as our own. (The Mesoamericans, in contrast, used a base-twenty counting system.) The lack of a knot in a significant place is used to signify the digit zero.

In addition to the numerical data contained in the knots, the *quipu* have the potential to convey information in a variety of ways, for instance in the spatial configurations of strings, the type of yarn used in each particular string, its color, its direction of twist, and so on. Unfortunately, we know next to nothing about how information was encoded and must try to reconstruct this from scratch. On the other hand, ethnohistory and archaeology do tell us something of the context in which some of the *quipu* were used. Accounts by early Spanish chroniclers, such as Felipe Guaman Poma de Ayala, attest to their use as tallying devices by local officials, perhaps recording the payment of tributes, or census information. Archaeologically, the fact that some *quipu* are found, like those already mentioned, among grave goods—evidently part of the personal paraphernalia of ordinary individuals—suggests that they were of significance to the individual or to their immediate household or local social group (*ayllu*). It is certainly plausible, then, that they contained a more permanent record of things of (local) cultural and historical relevance. Finally, and perhaps most intriguingly, we know that some of the native storytellers who recounted histories, chiefly genealogies and myths to the early chroniclers, did so by "reading from," or at least by interpreting in some way, certain *quipu*.

A few of the better-known *quipu* have been studied for many years. But recently new large-scale and comparative analyses have become possible, thanks to an extensive database being compiled at Harvard University by Gary Urton and his colleagues. This work has revealed, for example, common sequences of numbers on different *quipu*, implying perhaps that local tallies were copied by regional officials, further up the Inca hierarchical chain, and incorporated into summaries covering wider areas.

Some *quipu* appear to contain calendrical information. One of the largest of the thirty-two found at Lake of the Condors, known simply as "UR6," has 762 pendant strings, 730 of which are systematically organized into twenty-four sets, each containing twenty-nine, thirty, or thirty-one strings. An obvious interpretation is that each set represents a synodic month, that is, a month defined by the phase cycle of the moon, and that the whole represents a period of two years. However, this does not necessarily mean that the purpose of the *quipu* was to represent calendrical information. The strings themselves contain knots representing numbers. The obvious conclusion is that *quipu* UR6 contained information (as yet unknown in nature) organized calendrically. If this is true, it shows that while no *quipu* may in itself have been a calendar, we can nonetheless retrieve information about calendars in everyday use within the Inca empire from certain *quipu* indirectly.

To judge from a sample of over 350 *quipu* studied recently by Urton, the knot information on about a third of all *quipu* does not resolve obviously into base-ten numbers. An intriguing possibility is that these *quipu* (or at least some of them) may have been "narrative" *quipu,* somehow providing information from which a practiced interpreter could have generated, or reconstructed, a story or account. We may never know the particulars of the information encapsulated in this way on certain *quipu*, or the nature of the process by which narratives were extracted. And yet some of the first tentative steps in identifying some of the non-numerical information contained on *quipu* may now have been taken. A few numbers much larger than the rest appear on *quipu* UR6, but only in sequences of three together. The suggestion is that these were not actually number triples as such but *ayllu* labels or identifiers, tags that identified the *ayllu* to which associated numerical information referred. If this interpretation is correct, then we have made our first identification, and tentative interpretation, of non-numerical information contained on a *quipu*.

Beyond the fact that they recorded tallies and used a base-ten numbering system, interpretations of the purpose of the *quipu* and the nature of the information contained in them have, until recently, remained largely speculative. This is now changing, thanks to painstaking analyses of the numerical data combined with new archaeological evidence. The process of decoding these idiosyncratic recording devices may just have begun.

**See also:**
Lunar and Luni-Solar Calendars.
*Ceque* System.
Lunar Phase Cycle.

**References and further reading**
Ascher, Marcia, and Robert Ascher. *Code of the Quipu*. Ann Arbor: University of Michigan Press, 1981.
Quilter, Jeffrey, and Gary Urton, eds. *Narrative Threads: Accounting and Recounting in Andean Khipu*. Austin: University of Texas Press, 2002.
Urton, Gary. *The Social Life of Numbers*. Austin: University of Texas Press, 1997.
———. *Inca Myths*. Austin: University of Texas Press, and London: British Museum Press, 1999.
———. *Signs of the Inka Khipu: Binary Coding in the Andean Knotted-String Records*. Austin: University of Texas Press, 2003.

## Recumbent Stone Circles

A recumbent stone circle (RSC) is a distinctive type of stone circle containing one stone placed on its side. RSCs occur in just two regional groupings at opposite corners of the British Isles: one is in the Grampian region of northeastern Scotland around the modern city of Aberdeen, and the other in the southwestern corner of Ireland, in Counties Cork and Kerry. In both cases, the recumbent stone is generally found on the southwestern side of the circle. In the Scottish circles it is large and set between two tall uprights, known as *flankers,* that are normally the tallest stones in the circle. The heights of other stones tend to taper away as one goes round the circle, so that the smallest are found on the opposite, or northeastern, side. In the Irish circles, on the other hand, the tapering works in the other direction; the recumbent stone is small and placed in isolation, and the tallest stones are normally a pair called *portals* on the opposite northeastern side.

The clear similarity and obvious systematic difference between these two groups, together with the fact that they are separated from each other by sea and several hundred kilometers, have generated a long debate about the degree of relationship between them. It seems highly unlikely that such a similar idea could have developed independently in the two places. But was there necessarily direct communication between these two areas? Or was the idea carried by a single influential person or group of people, traders or emigrants, from one place to the other? Although some archaeologists, such as the British stone circle specialist Aubrey Burl, refer to both groups as RSCs, others, such as the Irish archaeologist Seán Ó Nualláin, prefer to avoid implying such a close relationship and refer to the Irish group as *axial stone circles*.

As a group, the Scottish recumbent stone circles present us with some of the clearest evidence from Neolithic or Bronze Age Britain or Ireland that the builders of a particular group of stone monuments carefully configured them in relation to a particular astronomical body or event. In this case the astronomical body concerned is not the sun, but the moon.

*The view looking out over the recumbent stone and flankers at the recumbent stone circle of Dyce, Aberdeenshire, Scotland. (Courtesy of Clive Ruggles)*

The RSCs are confined to a small geographical area, about a hundred kilometers (sixty miles) by sixty kilometers (forty miles), and the archaeological evidence, such as it is, suggests that they were constructed within a period of no more than a few centuries, probably in the later part of the third millennium B.C.E. However, it is their configuration that is of primary interest to archaeoastronomers. Their orientation is remarkably consistent. Over fifty RSCs are sufficiently well preserved for us to be able to determine the orientation, and there is a clear candidate for the principal orientation at each site: the axis of the circle through the recumbent stone. With not a single exception, the recumbent stone is placed within the quarter of the circle centered upon south-southwest, that is, between west-southwest and south-southeast. Some have argued that this orientation may simply reflect a relationship with the local topography or the prevailing wind, but this idea does not stand up to closer scrutiny: the topography of the area is varied and the patterns of wind flow are diffused around irregular hills and winding valleys. In fact, the reverse is true: the consistency in orientation that has been achieved over wide area despite the varied topography makes it clear that the sky was used in determining the orientation. At the very least, the builders must have determined the orientation in relation to the diurnal motion of the celestial bodies that defines the north-south axis.

But there is more to it. As viewed from within the circle, the recumbent stone and flankers frame a part of the horizon, so it is a natural supposition that this direction might be the one of significance (though most analyses have also checked the opposite direction). There is always at least a reasonably distant view beyond the recumbent, and quite often the stretch of horizon framed by the recumbent and flankers contains a conspicuous hill. This strongly suggests that it was important to stand inside the circle looking outwards toward the horizon. When one examines the declination ranges of the horizons, then the goal of such observations seems to be revealed: they seem to relate to the motions of the moon. Put simply, the full moon would pass low over the recumbent stone around midsummer each year.

It is tempting to imagine these monuments being used for ceremonies or rituals that took place by the light of the moon on a midsummer evening, when the moon was seen to reach its correct configuration with respect to the circle. Some archaeological evidence backs this up. Scatters of white stone and particularly quartz, which are perceived in some human communities as reflecting or encapsulating the light of the moon, have been found around the base of the recumbent stone at one excavated site, Berrybrae. At another recently excavated RSC, Tomnaverie, the recumbent stone itself contains more quartz than the other stones; at two others, Auchmallidie and North Strone, the recumbent stone is itself one huge block of quartz.

The other feature that reinforces the idea that the people who built and used these circles were interested in the moon is *cup marks*. There are few cup-marked stones at RSCs; cup marks occur only on the recumbent stone or flankers, or in one case only on the adjacent circle stone. They are never found in other parts of the circle. Remarkably, the orientations of these cup marks from the center correspond quite closely to specific rising and setting points of the moon, namely the most southerly limits, known as the major and minor standstill limits.

However, the image of people dancing by moonlight in a recumbent stone circle on a midsummer night may be a misleading one. Excavations undertaken by the British archaeologist Richard Bradley in 1999–2001 show that at least four RSCs were put up around an already existing feature, such as a funerary pyre or platform. They also reveal quartz scatters at different positions in the circle. Although there is also some evidence that the eventual addition of a stone circle was anticipated from the outset, it may be that we are seeing structures transformed from other materials, such as wood, into stone. This possibility does not detract from the importance of the lunar alignment, but it does alter how it can be interpreted. On the one hand, it reminds us that people must have identified a sacred place and perhaps used it for a variety of ritualistic purposes in order to make the necessary preliminary observations that enabled them to get the astronomical alignment right when it was set into stone. But it is also possible that the erection of the stones may have closed off rather than initiated a period of use, so that the alignment, once set up, was not used by people at all—or at least not by living people.

The Scottish recumbent stone circles are important to archaeoastronomers because they provide evidence of a consistent trend of astronomical orientation, one so strong that it is unthinkable that it could have arisen by chance. On the other hand, we will never know as much as we would like about why such alignments were set up or how (or even if!) they were subsequently used. The recent excavations show how, by looking more closely at stratigraphic, dating, and artifactual evidence, we can begin to gain a firmer idea of how particular monuments developed and what people did there at different times. We need this knowledge in order to put our tentative interpretations of symbolic relationships that we find encapsulated in monuments such as the RSCs into a more reliable context.

By comparing different groups of monuments we can also attempt to explore where cognitive principles, such as the interest in the moon that is so clearly manifested at the RSCs, first originated and where and how they were propagated. For example, evidence of lunar-related orientation can be found at the Clava cairns of Invernessshire to the northwest of the RSC area and also at many of the short stone rows found more widely in western Scotland and western Ireland. On present evidence, it seems that the concern with the moon that we see at the RSCs emerged in Britain and Ireland late in the Neolithic period and continued well on into the Bronze Age (from the late third until the mid or late second millennium B.C.E.). This contrasts with the solar orientation of certain monuments such as cursus monuments (Dorset Cursus, for example), passage tombs such as Maes Howe and Newgrange, and stone circles such as the group in Cumbria. Such practices extend back to some of the earliest farmers in Britain and Ireland in the fourth millennium B.C.E.

See also:
Orientation.
Axial Stone Circles; Clava Cairns; Cumbrian Stone Circles; Cursus Monuments; Maes Howe; Newgrange; Short Stone Rows; Stone Circles.
Declination; Diurnal Motion; Moon, Motions of.

**References and further reading**
Burl, Aubrey. *The Stone Circles of Britain, Ireland and Brittany,* 215–242. New Haven: Yale University Press, 2000.
Ruggles, Clive. *Astronomy in Prehistoric Britain and Ireland,* 91–99. New Haven: Yale University Press, 1999.

# Refraction

Atmospheric refraction, the bending of light downwards as it passes through the atmosphere, causes astronomical bodies to appear higher in the sky than they actually are. At high altitudes the effect is slight, but close to low horizons—when light rays have to travel through more of the earth's atmosphere to reach the observer—the apparent altitude can be as much as half a degree above the actual altitude. Certainly if something appears at the horizon, it is actually below it.

When we are calculating the declination of a celestial body following a field survey, it is necessary to take refraction into account. Because of the extremely high precision of the sightlines he was postulating, Alexander Thom went to pains to estimate atmospheric conditions at the time of supposed observations in order to calculate the appropriate refraction correction. However, as astronomers Bradley Schaefer and William Liller have since pointed out, day-to-day changes in atmospheric conditions effectively rule out sightlines precise to just a few arc minutes. For most purposes, a mean refraction correction (related only to the horizon altitude concerned) will suffice.

See also:
Thom, Alexander (1894–1985).
Declination; Precision and Accuracy.

**References and further reading**
Aveni, Anthony F. *Skywatchers*, 103–105. Austin: University of Texas Press, 2001.
Ruggles, Clive. *Astronomy in Prehistoric Britain and Ireland*, 23, 25. New Haven: Yale University Press, 1999.
Schaefer, Bradley E., and William Liller. "Refraction Near the Horizon." *Publications of the Astronomical Society of the Pacific* 102 (1990), 796–805.

# Ring of Brodgar
*See* Brodgar, Ring of.

# Roman Astronomy and Astrology
Unlike the temples in ancient Greece, Roman temples are not obviously consistent in their orientation. Nor did any great innovations in philosophical cosmology or mathematical astronomy emerge in the Roman world—at least none that are well known to modern historians of science. Roman astronomy, it seems, was more pragmatic in nature, often intimately bound up with prognostication and astrology.

One of the best-known manifestations of Roman astronomy is the Julian calendar. It evolved initially from a simple sequence of indigenous festivals related to the farmers' seasonal year, then became increasingly formalized as Roman society became increasingly urbanized. At first the civic calendar was based on the phase cycles of the moon, but it faced increasingly serious problems in getting out of step with the seasonal year because of political interference in the process of intercalation. All this culminated in 45 B.C.E. in the switch from a lunar to a solar calendar, dividing the year into twelve "months" that were in fact completely independent of the moon.

The civic calendar aside, there is little doubt that astronomy played a significant role in many aspects of Roman life. Land surveyors, employed all over the empire with simple instruments, needed good astronomical knowledge: this is clear from surviving surveyors' manuals. Other books instruct the navigators of commercial ships in navigating by the stars. Sundials were in common use for telling the hour of the day: over thirty were recovered in Pompeii alone.

Astrology, popular among the Romans' Etruscan predecessors as well as in Hellenistic Greece, had an uneasy relationship with the Roman state at first, but increased greatly in popularity in the second century C.E. with the rise of mystery cults such as those of Isis and Mithras. These were astral religions, and for many people the sun itself became an object of direct worship. Later, these cults came into direct conflict with Christianity.

**See also:**
Astrology; Lunar and Luni-Solar Calendars.
Julian Calendar; Mithraism; Pantheon; Temple Alignments in Ancient Greece.

**References and further reading**
Aveni, Anthony. *Empires of Time: Calendars, Clocks and Cultures,* 111–115. New York: Basic Books, 1989.
Krupp, Edwin C. *Skywatchers, Shamans and Kings,* 240–243. New York: Wiley, 1997.
Magini, Leonardo. *Astronomy and Calendar in Ancient Rome.* Rome: «L'Erma» di Bretschneider, 2001.
Walker, Christopher, ed. *Astronomy before the Telescope,* 92–97. London: British Museum Press, 1996.

# Rujm el-Hiri

The megalithic monuments of Britain and Ireland have been studied in great detail by archaeoastronomers. Numerous systematic investigations of the orientations of prehistoric tombs and temples throughout the western Mediterranean have been carried out. But comparable effort has rarely been directed towards similar sites elsewhere in the world. One exception is a curious circular construction built by Early Bronze Age peoples in the Golan Heights, close to the border between Israel and Syria. Its name, Rujm el-Hiri, means "stone pile of the wild cat." Built around 3000 B.C.E., it consists of a massive circular (actually slightly oval) wall some one hundred fifty meters (five hundred feet) across and over three meters (ten feet) thick with broad entrances to the northeast and southeast. Inside this are four further concentric and successively thinner walls with irregularly placed openings. Between the circular walls are curved corridors, some of them divided into segments by blocking walls built perpendicular to the circles (oriented radially out from the center), but there is no evident pattern to them. Most of the walls still stand to between two and three meters (seven to ten feet) high. The result resembles the sort of maze encased in plastic into whose center children try to guide ball bearings. Filling most of the central space, which measures some thirty meters (one hundred feet) across, is a cairn still standing to a height of almost five meters and built over a burial chamber oriented towards the northeast entrance. It is surrounded by three terraces with circular stone kerbs. The obvious conclusion—that the walls were built successively around the cairn—seems to be contradicted by the archaeological evidence, which suggests instead that the central cairn was actually built more than a millennium after the original construction, in around 1500 B.C.E. Rujm el-Hiri, in other words, appears to be a temple that later became a tomb.

Various astronomical alignments have been identified at the site, the clearest being the solstitial alignment of the northeast entrance, which is reflected in the orientation of the later burial chamber. Curiously, though, the southeast entrance is not aligned upon the other solstitial axis: it is too far to the south. On the other hand, two exceptionally large boulders on the eastern side could have referenced sunrise close to the equinox, and it has been suggested that the purpose of this was to indicate that the first spring rains were imminent, allowing final preparations to be made for water to be efficiently collected for irrigation. Statistical analyses of the alignments of over thirty radial wall segments suggest that they may have been used as stellar sighting devices.

What should we make of these alignments? It is almost certainly misleading to think of the site as an astronomical observatory. As a center for religious rituals, however, cosmic referencing would have been important. The location of the temple may have been significant in this regard, in that it resulted in the two most prominent mountains—Mount Hermon and Mount Tabor—appearing, respectively, almost due north and close to December solstice sunrise. By visually linking places of vital cosmic significance, some in the landscape and some in the sky, the builders would have reinforced the sacred power of the temple itself. More controversial, perhaps, is the idea that the radial walls could have functioned as stellar-alignment-fixing devices. But then, what other purpose could they have served? One possibility is that they simply served to restrict and confound access to the central space, the act of negotiating an asymmetric labyrinth serving to instill in the visiting pilgrim a heightened sense of the sacred power of the place. The approximately solstitial orientation of the (later) tomb was certainly nonfunctional but may have been deliberate and symbolic, following the alignment of the original temple.

Although Rujm el-Hiri has been hailed as a Stonehenge of the Levant, and its construction is certainly broadly contemporary with the earliest known (earth and timber) constructions at Stonehenge, the comparison may be misleading and inappropriate. It is a magnificent temple in its own right and gives us some special clues about the beliefs and practices of the Bronze Age cultures of this part of the Near East.

**See also:**
Cosmology; Solstitial Directions.
Megalithic Monuments of Britain and Ireland; Prehistoric Tombs and Temples in Europe; Stonehenge.

**References and further reading**
Aveni, Anthony F. *Skywatchers*, 323–326. Austin: University of Texas Press, 2001.
Aveni, Anthony F., and Yonathan Mizrachi. "The Geometry and Astronomy of Rujm el-Hiri, a Megalithic Site in the Southern Levant." *Journal of Field Archaeology* 25 (1998), 475–496.
Dgani, Avi, and Moshe Inbar, eds. *Eretz Hagolan [The Golan Heights and Mount Hermon]*, 403–412. Tel Aviv: Misrad Hafitahen, 1993. [In Hebrew.]

## Sacred Geographies

The term *sacred geography* stems from the realization that people in the past conceived the landscape as charged with meaning, some of which stemmed from the common worldview developed by their community as a whole, and some of which worked on a personal level. The significance assigned to particular places, routes through the landscape, and other (generally more visual or perceptual than physically marked) divisions, connections, and directions might well be a result of real events or memories. But it might equally well derive from mythical or sacred connotations that make little sense from a modern, Western perspective. Time is an important factor, too. The sacred power of a place was often thought to be enhanced at particular times, for example, when it was visually reinforced or revealed by a brief play of sunlight. Once we realize this, we are in a better position to appreciate conceptions of landscape in the context of worldviews—cosmologies—other than our own; conceptions that are unlikely to view landscape simply as a resource to be exploited and can give rise to actions that we would not necessarily perceive as rational.

At the same time, using the term *sacred geographies* (or *sacred landscapes*) to describe these conceptions has a major drawback. It implies that there was a clear separation between the sacred and the secular. That this is unhelpful is demonstrated by innumerable anthropological studies of historical and modern peoples. There are different degrees and types of sacredness; and while some sacred places were very special indeed, mundane places often had sacred aspects as well, helping to integrate the need for preserving worldly harmony into many different aspects of everyday life. Rites to ensure continuity and regeneration were often related to particular places, whether public temples or personal shrines, and tended as well to be related to natural (that is, seasonal) phenomena: accordingly, they were often carefully timed in relation to the cycles of the celestial bodies.

In all probability, true sacred landscapes, devoid of all mundane activity, simply did not exist. However, the sacred aspects of all landscapes were crucially important. The methods of archaeoastronomy are particularly important in their potential to reveal sacred aspects of prehistoric landscapes (for which, by definition, no historical evidence is available) where these have resulted in spatial patterning with respect to the heavenly bodies.

See also:
Archaeoastronomy; Cosmology; Landscape.
Aboriginal Astronomy; Aztec Sacred Geography; Hopi Calendar and Worldview; Lakota Sacred Geography.

**References and further reading**
Ashmore, Wendy, and Bernard Knapp, eds. *Archaeologies of Landscape: Contemporary Perspectives*. Oxford: Blackwell, 1999.
Carmichael, David, Jane Hubert, Brian Reeves, and Audhild Schanche, eds. *Sacred Sites, Sacred Places*. London: Routledge, 1994.
Ruggles, Clive. *Astronomy in Prehistoric Britain and Ireland*, 120–121. New Haven: Yale University Press, 1999.

## Sarmizegetusa Regia

The Carpathian mountains run through northwest Romania in a wide arc, enclosing an area that stretches down westwards into the Carpathian basin in Hungary. This region forms modern-day Transylvania, but in the centuries immediately before and after the birth of Christ it was home to the Dacians, a powerful kingdom whose later resistance to the Roman conquest is famously depicted in Trajan's Column in Rome. The rugged landscape to the south of the modern town of Orăştie contains a number of impressive Dacian monuments, including settlements, stone-built fortresses and "sanctuaries," assumed to have been places of religious worship. It is in this area, at an elevation of over 1,000 meters (3,300 feet), that one encounters Sarmizegetusa Regia, the ancient capital of the Dacians.

Sarmizegetusa Regia contained a citadel and living areas with dwellings and workshops, but it also contained a sacred zone. A number of rectangular temples were located there, the bases of their supporting columns still visible in regular arrays. Perhaps the most enigmatic construction at the site, however, is the *large circular sanctuary*. This comprised a "D"-shaped setting of timber posts surrounded by a timber circle, which was surrounded in turn by a low stone kerb. The original timber posts have long since rotted away, and the ones currently visible on the site are modern reconstructions in the original postholes. They help the modern visitor visualize the layout of the original construction, although the heights of the posts is completely speculative.

The layout of the timber settings bears a broad resemblance to the sarsen stone monument at Stonehenge in England, where five trilithons laid out in a horseshoe shape are surrounded by a complete circle. This has led

The large sanctuary at Sarmizegetusa Regia, Romania, with the "Andesite Sun" in the foreground. (Courtesy of Clive Ruggles)

some people to proclaim Sarmizegetusa Regia as a "Romanian Stonehenge" and to propose it as an ancient observatory comparable in importance to the British site. In fact, this logic is flawed on several grounds. First, the two sites are culturally unrelated and well separated in time: Sarmizegetusa Regia was not built until 1,500 years after Stonehenge had fallen into disuse. The importance of the Romanian sanctuary must be measured in the context of the Dacian cultural heritage. Second, Stonehenge is no longer considered by serious scholars to have been an ancient observatory. Its main axis was certainly aligned toward the solstices, and the same has been claimed of the large circular sanctuary. But in fact there are problems. The Romanian "horseshoe" (if one excludes the straight side of the "D") faces southeast rather than northeast as at Stonehenge, so it is the midwinter rather than midsummer sunrise upon which it is claimed to be aligned. This *would* be the direction of sunrise on the shortest day of the year if there were a distant, level horizon in this direction. But there is not. The horizon altitude in this direction is several degrees, and the midwinter sunrise would occur over ten degrees farther south.

Yet the circular sanctuary and other structures in the sacred zone may well have had astronomical or calendrical associations. It has been suggested, for example, that a large circular block of stone 1.5 meters (five feet) in diameter, laid flat and known popularly as the "Andesite Sun," was used as a sundial. It has been suggested that a circle of rectangular grooves cut into the stone might have been used to mark the length of the shadow from a central gnomon, but the existence of such a gnomon is pure speculation. Others have argued that the stone was simply used for sacrifices or other rituals. Yet the idea of as sundial is far from absurd, since there are known to have been influences on Dacian culture from Hellenistic Greece, and these may well have included ideas about geometry and astronomy.

**See also:**
Solstitial Alignments.
Stonehenge.
Altitude.

**References and further reading**
Frangopol, P. T., and V. V. Morariu, eds. *Archaeometry in Romania: 2nd Romanian Conference on the Application of Physics Methods in Archaeology* (vol. 2, 1–34.)
Glodariu, Ioan, Eugen Iaroslavschi, and Adriana Rusu. *Dacian Citadels and Settlements in the Orăştie Mountains,* 250–282. Bucharest: Editura Sport Turism, 1988.
Glodariu, Ioan, Adriana Rusu-Pescaru, Eugen Iaroslavschi, and Florin Stănescu. *Sarmizegetusa Regia: Pre-Roman Capital of the Dacians,* 277–310. Deva, Romania: Acta Musei Devensis, 1996.

# Saroeak

Basque stone octagons are symmetrical arrangements of eight small stones found scattered across the northern coastal parts of the Iberian peninsula but concentrated mainly in the Basque country—a culturally and ethnically distinctive region spanning four provinces in northern Spain and extending into southwestern France. The stone octagons represent an enigmatic class of monument constructed, apparently, using well-defined astronomical and geometrical principles. Each was of similar size, with a diameter of about 320 meters (1050 feet), and the stones were placed accurately on the cardinal and intercardinal points. (Sometimes an additional eight stones were placed at points in between, making sixteen in all.) In the center was a distinctive stone known as an *artamugarri*. The term *saroeak* (the plural, in the Basque language, of *saroe*), refers to the area demarcated by these stone arrangements. One closely studied area surrounds the coastal town of Donostia (San Sebastian) and the villages of Hernani and Urnieta to the south. Even though it is an area of only about forty square kilometers (fifteen square miles), nearly forty *saroeak* have been identified in this locality alone.

Until the 1990s, *saroeak* had been of little interest to archaeologists, who believed them to be a relatively modern phenomenon. However, it was

*The modest center stone at Akola* saroe *in the Basque Country, northern Spain. The distinctive carving on the top reflects the eightfold directional symmetry of the saroe itself. (Courtesy of Clive Ruggles)*

known that their distribution tended to coincide with places used for grazing by transhumant shepherding peoples as part of a strong pastoral tradition within the Basque Country. Then, during the 1980s and 1990s, regional cultural heritage projects combining documentary research and small-scale excavations revealed a quite remarkable continuity of material tradition extending over almost 2000 years. The historical documents showed the *saroeak* to be the material remnants of a system for organizing communal grazing lands dating back at least to the Early Medieval times—a system that, in some cases, remained in use until early in the twentieth century. The archaeological investigations showed that some center stones were constructed as far back as C.E. 200.

Some of the historical records actually specify the geometrical design of the *saroeak* and the methods that should be used in their construction, indicating that they were laid out using standard units of length, and confirming the importance of celestial orientation in relation to the cardinal and intercardinal directions. They also tell us that the *saroeak* could act as local seats of government and law and order; as meeting places; as safe grazing grounds; as farmstead boundaries; and as centers for religious rites. The importance of reflecting the perceived cosmic order in their design perhaps comes as no surprise, then, especially in view of the last of these.

The archaeological evidence shows that the design principles underlying the *saroeak* had already remained relatively constant for many centuries before the earliest documents relating to them were written. Furthermore, the earliest *saroeak* respected even earlier megalithic tombs and dolmens, in that they never enclosed them, although such monuments are sometimes found just outside a *saroe* boundary. Some have interpreted this as evidence of an even longer continuity of tradition dating back to the Neolithic or Bronze Age. Support for this view has been sought from the fact that the Basque language is unique in western Europe in not belonging to the Indo-European linguistic family, suggesting that its roots are more localized and pre-date the spread of Indo-European languages across the continent. If such a long continuity of tradition did indeed exist in the Basque country, then the Basque *saroeak* could be relevant to interpretations of later prehistoric monuments all over western Europe, including the famous Neolithic and Bronze Age stone circles of Britain and Ireland. This is because—despite efforts by some scholars to argue otherwise, for example in relation to the "megalithic" calendar—such interpretations are normally constrained by the fact that the only direct evidence available is archaeological.

Yet it is also possible that the earliest *saroeak* simply implemented sacred or symbolic principles that respected what were already, even by the early centuries C.E., conspicuous ancient features in the landscape, just as the Nahua (Aztecs)—in another continent and another millennium—respected the already ancient pyramids of Teotihuacan as works of the gods. However, if we become prepossessed by the possibility of even longer continuities of tradition, we divert attention away from the broader significance of the *saroeak* in a different respect. They demonstrate remarkably well that functions and meanings can vary considerably while a material tradition remains essentially unchanged. (Another, very different, example of this is the Mesoamerican cross-circle designs.) And they also provide an important case study concerning the ways in which this can happen in practice, over a period of nearly two millennia.

**See also:**
"Megalithic" Calendar.
Megalithic Monuments of Britain and Ireland; Mesoamerican Cross-Circle Designs; Stone Circles.

**References and further reading**
Blot, Jacques. *Montaña y Prehistoria Vasca*. Donostia [San Sebastian], Spain: Elkar, 1988. [In Spanish.]
Jaschek, Carlos, and F. Atrio Barandela, eds. *Actas del IV Congreso de la SEAC «Astronomía en la Cultura»*, 95–102. Salamanca: Universidad de Salamanca, 1997. [Article is in English.]

Markey, T. L., and A. C. Greppin, eds. *When Worlds Collide: The Indo-Europeans and Pre-Indo-Europeans,* 115–150. Ann Arbor: Karoma, 1990.

Ott, Sandra. *The Circle of Mountains: A Basque Shepherding Community* (2nd ed.). Reno: University of Nevada Press, 1993.

Ruggles, Clive, ed. *Archaeoastronomy in the 1990s,* 77–91. Loughborough, UK: Group D Publications, 1993.

Walker, Christopher, ed. *Astronomy before the Telescope,* 15–27. London: British Museum Press, 1996.

Zaldua, Luix Mari. *Saroeak Urnietan [Stone Octagons in Urnieta].* Urnieta, Spain: Kulturnieta, 1996. [In Basque, Spanish, and English.]

## Science or Symbolism?

In the early days of modern archaeoastronomy, in the 1970s and early 1980s, there was a long and occasionally rancorous argument about whether prehistoric astronomy, as manifested in astronomical alignments at British megalithic sites such as Ballochroy and Kintraw, was scientific or symbolic in nature. This question was intimately bound up with the issue of whether, as proposed by Alexander Thom, prehistoric people created deliberate sightlines on the rising and setting points of the sun, moon, and stars, making use of features on distant horizons as "foresights" to achieve pinpoint precision in their astronomical observations. The interpretation favored by many archaeologists was that the high-precision alignments picked out by Thom were fortuitous and that the people who built large stone monuments were aware only of rougher alignments, such as the famous midwinter sunrise alignment at the Irish passage tomb of Newgrange. They argued that these lower-precision alignments were more likely to be related to rituals and symbolism associated with seasonal activities, death, renewal, ancestor worship, and so on rather than used as precise "scientific" measuring instruments.

There are two ironies here. One is that "high precision" should not necessarily be equated with "scientific" and "low precision" with "symbolic." Many historical and modern instances demonstrate that human communities perform remarkably painstaking and exact astronomical measurements for what, ultimately, are "merely" (as we might mistakenly see it) ceremonial purposes. (A much-cited example is the traditional calendar of the Hopi in the southwestern United States.) Furthermore, the fact that prehistoric people went to the bother of encapsulating astronomical alignments, whether rough or more precise, in stone argues that they were *more* than mere observing instruments—that they had strong symbolic motivation and meaning.

The second irony in that "our" science—the modern Western way of understanding the world—is not likely to be the way that human communities understood the world in prehistory, any more than in indigenous communities today. One might well see other worldviews as representing alternative forms of "science," insofar as they explain what people see in the world (the cosmos) around them and often have a predictive capability. (A good example of this is the role of the Caterpillar Jaguar constellation of the Barasana people of the Columbian Amazon in relation to earthly caterpillars.) Such "alternative sciences" often have a strong ceremonial element, since appropriate observances help to keep human actions in tune with the cosmos as it is perceived and understood. Incorporating alignments in architecture to symbolize and reinforce the properties of the world is another. (This is evident, for example, in the design of the earth lodges of the Pawnee, or the hogans of the Navajo, to name but two of many examples.) If we are really interested in understanding more about prehistoric "science," then we should probably be looking for evidence of the kind of symbolical alignments and ceremonial activities that were seen by many back in the 1970s as its very antithesis.

See also:
Archaeoastronomy; Cosmology; Ethnocentrism; Megalithic "Observatories"; Palaeoscience; Symbols; Thom, Alexander (1894–1985).
Ballochroy; Barasana "Caterpillar Jaguar" Constellation; Hopi Calendar and Worldview; Kintraw; Navajo Hogan; Newgrange; Pawnee Earth Lodge.

**References and further reading**
Aveni, Anthony F. *Skywatchers*, 339–341. Austin: University of Texas Press, 2001.
Ruggles, Clive. *Astronomy in Prehistoric Britain and Ireland,* 1–81. New Haven: Yale University Press, 1999.

## Short Stone Rows

Among the most enigmatic of all the different types of megalithic monuments in Britain and Ireland are the short rows of up to six standing stones that are found scattered around northern and western parts of the islands, as well as in northwest France. Well over 250 examples are known, and the dates that exist suggest that the majority of them were erected at a relatively late stage as far as megalithic monuments go: in the Early to Middle Bronze Age, well into the second millennium B.C.E. Typically, the stones are one to two meters (three to six feet) in height and placed a few meters (up to about fifteen feet) apart, although there are some spectacular exceptions, such as the three-stone row at Gurranes in County Cork, Ireland, where the slender stones rise to over four meters (thirteen feet). Most of the rows stand in apparent isolation, though some are associated with burials or other megalithic structures such as stone circles. Particular concentrations of short stone rows are found in western Scotland—in Argyll on the mainland and on inner Hebridean islands such as Mull—and in southwestern Ireland, in Counties Cork and Kerry.

*The stone row at Gurranes, County Cork, Ireland. (Courtesy of Clive Ruggles)*

Unlike stone circles, stone rows do not enclose an area so cannot be interpreted as demarcating a central place or sacred space. They can, however, easily be envisaged as pointing at something, and hence have long evoked the interest of archaeoastronomers. Certain solstitially aligned rows have achieved considerable public notoriety, such as Gleninagh in Connemara, western Ireland. Alexander Thom certainly saw their potential as megalithic "observatories." He included many British examples in his lists of monuments aligned upon horizon features such as hilltops and notches that marked significant rising or setting positions of the sun, moon, and stars. The row at Ballochroy, on the west coast of the Kintyre peninsula in southern Argyll, Scotland, was one of his earliest published and most quoted solar observatory sites, but it also became one of the most important case studies in demonstrating why apparent astronomical alignments, and especially those of high precision, have to be treated with considerable care and caution.

Studies of repeated trends among regional groupings have provided the strongest evidence that many stone rows do have astronomical connections, though at a lower level of precision than that envisaged by Thom. In both of the strongest geographical concentrations, the clear consistency in the orientation of the rows is not a simple consequence of the local topography. It appears that the location of some of the monuments was carefully selected in order to follow the prevailing tradition of orientation (and the need for a distant view in certain directions) despite the general lie of the land. In southwestern Ireland there are just over fifty measurable rows, and their orientations are strongly clustered around northeast-southwest. The orientations of the western Scottish rows are just as consistent, but clustered around north-south. In both cases the evidence indicates a linkage with the moon, but for a stone row as opposed to a stone circle (the prime example being the Scottish recumbent stone circles), it is not obvious where in relation to the stones any moon-related ceremonials might have taken place. Only in the case of the handful of short stone rows on the northern part of the Isle of Mull in western Scotland does there seem to be a clearer answer to this question. These rows are all situated so that a prominent distant mountain is on the very limit of visibility—in other words, it is not clearly visible unless ones moves to one side of the row. This side, then, must be where any moon-related observances would have taken place. But this in itself doesn't help to explain why the rows should, in the first place, be so carefully located on the very limit of visibility.

In recent years, many archaeologists have started to question the assumption that, once farming spread to islands such as Britain and Ireland, marking the beginning of the Neolithic period, people settled into permanent villages, tending their crops and animals. Many peoples may have remained essentially mobile, with animal herders or perhaps whole communities following seasonal paths around the landscape. On an island such as Mull, where soils are poor and the agricultural potential is low, it is possible that few people lived there, even in the Bronze Age, and that they were essentially mobile. Could monuments such as the stone rows, then, mark places that were only visited occasionally? Or could they mark boundaries, social or symbolic? Such ideas do not imply that lunar alignments were not an integral part of these monuments but may suggest different ways in which they could be interpreted.

See also:
Thom, Alexander (1894–1985).
Ballochroy; Megalithic Monuments of Britain and Ireland; Recumbent Stone Circles; Stone Circles.

**References and further reading**
Burl, Aubrey. *From Carnac to Callanish: The Prehistoric Stone Rows and Avenues of Britain, Ireland and Brittany.* New Haven: Yale University Press, 1993.
Ruggles, Clive. *Astronomy in Prehistoric Britain and Ireland*, 102–124. New Haven: Yale University Press, 1999.

# Sky Bears

The constellations of Ursa Major and Ursa Minor, the Great Bear and Little Bear, circle the north celestial pole each night and have done so for several millennia, although the position of the pole gradually shifts owing to precession. These twin formations of stars, though identified in a variety of ways by different peoples, have formed a conspicuous feature of the skies for every human culture in the northern hemisphere—particularly from temperate latitudes northward, where they are circumpolar and thus remain visible on every clear night of the year.

Since the two constellations have similar shapes, it is not surprising that many peoples have perceived them as larger and smaller versions of the same thing. It is more mystifying why anyone would have conceived of them as bears. The idea comes down to us (along with many other constellation identifications) from ancient Greece. It does not require too much imagination to interpret their shapes as pans or ladles, wagons or ploughs, but it is a great deal more difficult to liken them to bears. This makes it all the more curious that a number of culturally unrelated groups also, apparently, saw these constellations as bears. Examples include several indigenous North American groups, such as the Iroquois and the Mi'kmaq of Nova Scotia. The desire not to dismiss this as a coincidence has led some people to propose a radical explanation.

The general idea of a bear, or bears, in the sky is widespread, especially in northerly latitudes. A variety of traditions—folk tales, public performances, and shamanic rituals—involving terrestrial bears and connected to a Celestial Bear exist among cultures scattered throughout Europe and the northernmost parts of Asia and the Americas, including the Basque country, Finland, Siberia, the Kamchatka peninsula of eastern Russia, Alaska, and eastern Canada. A common feature of many of these fragmentary traditions is that bears were seen as the ancestors of humans and suitably revered, the common ancestor of all being the Celestial Bear or Great Sky Bear.

In several parts of Europe, plays are still performed as part of seasonal festivities in the winter or spring in which the players enact the story of a bear who is hunted and killed, but subsequently returns. In Andorra, for example, the bear actor—or even a live bear—performs a dance accompanied by musicians and masked actors representing other animals. In some places older traditions have been reported in which an actual bear was hunted and sacrificed, then eaten at a banquet. The American linguist Roslyn Frank argues that such ceremonies represent the surviving fragments of a tradition in which the earthly bear was seen as being "sent home," its soul ascending up the "sky pole" to the ancestral bear in heaven, to whom it would report on its treatment in the earthly world. A positive report guaranteed health and well-being for another year.

Frank argues that many of these bear myths may have a common origin in an ancient worldview that predated by several millennia the great ancient civilizations, possibly dating back as far as the Upper Palaeolithic period over 10,000 years ago. If so, then Ursa Major and Ursa Minor might have been recognized as bears as long ago as this. And if this was the case, it might explain how such a conception could have arrived in the Americas—carried there by the earliest settlers entering Alaska across the land bridge that then connected it to Siberia. But such ideas are highly controversial, and it is better to separate them into three different strands.

The first is the suggestion that in modern European folk performances involving bears we may be glimpsing one aspect of a widespread worldview—prevalent at least as far back as Medieval times and existing alongside the dominant Christian tradition—in which bears, seen as human ancestors, played a key part in ensuring seasonal renewal through their connection with an ancestral bear spirit or spirits in the sky. The second is that folk tales and practices relating to sky bears in Europe relate directly (as opposed to bearing a fortuitous resemblance) to indigenous cosmological beliefs and rituals in other parts of the northern hemisphere, with the implication that this must represent a tradition whose origins stretch back to the very distant past. The final strand is that these practices relate specifically to the identification of one or two bears circling around the northern celestial pole: in other words, that the bear(s) in the sky can consistently be identified with Ursa Major and/or Ursa Minor. Each of these strands provides its own challenges in terms of the available evidence and its interpretation.

Nonetheless, if we do not accept Frank's scenario we are left with the open questions of why celestial bear myths are so commonplace, and why such an inherently improbable identification of these two groups of stars seems to crop up repeatedly in entirely different places. Finally, there is the mystery of why the celestial bear was perceived by the ancient Greeks as having a long tail, unlike terrestrial bears. There is no satisfactory explanation, and it has been suggested that this might itself be a consequence of the transfer into ancient Greece of a much older tradition.

See also:
Cosmology.
Celestial Sphere; Circumpolar Stars; Precession.

**References and further reading**
Koleva, Vesselina, and Dimiter Kolev, eds. *Astronomical Traditions in Past Cultures*, 116–142. Sofia: Bulgarian Academy of Sciences, 1996.
Ruggles, Clive, Frank Prendergast, and Tom Ray, eds. *Astronomy, Cosmology and Landscape*, 15–43, 133–157. Bognor Regis, UK: Ocarina Books, 2001.
Thurston, Hugh. *Early Astronomy*, 2–4. Berlin: Springer-Verlag, 1994.

# Solar Eclipses

It is a staggering coincidence—but a coincidence nonetheless—that, as seen from the surface of the earth, the apparent diameter of the sun is more or less the same as that of the moon. It is only because of this fluke of nature that total solar eclipses appear as spectacular as they do.

A total solar eclipse occurs when the moon passes directly in front of the sun, just covering its disc, but rendering visible the ethereal white light of the surrounding solar corona—the sun's outer atmosphere—and reducing the remainder of the sky to a dusk-like half-light. The same fluke of nature also ensures that total solar eclipses are rare, since if the alignment is not exactly right, then the disc of the sun will never be completely covered and the eclipse will only be partial. As long as only a small part of the sun's disc remains uncovered, its light will continue to light up the sky much as normal and the eclipse might easily pass unnoticed. The same is true of an annular eclipse, which occurs when the alignment is exact but the moon is slightly farther away from the earth than it is on average, so that its disc appears slightly smaller than usual; in this case, a thin ring of the sun's disc remains uncovered.

A total solar eclipse can be seen from somewhere on earth about once every year and a half on average, but each one is only visible from within a narrow track, typically about 160 kilometers (100 miles) in width, though stretching for several thousand kilometers across the surface of the planet. From a typical spot on earth, a total solar eclipse will only occur once every four hundred years on average. Fewer than one in four people in the past, lacking the benefit of modern scientific foreknowledge and travel possibilities, would have experienced a total solar eclipse during their lifetime.

Witnessing a total solar eclipse at first hand is an awe-inspiring experience. In a few moments the sun is transformed from something too bright to be directly visible into a black disc surrounded by a wispy white halo, and the remainder of the sky becomes as dark as early dawn or late dusk, with planets and bright stars visible. All around, nature reacts: the atmosphere cools, birds stop singing, and other animals are in confusion. Then, after no more than a few minutes, and equally rapidly, normality is restored. Even in the modern world, knowing what to expect, one is left with a profound sense of the power and immutability of nature. There can be little doubt that the psychological effect upon people and cultures in the past for whom such an event occurred entirely without warning—upsetting the rhythms of nature known and relied upon since time immemorial—was potentially devastating.

Despite this, hardly any eclipse sightings in prehistoric times have left a definite trace in the archaeological record. Most claims of depictions of eclipses in rock art are purely speculative. For example, on one panel in a large petroglyph field at Puʻu Loa on the Big Island of Hawaiʻi is a design consisting of a full circle with four adjacent crescents, increasing in size away from it. It has been suggested that these represent the stages of a solar eclipse, up to and including totality. However, the crescents vary in size rather than shape, as one would expect if they were literal depictions, and there are no symmetrical crescents on the other side of the circle, which one might expect to represent the phases after totality. Furthermore, the crescent sun cannot be viewed directly. It is more likely that these symbols have another meaning entirely. Similarly, a group of petroglyphs on a rock outcrop at Ekenburg in Sweden have been interpreted as a literal depiction of the sky during a total eclipse in 1596 B.C.E. However, this interpretation depends upon a number of unsupported assumptions. Given the likely impact of such an unexpected event on the people of the time, it seems unthinkable that, in less than three minutes, they could have recorded the appearance of the sky in enough detail to then reproduce it faithfully in stone.

Even in some of the great ancient empires and city-states, where literacy permitted detailed records to be kept of a variety of astronomical (and meteorological) phenomena, total solar eclipses could not be reliably predicted and remained greatly feared. In ancient China, for example, records of total eclipses of the sun go back at least to about 700 B.C.E., but even two complete millennia after this date there was no means to predict them. When they did happen, they were taken as particularly disastrous omens for the state, or for the Emperor himself. Babylonian astronomers were the first to discover that lunar eclipses tend to recur after certain fixed periods of time. This enabled them to predict lunar eclipses but not solar ones, which remained a sign of an impending calamity.

Across the world in pre-Columbian Mesoamerica, the Maya almanac subsequently known as the Dresden Codex contained a tabulated record of eclipses that, once again, could have been used to provide warnings of further lunar eclipses, with a reasonable (although probably not good) success rate—but not solar ones. The Maya city-states had collapsed by the time the Europeans arrived, but the Aztecs continued to record eclipses, even after the conquest. The tremendous fear felt by those natives who witnessed a total solar eclipse in 1531 is recorded in graphic detail by the chronicler Fray Bernardino de Sahagún.

The problem with predicting total solar eclipses is that regular cycles (such as the Saros cycle of eighteen years and eleven days, which the Babylonians identified) are only crude. They take no account of various additional, more complex irregularities in the motions of the earth and moon. These small irregularities are relatively unimportant in predicting eclipses of the moon, which affect half of the earth, but are critical if we need to predict the narrow path of totality of a solar eclipse.

Many people have been keen to demonstrate that prehistoric people could predict eclipses. Famously, the Aubrey Holes, a ring of fifty-six pits surrounding the sarsen circle at Stonehenge in England, was interpreted by astronomers Gerald Hawkins and Fred Hoyle in the 1960s as an eclipse-pre-

*Desperation among Peruvian villagers on the occurrence of a solar eclipse. Copper engraving, Bernard Picardi. (Bettmann/Corbis)*

dicting device. The two astronomers proposed slightly different methods, but both schemes involved wooden posts being moved around the holes in prescribed ways. Both methods would have successfully identified "eclipse danger periods"—periods when a solar *or* lunar eclipse *could* occur—at roughly six-month intervals. Eclipses of the moon could only occur at full moons, and eclipses of the sun at new moons, falling within these danger periods. However, even partial solar eclipses would only occur on average once every two and a half years, and a total solar eclipse would only happen on average once in several generations. Hawkins's scheme only successfully predicted a fraction of eclipses; Hoyle claimed that his own method would have successfully predicted about half. However, further debate about the relative merits of either scheme is unnecessary. The Aubrey Holes are just one of the several Neolithic pit circles known in Britain, of a range of sizes. It is archaeological nonsense to pick this one as a potential eclipse predictor while ignoring the rest.

Few would doubt that total solar eclipses, when they did occur in the past, could have made a huge impression on the people who witnessed them. Where historical or written records exist they give us concrete information about particular eclipse observations and their social impact. But we must beware of letting our eagerness to find notable accomplishments back in prehistory lead us to invent archaeological evidence that is simply not there.

**See also:**
Astrology; Eclipse Records and the Earth's Rotation; Lunar Eclipses; Power. Babylonian Astronomy and Astrology; Chinese Astronomy; Dresden Codex; Stonehenge; Swedish Rock Art.

**References and further reading**
Aveni, Anthony F. *Stairways to the Stars: Skywatching in Three Great Ancient Cultures,* 33–37. New York: Wiley, 1997.
———. *Skywatchers,* 26–29, 173–184. Austin: University of Texas Press, 2001.
Espenak, Fred. *NASA/Goddard Space Flight Center Solar Eclipse Page.* http://sunearth.gsfc.nasa.gov/eclipse/solar.html.
———. *Solar Eclipses of Historical Interest.* http://sunearth.gsfc.nasa.gov/eclipse/SEHistory/SEHistory.html.
Hoyle, Fred. *From Stonehenge to Modern Cosmology,* 19–54. San Francisco: Freeman, 1972.
Lee, Georgia, and Edward Stasack. *Spirit of Place: Petroglyphs of Hawai'i,* 97. Los Osos, CA: Easter Island Foundation, 1999.
Littmann, Mark, Ken Willcox, and Fred Espenak. *Totality—Eclipses of the Sun* (2nd ed.), 1–53. New York: Oxford University Press, 1999.
Schaefer, Bradley E. "Solar Eclipses that Changed the World." *Sky and Telescope* 87 (1994), 36–39.
Selin, Helaine, ed. *Astronomy across Cultures,* 450, 547. Dordrecht, Neth.: Kluwer, 2000.
Souden, David. *Stonehenge: Mysteries of the Stones and Landscape,* 125–126. London: Collins and Brown/English Heritage, 1997.
Thurston, Hugh. *Early Astronomy.* Berlin: Springer-Verlag, 1994.
Xu Zhentao, David Pankenier, and Jiang Yaotiao. *East Asian Archaeoastronomy: Historical Records of Astronomical Observations of China, Japan and Korea,* 13–19, 25–60. Amsterdam: Gordon and Breach, 2000.

# Solstices

If one were to go to a fixed spot every morning and watch the sun rise beyond a horizon full of distinctive features, such as distant mountains and valleys, it would soon become obvious that the rising position moves along the horizon day by day. During the months of January to May, the daily change is northward, becoming noticeably smaller in early June and finally almost imperceptible, until it comes to stop on or very close to June 21. After this, the motion is southward once again, at first imperceptible, but gradually hastening until by August the change is almost a whole solar diameter each day. By the end of November, the daily progress noticeably slows once again, until the southerly limit is reached on or about December

21. Exactly the same pattern would emerge if one were to watch sunsets rather than sunrises. The dates when the limits are reached, June 21 and December 21, are known as the solstices.

In a location north of the tropics, the farther north the sun rises and sets, the higher it climbs in the sky during the day. The June solstice occurs, as a consequence, in summer. Thus in the northern hemisphere, June 21 is generally known as the *summer solstice* and December 21 as the *winter solstice*. In the southern hemisphere the converse is true. However, as this can cause confusion, especially within the tropics, the terms *June solstice* and *December solstice* are sometimes preferable.

Outside the polar regions, the summer solstice also represents the longest day and shortest night (technically, this may actually occur a couple of days before or afterward, but the difference is negligible). The winter solstice represents the shortest day and longest night. Within the tropics, though, this makes little difference—as one approaches the equator all days in the year come to have the same length.

The solstices are widely recognized culturally. Historically, summer and winter solstice festivals are documented the world over. The winter solstice in particular is a time of death and renewal, a time when, for many human communities, the sun must be halted in its southward (or northward in the southern hemisphere) progress and brought back—for example, by shamanistic intervention—to enable a new growing season. The indigenous Chumash of California were just one of many peoples who held a major ceremony at this time because they believed human intervention to be necessary in order to maintain the delicate balance of nature. Similar principles may well have motivated a variety of rituals and ceremonies held at and around the time of winter solstice, such as the Hawaiian *makahiki* and the Incaic *Inti Raymi*. Pagan solstice festivals underlie the timing of the Christian festivals of the feast of St. John and Christmas.

See also:
Christianization of "Pagan" Festivals; Solstitial Alignments; Solstitial Directions.
Cusco Sun Pillars; Hawaiian Calendar.
Obliquity of the Ecliptic.

References and further reading
Krupp, Edwin C. *Skywatchers, Shamans and Kings*. New York: Wiley, 1997.
Ruggles, Clive, ed. *Archaeoastronomy in the 1990s*. Loughborough, UK: Group D Publications, 1993.

# Solstitial Alignments

The term *midsummer* is loosely applied: it derives from the common name, *Midsummer's Day*, for the feast of St. John, which was timed to coincide with the earlier pagan festivals associated with the summer solstice. *Midsummer sunrise*, then, really means "sunrise at the summer solstice." Likewise, *midsummer sunset* means "sunset at the summer solstice," and midwinter sunrise and sunset refer to sunrise and sunset at the winter solstice.

The direction of midsummer sunrise is always in the northeast quadrant of the horizon for an observer in the northern hemisphere (southeast in the southern hemisphere), but the exact azimuth depends upon the observer's latitude and the altitude of the horizon in that direction. The same applies to midsummer sunset (northwest in the northern hemisphere, southwest in the southern hemisphere), midwinter sunrise (southeast in the northern hemisphere, northeast in the southern hemisphere), and midwinter sunset (southwest in the northern hemisphere, northwest in the southern hemisphere). The only exceptions to this rule occur in the Arctic and Antarctic regions, where the midwinter sun never rises and the midsummer sun never sets.

A great many architectural alignments upon midwinter sunrise have been noted from a variety of places and times. The most famous is the alignment of the passage at Newgrange passage tomb in Ireland, built in the late fourth millennium B.C.E. Similar—in the sense that sunlight only penetrates into a dark space just after sunrise at the winter solstice—but very much older is the palaeolithic cave at Parpalló in eastern Spain. If deliberate, this alignment may date back as far as 19,000 B.C.E.

Midsummer sunrise alignments include what is arguably the most famous solstitial alignment of them all, along the axis of the sarsen monument at Stonehenge in England, toward the Heelstone. Ironically, though, it is quite likely that the alignment in the opposite direction, toward midwinter sunset, was of greater significance to the builders.

Alignments upon sunset at one or the other solstice seem to be no less common than those upon sunrise. A well-known example from North America is the Serpent Mound, a 180-meter- (600-foot)-long serpent-shaped effigy mound in southern Ohio, whose head faces the direction of midsummer sunset. Midwinter sunset at Kintraw in western Scotland was one of the earliest claimed examples of a high-precision sightline using a prominent notch in the natural horizon as a foresight. But that claim has not withstood the test of time.

Other controversial examples of solstitial alignments include the Bronze Age site at Brainport Bay, again in Scotland, where a fierce debate arose about whether the alignment of archaeological features, indubitably oriented upon midsummer sunrise, was precise and "calendrical" or less precise and "ceremonial." The circular sanctuary of timber posts at the Dacian sanctuary of Sarmizegetusa Regia in Romania (late first millennium B.C.E.) is said to face midwinter sunrise, but here the problem is that the alignment exists only in a theoretical sense, since the high horizon in the relevant direction displaces the actual direction of sunrise on the shortest day by some ten degrees along the horizon.

The significance of the solstitial directions is often cosmological rather than practical/calendrical, although the two are not mutually exclusive. In Hopi tradition, the main importance of the solstitial directions is that they define the fundamental divisions of the world into four parts. However, observations of the changing position of the rising and setting sun are also crucial in regulating their elaborate seasonal calendar. This includes, on the solstices themselves, laying prayer sticks on shrines that are themselves placed in the landscape along the solstitial directions.

A key issue is the extent to which we find solstitial alignments because we are looking for them. Nobody would deny that the directions of sunrise and sunset at the solstices are hugely significant in many cultures; the question is the extent to which we approach other cultural traditions armed with a "toolkit" that reflects our own predilections and preferences, and whether this can make us (perhaps unwittingly) selective with the evidence and hence biases our interpretation. To give just one example, the solstitial orientation of the Big Horn medicine wheel in Wyoming, when first announced in the 1970s, engendered much debate because of the suspicion that archaeoastronomers were simply extending the techniques and methods of British archaeoastronomy uncritically into North America. In general, such issues can only be satisfactorily addressed by paying careful attention to field methodology, the cultural context, and the broader archaeological/historical evidence of the site.

**See also:**
Christianization of "Pagan" Festivals; Methodology; Solstitial Directions.
Brainport Bay; Hopewell Mounds; Hopi Calendar and Worldview; Kintraw; Medicine Wheels; Newgrange; Sarmizegetusa Regia; Stonehenge.
Altitude; Azimuth; Solstices.

**References and further reading**
Aveni, Anthony F. *Skywatchers,* 301–303. Austin: University of Texas Press, 2001.
Cunliffe, Barry, and Colin Renfrew, eds. *Science and Stonehenge,* 203–229. London: British Academy/Oxford University Press, 1997.
Romain, William F. *Mysteries of the Hopewell,* 247–250. Akron: University of Akron Press, 2000.
Ruggles, Clive. *Astronomy in Prehistoric Britain and Ireland,* 12–19, 100–101, 127–131, 136–139. New Haven: Yale University Press, 1999.
Ruggles, Clive, ed. *Archaeoastronomy in the 1990s,* 40–43. Loughborough, UK: Group D Publications, 1993.
Ruggles, Clive, Frank Prendergast, and Tom Ray, eds. *Astronomy, Cosmology and Landscape,* 8–14. Bognor Regis, UK: Ocarina Books, 2001.
Ruggles, Clive, and Nicholas Saunders, eds. *Astronomies and Cultures,* 163–201. Niwot, CO: University Press of Colorado, 1993.

## Solstitial Directions

The direction of sunrise at the June solstice represents the northernmost limit of the range of horizon, centered upon east, where the sun will rise on different days through the year. Similarly, the direction of sunrise at the December solstice represents the southernmost limit of this range. Likewise, the directions of sunset at the June and December solstices represent, respectively, the northern and southern limits of the western range of horizon where the sun sets at different times of the year.

Everywhere on earth apart from the polar regions, the directions of June solstice sunrise and sunset, and of December solstice sunrise and sunset, fall respectively within the northeast, northwest, southeast and southwest quadrants of the horizon. These four directions divide the horizon into four distinct parts. The eastern part represents the range of horizon within which the sun always rises; everywhere in this range is a point where the sun will rise at some time during the year. Similarly, the western part represents the range within which the sun always sets. In northern temperate zones, the southern part represents the horizon the sun passes over each day; the direction of the sun in the middle of the day. Things facing in this direction will be open to the light of the sun. The north, on the other hand, is a "dark" direction. (The properties of the north and south parts are reversed in the southern hemisphere.)

Each of these four directions, in other words, has distinct properties and, in different human societies, can acquire distinct cosmological connotations. In one type of quadripartite cosmology (perceived division of the world, as viewed from some central place, into four parts), the boundaries between the different quarters of the world are formed by the solstitial directions. One example of this is the worldview of the Hopi people of Arizona. Another example is the layout of the Andean village of Misminay. It can be misleading to speak of a "quartering" of the world, since the four "quarters" are generally nowhere near equal in size: in fact, equal quarters are only possible (depending upon the horizon altitude) at latitudes of around fifty-five degrees (north and south). Closer to the equator, the east and west parts become narrower and the north and south parts wider. At the equator itself, the east and west parts are only about 47 degrees wide, and the north and south parts are 133 degrees wide.

As one approaches the polar regions, the northern and southern "quarters" rapidly shrink away, and in the polar regions themselves the quadripartite model is not applicable at all. Here, every part of the horizon is the rising or setting point of the sun at some time in the year. Indeed, there is a period around the winter solstice when the sun never rises above the horizon at all, and a period around the summer solstice when the sun never sets. The length of this period depends upon the latitude, but it is clearly a critical time for indigenous groups living in northern polar regions, such as the Inuit.

See also:
Cardinal Directions; Solstices; Solstitial Alignments.
Hopi Calendar and Worldview; Inuit Cosmology; Misminay.
Altitude.

**References and further reading**
Aveni, Anthony F. *Stairways to the Stars: Skywatching in Three Great Ancient Cultures,* 18–27. New York: Wiley, 1997.
———. *Skywatchers,* 55–67. Austin: University of Texas Press, 2001.

# Somerville, Boyle (1864–1936)

Henry Boyle Townshend Somerville was one of a handful of keen amateur astronomers and surveyors in the early part of the twentieth century who extended the work of Sir Norman Lockyer by carrying out their own surveys of megalithic monuments. Having risen to the rank of vice-admiral in the Royal Navy, he retired in 1919. After he returned to his native County Cork, he was responsible for the first published archaeoastronomical surveys and interpretations of a number of Irish chambered tombs, stone circles, and rows. It was he, for example, who first drew attention to the solstitial orientation of the stone circle at Drombeg.

Somerville is best known, however, for first drawing attention in 1912 to the possible lunar orientation of some of the standing stones at Callanish on the Isle of Lewis, Scotland. One consequence of his paper on Callanish, published in the *Journal of the British Astronomical Association,* was to trigger the interest of Alexander Thom, who went on to become one of the most influential twentieth-century figures in the field. More fundamentally, though, it contained the first suggestion that a prehistoric monument might have been aligned upon the horizon rising or setting point of the moon, not just of the sun or stars. This initiated a line of inquiry that is still bearing fruit today among the Bronze Age monuments of Scotland and Ireland, in particular the Scottish recumbent stone circles and some of the short stone rows found in western Scotland and western Ireland.

Somerville was murdered in 1936 by Irish republican activists, who saw him as encouraging Irish recruits into the British armed forces.

See also:
Lockyer, Sir Norman (1836–1920); Thom, Alexander (1894–1985).
Callanish; Drombeg; Recumbent Stone Circles; Short Stone Rows; Wedge Tombs.

**References and further reading**
Michell, John. *A Little History of Astro-Archaeology,* 40–42. London: Thames and Hudson, 1989.

## Son Mas

Scattered around Mallorca, the largest and most mountainous of the Balearic Islands in the western Mediterranean, are a number of walled sanctuaries built during the first millennium B.C.E. that seem to have been the setting for various ritual activities. Son Mas, near the northwest coast, is one of the largest of them but seems to have been in use for a far longer period: radiocarbon dates stretch back into the third millennium B.C.E. However, there is a curious gap around 1700 B.C.E., when the site seems to have been abandoned for several centuries for no reason that is apparent from the archaeology, before being rebuilt and reoccupied.

The reason that the earlier sanctuary was so precipitously abandoned remained unexplained until the site was studied archaeoastronomically. Like the taulas of Menorca, it faces south, but in the direction of a nearby valley flanked by steep hills. This is indicated by a curious, artificial groove cut into the side of a large upright boulder in front of the sanctuary. When the Southern Cross and Pointers appeared down low in the southern sky, passing almost horizontally, from east to west, they would have passed across this valley and hung spectacularly in the sky just above it. But due to precession they would have gotten gradually lower century by century until they were beginning to disappear in around 1700 B.C.E. The result, quite plausibly, would have been a crisis, and it was that which caused the sanctuary to be abandoned.

See also:
Taulas.
Precession.

**References and further reading**
Hoskin, Michael. *Tombs, Temples and Their Orientations*, 46–52. Bognor Regis, UK: Ocarina Books, 2001.
Hoskin, Michael, and William Waldren. *Taulas and Talayots*. Cambridge: Michael Hoskin, 1988.

## Space and Time, Ancient Perceptions of

Long before the invention of writing or the construction of observing instruments, people made sense of the world by linking objects, events, and cycles of activity, both on the land and in the sky. The regular cycles of the skies were often vital in developing notions of temporality, in the sense that they were used to regulate both practical and ritualistic aspects of human activity and played a crucial role in the development of calendars. But this is a very different matter from saying that astronomical periodicities were used to measure units of time, because this assumes an abstract notion of time that may not be appropriate in a non-Western worldview. The same can be said of an abstract notion of space.

For us, time progresses along a line and provides a backdrop against which we live our lives, constantly measurable from calendars and clocks that have become divorced from the physical, observable phenomena in the natural world that once generated them. Likewise, we measure our position in space using maps or GPS receivers that fit comfortably in the hand and, at the touch of a button, deliver our spatial coordinates to within a few meters.

Prehistoric peoples and others in the past who lacked such devices would have been no less aware of the passage of time. The basic needs of survival and subsistence would have forced them to keep their own actions in tune with various regularly recurring events going on in the natural world around them. It is also likely that the regular cycles of the heavens, starting with the most obvious one of them all—the phase cycle of the moon—provided a key point of reference for many and led to the eventual development of early calendars. But we must be careful not to approach the issue too naively. The passage of time (as we would think of it) might have been conceived in many different ways. One of the most fundamental conceptual distinctions is between circular and linear time—the one implying an endless perceived repetition of regular events and the other a single history and future. The ancient Maya are noted for having used both conceptions in parallel, in the form of the Calendar Round and the Long Count. Even the distinctions between past, present, and future break down differently in many other systems of thought.

Similarly, there are numerous ways of perceiving the world that differ from our view of things separated by empty space, from which stems our own idea of space as an abstract backdrop. Notions of the spatial interrelationship of things were inherently contextualized in places and paths with particular qualities and meanings. Furthermore, spatial and temporal attributes of things were intricately bound together in many ways, with the distinction between them (as we would see it) rather blurred. Thus, for the Lakota, the names and meanings of some places changed according to the time of year as they followed the buffalo through the landscape, mimicking the passage of the sun through the sky. The Aztecs, it has been claimed, linked the mountain dwelling of the rain god Tlaloc, the time when the sun rose in line with this mountain as viewed from an ancient pyramid (Cuicuilco), and the time when child sacrifices should be offered to Tlaloc in order to ensure rain and fertility in the coming growing season. And more generally, horizon calendars such as those of the Hopi, or those thought to have operated in central Mexico in pre-conquest times, function by directly associating time (a particular calendar date) with place (the point on the horizon where the sun rises or sets on the day in question).

Although such arguments give us a number of reasons for seeking astronomical alignments in the prehistoric context, they also provide a warning against pre-judging potential horizon targets of astronomical significance. Thus, while the solstitial directions have an obvious physical reality (as the limiting positions of sunrise or sunset on the horizon), the equinoxes, in general, do not. They are commonly taken as self-evident horizon targets, but in fact they only make sense as half-way points between the solstices in space and/or time, and this implies an abstract view of either or both. For a Neolithic Briton standing in a sacred stone circle, the date when (say) a crucial crop should be planted, or when the sun rose behind a particular sacred mountain—both very concrete associations (though of very different types)—are much more likely to have been significant than the abstract equinoxes. This makes it much more difficult for the modern archaeoastronomer to determine whether any stone circles intentionally incorporated symbolic alignments upon sunrise on significant days, since different days are likely to have been significant—and so the "targets" would have differed—from one stone circle to another. Only solstitial alignments could show up consistently in a statistical analysis of several sites taken together.

See also:
Calendars; Equinoxes; Methodology; Sacred Geographies; Solstitial Directions; Statistical Analysis.
Aztec Sacred Geography; Hopi Calendar and Worldview; Horizon Calendars of Central Mexico; Lakota Sacred Geography; Maya Long Count.
Lunar Phase Cycle.

References and further reading
Darvill, Timothy, and Caroline Malone, eds. *Megaliths from Antiquity*, 339–354. Cambridge: Antiquity Publications, 2003.
Goodman, Ronald. *Lakota Star Knowledge: Studies in Lakota Stellar Theology*. Rosebud, SD: Sinte Gleska College, 1990.
Ingold, Tim. *Companion Encyclopedia of Anthropology* (revised ed.), 503–526. New York: Routledge, 2002.
Lucas, Gavin. *The Archaeology of Time*. New York: Routledge, 2005.
Renfrew, Colin, and Paul Bahn. *Archaeology: Theories, Methods and Practice* (4th ed.), 408. London: Thames and Hudson, 2004.
———, eds. *Archaeology: The Key Concepts*, 268–273. Abingdon, UK: Routledge, 2005.
Ruggles, Clive, *Astronomy in Prehistoric Britain and Ireland*, 148–152. New Haven: Yale University Press, 1999.
Ruggles, Clive, and Nicholas Saunders, eds. *Astronomies and Cultures*, 275–283. Niwot: University Press of Colorado, 1993.
Tilley, Christopher. *A Phenomenology of Landscape: Places, Paths and Monuments*, 7–34. Oxford: Berg, 1994.

## Star and Crescent Symbol

A conjunction of the thin crescent moon and the planet Venus is one of the most striking sights in the sky—even visible within a modern city where the nighttime stars are all but forgotten. The spectacle of these two conspicu-

*The planet Venus (left) near a crescent moon. Earthshine illuminates the rest of the moon's disk. (Roger Ressmeyer/Corbis)*

ous objects hanging close together in a clear blue (and otherwise quite empty) sky, either in the early evening or the early morning, is likely to have given rise to what is one of the best known and most widespread celestial icons—the star and crescent symbol. In various forms it can be found on the flags of several modern Islamic countries, including Turkey, Pakistan, Algeria, and Azerbaijan; on modern coins from European countries such as Italy and Romania; on coins from ancient Rome, Greece, and Persia; and even in prehistoric rock art from various parts of the world.

The star and crescent is mostly widely recognized in the modern world as a symbol of the Islamic faith, although its origin in this context is much older. It originally passed into several Islamic countries from the Ottoman Empire, and its use within what is now Turkey can be traced back to the third millennium B.C.E. Although it is often claimed that the symbol (in this context) owes its origins to a particular astronomical event, several different candidates have been proposed, and there is really no firm evidence to support any of them.

Elsewhere, where the symbol arose in quite independent cultural contexts, the motivation is most likely to have been one or more sightings of a conjunction of the crescent moon and Venus. This was an event that was spectacular but not unduly rare. And one of the most famous rock art designs in North America including a star and crescent, the so-called "supernova petroglyph" on a rock overhang in Chaco Canyon, New Mexico, is actually a set of four symbols used to mark a sun-watching station.

See also:
Symbols.
Chaco Supernova Pictograph.
Inferior Planets, Motions of.

**References and further reading**
Ruggles, Clive, ed. *Archaeoastronomy in the 1990s*, 165–167. Loughborough, UK: Group D Publications, 1993.
Schaefer, Bradley E. "Heavenly Signs." *New Scientist* 132 (1991), 48–51.

## Star Compasses of the Pacific

Micronesian and Polynesian navigators used the stars in different ways to help them find their way across the open waters of the Pacific Ocean at night. One method was to learn the sequence of stars that would rise or set in the direction in which one wished to sail—the relevant *star path* or *kavienga* (as it was known in Tonga). The New Zealander David Lewis recorded a few surviving stone structures used for committing a given star path to memory. A single sighting stone in Tonga and two sets of sighting stones in the Gilbert Islands (Kiribati) farther to the north seem to have been set up to mark the directions to various distant islands. (The direction marked was usually somewhat displaced from the actual azimuth of the target island, so that landfall could be achieved by running with the prevailing wind and current.) By sighting along one of these lines sufficiently often and for sufficiently long periods, the novice navigator would come to recognize the succession of stars that marked a particular course. On the island of Beru in Kiribati, Lewis observed the continuing use of a *stone canoe* for a similar purpose.

When traveling greater distances, following more complex courses, or exploring rather than traveling to predetermined and familiar places, navigators needed a wider knowledge of the stars in order to set different courses from different locations in the ocean. The difficulty is that there are so many stars in the sky that to memorize the rising and setting positions of many hundreds of them would surely be impracticable. One Micronesian example of a *star compass*, which was analyzed in detail by the ethnographer Ward H. Goodenough in the 1950s, overcame this problem in a clever way. (It existed in a number of slightly different forms among the Caroline Islands of the western Pacific, although the underlying principles are common to all the variants.) The Carolinian navigator typically relied on just fifteen reference stars or star groups whose rising and setting positions marked thirty directions around the horizon. The north and south directions, in addition, were indicated by Polaris and the upright Southern Cross. The relative positions of these thirty-two guide points around the horizon were carefully memorized: the spacing was not particularly regular, but then it didn't need to be.

In practice, at a given time on a given night, few of the guide stars would generally be anywhere near to rising or setting, and many would not be visible in the sky at all. However, the Carolines are situated quite close to the equator, and the paths of the stars through the sky in this region are more or less vertical (this is only not the case in the vicinity of the celestial poles). This means that it was relatively easy to trace the paths of visible stars down to the horizon. This technique provided a number of reference points. The experienced navigator could then fill in the gaps and deduce the positions of all of the thirty-two guide directions that constituted his mental compass in order to orient himself.

No such detailed accounts exist of star compasses in Polynesia, although it is clear from the accounts of Captain James Cook and other early European explorers that Polynesians made extensive use of directional stars for navigation. There is, however, a suggestion that a navigational device in the form of a large hollowed-out gourd was used in Polynesia. The "sacred calabash" was carried in the canoe for use as a horizon compass or a sort of sextant for determining latitude. This interpretation has been hotly contested, but there is certainly evidence of the use of *navigation gourds* in the Hawaiian islands after European contact. In addition, at least one Hawaiian account recorded in the nineteenth century suggests that large gourds were used, back on land, as visual aids for teaching about the skies.

See also:
Navigation in Ancient Oceania.
Azimuth; Celestial Sphere; Star Rising and Setting Positions.

**References and further reading**
Finney, Ben. *Voyage of Rediscovery*, 55–57. Berkeley: University of California Press, 1994.
Gladwin, Thomas. *East Is a Big Bird: Navigation and Logic on Puluwat Atoll*, 148–160. Cambridge, MA: Harvard University Press, 1970.
Goodenough, Ward H. *Native Astronomy in the Central Carolines*. Philadelphia: University Museum, University of Pennsylvania, 1953.
Hodson, F. R., ed. *The Place of Astronomy in the Ancient World*, 137–139, 141–142. London: Royal Society, 1976.
Johnson, Rubellite K., and John K. Mahelona. *Nā Inoa Hōkū: A Catalogue of Hawaiian and Pacific Star Names*, 62–65, 70–74. Honolulu: Topgallant Press, 1975.
Lewis, David. *We the Navigators* (2nd ed.), 102–111, 283–284. Honolulu: University of Hawai'i Press, 1994.
Selin, Helaine, ed. *Astronomy across Cultures*, 106–108. Dordrecht, Neth.: Kluwer, 2000.

## Star Names

Individual stars and groups of stars have different names in different cultures, but in order to speak of these we need to identify them unambiguously using reference names acceptable to everyone. It makes sense to follow the nomenclature used by modern astronomers.

Many stars have generally agreed upon proper names, such as Sirius, Arcturus, Vega, Rigel, and Betelgeuse. However, just as the world is divided into countries, so the sky is now divided up into strictly defined regions—*modern constellations*. This means that all stars, whether named or not, belong to a modern constellation. The convention is to label the most prominent stars in each constellation, usually in order of brightness but sometimes according to their position, using the Greek letters—$\alpha$ (alpha), $\beta$ (beta), $\gamma$ (gamma), $\delta$ (delta), $\varepsilon$ (epsilon), $\zeta$ (zeta), and so on—combined with the constellation name (actually the Latin genitive): thus Sirius, the brightest star in Canis Major, is $\alpha$ Canis Majoris. One example best known by its name in this form is $\alpha$ Centauri.

Each constellation also has a three-letter abbreviation, for example Dra for Draco(nis), CMa for Canis Major(is), and Ori for Orion. Thus Betelgeuse is $\alpha$ Ori, Rigel is $\beta$ Ori, and Sirius is $\alpha$ CMa. Other conventions hold for fainter stars but are generally of no relevance to archaeoastronomy or ethnoastronomy.

**See also:**
Archaeoastronomy; Ethnoastronomy; Orion.

**References and further reading**
Allen, Richard H. *Star Names: Their Lore and Meaning.* New York: Dover, 1963.
Ridpath, Ian, ed. *Norton's Star Atlas and Reference Handbook* (20th ed.), 103–105. New York: Pi Press, 2004.

## Star of Bethlehem

According to the Christian scriptures—specifically, the Gospel of St. Matthew—wise men from the east arrived in Jerusalem to worship the King of the Jews, declaring that they had seen his star in the east. As they journeyed southward to Bethlehem the star went before them until it stopped over the place where the child was. Ever since the third century, scholars have struggled to find a concrete explanation for what has become known as the Star of Bethlehem. Twentieth-century efforts are reflected in scores of books and literally hundreds of articles in academic journals.

Although translations differ, one reading of Matthew's account, assuming that it does indeed have its basis in observed fact, interprets "star in the east" as "star at its rising." We must then assume that the journey to

Bethlehem was made when the star was high in the sky ahead of the Magi, to the south. Clearly it could not have stopped dead in the sky, so "stopped over" is then taken to mean "set behind."

Several types of spectacular astronomical phenomena have been identified as leading candidates. The historical context, which fixes the date of Christ's birth to around 6 B.C.E., provides a strong constraint. On the assumption that the Star of Bethlehem was literally a star, but an unfamiliar one that suddenly appeared in the sky at the time, an obvious possibility is a nova or supernova. The latter is a dying star that gains hugely in brightness for a short period, typically just a few weeks. Some supernovae are even visible by day, so we would not be forced to speculate that the Magi traveled by night. Unfortunately, if a supernova had occurred at the time we would expect Chinese astronomers to have recorded a "guest star." They did not, unless one counts a "tailed object," more probably a comet, seen in the spring of 5 B.C.E. Another possibility is a conjunction (close coming together) of two planets, each themselves among the brightest objects in the sky. In the summer and autumn of 7 B.C.E., the planets Jupiter and Saturn passed each other three times. Nonetheless, eye-catching as this would have been, they would still have been seen as two "stars" and would only have led the way at night. Some have argued that it was the succession of several rather unusual astronomical phenomena that augured great events, but this stretches credulity even further.

A serious weakness in all of these arguments is that they fail to take account of the Magi's own worldview. The Magi were astrologers, probably from Persia. We know that astrological prognostications of the time derived from a complex set of interrelationships between the planets and the zodiacal constellations they appeared in. We also know that each constellation had terrestrial associations, and that Aries or Pisces was associated with Judaea. American astronomer Michael Molnar has recently argued from this standpoint that an occultation of the planet Jupiter by the moon (in other words, the moon passing directly in front of the planet) that occurred in the constellation of Aries in the spring of 6 B.C.E. was an astrological portent of enormous significance and pointed directly to the birth of a new king in Judaea. In Molnar's reading, Matthew's "rising in the east" refers to a planetary heliacal rising, and "stopping" to retrograde motion. Aspects of this interpretation continue to be strongly debated. Most controversially, the occultation in question would actually have been rendered invisible by the light of the sun. This would not have diminished its astrological significance, but the question remains of whether the astrologers of the time would have been able to calculate it.

Be this as it may, it seems obvious in retrospect that the solution to the Star of Bethlehem mystery lies not in events that would be considered conspicuous by modern astronomers but in events that would have stood out in terms of the horoscopes generated by the astrologers of the time.

**See also:**
Astrology.
Babylonian Astronomy and Astrology.
Superior Planets, Motions of.

**References and further reading**
Gingerich, Owen, ed. "Review Symposium: The Star of Bethlehem." *Journal for the History of Astronomy* 33 (2002), 386–394.
Kidger, Mark. *The Star of Bethlehem: An Astronomer's View.* Princeton: Princeton University Press, 1999.
Molnar, Michael. *The Star of Bethlehem: The Legacy of the Magi.* New Brunswick, NJ: Rutgers University Press, 1999.

## Star Rising and Setting Positions

Any given star, viewed from a given place, traces the same path across the celestial dome night after night. Though it always rises and sets in the same place, the time of day when these two events occur alters constantly, getting earlier by about four minutes each night. Each of them will occur during daylight for about half of the year, and hence be invisible. So also will all or part of the star's daily passage across the sky, whenever the sun is in the sky at the same time. When people in the past observed the rising of certain stars and aligned sacred structures upon them, the event of particular significance was often the heliacal rise, the annual event when the star first rose sufficiently long before the sun to become visible in the predawn sky.

Observations of the rising and setting of stars are especially useful to the navigator wishing to set or maintain a course. Micronesian and Polynesian navigators, for example, had an excellent knowledge of the stars associated with various directions, and this served them well during their long voyages across the Pacific. In practice, such observations are hampered by extinction, the dimming of a star's light due to absorption in the intervening atmosphere. This problem is most severe when a star is close to the sea horizon because its light has had to travel through a great deal more of the earth's atmosphere before it reaches the observer than when it is higher in the sky. This is a potential headache for the navigator, as all but the brightest stars may need to be considerably above the horizon to be visible at all. However, it is less of a problem close to the equator, where the celestial bodies rise and set almost vertically.

Over a long timescale, the slow tumbling of the whole celestial sphere due to precession gradually causes the rising and setting positions of stars to drift. As a result, the place on the horizon where a star rises or sets in the present day may be several degrees away from where it rose or set at the time of interest (for example when a monumental alignment was built) several centuries or more in the past. Some stars that (for locations in the northern hemisphere) rise and set close to due south (or close to due north in the southern hemisphere) may not have appeared above the horizon at all at certain times in the past, while others visible then may since have disappeared.

Precession limits our confidence when we try to identify putative stellar alignments at ancient monuments, since there is invariably some (and often considerable) uncertainty about the date of construction and use. Extinction adds a further complication to the interpretation of such alignments, especially at high latitudes where stars rise and set at a considerable angle to the vertical, since it is difficult to be sure about exactly where a given star was when it appeared or disappeared.

See also:
Astronomical Dating; Methodology; Navigation.
Star Compasses of the Pacific.
Celestial Sphere; Declination; Extinction; Heliacal Rise; How the Sky Has Changed over the Centuries; Precession.

**References and further reading**
Ridpath, Ian, ed. *Norton's Star Atlas and Reference Handbook* (20th ed.), 5–7. New York: Pi Press, 2004.

# Statistical Analysis

The need for statistical analysis arises in archaeoastronomy because astronomical patterns we perceive in the archaeological record may have come about purely by chance, or to be more precise, as a result of factors quite unrelated to astronomy. Doubts about intentionality can apply to a whole range of types of evidence concerning ancient astronomy: monuments aligned upon the rising and setting positions of astronomical bodies; displays of sunlight and shadow only visible on rare occasions; groups of monuments placed so that their positions on the ground mimic the shape of a constellation; patterns of tally marks carved on rocks or portable artifacts that include could have functioned as calendars; and so on.

In the formative days of archaeoastronomy in the 1960s to 1980s, statistical analysis was most extensively applied in assessing monumental alignments. Indeed, this approach came to characterize the "green," or European, approach to archaeoastronomy, which focused on prehistoric monuments without the benefit of related historical or written evidence. Broadly speaking, the approach taken was that pioneered by the Scottish engineer Alexander Thom, which was to accumulate data from many alignments and determine the declination of the horizon point "indicated" by each alignment, and then to use statistical methods to determine whether observed declination "peaks"—repeated occurrences of particular declinations—were in fact significant. If so, then a consistent astronomical purpose was evident and the question could be asked, which astronomical body the given declination might have corresponded to.

Devising a suitable statistical test is by no means straightforward. Sometimes one can only resort to *Monte-Carlo testing,* a method by which one generates, for example, randomly oriented sets of monuments in the computer and sees in what proportion of cases they contain at least as many of a particular type of alignment as were found in the "real" data. If the answer is, say, only once in every thousand runs, then one can conclude that the chances of the real data having arisen fortuitously are only one in a thousand; as this probability is so small, it is fair to conclude that the alignments were deliberately intended. However, there are a number of dangers and pitfalls with this method. An obvious one is that the data must be fairly selected. Even leaving out one site simply because it seems to be pointing in a different direction from all the rest (and it is often easy enough to find retrospective reasons for doing this) can badly distort the results and so undermine any statistical conclusions.

A less obvious but more fundamental difficulty relates to the whole nature of hypothesis testing in this way. Many statistical procedures assume that the hypothesis preceded the data, whereas in an archaeological context, a particular interpretation is generally first suggested by the data (or most of it), since archaeologists do not have the opportunity for repeated experiments. The hypothesis that we choose to test may itself be selected from innumerable possibilities. To take an example, it is a well-known though astonishing fact that if twenty-three people are selected randomly, there is an even chance that two of them will have the same birthday. However, if we took such a group, observed that two of them had birthdays (say) on July 27, and on the basis of that chose to assess the chances (test the hypothesis) that two July 27 birthdays could have arisen among the group purely fortuitously, we would obtain the result "highly unlikely." The reason, of course, is that we chose to test the "July 27" hypothesis (rather than any one of 364 others) because the data contained that pattern in the first place. In short, statistical methods must be used with considerable care, especially where they purport to assign *significance levels* to particular results.

Nowadays there is a much greater reliance on simply displaying the data. If certain trends are strong enough, then their intentionality is obvious, as is the case for the orientations of numerous local groups of later prehistoric tombs and temples in western Europe. Where we need to assess sets of measured declinations, Alexander Thom's method of *curvigrams*—graphs showing accumulations of probability—seems as good a method of visualizing the likely significance of repeatedly occurring declinations as any. The fair selection of data remains a much more crucial issue.

In recent years archaeoastronomers have begun to recognize the potential of a rather different statistical approach: the *Bayesian paradigm*. This approach allows one to assess data repeatedly against different "prior" ideas and to take into account background information. It is especially useful where we are trying to integrate different types of evidence, such as historical records.

However, we must look beyond statistical analysis if we wish to do more than establish the intentionality of astronomical alignments or other astronomically related patterns evident in the archaeological record. No amount of statistical analysis can help with the interpretation of their significance and meaning to the people who built and used them.

### See also:
Alignment Studies; Constellation Maps on the Ground; "Green" Archaeoastronomy; Methodology; Orientation; Thom, Alexander (1894–1985).
Prehistoric Tombs and Temples in Europe.
Declination.

### References and further reading
Heggie, Douglas, ed. *Archaeoastronomy in the Old World*, 83–105. Cambridge: Cambridge University Press, 1982.
Iwaniszewski, Stanisław, Arnold Lebeuf, Andrzej Wierciński, and Mariusz Ziółkowski, eds. *Time and Astronomy at the Meeting of Two Worlds*, 497–515. Warsaw: Centrum Studiów Latynoamerykańskich, 1994.
Ruggles, Clive. *Megalithic Astronomy: A New Archaeological and Statistical Study of 300 Western Scottish Sites*. Oxford: British Archaeological Reports (British Series, 123), 1984.
———. "The General and the Specific," *Archaeoastronomy: The Journal of Astronomy in Culture* 15 (2000), 151–177.
———. *Astronomy in Prehistoric Britain and Ireland*, 159–162. New Haven: Yale University Press, 1999.

# Stone Circles

The stone circle is a type of later prehistoric monument that is commonly found in northwestern Europe but is also found elsewhere in Europe and also in certain other parts of the world. Over 1,300 examples survive in Britain, Ireland, and Brittany, ranging from a few meters to over 350 meters in diameter and containing from five to over thirty standing stones. The stones at some stone circles are mere stumps and rounded boulders; at others they are great monoliths several meters in height.

Stone circles are not always found in isolation. They can be found surrounding tombs, as at Newgrange, or within the banks and ditches of henges, as at Avebury; they can also be connected to lines or avenues of stones, as at Callanish and Beaghmore. Some of the most complex arrangements of several circles and stone alignments are the stone monuments of

One of a complex of Bronze Age stone circles and rows at Aughlish, County Derry, Northern Ireland. (Courtesy of Clive Ruggles)

mid-Ulster. The designs of individual circles are also quite diverse, with a number of regional variants. One of the most distinctive of these is the recumbent stone circle, where one stone is placed on its side. Others include stone circles with entrances, circles with outlying standing stones, embanked circles, and "circles" with distinctive shapes such as ellipses or flattened circles. (Since many are nowhere near true circles, some have suggested that stone circles should more correctly be called "stone rings.")

It is difficult to date a stone circle, but most northwest European examples are thought to have been built in the Later Neolithic or earlier part of the Bronze Age, broadly in the third and second millennia B.C.E. Stones were sometimes transported several kilometers to the site; famously (although exceptionally) the Stonehenge bluestones, weighing up to four tons each, were transported over sea and land some three hundred kilometers (two hundred miles) from the Preseli mountains in southwest Wales.

The purpose of stone circles has long been a mystery. Only a minority are directly associated with tombs. Cists containing small burials are found in some of them, but these are often late additions, perhaps placed at what was already an ancient hallowed or ancestral place whose original purpose was long forgotten. Other than this, archaeological excavations tend to find little in the way of stratigraphy and artifacts, so that for many years stone circles were largely ignored by archaeologists—so much so that the British archaeologist Aubrey Burl once referred to them as the "*personae non gratae* of British prehistory" (Ruggles 2002, p. 176). Yet an obvious assumption—as for any open ring, whether built of earth, timber, or stone—is that they demarcated a piece of special or sacred ground, perhaps a place used for a ceremonial gathering or ritual activity. This means it makes sense to study the form of the rings, their position in the landscape, and their relationship to visible features in the surrounding environment, including objects in the sky. Such investigations are now carried out by archaeologists, but until late in the twentieth century they were the almost exclusive preserve of engineers, astronomers, and others from outside the discipline.

One of the most famous of these was the engineer Alexander Thom, who proposed that many British stone circles were laid out using a standard unit of measurement, the "megalithic yard," of 0.83 meters (2.72 feet). He also claimed that many evidently noncircular rings represent precise geometrical constructions, such as ellipses and various types of "flattened circle," laid out using Pythagorean triangles more than a millennium before they were (re)discovered by Pythagoras. Given all this, Thom's other main idea—that many stone circles incorporated remarkably precise alignments upon the rising and setting positions of the sun, moon, and certain stars—seemed less surprising.

However, time has dealt harshly with the specifics of "megalithic mensuration," "megalithic geometry," and "megalithic astronomy," even though Thom himself is remembered as an important pioneer in getting people to think about such issues. Statistical reappraisals of the diameters of stone circles, given the large size of the stones and their often poor state of preservation, only gives marginal support for the notion that some common unit (perhaps related to the average human pace) was used in setting out many of them; there is no sustainable evidence whatsoever for the idea of a precise measuring stick being carried the length and breadth of later prehistoric Britain. Similarly, the observed shapes of stone circles can all be adequately explained by the use of simpler construction techniques: there is no need to resort to Thom's geometrical constructions. Some have even argued that these monuments may all simply be poor attempts at circles, or that the precise shape didn't matter and was simply laid out by eye.

A major problem with studying possible astronomical alignments at stone circles is the number of different ways in which such an alignment might have been set up: in the direction of the entrance where one exists; in the direction of the tallest stone; from the circle center to an outlier where one exists, or vice versa; (either way) along the main axis where one can be identified (e.g., in the case of an elliptical ring); (either way) between the tallest stones; and so on. At many stone circles, many potential alignments of various types may exist, making it quite likely—even, in some cases, highly probable—that one or more of them will align purely by chance upon a prominent feature in the landscape or a major celestial body or event. This means that in order to have some degree of confidence that any alignment we spot was actually intentional and meaningful, we have to pay particular attention to questions of methodology.

In just a few types of stone circle, one direction of alignment stands out from any others as possibly significant. Where this occurs we find some of the strongest evidence of an intense interest in the sun and/or moon, although not consistently. In this context the Scottish recumbent stone circles and the Irish axial stone circles are particularly important. However, as with mensuration and geometry, the high precision astronomical alignments envisaged by Thom have not withstood the test of time. They simply cannot be convincingly demonstrated on the basis of the evidence available. Though absence of evidence is not evidence of absence, until and unless such evidence emerges, we have no need to fit chiefs, shamans, or priests obsessed with the minutiae of the moon's motions over several generations into our models of later prehistoric societies in Britain and Ireland.

On the other hand, solar and lunar alignments of less than pinpoint accuracy are nonetheless highly fascinating. They suggest, for example, that circles may have become charged with meaning at certain times, when the celestial bodies reached particular configurations in relation to the stones of the circle. At some sites, other forms of archaeological evidence reinforce the idea of ceremonies taking place when the celestial configuration was right. A good example of this is the scatters of moon-reflecting white stones such as quartz that are found around the base of the recumbent stone at some Scottish recumbent stone circles, reinforcing the idea that ceremonies took place here by the light of the full moon on a midsummer evening when it appeared low over the recumbent stone. Such ideas must be considered in the broader context of cosmological ideas that could have viewed these monuments as microcosms or placed them conceptually at the very center of the world.

**See also:**
Methodology; Palaeoscience; Thom, Alexander (1894–1985).
Avebury; Axial Stone Circles; Callanish; Circles of Earth, Timber, and Stone; Megalithic Monuments of Britain and Ireland; Newgrange; Recumbent Stone Circles; Stonehenge.

**References and further reading**
Burl, Aubrey. *A Guide to the Stone Circles of Britain, Ireland and Brittany.* New Haven: Yale University Press, 1995.
———. *Prehistoric Stone Circles.* Princes Risborough, UK: Shire, 1997.

———. *Great Stone Circles*. New Haven: Yale University Press, 1999.
———. *The Stone Circles of Britain, Ireland and Brittany*. New Haven: Yale University Press, 2000.
Ruggles, Clive, ed. *Records in Stone: Papers in Memory of Alexander Thom*, 175–205. Cambridge: Cambridge Univerity Press, 2002.

# Stonehenge

Stonehenge, one of the most famous archaeological sites in the world, has long been associated with astronomy. Situated on the chalk uplands of the Wessex region in southern England, in the county of Wiltshire, the monument is characterized by its huge sarsen stones, some of which stand over seven meters (twenty-two feet) tall. The tallest are the remains of five *trilithons* (two uprights with a horizontal *lintel* placed across their tops) arranged in the shape of a horseshoe, and these are surrounded by what was once a continuous ring of only slightly less tall uprights and lintels. Almost unimaginably, these huge stones, weighing up to forty tons apiece, were dragged here, down- and uphill, over a distance of some thirty kilometers (eighteen miles) from the Marlborough Downs to the north. Where the stones have collapsed it is possible to see some of the carefully carved mortice-and-tenon and tongue-and-groove joints (joints that would normally be associated with woodworking) that were used to hold them together. The horseshoe is oriented northeastwards. Looking out in this direction through a well-preserved section of the outer circle of sarsens, one sees a much rougher stone, leaning over and badly weathered. This is the famous Heelstone, which stands some sixty meters outside the outer ring.

The purpose and function of the great stone monument at Stonehenge has been a source of speculation for centuries. That it was a place of great power is self-evident. Regarding astronomy, it has been portrayed variously as a cosmic temple, a calendrical device, an observatory, and an eclipse calculator. Many of these ideas have attracted widespread public interest.

However, the majority of astronomical theories concerning Stonehenge have little hard evidence to support them. Many are based on the idea that, at one stage or another, the monument incorporated deliberate architectural alignments upon the rising and setting positions of celestial bodies, particularly the sun or moon. The problem is that every architectural alignment must point somewhere, and many factors might have influenced any particular orientation. Just because a pair of stones is aligned upon, say, the position of sunrise at one of the solstices, does not mean that this was deliberate or that it meant anything at all to the people who built and used the monument. We must be particularly cautious if we only find a modest number of potential astronomical alignments among a great many possibilities: and there are hundreds of ways of aligning different pairs of stones at Stonehenge.

The view out along the axis of the sarsen monument at Stonehenge, in the direction of midsummer sunrise. The Heelstone is visible in the distance to the right of the axial alignment; another stone, probably similar, once stood symmetrically to the left. (Courtesy of Clive Ruggles)

One of the most famous astronomical speculations about Stonehenge is that it functioned as a calculating device for predicting eclipses. Put forward in the early 1960s in a popular book by Gerald Hawkins entitled *Stonehenge Decoded,* and later elaborated by the astronomer Fred Hoyle, the idea is basically that a small number of wooden posts were moved around the Aubrey Holes according to strictly defined rules (the two authors suggested different methods). When certain configurations were reached, then it was a "danger period" when eclipses could occur.

The Aubrey Holes are a ring of fifty-six pits surrounding the sarsen monument, named after the antiquarian John Aubrey who discovered them in the seventeenth century. Concrete markers mark their positions for the modern visitor. Those who challenged the theory were often asked "why *were* there fifty-six Aubrey Holes, then?" But to try to answer this question is to miss the point. There has to be some number of holes in the ring, even if it was created for some other purpose; and we may never know with any certitude what that purpose was. Some astronomers in the 1960s showed that the number fifty-six is not that special with regard to predicting eclipses: similar post-moving schemes can be devised with other numbers of holes in the circle. Archaeologists know of many similar pit circles in Britain, varying in number from less than twenty to over a hundred. The number fifty-six is not special, then, either archaeologically or astronomically. Finally, to further confound the Hawkins/Hoyle theory, the archaeological evidence shows that at least some of the Aubrey Holes at Stonehenge held permanent timber posts, albeit possibly after being left open for a time.

To be plausible, astronomical theories at Stonehenge must be consistent with what we know of the chronological development of the monument from the many archaeological investigations that have taken place there, particularly during the twentieth century. The great sarsen circle and trilithons, constructed in about 2400 B.C.E., were not the first construction that left its trace at this spot. The smaller bluestones had been brought there all the way from the Preseli mountains in South Wales, a distance of some 300 kilometers (200 miles), perhaps a century earlier. And some four or five centuries earlier still, an earthen enclosure, consisting of a circular ditch and inner bank, was built here, probably broadly contemporary with the Aubrey pits/posts. Yet, more remarkably, the postholes of five huge timber poles, erected several millennia earlier, in about 8000 B.C.E., were discovered during the construction of the Stonehenge parking lot in the 1970s.

During the first half of the third millennium B.C.E., the earthen enclosure seems to have been left to decay. The ditch silted up and the timber posts eventually rotted away, leaving only the Aubrey Holes. People did come, though, and carefully placed offerings such as articulated animal bones and later animal and human cremations, in the ditch and in the Aubrey Holes themselves. The distribution of these offerings around the site is far from haphazard and does seem to have been related to the motions of the sun and moon, perhaps reflecting the cosmological beliefs of the visitors to what may already have been a sacred ancestral site. This recent discovery is important, because it demonstrates that in order to understand better how and why people's activities related to the celestial bodies in the past we must try to look beyond the architecture of monuments. Archaeological excavations, which can reveal evidence of ritualistic activities that have no apparent "rational" explanation in our terms, may give us some important insights.

When stones were first brought to Stonehenge, its axis was changed. It was now aligned in the northeast upon midsummer sunrise. In other words, at dawn on the longest day of the year (and for some days before and after) the sun would have risen squarely between the Heelstone and a companion that has since disappeared, a beam of light shining along a corridor of stones and into the interior of the sarsen ring. The axis was also aligned, in the opposite direction, upon sunset on the shortest day. It is likely that this was the real focus of the monument, since the ceremonial approach was along an avenue from the northeast, so that people would have approached facing the southwest.

*Modern druids participating in a ceremony at Stonehenge, England, on the morning of the summer solstice in 1976. (Courtesy of Clive Ruggles)*

Among many other astronomical alignments claimed at Stonehenge, one of the most intriguing concerns the four *station stones*. Two of these remain in situ, just inside the ditch where the timber circle once stood. They appear to be broadly contemporary with the first appearance of other stones at the site (though this is unconfirmed archaeologically) and formed a rectangle. The lengths of its sides are in the ratio 5:12, a fact that has led some people to speculate that the builders had knowledge of Pythagorean geometry, since this ratio gives a diagonal whose length is a whole number (thirteen) of the same units. The shorter sides of the rectangle are aligned along the main axis of the monument, on midsummer sunrise and midwinter sunset, while its longer sides are aligned toward the southeast upon a point close to the most southerly rising point of the moon. This alignment has led to claims that the latitude of Stonehenge was carefully chosen to optimize this arrangement. However, such an argument is flawed on a number of grounds, most obviously because the earlier monument had already existed on this spot for several centuries. In any case, the optimal latitude would actually be well to the south, somewhere in the English Channel.

In comparison, there is little doubt that the solstitial alignment of the main axis of Stonehenge in its later phases was intentional. Its purpose may have been as a display of power, showing that this monument, erected with a huge input of human labor from exotic materials (stone brought from great distances), was in harmony with nature. Its power would reinforce the power of those who were seen to be in control of it. Certainly by the middle of the third millennium B.C.E., Wessex cheftains were powerful and commanded great long-distance exchange networks. During the ensuing millennium, dozens of rich and powerful chieftains were buried along with their personal treasures in round barrows within sight of Stonehenge, relating in some way to the power of this great ancestral monument.

See also:
Cosmology; Lunar Eclipses; Solar Eclipses; Solstitial Alignments.
Bush Barrow Gold Lozenge; Crucuno; Megalithic Monuments of Britain and Ireland.
Moon, Motions of; Solstices.

**References and further reading**
Burl, Aubrey. *The Stone Circles of Britain, Ireland and Brittany*, 349–375. New Haven: Yale University Press, 2000.
Cleal, Rosamund, Karen Walker, and R. Montague. *Stonehenge in Its Landscape: Twentieth-Century Excavations*. London: English Heritage, 1995.
Cunliffe, Barry, and Colin Renfrew, eds. *Science and Stonehenge*. London: British Academy/Oxford University Press, 1997.
Ruggles, Clive. *Astronomy in Prehistoric Britain and Ireland*, 35–41, 136–139. New Haven: Yale University Press, 1999.
Souden, David. *Stonehenge: Mysteries of the Stones and Landscape*. London: Collins and Brown/English Heritage, 1997.

# Sun, Motions of

On any particular day of the year, the sun occupies a certain position among the stars. As the celestial sphere turns in the course of its daily (diurnal) rotation, the sun—like all the stars—appears to trace out a line of constant "latitude" (for which the technical term is *declination*) on that rotating sphere. Unlike the stars, though, during the course of a year—as the sun moves gradually along its annual path through the stars, known as the *ecliptic*—its declination changes. It is at its farthest north at the June solstice, when its declination is about +23.5 degrees, corresponding to the latitude of the Tropic of Cancer on the earth. Conversely, it is at its farthest south at the December solstice, when its declination is about –23.5 degrees, corresponding to the terrestrial latitude of the Tropic of Capricorn. Over the years its declination varies continuously between these limits in a smooth, wave-like manner. In the same way, its rising position moves continually to and fro along the eastern horizon, reaching its limits at the solstices; and the same is true of its setting position in the west.

If we have an alignment that we suspect relates to the rising or setting position of the sun, then we only have to determine the declination of the relevant point on the horizon in order to deduce when in the year the sun will rise or set there. There are always two such times: one when the sun is moving north (in the northern hemisphere spring or southern hemisphere autumn) and the other when it is moving south. There will be a slight uncertainty of up to a day in each case, mainly because of the leap-year cycle. It is convenient to express the resulting dates in terms of the modern Gregorian calendar, but when we interpret our alignments, we must bear in mind that the people who set them up would have used different calendars. For example, when medieval churches were constructed, the Julian calendar was in use. The other complication is that if we delve deeply into the past, we must make a slight correction (only normally significant on a time scale of millennia) to take into account an effect known as the change in the obliquity of the ecliptic.

**See also:**
Church Orientations; Gregorian Calendar; Julian Calendar; "Megalithic" Calendar; Solstitial Directions.
Celestial Sphere; Declination; Diurnal Motion; Ecliptic; Obliquity of the Ecliptic; Solstices.

**References and further reading**
Ruggles, Clive. *Astronomy in Prehistoric Britain and Ireland,* 24–25. New Haven: Yale University Press, 1999.

## Superior Planets, Motions of

The planets Mars, Jupiter, and Saturn revolve around the sun on orbits outside that of Earth. As a result, they travel round the sun more slowly. Suppose for a moment that the earth simply vanished, leaving a sentient space monster orbiting the sun in its place, its head upward to the north. For want of anything better to do, the monster decides to ignore everything else in the sky but the sun and Mars, keeping its face toward the sun and watching Mars gradually progressing around it. At intervals it sees Mars pass slowly by the sun on the far side (only rarely will it pass directly behind it): this is known as *conjunction*. Mars is actually traveling to the monster's left, but the monster is traveling faster and in the opposite direction, so Mars appears to emerge from behind the sun traveling right. As Mars continues its journey round the sun, its angle rightward from the sun steadily increases, slowly at first, but more rapidly as the monster catches it up. Eventually, it passes round behind the monster, an event known as *opposition,* and the sequence is then repeated in reverse on the monster's left until Mars eventually approaches the sun again from the left prior to another conjunction.

The sequence of events for an Earth-based observer reflects those just described, except that we view them from the surface of a spinning planet.

An observer in the northern hemisphere perceives the celestial sphere to be turning to the right. If Mars is to the right of the sun, it rises earlier, passes across the sky to the right (west) of the sun, and sets earlier. The directions are reversed for an observer in the southern hemisphere, whose perceptions can be modeled by turning the space monster upside-down.

About sixty days after conjunction, then, Mars appears in the eastern sky before sunrise. This event is known as the *heliacal rise*—a similar concept to the heliacal rise of stars. As it gradually draws away westward from the sun, it rises earlier and climbs higher in the sky by sunrise. It also increases in brightness as it comes closer. The changes become more rapid as the planet approaches opposition, when it is at its closest and brightest. At this time, it rises at about sunset and can be seen in the sky all night before setting at about sunrise. The sequence is now repeated in reverse and in mirror image. The planet increases in distance and decreases in brightness, rapidly at first. It is now visible only in the evenings and gradually approaches the sun, appearing ever lower in the sky at sunset and setting earlier. Eventually, it ceases to be visible (this is known as its *heliacal setting*). This is followed by another conjunction, and the cycle is repeated.

Returning to the monster for a moment, suppose that it turns its attention to the apparent motion of Mars against the background stars. Suppose that it now keeps its face toward Mars, watching it pass steadily leftward (eastward) through the stars as the weeks and months go past. It probably expects this to be fairly uneventful, but it is in for a surprise. As it approaches opposition, Mars appears to stop and go into reverse. (Actually, it does not stop, because there is a slight up-and-down motion as well, so that the path traced out against the stars is a U-shape or loop.) At opposition, it appears to be going backward (westward), but then, some days later, it stops a second time and resumes its forward (eastward) progress. This phenomenon is known as *retrograde motion*. Around the time Earth is overtaking Mars, it is traveling faster than Mars in roughly the same direction, so Mars must appear to be moving backward relative to the distant stars. At other times the two planets are heading in very different directions, so the component of the earth's motion in the same direction as Mars is either much smaller or, for much of the time, in the opposite direction. This means that Mars will still appear to be moving forward in relation to the distant backdrop of stars.

The period between successive occurrences of the same configuration (the *synodic period*) of Mars is 780 days. Invisibility around conjunction lasts for about 120 days, and retrograde motion around opposition for about 75 days (with wide variations). Jupiter and Saturn, being much farther from the sun, progress much more slowly around their orbits, taking several years to complete even one circuit. This means that their synodic periods are shorter, since these are mainly determined by the rate of the earth's rotation around the sun. However, the essential characteristics of their motions are the same as for Mars. The synodic period of Jupiter is 399 days, disappearance around conjunction lasts for about 32, and retrograde motion for around 120. For Saturn the respective figures are 378, 25, and 140 days.

Retrograde motion was an enigma to pre-Copernican astronomers. Some of their solutions—such as the Greek epicycles—were extremely elegant, but they were ultimately doomed to failure.

**See also:**
Dresden Codex; Star of Bethlehem.
Celestial Sphere; Heliacal Rise; Inferior Planets, Motions of.

**References and further reading**
Aveni, Anthony F. *Conversing with the Planets*, 24–33. New York: Times Books, 1992.
———. *Stairways to the Stars: Skywatching in Three Great Ancient Cultures*, 37–40. New York: Wiley, 1997.
———. *Skywatchers*, 80–94. Austin: University of Texas Press, 2001.
Department of Physics and Astronomy, University of Tennessee. *Astronomy 161: The Solar System.* http://csep10.phys.utk.edu/astr161/lect/retrograde/copernican.html and http://csep10.phys.utk.edu/astr161/lect/solarsys/revolution.html.
Hoskin, Michael, ed. *Cambridge Illustrated History of Astronomy*, 29–47. Cambridge: Cambridge University Press, 1997.

## Supernovae
*See* Comets, Novae, and Meteors.

## Surveying
*See* Field Survey.

## Swedish Rock Art

Scattered among the fertile plains of central and southern Sweden, and found on boulders, outcrops, and flat rock surfaces, are a remarkable quantity of prehistoric rock carvings. Mostly attributed to the Bronze Age, and thus dating to the second and the early part of the first millennia B.C.E., they include depictions of people, boats, horses, and other animals; wagons and wheels; the soles of feet, resembling footprints, either barefoot or wearing sandals; and more abstract elements such as spirals, discs, and cup marks. Particularly impressive are those sites where we find combinations or accumulations of scores, or even hundreds, of rock carvings in a single place. One small area to the northwest of Stockholm, covering just nine parishes, was studied recently by British archaeologist John Coles. It alone was found to contain 840 rock art sites containing over 1,500 boats, 350 people and animals, and 300 soles of feet, together with around 19,000 cup marks.

Some Swedish rock carvings have been interpreted as representations of ancient constellations and calendars, along with depictions of spectacular events such as comets, supernovae, and eclipses. These ideas, developed in great detail during the 1980s and early 1990s by the Swedish astronomer Göran Henriksson, received a good deal of publicity within Sweden, but also proved controversial and even contentious. The main criticism leveled at them was that they were highly subjective. Attributing meaning to rock art, where there are no "informants" to give us any sort of clue, is notoriously difficult—many would say impossible—since meaning is highly dependent upon cultural context. Archaeologists, for this reason, normally tend to limit themselves to formal analyses of the frequency of certain designs, the location of rock art sites in the landscape, and similar questions, and to broad interpretations relating to the cultural setting and to ancient perceptions of the environment. Attributing specific meaning to particular carvings is unavoidably speculative and especially dangerous, since there is very wide scope for fitting complex explanations to complex data.

A particular scene found on a rock panel at Ekenburg near the town of Norrköping, about 140 kilometers (90 miles) southwest of Stockholm, has been claimed to depict the sky as seen during a total solar eclipse. The event in question took place in 1596 B.C.E. For just under three minutes, as the sun was covered by the moon, people would have seen four bright planets and a spectacular comet—Comet Encke—close to it in the daytime sky. A set of concentric rings with a dagger-like object sticking into one side is interpreted as the eclipsed sun with the tail of Comet Encke adjacent to it. Four nearby cup marks are interpreted as the planets, and a nearby boat is interpreted as a constellation comprising parts of Gemini and Taurus.

Attractive as it might seem, there are a number of problems with this interpretation. For one thing, some astronomers dispute the calculation of the time of the eclipse, which depends on certain assumptions about the slowing of the rate of rotation of the earth. Others point out that the orbit of Encke's Comet cannot be calculated prior to C.E. 1100 owing to a close encounter with Jupiter. In any case, the attribution of this particular astronomical event is speculative; the choice of which carvings are included while other adjacent ones are dismissed as not part of the eclipse scene is arbitrary (and influenced by the interpretation being put forward); and the goodness of fit—in other words, the positions of (for example) the cup marks compared with the positions of the actual planets in the sky at the time—is not good.

The lesson to be learned here is a methodological one. Enthusiasm in fitting an astronomical explanation to an archaeological phenomenon is no substitute for fairly assessing the data. Fitting a complex astronomical explanation to complex archaeological data brings the danger that any number of different, equally complex theories might fit the data equally well. If this is so, then we cannot claim that the evidence supports our theory in particular. How, then, do we estimate how many other complex explanations could have fit equally well? This is not an easy question to answer and not always of immediate interest to those who see that the data appear to fit their favored theory well enough; but a scientific approach demands that we be prepared to consider this question, and that we make a serious attempt to assess it fairly.

See also:
Comets, Novae, and Meteors; Eclipse Records and the Earth's Rotation; Methodology; Solar Eclipses.
Presa de la Mula.

**References and further reading**
Bradley, Richard. *Rock Art and the Prehistory of Atlantic Europe.* London: Routledge, 1997.
Chippindale, Christopher, and Paul Taçon, eds. *The Archaeology of Rock Art.* Cambridge: Cambridge University Press, 1998.
Coles, John, in association with Bo Gräslund. *Patterns in a Rocky Land: Rock Carvings in South-West Uppland, Sweden.* Two volumes. Uppsala: Department of Archaeology and Ancient History, 2000.
Lebeuf, Arnold, and Mariusz Ziółkowski, eds. *Actes de la Vème Conférence Annuelle de la SEAC,* 155–173. Institute of Archaeology, Warsaw: University of Warsaw, 1999.
Lindström, Jonathan, and Curt Roslund. "Arkeoastronomi och Göran Henrikssons nonsensforskning." *Folkvett* 3–4 (2000), 7–30. [In Swedish.]

# Symbols

Many archaeologists now believe that it is possible to reconstruct elements of human cognition from material remains. But in order to do so, it is crucial to recognize a means of expression unique to human beings: the use of symbols. The digit "2," for example, is a symbol used by Europeans to define a certain numerical concept. But if a hypothetical future archaeologist, ignorant of all other aspects of twentieth-century European culture, were to try to interpret the digit "2" carved, say, on an isolated stone, they could not deduce its meaning from the form of the symbol alone. On the other hand, they might be able to deduce something from the fact that stones marked "1 km," "2 km," "3 km," etc. occurred at intervals a fixed distance apart by the side of a road. The context of the symbols—their association with one another, as well as their association with the road, which seems relevant to the way in which they were used, and their spatial distribution—clearly does suggest possible meanings. A complication is that symbols may mean several different things at once; they may convey different meanings to different people, some of which may be very different from those intended; and the meanings may well change through time. But just as the Stars and Stripes would be recognized by virtually everyone in the modern world as a symbol of the United States, people sharing a common cultural background are likely to perceive common sets of meanings. It is this set of shared meanings that gives us some hope of retrieving meanings that reflect certain worldviews.

Designs found in prehistoric rock art, especially those that do not obviously "literally" represent something like an animal, may have had symbolic meaning, and it is possible to try to deduce something about that meaning from its context. For example, it has been suggested that many circles and rayed circles in megalithic art all round the Atlantic fringe of Europe may be sun symbols or representations. Such a design, along with other decorations, was carved onto the back corbel of the roof-box above the entrance of the Neolithic passage tomb at Newgrange, Ireland. The tomb was apparently designed so that shortly after sunrise on days around winter solstice, the sun would shine through the roof-box and down the nineteen-meter-long passage, reaching the chambers in the interior of the tomb.

Can we do more than simply recognize possible symbols and possible meanings? Can we use them to reconstruct elements of other worldviews? What gives us hope is the human propensity to express and reinforce the nature of the cosmos, as a given group of people perceives it, in many different ways in order to keep their own lives and activities in harmony with the world. Verbal or visual metaphors might be expressed in myth, exhibited in activities that we might interpret as either ritualistic or mundane (or both), or displayed in architecture. For example, hogans, traditional round or octagonal buildings built by Navajo families in the southwestern United States, reflect the harmony of the cosmos in various ways. Their single door faces the rising sun, and the space within is conceptually divided into four quarters used in different ways, reflecting the four directions of the cosmos and their associated properties and qualities. Movement within the hogan is always clockwise, reflecting the motion of the sun around the sky.

Where we have only the material record to go by, associations between symbols, if we can fathom their meaning, may give us insights into correspondences that form part of a worldview. We might notice, for example, the consistent orientation of hogans toward the solar rising arc in the east; if in addition we consistently found material evidence of different activities in the four quarters of the hogan, we might correctly deduce the quadripartite structure of the Navajo cosmos.

**See also:**
Cognitive Archaeology; Cosmology; Science or Symbolism?; Star and Crescent Symbol.
Mesoamerican Cross-Circle Designs; Navajo Hogan; Newgrange.

**References and further reading**

Green, Miranda. *The Sun-Gods of Ancient Europe*, 20–28. London: Batsford, 1991.

Hodder, Ian. *Symbols in Action: Ethnoarchaeological Studies of Material Culture*. Cambridge: Cambridge University Press, 1982.

Renfrew, Colin, and Ezra Zubrow, eds. *The Ancient Mind: Elements of Cognitive Archaeology*, 3–12. Cambridge: Cambridge University Press, 1994.

Renfrew, Colin, and Paul Bahn. *Archaeology: Theories, Methods and Practice* (4th ed.), 497–500. London: Thames and Hudson, 2004.

# Synodic Month

*See* Lunar Phase Cycle.

## Taulas

Taulas are a distinctive type of later prehistoric monument found exclusively on the Mediterranean island of Menorca. Attributed to the talayotic culture—so called because it is characterized by a type of dry-stone tower known as a *talayot*—they date broadly to the late second and early first millennia B.C.E. Taulas consist of two large rectangular blocks of stone, one standing upright and the other placed across the top, balanced centrally, so as to form a shape resembling the letter "T." These were placed within stone-walled precincts. One side of the "T" was worked flat and smooth. This side faced the visitors as they came in through the single entrance, as well as the wider landscape beyond. Each major village seems to have had its taula, and some twenty-five examples are known, in various states of preservation, scattered about the small island.

The orientations of taulas (and the precinct entrances) are not random; they face generally southwards, within a range from southeast to south-southwest. In every case there is an uninterrupted view out to a distant horizon, suggesting that it was important to have a clear view of the sky in that direction. Yet this direction does not correspond to that of the rising or setting position of the sun, moon, or planets, leaving only the possibility that the taulas were aligned on certain stars. It is possible that the asterism of interest was the Southern Cross and Pointers, a very conspicuous group of bright stars appearing low in the southern sky at this latitude in the second millennium B.C.E. This might also explain the puzzling fact that no taulas are found on neighboring Mallorca, despite the fact that talayots are equally abundant there. Mallorca, unlike Menorca, is mountainous and the landscape around the principal Mallorcan villages (concentrated in the fertile northern part of the island) would not have permitted a clear view of the southern sky. In other words, people living on Mallorca had no possibility of seeing the Southern Cross and so did not build taulas. However, if the Menorcan taulas were indeed linked inextricably with the Southern Cross and Pointers, then the passage of time would have presented the ancient

*The taula at Talatí de Dalt, Menorca, viewed from the entrance of its enclosure. Leaning against it is a pillar, with a small capstone of its own, which incredibly became wedged when it fell from the vertical. (Courtesy of Clive Ruggles)*

Menorcans with a problem. Owing to precession, these stars dropped lower and lower in the southern skies as the centuries progressed, until by the onset of the first millennium B.C.E. they were hardly visible at all.

Taulas certainly had sacred associations, as is clear from broken statues and sacrificed animal bones strewn about the precincts. In one case these included a statue of the Egyptian god of medicine, Imhotep, complete with a hieroglyphic inscription; at another, excavations uncovered a hoof from a bronze statue, evidently of a horse (or centaur), the remainder presumably having been plundered. Since there is no other evidence of local horse gods, this again suggests links with the eastern Mediterranean. In ancient Egypt, it was common to associate gods and goddesses with stars and constellations, and in Greek mythology the god of medicine, Asclepius (counterpart to the Egyptian Imhotep), was tutored by a centaur—Chiron. Tentatively weaving together these threads, the British historian of astronomy Michael Hoskin suggested that the taula sanctuaries may have been places of healing and were oriented upon the constellation of the centaur—Centaurus. The Southern Cross and Pointers are at the feet of the centaur in the sky, the Pointers ($\alpha$ and $\beta$ Centauri) forming the two brightest stars of the modern constellation.

See also:
Son Mas.
Precession.

**References and further reading**
Hochseider, Peter, and Doris Knösel. *Les Taules de Menorca: Un Estudi Arqueoastronòmic.* Maó (Mahón), Menorca: Govern Balear, 1995. [In Catalan.]
Hoskin, Michael. *Tombs, Temples and Their Orientations,* 38–46. Bognor Regis, UK: Ocarina Books, 2001.
Hoskin, Michael, and William Waldren. *Taulas and Talayots.* Cambridge: Michael Hoskin, 1988.
Pasarius, J. Mascaró. "Las Taulas: Testimonio de la fe religiosa y de la capacidad creadora de los paleo-menorquines." *Revista de Menorca,* número extraordinario (1968), 215–330. [In Spanish.]
Waldren, W. H., J. A. Ensenyat, and R. C. Kennard, eds. *Recent Developments in Western Mediterranean Prehistory: Archaeological Techniques, Technology and Theory, Volume II: Archaeological Technology and Theory,* 217–236. Oxford: Tempus Reparatum (BAR International Series S574), 1991.

# Temple Alignments in Ancient Greece

The first systematic study of temple orientations in ancient Greece was carried out by Heinrich Nissen in the nineteenth century. William Dinsmoor, carrying out another study in the 1930s, concluded that almost three quarters (73 percent) of them were oriented eastward within the range of sunrise. Some were oriented on sunrise on the feast day of the deity to whom the temple was dedicated. One example is the Parthenon itself, dedicated to Athena. It is not only aligned upon the sunrise on the feast day of Athena but also upon a prominent hill—Mount Hymettos. How widespread or long-lasting this practice was is a hard question to answer. The evidence is complicated by the fact that many temples were reconstructed on a different orientation or their dedication was changed. Historical sources also make it clear that sunrise was not always the principal concern. The Greek writer Vitruvius, in his work *De Architectura,* indicates that there was sometimes a preference for facing nearby cities or directions of approach along rivers or roads.

It has been suggested that the predominantly easterly orientation of Greek temples may reflect broader orientation practices extending back as far as the Bronze Age. The Minoan culture survived on the island of Crete until the middle of the second millennium B.C.E., after which the Mycenaeans from the Greek mainland took control. Elements of both Mycenaean and Minoan orientation practices may have come to influence Archaic and, in due course, Classical Greek temples, but the practice of easterly orientation seems to derive from the Minoans. Studies of the orientations of over three

The Temple of Poseidon at Sounio, Greece, viewed toward the entrance. Like many such temples it faces within a few degrees of east. (Courtesy of Clive Ruggles)

hundred Bronze Age passage tombs on Crete, dating back to the mid-third millennium B.C.E., show a very strong easterly trend, with over 80 percent of them lying within the solar arc. An even higher proportion (about 87 percent) are within the slightly wider lunar arc. Over half the data are derived from a single cemetery, Armenoi in western Crete. Each of the more than two hundred tombs there is oriented within the range of moonrise; and it is surely not coincidental that three kilometers (two miles) to the east is Mount Vrysinas, upon whose summit was a peak sanctuary apparently, although not certainly, dedicated to a moon goddess.

The development of mathematical astronomy in ancient Greece from around 400 B.C.E. onward seems to have derived in large part from ancient Babylonia. However, in constructing large stone temples as far back as the seventh century B.C.E., the ancient Greeks acquired both inspiration and technological expertise from ancient Egypt. This linkage opens up the possibility that some of the more symbolic aspects of Greek temple construction, such as architectural alignments upon celestial objects, also derived from ancient Egypt, where there is cultural continuity stretching back into the third millennium B.C.E. Studying the common aspects of astronomical symbolism incorporated in Greek and Egyptian temples may throw more light on this possibility in the future.

See also:
Hesiod (Eighth Century B.C.E.); Nissen, Heinrich (1839–1912).
Egyptian Temples and Tombs; Minoan Temples and Tombs.
Moon, Motions of.

**References and further reading**
Dinsmoor, William. "Archaeology and Astronomy." *Proceedings of the American Philosophical Society* 80 (1939), 95–173.
Hoskin, Michael. *Tombs, Temples and Their Orientations*, 217–222. Bognor Regis, UK: Ocarina Books, 2001.
Ruggles, Clive, Frank Prendergast, and Tom Ray, eds. *Astronomy, Cosmology and Landscape*, 72–91. Bognor Regis, UK: Ocarina Books, 2001.

# Teotihuacan Street Grid

Teotihuacan was the first pre-industrial city in the New World. Rising to prominence in about 100 B.C.E., it subsequently reached massive proportions, supporting at its height (between about C.E. 350 and 650) a population of around 200,000 and covering some twenty square kilometers (eight square miles). This entire area was partitioned by a regular street grid into about two thousand rectangular "apartment compounds," some containing groups of occupational specialists and others housing different social classes or ethnic groups. Most remarkable of all is the ceremonial center, its wide main street (the so-called Street of the Dead) lined with splendid courtyards, palaces, and temples. It also contained the two great pyramids that have subsequently become known as the Pyramid of the Sun and the Pyramid of the Moon.

Like many other Mesoamerican cities, the street grid is precise and oriented slightly askew from the cardinal directions. There are two different parts to the city, one around the Pyramid of the Sun, which deviates from the cardinal directions by about 15.5 degrees clockwise, and another farther to the south that deviates by about 16.5 degrees. Curiously, quite a large area in between seems to follow one north-south convention but the other an east-west one, resulting in street intersections, and even individual buildings, that have corners of around eighty-nine degrees and ninety-one degrees.

In all likelihood, one or two different sightings fixed the original alignments in each direction and were then propagated as more streets were constructed. But what was being sighted upon? One possibility is that the north-south direction reflects the orientation of the Pyramid of the Sun directly upon the hill Cerro Gordo to the north. Indeed, when viewed from the south, the

*The Temple of the Sun at Teotihuacan, Mexico, reflecting the shaping of the hill, Cerro Gordo, to the north. (Courtesy of Clive Ruggles)*

pyramid does seem to "reference" the hill behind by mimicking its shape. Others have argued that the focus of interest was topographic features on the eastern horizon. But it may be misleading to consider the Teotihuacan street grid in isolation. Systematic studies of the orientations of the street grids of Mesoamerican cities show that deviations of between fifteen degrees and eighteen degrees clockwise from north-south-east-west are surprisingly common; indeed, they form a distinctive orientation group. This characteristic strongly implies that they were fixed using a common astronomical determinant.

Yet no overwhelmingly convincing possibilities emerge, especially among celestial objects rising to the east. A possible explanation of the east-west direction is that it aligned in the west with the horizon setting position of the Pleiades, but if so it is curious that the orientations of structures spanning many centuries remained on the same orientation, while the setting direction shifted owing to precession. Another possibility is that the direction corresponds to sunset on April 30 and August 13, dates significant because the longer interval between them is exactly 260 days, or the length of the sacred cycle that was a defining feature of all Mesoamerican calendars.

See also:
Horizon Calendars of Central Mexico; Mesoamerican Calendar Round; Mesoamerican Cross-Circle Designs.

*The Street of the Dead at Teotihuacan, Mexico, viewed from the Temple of the Moon at its northern end. (Courtesy of Clive Ruggles)*

### References and further reading
Aveni, Anthony F. *Skywatchers*, 223–235. Austin: University of Texas Press, 2001.

Millon, René, ed. *The Teotihuacan Map* (in two parts). Austin: University of Texas Press, 1973.

Šprajc, Ivan. "Astronomical Alignments at Teotihuacan, Mexico." *Latin American Antiquity* 11 (2000), 403–415.

———. *Orientaciones Astronómicas en la Arquitectura Prehispánica del Centro de México*, 209ff. Mexico City: Instituto Nacional de Antropología e Historia (Colección Científica 427), 2001. [In Spanish.]

## Theodolite Surveys

The theodolite, sometimes referred to in the United States as a *surveyor's transit*, is the surveying instrument that is most useful in determining the astronomical potential of structural alignments at archaeological sites such as prehistoric monuments. Although the designs of theodolites vary considerably, as they have throughout the century or so that they have been manufactured, their fundamental components remain invariable. These consist of a telescope that can be pointed in any direction, together with two graduated circles, mounted horizontally and vertically, which enable that direction to be defined quantitatively. In order to determine the astronomical potential of (say) a point on the horizon as viewed from a given observing position—in other words, to determine the apparent positions of the distant celestial bodies in relation to that closer feature—the direction of that point from the observer is precisely what we need.

Any direction can be defined in terms of two angles: a horizontal angle measured around from some suitable zero point, and a vertical angle measured up or down from level. Determining the second of these is straightforward. When a theodolite is first set up on its tripod, a leveling process involving various bubbles ensures that the vertical circle is properly zeroed and will give us the required angle, which is known as the *altitude*. The situation regarding the horizontal angle is more complicated. The leveling process ensures that the horizontal circle is truly horizontal, but the zero point on this scale will be pointing in an arbitrary direction. What we actually want is the angle measured clockwise around from true north, which is known as the *azimuth*. However, generally we can't know the direction of true north (at least, to the required accuracy) in advance. What is normally done in practice is to measure what is called the horizontal *plate bearing*: the clockwise angle around from the arbitrary zero point. We must make the adjustment from plate bearings to azimuths in retrospect. This will involve making the same correction to each reading, by adding the actual azimuth of the zero point, sometimes known as *plate bearing zero* or PBZ.

But how do we determine PBZ once we have set up the theodolite? One way is to take a series of timed observations of the sun (it is very important not to look through the telescope when it is pointing at the sun, but the sun's image can be projected on to a piece of paper). The plate bearing of each reading can then be compared with the sun's actual azimuth as determined using published or on-line ephemerides. Each direction we measure (e.g., of a point on the horizon) will eventually be defined in terms of azimuth and altitude, from which we can calculate the *declination* and hence know what rises and sets there, or would have risen and set there at any given era in the past.

A modern theodolite will typically measure the azimuth and altitude of a distant object to a precision of twenty arc seconds or better. For work that does not warrant anything like this sort of accuracy, it may be possible to obtain the necessary information on site using a magnetic compass and a clinometer, or by using large-scale topographic maps or digital topographic data.

In addition to determining the directions of horizon points, etc., it is important to fix the position of the theodolite in space, for example, in relation to a monument. This can be done using a tape, but modern surveying instruments are very useful for this. Theodolites incorporating an electronic distance measurement device (EDM) provide a convenient way of determining the distance between the instrument and a reflector placed within a range of distances appropriate for most site surveys. Total Stations combine this technology with a computer chip that can convert direction-and-distance determinations of measured points into positions in space expressed in a chosen coordinate system. They can also store measurements for downloading directly to a computer, which not only allows the surveyor to dispense with the task of recording measurements manually on site, but also allows for a quick display of the results, using appropriate software.

**See also:**
Compass and Clinometer Surveys; Field Survey; Precision and Accuracy. Altitude; Azimuth; Declination.

**References and further reading**
Aveni, Anthony F. *Skywatchers,* 120–124. Austin: University of Texas Press, 2001.
Ruggles, Clive. *Astronomy in Prehistoric Britain and Ireland,* 165–169. New Haven: Yale University Press, 1999.

# Thom, Alexander (1894–1985)

Alexander Thom is widely acknowledged as one of the founding fathers of modern archaeoastronomy. An engineer by profession, he was also a keen amateur astronomer, and he was aware from an early age of the writings of Sir Norman Lockyer on the possible astronomical significance of ancient stone monuments in his native Scotland and elsewhere in the British Isles. His own active interest in the megalithic monuments of Britain, and particularly in his native Scotland, is said to have been aroused in 1933 by the sight, one evening as he was sailing in the Outer Hebrides looking for an anchorage, of one of the stone circles at Callanish silhouetted against the rising moon. Twenty years and many dozens of site surveys later, he began to publish research papers in which he proposed three ideas that, during the 1960s and 1970s, would shake the archaeological world.

The first of Thom's ideas was that stone circles were laid out using a precisely defined unit of measurement, which he called the "megalithic yard." The second, and related, idea was that each circle was laid out according to one of a small number of particular geometrical designs using techniques based upon Pythagorean triangles. The third idea was that a variety of "megalithic sites" incorporated alignments of remarkable precision upon the rising and setting positions of various celestial bodies. Taken together, these showed that "megalithic man," to use Thom's own term, was a competent engineer with considerable technical knowledge, whose accomplishments included sophisticated numeration, geometry, and astronomy—three things that would come to be known together as "megalithic science." Thom's ideas first appeared in book form in 1967, in a slim but highly technical volume called simply *Megalithic Sites in Britain*, which contained the detailed analysis of survey results from over 250 standing stone monuments.

Thom's second book, *Megalithic Lunar Observatories*, published in 1971, set the archaeological world into a turmoil that did not calm down fully until more than a decade later. His claim was that many standing stone monuments were deliberately aligned upon "notches" and other features in distant mountainous horizons, and that these horizon points marked extreme rising and setting points of the moon with incredible precision. To attain that level of precision, prehistoric astronomers would have had to carry out meticulous programs of observation, following them consistently for years if not decades (in other words, over generations). The problem was that these conclusions were completely unpalatable to most prehistorians at the time. Their social models, informed by a variety of archaeological evidence, simply did not fit with such ideas. Yet Thom's results could not easily be refuted: they were based on careful fieldwork and supported by rigorous statistical analysis.

When detailed reappraisals of Thom's ideas on "megalithic astronomy" were eventually completed in the 1980s, it emerged that the supposed astronomical sightlines of very high precision could all be explained away as chance occurrences. On the other hand, certain groups of monuments, such as the recumbent stone circles of eastern Scotland and the short stone rows of western Scotland and southwest Ireland, definitely did contain consistent lunar alignments, though at a much lower degree of precision, that could not be ignored. Similarly, although the idea of a precise "megalithic yardstick" being used all over Britain is not sustainable as Thom conceived of it, the idea that there was a widespread practice of laying out stone circles and other monuments using broadly consistent measurements based upon the human body (such as arm lengths or paces) remains a distinct possibility.

In hindsight, even though Thom's theories have been largely superseded, his work led archaeologists to start thinking about issues and facing problems that they had largely avoided. In particular, Thom's work motivated others to investigate intentional astronomical alignments at prehistoric monuments and eventually resulted in this area of study becoming archaeologically informative and respectable. Finally, Thom pioneered ways of carrying out site surveys using a theodolite, analyzing the data, and presenting the results that are now a fundamental part of field methodology within archaeoastronomy.

**See also:**
Archaeoastronomy; Field Survey; Lockyer, Sir Norman (1836–1920); "Megalithic Astronomy."
Megalithic Monuments of Britain and Ireland; Recumbent Stone Circles; Short Stone Rows; Stone Circles.

**References and further reading**

Ruggles, Clive. *Astronomy in Prehistoric Britain and Ireland*, 49–67. New Haven: Yale University Press, 1999.

———, ed. *Records in Stone: Papers in Memory of Alexander Thom*. Cambridge: Cambridge University Press, 2002.

Thom, Archibald S. *Walking in All the Squares: A Biography of Alexander Thom*. Glendaruel, Scotland: Argyll Publishing, 1995.

# Thornborough

Situated on a gravel plateau near Thornborough in Yorkshire, northern England, is a remarkable monument consisting of not just one, but an alignment of three large, round Neolithic earthwork enclosures with high outer banks, known as henge monuments. The three henges are almost identical in size and form, each being some two hundred meters across with two entrances on opposite sides, and they are more or less equally spaced some five hundred meters apart, forming an alignment over 1.5 kilometers (1 mile) long. The entrances are oriented along the alignment. Surrounding the henges are a number of other prehistoric features, including two large pits, which appear to have held sizeable wooden posts, aligned upon a double ring-ditch that represents the remains of a ploughed-out barrow.

What makes this monument fascinating from an astronomical point of view is the possibility that it was a pilgrimage center whose astronomical associations tell us not only the time of year when the major ceremonies took place here, but also reveal how people from far afield may have known when to set out in order to arrive in time for those ceremonies.

A number of factors support the pilgrimage center idea. First, the henges were constructed along what is known from other archaeological evidence to have been an important routeway. Second, the formal design of the monuments was evidently important, and it developed over time. The henges are embellishments of earlier, smaller earthwork enclosures, and the central one was built over the central part of an earlier linear earthwork known as a cursus monument, itself over two kilometers (1.5 miles) long. Third, the high outer banks would have obscured the view of the distant skyline (something that is still evident from what remains of the banks today), and this suggests that the henge interiors were exclusive places, cut off from the outside world. Fourth, archaeological field-walking has revealed an almost complete absence of surface scatters of flints from the area immediately around the henges, strongly implying that this was not a place where people routinely lived and worked. Finally, and most extraordinarily, about one kilometer (half a mile) to the east is a large concentration of surface flint, but a good deal of it is non-local and from scattered sources, and lightly worked. It appears that people came here from some distance and brought the stone with them, which implies in turn that this may have been a visitors' encampment, rather than a permanent settlement.

There are a number of locations at the site where prominent features of the monument seem to frame parts of the horizon. Intriguingly, several of these align upon the same asterism, namely the three stars of Orion's belt. It is worryingly easy to identify potential astronomical, and particularly stellar, alignments at almost any site, whether or not they were intentional. But this consistency argues in favor of the Thornborough alignments being deliberate. Viewing along the cursus to the west, the three stars would have been framed by the western terminal as they set. Viewing along the line of the two large timber posts, toward the barrow, an observer would have seen them rising in the alignment. And by the time the bright star Sirius rose, in roughly the same alignment, Orion's belt would have moved up and to the right, to be in line with the henge alignment itself and visible from the henge interiors over their southern entrances. A final, more speculative suggestion is that the three henges themselves formed a representation on the ground of Orion's belt. Because of precession, the rising and setting positions of any star or group of stars varies significantly over a time scale of centuries. Though it is highly inadvisable to try to date sites astronomically by fitting stars to alignments, precession serves at Thornborough to reinforce the Orion's belt alignments, in that the "best fit" dates are wholly compatible with the archaeological estimates of the dates of different features at the monument.

If Thornborough really was a pilgrimage center, then two questions arise. Were there particular times when people congregated here, perhaps for large ceremonies? And if so, and if people had to come from some distance, how did they know when to leave in order to arrive on time?

The astronomical associations of the site may provide answers to both questions. The annual first appearance of Orion's belt in the eastern sky before dawn—its heliacal rise—would have occurred in the late summer. The annual reappearance of Sirius—the brightest star in the sky—would have occurred two or three weeks later. Once Orion's belt had appeared in the predawn sky, it could have provided the necessary reminder for surrounding peoples that a ceremony was imminent. The actual ceremony could then have taken place somewhat later, at the time of the dawn reappearance of Sirius, with Orion's belt now prominently visible above the southern entrance of each henge just before dawn. Such a ceremony would have taken place in early autumn, when the harvests were in and the seasonal workload had eased up for a while.

This interpretation of the Thornborough henges was made possible using a combination of sky visualization software and virtual reality software. It shows the potential of this type of approach in contrast to the more traditional procedure, which is to survey horizon alignments and then calculate declinations.

> **See also:**
> Astronomical Dating; Orion; Pilgrimage.
> Circles of Earth, Timber, and Stone; Cursus Monuments.
> Heliacal Rise; Precession.
>
> **References and further reading**
> Harding, Jan. "Late Neolithic Ceremonial Centres, Ritual and Pilgrimage: the Monument Complex of Thornborough, North Yorkshire." In Anna Ritchie, ed., *Neolithic Orkney in Its European Context,* 31–46. Cambridge: McDonald Institute for Archaeological Research, 2000.
> Ruggles, Clive. "Landscape Archaeology and the Archaeology of Pilgrimage: A View from across the Atlantic." In John Carlson, ed., *Pilgrimage and the Ritual Landscape in Pre-Columbian America.* Washington, DC: Dumbarton Oaks, forthcoming.

## Total Stations
*See* Theodolite Surveys.

## Tri-Radial Cairns

Tri-radial cairns are a distinctive type of Bronze Age monument found on the moorlands of northern England, comprising three radial "arms" of stones, typically no more than two meters (six feet) long and half a meter (two feet) high, extending out from a common center. Modest in size and easily mistaken for sheep shelters or other more modern constructions, their status as prehistoric monuments was only recognized in the 1990s. Over twenty examples are now known, with a major concentration of eight at Lordenshaws in Northumberland.

The state of preservation of the known tri-radial cairns varies. It is often difficult to determine with any great precision the original direction of each of the arms, even assuming that they were straight in the first place. Nonetheless, there is a clear general consistency in the orientations, and this invites an astronomical explanation. In each case one of the arms points approximately due north, always to within about thirty degrees and in the majority of cases to within ten degrees. The remaining two arms point southeastward and southwestward respectively, with a similar scatter in the "best-fit" directions.

A group from the Border Archaeological Society who discovered the cairns has suggested that they might have been constructed by making observations of the sun around the time of the summer solstice. The technique would have been to erect a vertical post and to build the three arms in the direction of the shadow at sunrise, noon (when the shadow is shortest), and sunset. This might not have been done to any great precision, so we might expect a considerable scatter in the resulting orientations even at the time of construction. (It would also require clear skies, but given the tiny change in the sun's motions on days near the solstice, there would be many opportunities to take advantage of suitable weather conditions.) The result would be radial orientations concentrated around azimuths of 0 degrees, 135 degrees, and 225 degrees respectively.

But there is a simpler explanation that fits the data just as well—in fact, better. Suppose that one arm was oriented to the north, either by using the sun's shadow at noon (on any day of the year), or by reference to the diurnal motions of the stars at night, the other two arms being simply placed at regular intervals round the cairn. This would result in radial orientations concentrated around azimuths of 0 degrees, 120 degrees, and 240 degrees. Of course this begs the question of why the chosen number of arms was three, but the explanation could well have nothing to do with astronomy.

There is no conclusive evidence to support either of these explanations, but the fact that they are both viable shows the dangers of fitting particular astronomical orientations too readily, and not considering equally plausible alternatives. It also demonstrates the principle known as *Occam's razor*, which advises us that there is no need to resort to a complex hypothesis to explain some data if a simpler one will explain the data just as well. Here, the "regular spacing" explanation is arguably the simpler one and more consistent with the data.

See also:
Azimuth; Diurnal Motion.

**References and further reading**
Ford, Bill, Philip Deakin, and Manuella Walker. "The Tri-radial Cairns of Northumberland." *Current Archaeology* 182, 16(2), Nov. 2002, 82–85.

# U

## Uaxactún, Group E
*See* Group E Structures.

## Uxmal, Governor's Palace
*See* Governor's Palace at Uxmal.

# Venus

*See* Inferior Planets, Motions of; Venus in Mesoamerica.

# Venus in Mesoamerica

The planet Venus is the third brightest regularly visible object in the skies, behind the sun and moon. Its cycle of morning and evening appearances was well known to the ancient Mesoamericans, partly because it seemed to fit naturally with other astronomical cycles, in that five Venus cycles exactly equaled eight year-cycles of 365 days. In fact, five Venus cycles do not quite equal eight years, and we would view the near-coincidence of these two periods merely as an accident of nature. The ancient Mesoamericans, on the other hand, saw it as a significant property of the cosmos and this had two far-reaching consequences. The first was that they, and particularly the Maya, could integrate cycles of Venus with ease into the other intermeshing cycles of their calendar. Direct evidence of the care with which the synodic cycles of Venus were observed by the Maya comes from the Dresden Codex, where the motions of the planet are carefully tabulated and correlated with other astronomical cycles. The heliacal rising of Venus was probably the most significant event to the Maya, and certainly a time of dread for the Aztecs.

The second consequence was that Venus appeared to have a relationship with the seasons, even though a particular seasonal configuration of Venus would only recur after eight years, and so Venus became strongly associated with rain, fertility, and maize. It also appears to have become strongly associated with warfare and ritual sacrifice, with military campaigns carefully timed according to Venus-related prognostications. The two are related, since one of the main purposes of human sacrifice was to petition the gods for rain. There was also a clear association between Venus and the sacred ballgame played throughout Mesoamerica; this is evident from the common occurrence of Venus symbols both in depictions of the game in codices and on the garments and body decoration worn by the players themselves.

**See also:**
Cacaxtla; Dresden Codex; Caracol at Chichen Itza; Governor's Palace at Uxmal; Kukulcan.
Inferior Planets, Motions of.

**References and further reading**
Aveni, Anthony F. *Stairways to the Stars: Skywatching in Three Great Ancient Cultures*, 37–46, 93–146. New York: Wiley, 1997.
———. *Skywatchers*, 166–167. Austin: University of Texas Press, 2001.
Milbrath, Susan. *Star Gods of the Maya: Astronomy in Art, Folklore, and Calendars*. Austin: University of Texas Press, 1999.
Ruggles, Clive, and Nicholas Saunders, eds. *Astronomies and Cultures*, 202–252. Niwot: University Press of Colorado, 1993.
Šprajc, Ivan. *La Estrella de Quetzalcóatl: El Planeta Venus en Mesoamérica*. Mexico City: Editorial Diana, 1996. [In Spanish.]
Whittington, E. Michael, ed. *The Sport of Life and Death: The Mesoamerican Ballgame*, 42–45. London: Thames and Hudson, 2001.

# Wedge Tombs

Wedge tombs are a distinctive type of megalithic passage tomb found all over Ireland, but mainly in the west. Over five hundred examples are known, and while they vary widely in size, they do have a defining characteristic: a trapezoidal central chamber, its sides formed by two lines of large, upright stones (orthostats), which is wider and higher toward the entrance end, forming a wedge shape. The bulk of wedge tomb construction took place in the second half of the third millennium B.C.E., placing them chronologically toward the end of a rich tradition of Neolithic tomb construction in Ireland.

The other property that confirms the wedge tombs as a useful category of monuments is the strong consistency in their orientations. Nearly all known examples face the western arc of the horizon, with the majority facing southwest. It is unusual to have such a clear preference for westerly orientation among a group of Neolithic tombs, let alone such a substantial one. The only other example in Britain or Ireland is the few dozen Clava cairns of northeast Scotland. The fact that the Clava cairns also seem to represent a late development suggests that a strong predilection for westerly/southwesterly orientations only developed toward the end of the Neolithic. A similar preference is evident in the orientations of the axial stone circles of southwest Ireland, a region where there is a major concentration of wedge tombs, as well as among the recumbent stone circles of northeast Scotland, which occupy an area close to that where the Clava cairns are found.

What is the astronomical significance of the wedge tombs? Viewed in the context of groups of later prehistoric temples and tombs found all over western Europe, their pattern of orientation fits the "sun descending or setting" model. In other words, each tomb was oriented upon a position where the sun was seen either to set, or to be descending in the sky, on a significant day—perhaps the day on which construction was begun. However, there is one well-studied example that provides tantalizing evidence of more particular concerns. This is the tomb of Altar, situated on the south coast of the Mizen peninsula at the very extreme of southwest Ireland, close to the village of An Tuar Mór (Toormore).

The Altar tomb commands a clear view down toward the tip of the Mizen peninsula, and is directly oriented upon the conspicuous Mizen Peak, which forms a distinctive pyramid shape on the horizon some twelve kilometers (eight miles) away. Added to this, the tomb's orientation may have had a calendrical significance. As was first pointed out by Boyle Somerville, who surveyed the monument in 1931, it is oriented upon the position of sunset within a couple of days of February 4 and November 5. These are two of the mid-quarter days, which along with the solstices, equinoxes, and the other two mid-quarter days divide the year into eight equal parts. This interested Somerville because the November date corresponded to the festival of Samhain in the Celtic calendar.

The problems with this interpretation apply to all supposed equinoctial and mid-quarter-day alignments, many of which were highlighted in reassessments of the "megalithic" calendar proposed in Britain by Alexander Thom. On the other hand, an intriguing aspect of Altar is the existence of archaeological evidence that suggests a sequence of ritual activity stretching from around 2000 B.C.E. up to at least the first century C.E. This evidence might support the argument of continuity of tradition into "Celtic" times.

However, Altar does not seem to be typical even of the wedge tombs in its immediate vicinity, at least to judge from an analysis of nine other wedge tombs in the Mizen peninsula. Even the impressive topographic alignment appears to be a "one-off." This means that, as with other single instances among groups such as the solstitially aligned axial stone circle at Drombeg, one must continue to wonder whether the alignments could have arisen fortuitously rather than intentionally.

See also:
Celtic Calendar; Equinoxes; "Megalithic" Calendar; Mid-Quarter Days; Prehistoric Tombs and Temples in Europe; Somerville, Boyle (1864–1936); Thom, Alexander (1894–1985).
Axial Stone Circles; Clava Cairns; Drombeg; Recumbent Stone Circles.

**References and further reading**
O'Brien, William. *Sacred Ground: Megalithic Tombs in Coastal South-West Ireland.* Department of Archaeology, Galway: National University of Ireland Galway, 1999.

# Worldview
*See* Cosmology.

## Xochitecatl
*See* Cacaxtla.

## Years B.C.E. and Years before 0

There is great potential for confusion between the ways that archaeologists and astronomers reckon dates before the birth of Christ. For the archaeologist, the year before C.E. 1 is 1 B.C.E. In other words, there was no "year 0." The convention amongst astronomers, on the other hand, is to count years backward on a linear scale, so that the year before the year 1 was the year 0, the year before that was –1, and so on. This means, for example, that the archaeologist's year 331 B.C.E. is the astronomer's year –330.

This issue becomes crucially important, for example, at the Iron Age causeway site of Fiskerton in England, where it has been proposed that episodes of reconstruction may have been associated with lunar eclipses. In order to test this idea, archaeologically determined tree-ring dates need to be compared with astronomically determined eclipse dates.

**See also:**
Fiskerton.

**References and further reading**
Ruggles, Clive. *Astronomy in Prehistoric Britain and Ireland*, ix. New Haven: Yale University Press, 1999.

## Yekuana Roundhouses

The Yekuana live amid the rainforests of southern Venezuela, along the banks of the river Orinoco and its tributaries. Their traditional dwellings are communal roundhouses large enough to house several families. They have walls of wattle and daub and a huge conical roof, thatched over, through which the central pole protrudes, pointing directly upwards at the heavens.

According to Yekuana myth, the design of the roundhouse was handed down by the Sun (Wanadi) himself, the creator of the world, and reflects the properties of the cosmos in a variety of ways. A circular inner room represents the sea (this is used as sleeping space for bachelor men as well as a place for rituals and ceremonials presided over by the village shaman), and the outer ring, divided radially into compartments where different families live, represents the land. The roof is seen as a reflection of the celestial dome: it is supported by upright posts known as "pillars of the stars," and the horizontal crossbeam supporting it, always placed in a north-south orientation, has the name of the Milky Way. The center post reaches down into the underworld and up, beyond the roof, to the heavens. Placed centrally within the roof space, about three meters (ten feet) above the floor, is a set of tie beams forming a rectangular structure. This has a practical purpose in helping to support the crossbeam and providing a frame from which the men sleeping in the central space can hang their hammocks. However, its precise rectangular shape also has a cosmological significance: the directions of the four corners from the center pole are the directions of sunrise and sunset at the solstices and, together with the pole itself, represent the four corners and center of the world.

A trapezoidal opening in the western or southwestern side of the roof doubles as a vent for smoke and as a source of light. It is closed in the wet season (when the roundhouse fills unpleasantly with smoke), but opened up in the dry season; at this time its orientation, well away from the southerly or southeasterly direction of the prevailing wind, ensures that only rarely will it catch the wind thus blocking the escaping smoke. However, this does not explain why a northerly or northeasterly orientation could not equally well be chosen in many cases. The reason for this may be that the opening originally had a third purpose, as an observation window. The beam of the afternoon sun crosses the interior of the house, its movement marking out the time of day and the time of year. The interior of the house is richly decorated with paintings depicting scenes and figures important in Yekuana myth, some of which are lit up by the beam of sunlight at certain times. There are structural alignments, too: thus on the winter solstice, the sunbeam reaches the northeastern corner of the internal rectangle.

The Yekuana roundhouse was truly a model of the cosmos. It may also, in some senses, have functioned as an observatory, in the sense that observations of sunlight through the roof opening could have been used as a clock or calendar. It may even have been used to observe the night sky. There is no ethnographic evidence to support this claim directly, although the same is known to have been the case among some indigenous North America groups, such as the Pawnee.

The Yekuana creation myth, and a very specific set of principles of construction conceived within the framework of that myth, ensure that the traditional roundhouse not only incorporates sound structural principles and pragmatics for living, but also reflects and reinforces the prevailing worldview. It is interesting to speculate on how much of the cosmological symbolism incorporated just in the overall structure of the roundhouse could have been guessed at by an archaeologist of the future if all historical records had been lost. The answer is probably very little, a sobering thought when we struggle to interpret, for example, roundhouses surviving in Europe from the Iron Age.

**See also:**
Cosmology; Solstitial Directions.
Iron Age Roundhouses; Navajo Hogan; Pawnee Earth Lodge.
Solstices.

**References and further reading**
Aveni, Anthony F. *Ancient Astronomers,* 146. Washington, DC: Smithsonian Books, 1993.
Ruggles, Clive, and Nicholas Saunders, eds. *Astronomies and Cultures,* 296–328. Niwot: University Press of Colorado, 1993.
Wilbert, Johannes. "Warao Cosmology and Yekuana Roundhouse Symbolism." *Journal of Latin American Lore* 7 (1981), 37–72.

## Zenith Passage of the Sun

If you are within the tropics—but only if you are within the tropics—you will have the opportunity to see the sun pass directly overhead. The highest point in the sky is known as the zenith, so this event is known technically as *solar zenith passage*. Solar zenith passage will happen at local noon on two separate days in the year, but what those days are will depend upon your latitude on the earth. Right on the Tropics themselves, it happens on just one day in the year: the June solstice at the Tropic of Cancer in the northern hemisphere, and the December solstice at the Tropic of Capricorn in the south. A little south of the Tropic of Cancer, the dates of solar zenith passage will be a little before and after the June solstice: the farther south you are, the further apart these dates will be. At the equator, the dates occur exactly six months apart, at the two equinoxes. Moving farther south again, the two dates gradually converge on the December solstice.

One way of visualizing this is to imagine a tightly coiled spring wrapped around the earth with its open ends on the two Tropics, and its 183 coils not quite following lines of latitude. There is always some point on the earth where the sun is currently overhead, and from the June solstice to the December solstice this point will trace slowly along the coils of the spring, going around once every twenty-four hours and moving inexorably southwards along the coils, until the December solstice is reached and the sun starts to move back northwards again.

Times when the sun passes close to the zenith stand out, because people and other upright objects cease to have shadows. Such times have sacred and practical significance in a variety of cultures within the tropics. A number of legends identify it as a time when the way is open into the upper world. In some Guatemalan villages, for example, the two dates of zenith passage coincide with springtime and August rains, and are marked by ritual observances that still live on in Christian tradition in the form of parades. In ancient Hawai'i, the moment of solar zenith passage was a time with great *mana,* or sacred power. It was a time when a person's shadow was no longer visible and was thought to have retreated directly into the brain through the top of the head; a person's spirit could exit at this time.

The coming of the days when the sun will pass through the zenith may be recognized in various ways, for example, by observing the rising or setting position of the sun on the horizon, or by watching for the heliacal rise or set of certain stars or asterisms. Sometimes we have evidence of devices that marked the actual moment of solar zenith passage, often in a dramatic way. Zenith tubes, which allowed the light of the sun to pass down into a dark place only at the time of zenith passage, are known at the Mesoamerican cities of Monte Alban and Xochicalco.

The zenith sun may hold one of the keys to the mystery of why Necker Island, a remote and uninhabitable rock lying well beyond the larger islands of the Hawaiian chain, is covered in temple platforms. It may have been an especially sacred place because it lay right on the Tropic of Cancer, at the very edge of the region where the sun reaches the zenith.

**See also:**
Antizenith Passage of the Sun; Equinoxes; Zenith Tubes.
Necker Island.
Heliacal Rise; Solstices.

**References and further reading**
Aveni, Anthony F. *Stairways to the Stars: Skywatching in Three Great Ancient Cultures,* 21–25. New York: Wiley, 1997.
———. *Skywatchers.* Austin: University of Texas Press, 2001.
Pukui, Mary K., E.W. Haertig, and Catherine A. Lee. *Nānā I Ke Kumu (Look to the Source), Volume I,* 123–124. Honolulu: Hui Hānai, 1972.

## Zenith Stars in Polynesia

The paths of the stars across the sky depend upon the observer's latitude. At any given latitude, only certain stars will pass directly overhead. Conversely, any given star will only be seen to pass overhead from one particular latitude on earth. For this to happen, the observer's latitude has to be equal to the "latitude" of the star on the celestial sphere—in other words, its declination.

The New Zealander David Lewis, in his efforts to trace surviving fragments of traditional methods of navigation in the Pacific, encountered a number of references to "the star on top," "the overhead star," "the star that points down upon an island," and the like. Taken together, these suggest the former existence of a Polynesian tradition in which different islands were seen to have distinctive "marker" stars. Most intriguingly, a Tahitian chant recorded in 1818 listed eight "pillars that hold up the heavens." The declinations of several of these stars, Lewis found, corresponded in C.E. 1000 (since they change slightly over the centuries, owing to precession) to the latitudes of various southern Polynesian islands, from the Line Islands in the north down to New Zealand in the south. The implication is that the overhead stars, marker stars, and pillar stars could have been zenith stars for particular islands.

The extent to which zenith stars were used by Pacific navigators is another question. Zenith stars cannot be used to plot a course, merely to determine the current latitude. Even to do the latter at all successfully, one has to identify the upward direction with reasonable accuracy, which is of course difficult on a swaying canoe. On the other hand, Lewis himself did achieve this reasonably satisfactorily by siting up the mast, and there are various documented references to zenith stars being used for navigation, particularly in Micronesia. However, even with zenith star observations as a back-up, Oceanic navigators would still have needed to use other methods (such as a combination of star compasses and dead reckoning) in order to locate distant islands.

**See also:**
Navigation in Ancient Oceania.
Celestial Sphere; Declination; Precession.

**References and further reading**
Finney, Ben. *Voyage of Rediscovery,* 62–63. Berkeley: University of California Press, 1994.
Hodson, F. R., ed. *The Place of Astronomy in the Ancient World,* 142–143. London: Royal Society, 1976.
Lewis, David. *We the Navigators* (2nd ed.), 278–290. Honolulu: University of Hawai'i Press, 1994.
Makemson, Maud. *The Morning Star Rises: An Account of Polynesian Astronomy,* 13–14. New Haven: Yale University Press, 1941.
Selin, Helaine, ed. *Astronomy across Cultures,* 105–106. Dordrecht, Neth.: Kluwer, 2000.

# Zenith Tubes

Most people who lived in the tropics in the past would have been broadly aware of the two times of year when the sun passed more or less directly overhead at noon and they ceased to have shadows. If they wished to determine the precise dates when zenith passage occurred, though, they would have needed to construct an instrument. A vertical, straight-sided post would have sufficed—a gnomon whose shadow would have dwindled to nothing at the appropriate time. So runs a plausible argument; but to reason in this way is to approach the issue from a modern Western perspective. If people had wished instead to harness the sacred power of such a moment, for whatever ideological or political ends, they would surely have created a hierophany. One way to achieve this would have been to construct a long vertical tube opening into a dark chamber below. This would create a breathtaking effect when the light of the zenith sun suddenly passed directly down the tube and, for a few precious moments, brought a sudden transformation to the normally dark space.

Two of the most discussed examples of possible zenith tubes are found in Mesoamerica. One is at the Early Classic site of Monte Alban, a little under four hundred kilometers (250 miles) southeast of Mexico City, close to the modern city of Oaxaca. Monte Alban was the capital city of the Zapotec state, which formed in the valley of Oaxaca around 500 B.C.E. and grew steadily in power and influence until it reached its apex around C.E. 300–600. The massive main plaza, built on a flattened hilltop, was its ceremonial heart. Visitors cannot fail to be impressed by the sheer size of this plaza, which measures more than three hundred meters by two hundred meters (1,000 feet by 650 feet). Nor can they fail to notice something extremely strange about Building J, situated on the central axis towards one end: it is wildly skewed with respect to all the other buildings. Rather than being oriented along the plaza like all the others, it is aligned directly upon Building P, one of several stepped buildings along the eastern side. And halfway up the steep steps of Building P is a curious feature. Here, reached through a narrow entrance, is a small room built directly under the steps, and at the back of it is a bench. Above the bench is a narrow vertical tube about 1.5 meters (five feet) high leading up to an opening higher up the staircase. Some have suggested that this tube could be merely a chimney. But several other features in the alignments at Monte Alban suggest that it could have been an astronomical observing device.

The other Mesoamerican zenith tube, at Xochicalco, is on an altogether different scale from the one at Monte Alban. Located in the state of Morelos, some seventy kilometers (forty-five miles) south of Mexico City and about fifteen kilometers southwest of the city of Cuernavaca, Xochicalco's situation—draped over a series of artificially flattened and terraced hilltops overlooking a deep valley over a hundred meters (330 feet) below—is nothing short of stunning. This city-state flowered early in the Epiclassic period, c. C.E. 700–900 following the collapse of Teotihuacan to the north and Monte Alban to the south. The tube here is over forty centimeters (sixteen inches) wide and descends for more than five meters (sixteen feet) before opening out into the roof in the far corner of a cave chamber almost twenty meters (sixty-five feet) long and twelve meters (forty feet) wide, dug out of the living rock. The chamber is supported by three rock pillars and reached by a network of subterranean passages.

The effect of the midday sun shining down the Xochicalco zenith tube is spectacular indeed, sending a vertical beam of bright sunlight down into the dark cave. But because of the width of the tube, this phenomenon is not confined to the days of solar zenith passage. In fact, the sun first shines di-

*Building J at Monte Alban, Mexico, showing its anomalous orientation, as viewed from Building P, upon which it is aligned. (Courtesy of Clive Ruggles)*

rectly down the tube, producing a brief spot of sunlight on the chamber floor, at local noon on April 30. As the sun passes closer to the true zenith at noon on successive days, the length of time when its light shines directly onto the floor increases, as does the proportion of its disc that is visible at maximum, until May 15, when it passes directly across the zenith at noon. After this, the length and proportion visible start to decrease again until the solstice is reached, when the noonday sunbeam has almost disappeared. The sequence is then repeated in reverse: zenith passage is reached on July 29, and the last spot of direct sunlight is seen on around August 12. A number of early descriptions dating back to the mid-nineteenth century reassure us that no dubious reconstructions have taken place and that these alignments are reliable.

Since it was first investigated astronomically, there have been many further suggestions about the Xochicalco tube. One is that this is just an approximate zenith device, but others argue that the dates of appearance and reappearance of noon sunlight are significant, since they fall exactly fifty-two days before and after the solstice respectively, and the fifty-two-day interval is calendrically significant. Furthermore, the length of the period of darkness—before the sun starts to appear at noon again—is (within a day or two of) 260 days, exactly the length of the sacred calendar cycle or *tonalpohualli*. Still others suggest that the moon might have been at least as important as the sun at Xochicalco, since at minor standstill it passes directly through the zenith at this latitude. The idea is given some credence by the discovery, in excavations during the 1980s, of a unique crescent-moon-shaped stele.

We cannot be certain that the Monte Alban and Xochicalco zenith tubes were really used for the purposes of astronomical observation, but a good deal of circumstantial evidence supports this view. Furthermore, the existence of solar zenith passage devices like this seems plausible enough in the context of Mesoamerican worldview, in which (with variations) the up-down axis is a fundamental means of orientation, together with the four cardinal directions. The importance of the zenith sun is also well attested in historical accounts, iconography, and ethnography. The effect of sunlight entering the underground chambers is certainly spectacular, and it is hardly surprising that the Xochicalco "observatory" has become a popular tourist attraction. If we are lucky, other examples still wait to be discovered in the future, in Mesoamerica and elsewhere.

See also:
Zenith Passage of the Sun.
Mesoamerican Calendar Round; Teotihuacan Street Grid.
Moon, Motions of.

**References and further reading**
Aveni, Anthony F. *Skywatchers,* 262–271. Austin: University of Texas Press, 2001.
———, ed. *World Archaeoastronomy,* 167–179. Cambridge: Cambridge University Press, 1989.
———. "Zapotec Astronomy: Reconsideration of an Earlier Study." *Archaeoastronomy: The Journal of Astronomy in Culture* 18 (2004), 26–31.
Aveni, Anthony F., and Horst Hartung. "The Observation of the Sun at the Time of Passage through the Zenith in Mesoamerica." *Archaeoastronomy* 3 (supplement to *Journal for the History of Astronomy* 12 [1981], S51–S70.
De la Fuente, Beatriz, Silvia Garza Tarazona, Norberto González Crespo, Arnold Lebeuf, Miguel León Portilla, and Javier Wimer. *La Acrópolis de Xochicalco,* 210–287. Mexico City: Instituto de Cultura de Morelos, 1995. [In Spanish.]

# Zodiacs
*See* Ecliptic.

# Bibliography

This general bibliography contains the principal sources used in compiling the entries, together with a number of additional works. Some focus upon broad themes and issues in cultural astronomy and typically include a range of case studies; some relate to particular cultures with a greater or lesser emphasis on practices relating to the sky; and some serve to put such astronomy in a broader cultural context. Others provide essential information for those wishing to visit and investigate particular monuments first hand: this category includes serious guides and gazetteers. The list also includes a small number of works that introduce background concepts and information, both from astronomy and from the human sciences. Journal articles are included only where they address broad themes and include material of which summaries cannot readily be obtained in any available book. Likewise, web sites are cited only in exceptional cases, because they do not constitute a permanent medium. A small number of the cited works describe the development of this field of study, or themselves form landmarks within that development. The cited works are in English unless otherwise stated.

The reader should approach the literature on ancient astronomy with caution. Because of its interdisciplinary nature, peppered with popular misconceptions, the subject is particularly prone to sensationalism and uncritical scholarship. These creep all too frequently into academic publications from which they can easily become propagated further in books directed at a wider audience. Collective works and conference proceedings may, likewise, contain articles varying widely in content, approach, and quality. Short notes are included against many of the works listed in the bibliography, in an attempt to help the reader navigate this minefield and to select reliable and authoritative sources, or else to approach the wider available literature with a duly critical eye. Various encyclopedia entries elaborate further on some of the main problems and issues.

## Journals and Series

There are a small number of journals devoted to cultural astronomy. The following lists the principal ones. Academic papers on ancient astronomy also regularly appear in a broad variety of academic journals spanning a wide range of disciplines.

*Archaeoastronomy*, the bulletin, subsequently journal, of the Center for Archaeoastronomy, College Park, MD. Published by the Center for Archaeoastronomy. Vols. 1 (1977)–11 (1993). [Vols. 12–13 (1996) published as *Songs from the Sky* (see Chamberlain et al. 2005 below). Superseded by the University of Texas Press journal, see immediately below.]

*Archaeoastronomy: The Journal of Astronomy in Culture*. Vols. 14 (1999) to date. Published by the University of Texas Press.

*Archaeoastronomy*, the supplement to the *Journal for the History of Astronomy*. Nos. 1 (1979)–27 (2002). Published by Science History Publications, Cambridge, England. [Archaeoastronomical articles may also be found in the parent *Journal for the History of Astronomy*, particularly before 1979 and since 2003.]

*Astronomie et Sciences Humaines*, Nos. 1 (1988)–7 (1991). Published by l'Observatoire Astronomique de Strasbourg. [In French. Superseded by Jaschek 1992 and annual publications from SEAC meetings.]

*Culture and Cosmos*. Vols. 1 (1997) to date. Published by Culture and Cosmos, Bristol, England. [Includes archaeoastronomy and historical astronomy, but the main emphasis is upon the history of astrology. Vol. 8 (2004) published as *The Inspiration of Astronomical Phenomena* (see Campion 2005 below).]

*Journal for the History of Astronomy*. Vols. 1 (1970) to date. Published by Science History Publications, Cambridge, England. [Includes archaeoastronomy, but between 1979 and 2002, archaeoastronomical articles are mainly to be found in the supplement *Archaeoastronomy*.]

*Rivista Italiana di Archeoastronomia*, Vols. 1 (2003) to date. Published by Edizioni Quasar di Severino Tognon, Rome. [In Italian.]

*The "Oxford" series of international symposia on archaeoastronomy are held at intervals of three to five years and are the principal conferences in the field. These have been published as follows (for full details cross-refer to the general bibliography below):*

Oxford, England, 1981 (see Heggie 1982 and Aveni 1982).
Mérida, Mexico, 1986 (see Aveni 1989).
St. Andrews, Scotland, 1990 (see Ruggles and Saunders 1993; Ruggles 1993).
Stara Zagora, Bulgaria, 1993 (as yet unpublished).
Santa Fe, New Mexico, USA, 1996 (see Fountain and Sinclair 2005).
La Laguna, Tenerife (Spain), 1999 (see Esteban and Belmonte 2000; a number of keynote papers were published separately in vol. 15 of *Archaeoastronomy: The Journal of Astronomy in Culture*).
Flagstaff, Arizona, USA, 2004 (as yet unpublished).

*In Europe, the publications of annual conferences on archaeoastronomy since 1988 have formed a diverse but as yet unbroken series. Since 1993 these have been produced under the auspices of the European Society for Astronomy in Culture (SEAC). For full details cross-refer to the general bibliography that follows.*

Dobrič [then Tolbukhin], Bulgaria, 1988 (see Valev 1989).
Venice, Italy, 1989 (see Romano and Traversari 1991).
Warsaw, Poland, 1990 (see Iwaniszewski 1992).
Székesfehérvár, Hungary, 1991 (see Pásztor 1995).
Strasbourg, France, 1992 (see Jaschek 1992).
Smolyan, Bulgaria, 1993 (see Koleva and Kolev 1996).
Bochum, Germany, 1994 (see Schlosser 1996).
Sibiu, Romania, 1995 (see Stănescu 1999).
Salamanca, Spain, 1996 (see Jaschek and Atrio 1997).
Gdańsk, Poland, 1997 (see Lebeuf and Ziółkowski 1999).
Dublin, Ireland, 1998 (see Ruggles, Prendergast, and Ray 2001).
La Laguna, Tenerife (Spain), 1999 (see Esteban and Belmonte 2000; a number of keynote papers were published separately in vol. 15 of *Archaeoastronomy: The Journal of Astronomy in Culture*).
Moscow, Russia, 2000 (see Potyomkina and Obridko 2002).
Stockholm, Sweden, 2001 (see Blomberg, Blomberg, and Henriksson 2003).

Tartu, Estonia, 2002 (see Kõiva, Mürk, and Pustõlnik in press).
Leicester, England, 2003 (will be published jointly with the 2004 conference).
Kecskemét, Hungary, 2004 (as yet unpublished).
Isili, Sardinia (Italy), 2005 (as yet unpublished).

## General Bibliography

Aaboe, Asger. *Episodes from the Early History of Astronomy*. New York: Springer, 2001.

Aitken, Michael. *Science-Based Dating in Archaeology*. London: Longman, 1990. [An introduction to archaeological dating methods aimed at students in archaeological science.]

Åkerblom, Kjell. *Astronomy and Navigation in Polynesia and Micronesia*. Stockholm: Stockholm Ethnographical Museum, 1968.

Allen, Richard H. *Star Names: Their Lore and Meaning*. New York: Dover, 1963. (First published 1899.) [Compendium of star-name information from all over the globe. Of historical interest but also still useful.]

Altuna, Jesús, Angel Armendariz, Luis del Barrio, Francisco Etxeberria, Koro Mariezkurrena, Xavier Peñalver, and Franciscop Zumalabe. *Gipuzkoa Karta Arkeologikoa: I. Megalitoak*. Donostia [San Sebastián], Spain: MUNIBE, 1990. [Comprehensive set of maps, plans, and descriptions of megalithic monuments in the Gipuzkoa region of the Basque country. In Basque and Spanish.]

Antequera, Luz, Antonio Aparicio, Juan A. Belmonte, José R. Belmonte, César Estéban, Michael Hoskin, and Amador Rebullida. *Arqueoastronomía Hispana*. Madrid: Equipo Sirius, 1994. [A survey of archaeoastronomy in various regions of Spain. In Spanish.]

Armit, Ian. *The Archaeology of Skye and the Western Isles*. Edinburgh, Edinburgh University Press, 1996. [A good archaeological background on the Isle of Skye and the Outer Hebrides in Scotland.]

Ascher, Marcia, and Robert Ascher. *Code of the Quipu*. Ann Arbor: University of Michigan Press, 1981. [An early classic in the study of the knotted-string devices used by the Incas.]

Ashmore, Patrick. *Calanais: The Standing Stones*. Stornoway, Isle of Lewis, Scotland: Urras nan Tursachan, 1995. [A short introduction to the standing stones of Calanais/Callanish, including mention of astronomy, with nice illustrations.]

———. *Neolithic and Bronze Age Scotland*. London: Batsford/Historic Scotland, 1996. [An authoritative but accessible account organized around the best dating evidence available at the time of writing. It is divided into 500-year chunks running from 4000 B.C.E. to 1000 B.C.E., plus a final 250-year chunk at the threshold of the Iron Age.]

Ashmore, Wendy, and Bernard Knapp, eds. *Archaeologies of Landscape: Contemporary Perspectives*. Oxford: Blackwell, 1999. [A wide-ranging collection of papers exploring cultural perceptions of landscape through archaeological, historical, and ethnographic evidence.]

Atkinson, Richard J. C. *Stonehenge*. Harmondsworth: Penguin Books, 1979. [This was the final reprint of Atkinson's popular account of his excavations at Stonehenge, first published in 1956, which influenced all the astronomical interpretations in the 1960s. The full account of Atkinson's excavation was eventually published within Cleal et al. 1995.]

Aveni, Anthony F. *Empires of Time: Calendars, Clocks and Cultures*. New York: Basic Books, 1989. [An excellent thematic overview.]

———. *Conversing with the Planets.* New York: Times Books, 1992. [A wide-ranging account of mythology, astronomy, and astrology relating to the planets for a general audience.]

———. *Ancient Astronomers.* Washington, DC: Smithsonian Books, 1993. [A broad range of examples, richly illustrated.]

———. *Stairways to the Stars: Skywatching in Three Great Ancient Cultures.* New York: Wiley, 1997. [An introductory text contrasting three case studies—British megaliths, ancient Maya astronomy, and the Venus cult—and astronomy in the Inca empire. In doing so it introduces a number of key issues and themes. It also contains a useful initial chapter on terms and concepts and an appendix on fieldwork techniques. The "megalithic" chapter contains much that has now been superseded.]

———. *Between the Lines: The Mystery of the Giant Ground Drawings of Ancient Nasca, Peru.* Austin: University of Texas Press, 2000. (Published in the UK as *Nasca: The Eighth Wonder of the World.* London: British Museum Press, 2000.) [An account of recent work at Nasca directed at a general audience.]

———. *Skywatchers.* Austin: University of Texas Press, 2001. [A comprehensive and up-to-date survey of Mesoamerican archaeoastronomy, with briefer sections dealing with other parts of the Americas and elsewhere. Extensively revised version of *Skywatchers of Ancient Mexico,* published in 1980.]

———. *The Book of the Year: A Brief History of Our Seasonal Holidays.* Oxford: Oxford University Press, 2003.

———, ed. *Archaeoastronomy in Pre-Columbian America.* Austin: University of Texas Press, 1975. [A collection of papers deriving from an early conference on archaeoastronomy in the Americas.]

———, ed. *Native American Astronomy.* Austin: University of Texas Press, 1977. [A collection of papers deriving from an early conference on archaeoastronomy in the Americas.]

———, ed. *Archaeoastronomy in the New World.* Cambridge: Cambridge University Press, 1982. [A collection of "new world" papers from the first "Oxford" international symposium on archaeoastronomy held in England in 1981.]

———, ed. *World Archaeoastronomy.* Cambridge: Cambridge University Press, 1989. [A collection of papers from the second "Oxford" international symposium on archaeoastronomy held in Mexico in 1986.]

———, ed. *The Lines of Nazca.* Philadelphia: American Philosophical Society, 1990. [A collection of academic papers deriving from the fieldwork of Aveni et al. at Nasca in the early 1980s.]

———, ed. *The Sky in Mayan Literature.* New York: Oxford University Press, 1992. [A collection of papers from a workshop held in New York in 1989.]

Aveni, Anthony F., and Gordon Brotherston, eds. *Calendars in Mesoamerica and Peru: Native American Computations of Time.* Oxford: British Archaeological Reports (International Series 174), 1983. [A collection of papers from a conference held in Manchester, England, in 1982.]

Aveni, Anthony F., and Horst Hartung. *Maya City Planning and the Calendar.* Philadelphia: American Philosophical Society (Transactions, vol. 76, part 7), 1986.

Aveni, Anthony F., and Gary Urton, eds. *Ethnoastronomy and Archaeoastronomy in the American Tropics.* New York: New York Academy of Sciences, 1982. [A collection of papers arising from a conference held in New York in 1981.]

Bahn, Paul, ed. *The Cambridge Illustrated History of Archaeology.* Cambridge: Cambridge University Press, 1996.

Bailey, Mark, Victor Clube, and Bill Napier. *The Origin of Comets.* London: Pergamon Press, 1990. [Astronomers argue that the earth has had periodic encounters with large comets.]

Baillie, Mike. *From Exodus to Arthur: Catastrophic Encounters with Comets.* London: Batsford, 1999. [A leading dendrochronologist argues that cometary encounters may explain "cold summers" evident from tree-ring evidence.]

Barba de Piña Chán, Beatriz, ed. *Caminos Terrestres al Cielo.* Mexico City: INAH, 1998. [A collection of papers on pilgrimage in Mexico spanning the pre-Hispanic, colonial, and modern eras. In Spanish.]

Barclay, Alastair, and Jan Harding, eds. *Pathways and Ceremonies: The Cursus Monuments of Britain and Ireland.* Oxford: Oxbow Books, 1999. [A collection of archaeological papers concerning the Neolithic cursus monuments found in Britain and Ireland, providing an essential background to discussions of their possible astronomical significance.]

Barclay, Gordon. *Farmers, Temples and Tombs: Scotland in the Neolithic and Early Bronze Age.* Edinburgh: Canongate/Historic Scotland, 1998. [An excellent short introduction, nicely illustrated.]

Barclay, Gordon, and Gordon Maxwell. *The Cleaven Dyke and Littleour: Monuments in the Neolithic of Tayside.* Edinburgh: Society of Antiquaries of Scotland, 1998. [A broad-ranging archaeological investigation of two Neolithic monuments in central Scotland, including the possible astronomical significance of the linear earthwork known as the Cleaven Dyke.]

Barlai, Katalin, and Ida Bognár-Kutzián, eds. *"Unwritten Messages" from the Carpathian Basin.* Budapest: Konkoly Observatory, 2002. [Papers from a small conference held in Hungary in 2000, with a particular focus on grave and church orientations in central Europe.]

Barnatt, John. *Prehistoric Cornwall: The Ceremonial Monuments.* Wellingborough, UK: Turnstone Press, 1982. [A field guide to megalithic monuments in Cornwall, England, including many useful site plans.]

———. *Stone Circles of Britain: Taxonomic and Distributional Analyses and a Catalogue of Sites in England, Scotland and Wales* (in two volumes). Oxford: British Archaeological Reports. (BAR British Series 215), 1989. [Detailed tabulated data on British stone circles.]

Barrère, Dorothy B., Mary K. Pukui, and Marion Kelly. *Hula: Historical Perspectives.* Honolulu: Bishop Museum Press, 1980. [Contains background information on hula performances and on the hula platform at Ke-Ahu-a-Laka, Kauaʻi, relevant to the interpretation of the Nā Pali chant.]

Barrett, John, Richard Bradley, and Martin Green. *Landscape, Monuments and Society.* Cambridge: Cambridge University Press, 1991.

Barthel, Thomas. *The Eighth Land: The Polynesian Discovery and Settlement of Easter Island.* Honolulu, University Press of Hawaiʻi, 1978. [A much-criticized attempt to reconstruct how Polynesians arrived on Easter Island.]

Bauer, Brian. *The Sacred Landscape of the Inca: The Cusco Ceque System.* Austin: University of Texas Press, 1998. [Presents the results of a project to document the location in the landscape of the Cusco *ceques* and the sacred places (*huacas*) associated with them, and thereby to learn more about the spatial aspects of social organization in Incaic Cusco.]

Bauer, Brian, and David Dearborn. *Astronomy and Empire in the Ancient Andes.* Austin: University of Texas Press, 1995. [Inca cosmology and calendar explored by an archaeologist and an astronomer working together.]

Bauval, Robert, and Adrian Gilbert. *The Orion Mystery: Unlocking the Secrets of the Pyramids.* London: Heinemann, 1994. [Speculative theory that has been met with extensive criticism from academics.]

Beaglehole, J. C. *The Life of Captain James Cook.* Stanford, CA: Stanford University Press, 1974. [A classic account of Cook's life and voyages.]

Beckwith, Martha W. *Kepelino's Traditions of Hawaii*. Honolulu: Bishop Museum Press, 1932.

———. *Hawaiian Mythology*. Honolulu: University of Hawai'i Press, 1970. Originally published in 1940 by Yale University Press.

———. *The Kumulipo: A Hawaiian Creation Chant* (repr.). Honolulu: University of Hawai'i Press, 1972. [The classic translation of and commentary upon a sacred Hawaiian creation chant of extreme cultural importance, first published in 1951.]

Bellwood, Peter. *The Polynesians: Prehistory of an Island People* (rev. ed.). London: Thames and Hudson, 1987. [A synthesis of Polynesian prehistory now largely superseded by Kirch 2000.]

Belmonte Avilés, Juan Antonio. *Las Leyes del Cielo: Astronomía y Civilizaciones Antiguas*. Madrid: Ediciones Temas de Hoy, 1999. [A wide-ranging thematic introduction to ancient astronomy. In Spanish.]

———. *Tiempo y Religión: Una Historia Sagrada del Calendario*. Madrid: Ediciones del Orto, 2005. [In Spanish.]

Belmonte Avilés, Juan Antonio, and Michael Hoskin. *Reflejo del Cosmos: Atlas de Arqueoastronomía del Mediterráneo Antiguo*. Madrid: Equipo Sirius, 2002. [A very useful guide to archaeoastronomical sites around the western Mediterranean. In Spanish.]

Belmonte Avilés, Juan Antonio, and Margarita Sanz de Lara Barrios. *El Cielo de los Magos: Tiempo, Astronómico y Meteorológico en la Cultura Tradicional del Campesinado Canario*. Santa Cruz de Tenerife and Las Palmas de Gran Canaria: La Marea, 2001. [Study of traditional knowledge of the skies among old people in the Canary Islands. In Spanish.]

Benigni, Helen, Barbara Carter, and Éadhmonn Ua Cuinn. *The Myth of the Year: Returning to the Origin of the Druid Calendar*. Lanham, MD: University Press of America, 2003. [A reconstruction of Celtic myth and calendar by a writer, an astrologer, and a sculptor.]

Bertola, Francesco. *Via Lactea: Un Percorso nel Cielo e nella Storia dell'Uomo*. Cittadella: Biblos, 2003. [Beautifully illustrated coffee-table book. In Italian and English.]

Bertola, Francesco, et al., eds. *Archeologia e Astronomia: Esperienze e Prospettive Future*. Rome: Accademia Nazionale dei Lincei (Atti dei Convegni Lincei, 121), 1995. [A collection of papers presented at a convention held at the Lincean Academy, Rome, in 1994. Contains articles in Italian, English, and French.]

———. *Archeoastronomia, Credenze e Religioni nel Mondo Antico*. Rome: Accademia Nazionale dei Lincei (Atti dei Convegni Lincei, 141), 1998. [A collection of papers presented at a convention held at the Lincean Academy, Rome, in 1997. Contains articles in Italian and English.]

———. *L'Uomo Antico e il Cosmo*. Rome: Accademia Nazionale dei Lincei (Atti dei Convegni Lincei, 171), 2001. [A collection of papers presented at a convention held at the Lincean Academy, Rome, in 2000. Contains articles in Italian and Spanish.]

Best, Elsdon. *The Astronomical Knowledge of the Maori*. Wellington: Dominion Museum (Monograph, no. 3), 1922. [A classic that remains one of the most important sources of information on Maori astronomical knowledge compiled in the early twentieth century.]

———. *The Maori Division of Time*. Wellington: Dominion Museum (Monograph, no. 4), 1922. [A classic, still one of the most important sources of information on Maori calendrical knowledge as it remained in the early twentieth century.]

Bialas, Volker. *Astronomie und Glaubensvorstellungen in der Megalithkultur: Zur Kritik der Archäoastronomie*. München: Verlag der Bayerischen Akademie der

Wissenschaften, 1988. [A critique of "megalithic astronomy" as practised up to the mid-1980s. In German.]

Birmingham, Robert, and Leslie Eisenberg. *Indian Mounds of Wisconsin*. Madison: University of Wisconsin Press, 2000. [A broad survey and suggested framework for interpretation of the mound sites in the state of Wisconsin, including a list of selected sites open to the public.]

Blackburn, Bonnie, and Leofranc Holford-Stevens. *The Oxford Companion to the Year: An Exploration of Calendar Customs and Time-Reckoning*. Oxford: Oxford University Press, 1999.

Blomberg, Mary, Peter Blomberg, and Göran Henriksson, eds. *Calendars, Symbols and Orientations: Legacies of Astronomy in Culture*. Uppsala: Uppsala Astronomical Observatory, 2003. [A collection of papers from the SEAC (European Society for Astronomy in Culture) meeting held in Stockholm in 2001. Contains articles in English and French.]

Blot, Jacques. *Montaña y Prehistoria Vasca*. Donostia [San Sebastián], Spain: Elkar. [A nicely illustrated overview of prehistory in the Basque country, including descriptions of the principal monuments. In Spanish.]

Boccas, Maxime, Johanna Broda, and Gonzalo Pereira, eds. *Etno- y Arqueoastronomía en las Américas*. Santiago, Chile: Congreso Internacional Americanista, 2004. [A collection of papers from a session at the International Congress of Americanists in Chile in 2003. In Spanish.]

Bol, Marsha C., ed. *Stars Above, Earth Below: American Indians and Nature*. Niwot, CO: Roberts Rinehart/Carnegie Museum of Natural History, 1998. [Includes an overview of American Indian astronomy.]

Boone, Elizabeth H. *The Aztec Templo Mayor*. Washington DC: Dumbarton Oaks, 1987. [A collection of scholarly papers from a 1983 symposium, including a number focusing on ritual and cosmology.]

Bradley, Richard. *Altering the Earth*. Edinburgh: Society of Antiquaries of Scotland, 1993. [A series of lectures by a leading British archaeologist concerning European later prehistoric monuments and their interpretation, and including relationships to the sky.]

———. *Rock Art and the Prehistory of Atlantic Europe: Signing the Land*. London: Routledge, 1997. [An interpretation of the perceived meaning and purpose of prehistoric rock carvings in relation to their landscape setting.]

———. *The Significance of Monuments: On the Shaping of Human Experience in Neolithic and Bronze Age Europe*. London: Routledge, 1998. [A set of linked essays developing ideas concerning the development of tombs and ceremonial monuments in northwest Europe during the Neolithic and Early Bronze Age, the latter part focusing on a discussion of the ubiquitous circular form.]

———. *An Archaeology of Natural Places*. London: Routledge, 2000. [An exploration by a leading British prehistorian of the significance to the people in the past of natural, unaltered places in the landscape.]

———. *The Good Stones: A New Investigation of the Clava Cairns*. Edinburgh: Society of Antiquaries of Scotland (Monograph Series 17), 2000. [An archaeological investigation of the Clava cairns, providing an essential background to discussions of their possible astronomical significance.]

Brandt, John C., and Robert D. Chapman. *Introduction to Comets*, 1–54. Cambridge: Cambridge University Press, 2004. [The first part of this book is an account of historical perceptions and sightings of comets.]

Brennan, Martin. *The Stones of Time: Calendars, Sundials and Stone Chambers of Ancient Ireland*. Rochester, VT: Inner Traditions International, 1994. (Originally published in 1983 as *The Stars and the Stones: Ancient Art and Astronomy in*

*Ireland* by Thames and Hudson, London.) [Identifies numerous astronomical associations in Irish monuments and megalithic art, but has drawn severe scholarly criticism for being methodologically uncritical.]

Briard, Jacques, Maurine Gautier, and Gilles Leroux. *Les Mégalithes de Saint-Just.* Luçon, France: Éditions Jean-Paul Gisserot, 1993. [Colorful and helpful short guide to monuments in the St. Just area of Brittany. In French.]

Broda, Johanna, Davíd Carrasco, and Eduardo Matos Moctezuma. *The Great Temple of Tenochtitlan: Center and Periphery in the Aztec World.* Berkeley: University of California Press, 1987. [An excellent analysis of the Aztec Templo Mayor examined from three different scholarly perspectives.]

Broda, Johanna, Stanisław Iwaniszewski, and Lucretia Maupomé, eds. *Arqueoastronomía y Etnoastronomía en Mesoamérica.* Mexico City: Universidad Nacional Autónoma de México, 1991. [An extensive collection of papers covering both Meso- and North America, from a conference held in Mexico in 1989. Contains articles in Spanish and English.]

Broda, Johanna, Stanisław Iwaniszewski, and Arturo Montero, eds. *La Montaña en el Paisaje Ritual.* Mexico City: CONACULTA/INAH, 2001. [A general collection of papers on mountains and sacred geography, largely concentrated in the Basin of Mexico, including some articles that focus upon connections with calendrics and astronomy. In Spanish.]

Brown, Dayle L. *Skylore from Planet Earth: Stories from Around the World . . . Orion.* Bloomington, IN: AuthorHouse, 2004. [A collection for children of illustrated stories from around the world concerning the constellation that we call Orion. Supplemented with parent/teacher notes.]

Brunod, Giuseppe, Walter Ferreri, and Gaudenzio Ragazzi. *La Rosa di Sellero e la Svastica: Cosmologia, Astronomia, Danze Preistoriche.* Savigliano, Italy: I Quaderni di «Natura Nostra», 1999. [Highly speculative and methodologically questionable interpretations of Italian rock art. In Italian.]

Buck, Peter H. [Te Rangi Hiroa]. *Ethnology of Mangareva.* Honolulu: Bishop Museum Press, 1938.

Burl, Aubrey. *The Stone Circles of the British Isles.* New Haven: Yale University Press, 1976. [A detailed but highly readable account of British and Irish stone circles, aimed at a mixed audience, and including a comprehensive gazetteer. Achieved remarkable sales in the 1970s and 1980s. Now superseded by Burl 2000.]

———. *Prehistoric Avebury.* New Haven: Yale University Press, 1979. [Eminently readable but now superseded by more recent accounts.]

———. *Rings of Stone: The Prehistoric Stone Circles of Britain and Ireland.* London: Frances Lincoln, 1979. [Popular but now dated account of British stone circles including some discussion of orientation and astronomy. Includes the only reasonably detailed account of an excavation at Berrybrae recumbent stone circle, where scatters of white quartz fragments found near the recumbent stone reinforce the idea of a lunar significance.]

———. *Rites of the Gods.* London: Dent, 1981. [Popular account of ritual and religion in later prehistoric Britain, including some discussion of monumental orientation and astronomy. Now rather dated.]

———. *Prehistoric Astronomy and Ritual.* Princes Risborough, UK: Shire, 1983. [A short and easy-to-read introduction to British "megalithic astronomy," now somewhat dated.]

———. *Megalithic Brittany.* London: Thames and Hudson, 1985. [By far the best guidebook available in the English language.]

———. *Four-Posters: Bronze Age Stone Circles of Western Europe.* Oxford: British Archaeological Reports (BAR British Series 195), 1988. [Gazetteer containing detailed data of a distinctive type of British megalithic monument identified as such by Burl himself. Also includes possible examples in Ireland and Brittany.]

———. *From Carnac to Callanish: The Prehistoric Stone Rows and Avenues of Britain, Ireland and Brittany.* New Haven: Yale University Press, 1993. [The "stone rows" equivalent of Burl's highly successful *The Stone Circles of the British Isles* (see Burl 1976).]

———. *A Guide to the Stone Circles of Britain, Ireland and Brittany.* New Haven: Yale University Press, 1995. [Excellent guidebook.]

———. *Prehistoric Stone Circles.* Princes Risborough, UK: Shire, 1997. [Short introduction for a general audience.]

———. *Great Stone Circles.* New Haven: Yale University Press, 1999. [An account of a selection of English stone circles aimed at a general audience and beautifully illustrated.]

———. *The Stone Circles of Britain, Ireland and Brittany.* New Haven: Yale University Press, 2000. [An extensively revised version of Burl 1976, extended to include Brittany.]

Cairns, Hugh, and Bill Yidumduma Harney. *Dark Sparklers: Yidumduma's Wardaman Aboriginal Astronomy.* Merimbula, NSW: H.C. Cairns, 2003. [A remarkable account of celestial knowledge in a modern Aboriginal community in Australia's Northern Territory, taught by means of "songlines" that cross the sky.]

Calledda, Pino, and Giorgio Murru, eds. *Archeologia e Astronomia: Esperienze e Confronti.* Cagliari, Sardinia, Italy: Cooperativa Universitaria Editrice Cagliaritana, 2000. [A small collection of papers from a conference held in 1998, focusing mainly on Sardinia. In Italian; some abstracts in English.]

Campion, Nicholas. *The Great Year: Astrology, Millenarianism and History in the Western Tradition.* London: Arkana/Penguin, 1994.

———, ed. *The Inspiration of Astronomical Phenomena: Proceedings of the Fourth Conference on the Inspiration of Astronomical Phenomena, Magdalen College, Oxford, 3–9 August 2003.* Bristol: Cinnabar Books, 2005. [A collection of papers on a broad sweep of topics, including cultural astronomy.]

Carlson, John B. "America's Ancient Skywatchers." *National Geographic Magazine,* 177 (3) (1990), 76–107.

———, ed. *Pilgrimage and the Ritual Landscape in Pre-Columbian America.* Washington DC: Dumbarton Oaks, in press.

Carlson, John B., and W. James Judge, eds. *Astronomy and Ceremony in the Prehistoric Southwest.* Albuquerque, NM: Maxwell Museum of Anthropology, 1987.

Carmichael, David, Jane Hubert, Brian Reeves, and Audhild Schanche, eds. *Sacred Sites, Sacred Places.* London: Routledge, 1994.

Carr, Christopher, and D. Troy Case, eds. *Gathering Hopewell: Society, Ritual, and Ritual Interaction.* New York: Kluwer, 2004.

Carrasco, Davíd. *Religions of Mesoamerica.* San Francisco: Harper and Row, 1990. [Overview of Mesoamerican history and cosmovisión from the perspective of a leading historian of religions.]

———, ed. *To Change Place: Aztec Ceremonial Landscapes.* Niwot: University Press of Colorado, 1991. [A collection of papers from a 1989 symposium.]

———, ed. *The Oxford Encyclopedia of Mesoamerican Cultures: The Civilizations of Mexico and Central America* (3 vols.). New York: Oxford University Press, 2000. [This excellent encyclopedia contains several useful articles on, or relevant to, Mesoamerican archaeoastronomy. The entries on "Astronomy" and "Festivals and Festival Cycles" make good starting points.]

Castleden, Rodney. *The Making of Stonehenge.* London: Routledge, 1993. [Contains a number of controversial ideas.]

Chamberlain, Von Del. *When Stars Came Down to Earth: Cosmology of the Skidi Pawnee Indians of North America.* Los Altos, CA, and College Park, MD: Ballena Press/Center for Archaeoastronomy, 1982. [A detailed ethnoastronomical study by a professional astronomer.]

Chamberlain, Von Del, John Carlson, and Jane Young, eds. *Songs from the Sky: Indigenous Astronomical and Cosmological Traditions of the World.* Bognor Regis, UK: Ocarina Books, and College Park, MD: Center for Archaeoastronomy, 2005. [This extensive collection of papers derives ultimately from the first, and to date only, international conference on ethnoastronomy of worldwide scope, held in Washington, DC, in 1983.]

Champion, Timothy, and John Collis, eds. *The Iron Age in Britain and Ireland: Recent Trends.* Sheffield, England: Sheffield Academic Press, 1996.

Chapman, Robert. *Emerging Complexity: The Later Prehistory of South-East Spain, Iberia and the West Mediterranean.* Cambridge: Cambridge University Press, 1990.

Charles, Mary. *Winin: Why the Emu Cannot Fly.* Broome, Australia: Magabala Books Aboriginal Corporation, 2000. [A children's book describing the emu and other Aboriginal constellations.]

Chippindale, Christopher. *Stonehenge Complete* (rev. ed.). London: Thames and Hudson, 1994. [A compendium of historical interpretations and anecdotes concerning Stonehenge. Consigns nearly all astronomical interpretations to the historical dustbin.]

Chippindale, Christopher, and Paul Taçon, eds. *The Archaeology of Rock Art.* Cambridge: Cambridge University Press, 1998. [A collection of papers themed by the division between "informed" and "formal" approaches to the interpretation of rock art, the latter applying where there exist no insights from living or historical informants.]

Clagett, Marshall. *Ancient Egyptian Science: A Source Book, Vol. 2: Calendars, Clocks, and Astronomy.* Philadelphia: American Philosophical Society, 1995. [One of a set of books by this author addressing different aspects of ancient Egyptian science.]

Cleal, Rosamund, Karen Walker, and R. Montague. *Stonehenge in Its Landscape: Twentieth-Century Excavations.* London: English Heritage, 1995. [The definitive report on excavations at Stonehenge.]

Coe, Michael D. *The Maya* (4th ed). London: Thames and Hudson, 1987. [A classic introduction to Maya history and culture.]

———. *Breaking the Maya Code* (rev. ed.). New York: Thames and Hudson, 1999. [The enthralling story of the deciphering of Maya hieroglyphic script.]

———. *Angkor and the Khmer Civilization.* London: Thames and Hudson, 2003. [A highly readable and informative overview by a much respected scholar.]

Coles, John, in association with Bo Gräslund. *Patterns in a Rocky Land: Rock Carvings in South-West Uppland, Sweden* (in two volumes). Uppsala: Department of Archaeology and Ancient History, 2000. [The second volume contains meticulous drawings of many scores of rock carvings in the study area.]

Collis, John. *The European Iron Age.* London: Routledge, 1997. [Good background text.]

Constanza Ceruti, María. *Cumbres Sagradas del Noroeste Argentino.* Buenos Aires: Editorial Universitaria de Buenos Aires, 1999. [An astonishing account of sacred offerings including infant burials discovered on high mountain summits in the Andes. In Spanish.]

Cooney, Gabriel. *Landscapes of Neolithic Ireland.* London, Routledge, 2000. [A study by one of Ireland's leading prehistorians of the Irish Neolithic as evidenced through material artifacts, houses, monuments, and landscapes.]

Cooney, Gabriel, and Eoin Grogan. *Irish Prehistory: A Social Perspective*. Dublin: Wordwell, 1994. [A very useful overview of Irish prehistory with a particular emphasis on landscape.]

Cornell, James. *The First Stargazers: An Introduction to the Origins of Astronomy*. London: Athlone, 1981. [One of many books produced in the 1970s and early 1980s to introduce archaeoastronomy to a general audience. Badly dated in places.]

Coyne, George V., S. J. Sinclair, and Rolf M. Sinclair, eds. "The Inspiration for Astronomical Phenomena." *Vistas in Astronomy* 39 (4), 1995. [A special issue of the journal devoted to papers from the first conference on The Inspiration of Astronomical Phenomena, held in Castelgandolfo, Italy, in 1994. The remit of the INSAP conferences is very broad but includes aspects of cultural astronomy.]

Crump, Thomas. *The Anthropology of Numbers*. Cambridge: Cambridge University Press, 1990.

Cumont, Franz. *The Mysteries of Mithra* [trans. Thomas J. McCormack]. New York: Dover, 1956.

Cunliffe, Barry. *The Ancient Celts*. Oxford: Oxford University Press, 1997.

———. *Facing the Ocean: The Atlantic and Its Peoples, 8000 BC–AD 1500*. Oxford: Oxford University Press, 2001. [One of Europe's most eminent prehistorians argues lucidly that the inhabitants of Atlantic Europe have, over several millennia, developed a distinctive mindset—common elements of worldview—related to their proximity to the ocean.]

———, ed. *The Oxford Illustrated Prehistory of Europe*. Oxford: Oxford University Press, 1994. [Broad survey of European prehistory aimed both at students and the general reader.]

Cunliffe, Barry, and Colin Renfrew, eds. *Science and Stonehenge*. London: British Academy/Oxford University Press, 1997. [A collection of papers applying a variety of scientific approaches to the interpretation of Stonehenge.]

Cuppage, Judith. *Archaeological Survey of the Dingle Peninsula*. Ballyferriter, Ireland: Oidhreacht Chorca Dhuibne, 1986. [Inventory of monuments.]

D'Altroy, Terence. *The Incas*. Oxford: Blackwell, 2002.

Darvill, Timothy, and Caroline Malone, eds. *Megaliths from Antiquity*. York: Antiquity Publications, 2003. [A collection of classic papers on megaliths that originally appeared in the journal *Antiquity*.]

De la Fuente, Beatriz, Silvia Garza Tarazona, Norberto González Crespo, Arnold Lebeuf, Miguel León Portilla, and Javier Wimer. *La Acrópolis de Xochicalco*. Mexico City: Instituto de Cultura de Morelos, 1995. [In Spanish.]

DeMallie, Raymond J., and Douglas R. Parks, eds. *Sioux Indian Religion*. Norman: University of Oklahoma Press, 1987.

Depuydt, Leo. *Civil Calendar and Lunar Calendar in Ancient Egypt (Orientalia Lovaniensia Analecta)*. Leuven, Belgium: Departement Oosterse Studies, 1977.

De Santillana, Giorgio, and Hertha von Dechend. *Hamlet's Mill: An Essay on Myth and the Frame of Time*. London: Macmillan, 1970. [Highly provocative theory claiming that myths from all over the world encapsulated, metaphorically, a universal awareness of precession. Now discredited and of historical interest only.]

Dinsmoor, William. "Archaeology and Astronomy." *Proceedings of the American Philosophical Society* 80 (1939), 95–173. [Of significance in regard to the historical development of archaeoastronomy.]

Donald, Merlin. *Origins of the Modern Mind: Three Stages in the Evolution of Culture and Cognition*. Cambridge, MA: Harvard University Press, 1991.

Edmonson, Munro S. *The Book of the Year: Middle American Calendar Systems*. Salt Lake City: University of Utah Press, 1988.

Edwards, Iorwerth E. S. *The Pyramids of Egypt* (rev. ed.). London: Penguin, 1991.

Edwards, Kevin J., and Ian B. M. Ralston, eds. *Scotland: Environment and Archaeology, 8000 BC–AD 1000*. Chichester: Wiley, 1997. [Good general introduction with chapters by leading archaeologists and palaeoenvironmentalists.]

Emerson, Nathaniel B. *Unwritten Literature of Hawaii: The Sacred Songs of the Hula*. Honolulu: 'Ai Pōhaku Press, 1997. Originally published in 1909 by Government Printing Office, Washington, DC.

Emory, Kenneth P. *Archaeology of Nihoa and Necker Islands* (repr.). Honolulu: Mutual Publishing/Bishop Museum Press, 2002. [Originally published in 1928, this remains the only widely available gazetteer of archaeological sites on these two small islands at the northwestern end of the Hawaiian chain.]

Eogan, George. *Knowth and the Passage-Tombs of Ireland*. London: Thames and Hudson, 1986. [Excavation report providing essential background information relevant to discussions of the possible astronomical significance of the site.]

Espenak, Fred. *Lunar Eclipses of Historical Interest*. http://sunearth.gsfc.nasa.gov/eclipse/LEHistory/LEHistory.html.

———. *Solar Eclipses of Historical Interest*. http://sunearth.gsfc.nasa.gov/eclipse/SEHistory/SEHistory.html.

Esteban, César, and Juan Antonio Belmonte Avilés, eds. *Oxford VI and SEAC99: Astronomy and Cultural Diversity*. La Laguna, Spain: Organismo Autónomo de Museos del Canildo de Tenerife, 2000. [A collection of papers from the sixth "Oxford" international symposium on archaeoastronomy and SEAC (European Society for Astronomy in Culture) meeting held jointly in Tenerife in 1999.]

Evans, James. *The History and Practice of Ancient Astronomy*. New York: Oxford University Press, 1998. [Focuses mainly upon literate cultures and the history of astronomy.]

Evans, Susan T., and David L. Webster, eds. *Archaeology of Ancient Mexico and Central America: An Encyclopedia*. New York: Garland, 2001. [Contains a number of entries relating to Mesoamerican archaeoastronomy. The entry on "Astronomy, Archaeoastronomy, and Astrology" is a good starting point.]

Fabian, Stephen M. *Patterns in the Sky: An Introduction to Ethnoastronomy*. Prospect Heights, IL: Waveland Press, 2001. [The title is misleading; this is really a short (but certainly useful) introduction for anyone seeking a sufficient knowledge of celestial objects and phenomena to do ethnoastronomical fieldwork.]

———. *People of the Earth: An Introduction to World Prehistory* (11th ed.). Upper Saddle River, NJ: Prentice Hall, 2003. [A truly global overview aimed at students taking their first courses in prehistory.]

———, ed. *The Oxford Companion to Archaeology*. Oxford: Oxford University Press, 1996. [Enormously useful reference book.]

Farrer, Claire. *Living Life's Circle: Mescalero Apache Cosmovision*. Albuquerque: University of New Mexico Press, 1991. [An ethnoastronomical study of a nomadic Apache community.]

Faulkner, Raymond. *The Ancient Egyptian Pyramid Texts*. Oxford: Oxford University Press, 1969.

———. *The Ancient Egyptian Coffin Texts*. Oxford: Aris and Phillips, 2004. (Originally published in three volumes, 1973, 1977, and 1978.)

Ferguson, William M., and Richard E. W. Adams. *Mesoamerica's Ancient Cities* (rev. ed.). Albuquerque: University of New Mexico Press, 2001.

Field, Naomi, and Michael Parker Pearson, eds. *Fiskerton: An Iron Age Timber Causeway with Iron Age and Roman Votive Offerings*. Oxford: Oxbow Books, 2003.

Finney, Ben. *Voyage of Rediscovery.* Berkeley: University of California Press, 1994. [A key contribution to debates about Pacific navigation based on "experimental voyaging" in a reconstructed Polynesian voyaging canoe, the Hōkūle'a.]

Fischer, Steven Roger, ed. *Easter Island Studies.* Oxford: Oxbow Books, 1993.

Fitzpatrick, Andrew. *Who Were the Druids?* London: Weidenfeld and Nicolson (Mysteries of the Ancient World series), 1997. [A short but serious introduction by a respected archaeologist.]

Flenley, John, and Paul Bahn. *The Enigmas of Easter Island.* Oxford: Oxford University Press, 2003. [Describes how the inhabitants of isolated Easter Island brought an environmental catastrophe upon themselves. Revised version of *Easter Island, Earth Island* by Bahn and Flenley, published in 1992.]

Flood, Josephine. *Rock Art of the Dreamtime.* Sydney: HarperCollins, 1997. [An interpretative survey by an archaeologist of Aboriginal rock art in Australia.]

Foncerrada de Molina, Marta. *Cacaxtla: La Iconografía de los Olmeca-Xicalanca.* Mexico City: Universidad Nacional Autónoma de México, 1993. [A beautifully illustrated account of the Cacaxtla murals. In Spanish.]

Fountain, John W., and Rolf M. Sinclair, eds. *Current Studies in Archaeoastronomy: Conversations across Time and Space.* Durham, NC: Carolina Academic Press, 2005. [A collection of papers from the fifth "Oxford" international symposium on archaeoastronomy held in Santa Fe in 1996.]

Fraser, David. *Land and Society in Neolithic Orkney* (in two volumes). Oxford: British Archaeological Reports (BAR British Series 117), 1983. [An early example of an investigation of a set of monuments (the chambered cairns of the Scottish Orkney Islands) in relation to their landscape, including orientation in relation to the visible topography.]

Friedel, David, Linda Schele, and Joy Parker. *Maya Cosmos: Three Thousand Years on the Shaman's Path.* New York: William Morrow, 1995. [Controversial but challenging interpretation of Maya religious thought and sky knowledge, arguing a long-term continuity lasting into the modern era.]

Galindo Trejo, Jesús. *Arqueoastronomía en la América Antigua.* Madrid: Equipo Sirius, 1994. [A broad introduction from the viewpoint of an astronomer. In Spanish.]

Galindo Trejo, Jesús, et al. "Arqueoastronomía Mesoamericana." *Arqueología Mexicana* 7 (41) 2000. [A set of articles providing a broad introduction to Mesoamerican archaeoastronomy. In Spanish.]

Garnham, Trevor. *Lines on the Landscape: Circles from the Sky: Monuments of Neolithic Orkney.* Stroud, England: Tempus, 2004. [A range of ideas concerning the Neolithic monuments and landscape of the Scottish Orkney Islands.]

Gaspani, Adriano. *La Cultura di Golasecca: Cielo Luna e Stelle dei Primi Celti d'Italia.* Aosta, Italy: Keltia Editrice, 1999. [Speculative interpretation of monuments in the Golasecca region of northern Italy based on the identification of various putative solar, lunar, and stellar alignments. In Italian.]

Gibson, Alex. *Stonehenge and Timber Circles.* Mount Pleasant, SC: Tempus, 1998. [A good general introduction to European prehistoric timber circles. Includes Stonehenge interpreted in the light of this tradition.]

Gibson, Alex, and Derek Simpson, eds. *Prehistoric Ritual and Religion: Essays in Honour of Aubrey Burl.* Stroud, UK: Sutton, 1998. [A collection of essays on prehistoric northwest Europe dealing with topics from Neolithic monuments and landscapes to Iron Age calendars.]

Gifford, Edward W. *Tongan Society.* Honolulu: Bishop Museum Press, 1929.

Gillings, Mark, and Joshua Pollard. *Avebury.* London: Duckworth, 2004. [New descriptions and interpretations drawing upon recent excavations.]

Gingerich, Owen, ed. "Review Symposium: The Star of Bethlehem." *Journal for the History of Astronomy* 33 (2002): 386–394.

Giot, Pierre-Roland. *La Bretagne des Mégalithes*. Rennes, France: Éditions Ouest-France, 1997. [Comprehensive and definitive guide to Breton megaliths, more selective only in the Carnac area where the concentration of monuments is particularly great. In French.]

Gladwin, Thomas. *East Is a Big Bird: Navigation and Logic on Puluwat Atoll*. Cambridge MA: Harvard University Press, 1970.

Glodariu, Ioan, Eugen Iaroslavschi, and Adriana Rusu. *Cetăţi şi Asezări Dácice în Munţii Orăştiei [Dacian Citadels and Settlements in the Orăştie Mountains]*. Bucharest: Editura Sport Turism, 1988. [An archaeological overview, including an account of the use of the sundial and gnomon. In Romanian with a summary in English.]

Glodariu, Ioan, Adriana Rusu-Pescaru, Eugen Iaroslavschi, and Florin Stănescu. *Sarmizegetusa Regia: Capitala Daciei Preromane [Sarmizegetusa Regia: Pre-Roman Capital of the Dacians]*. Deva, Romania: Acta Musei Devensis, 1996. [An archaeological overview of the site, including an account of its possible astronomical significance. In Romanian with a summary in English.]

Gomes, Carlos J. Pinto, et al. *Paisagens Arqueológicas Aoeste de Évora*. Évora: Câmara Municipal de Évora, 1997. [Provides useful archaeological background on the megalithic monuments of the western Évora region in central Portugal. Contains articles in Portuguese and English.]

González Reimann, Luis. *Tiempo Cíclico y Eras del Mundo en la India*. Mexico City: El Colegio de México, 1988. [In Spanish.]

Goodenough, Ward H. *Native Astronomy in the Central Carolines*. Philadelphia: University Museum, University of Pennsylvania, 1953.

Goodman, Ronald. *Lakota Star Knowledge: Studies in Lakota Stellar Theology*. Rosebud SD: Sinte Gleska College, 1990.

Green, Miranda. *The Sun-Gods of Ancient Europe*. London: Batsford, 1991. [An archaeologist traces the origins of sun cults in Iron Age Europe back into the Neolithic and Bronze Age.]

Griffin-Pierce, Trudy. *Earth Is My Mother, Sky Is My Father: Space, Time and Astronomy in Navajo Sandpainting*. Albuquerque: University of New Mexico Press, 1992. [A valuable study of Navajo cosmology and sky knowledge and its expression through the medium of sandpainting.]

Grimble, Arthur. "Gilbertese Astronomy and Astronomical Observances." *Journal of the Polynesian Society* 40 (1931): 197–224.

Gwilt, Adam, and Colin Haselgrove, eds. *Reconstructing Iron Age Societies*. Oxford: Oxbow Books, 1997.

Hadingham, Evan. *Circles and Standing Stones*. London: Heinemann, 1975. [One of many books produced in the 1970s and early 1980s to introduce archaeoastronomy to a general audience. Badly dated in places.]

———. *Early Man and the Cosmos*. London: Heinemann, 1983. [A broad introduction to archaeoastronomy for a general audience. Much has now been superseded.]

———. *Lines to the Mountain Gods: Nazca and the Mysteries of Peru*. London: Harrap, 1987. [A popular account of investigations at the Nasca lines up to the mid-1980s including the work of Gerald Hawkins, Maria Reiche, and Anthony Aveni and his team.]

Haleʻole, S. N. *Ka Moʻolelo o Lāʻieikawai: The Hawaiian Romance of Lāʻieikawai*, translated by Martha W. Beckwith. Honolulu: First People's Productions, 1997. [First published in translation in 1919 by the Smithsonian Institution, Washington, DC.]

Hamilton, Roland, ed. and trans. *History of the Inca Empire* [a translation of books 11 and 12 of Cobo's *Historia del Nuevo Mundo*, 1653]. Austin: University of Texas Press, 1979.

Hancock, Graham, and Santha Faiia. *Heaven's Mirror: Quest for the Lost Civilization*. London: Penguin, 1998. [Highly controversial theory that has been met with extensive criticism from academics.]

Harding, Jan. *Henge Monuments of the British Isles*. Mount Pleasant, SC: Tempus, 2003. [An excellent general introduction to this type of monument, providing a useful background to discussions of their possible astronomical significance.]

Hawkes, Jacquetta. *Man and the Sun*. London: Cresset Press, 1962.

Hawkins, Gerald (with John B. White). *Stonehenge Decoded*. New York: Doubleday, 1965. [Interpretation of Stonehenge as an astronomical computer. Heavily criticized by academics at the time and now seen as an example of the dangers of misinterpretation.]

Hawkins, Gerald. *Beyond Stonehenge*. New York: Harper and Row, 1973. [More on Stonehenge and a wider set of examples, written in a journalistic style for a popular audience.]

Heggie, Douglas. *Megalithic Science: Ancient Mathematics and Astronomy in Northwest Europe*. London: Thames and Hudson, 1981. [A landmark as the first serious critique of the "megalithic astronomy" of Gerald Hawkins and Alexander Thom by an astronomer as opposed to an archaeologist.]

———, ed. *Archaeoastronomy in the Old World*. Cambridge: Cambridge University Press, 1982. [A collection of "old world" papers from the first "Oxford" international symposium on archaeoastronomy held in England in 1981.]

Heilbron, John. *The Sun in the Church: Cathedrals as Solar Observatories*. Cambridge MA: Harvard University Press, 1999. [An intriguing exploration by a highly respected historian of science of the use of medieval churches and cathedrals as solar observatories.]

Hetherington, Norriss S. *Ancient Astronomy and Civilization*. Tucson, AZ: Pachart, 1987. [An overview from the perspective of a historian of science.]

———. *Science and Objectivity: Episodes in the History of Astronomy*. Ames: Iowa State University Press, 1988.

Higham, Charles. *The Archaeology of Mainland Southeast Asia: From 10,000 BC to the Fall of Angkor*. Cambridge: Cambridge University Press, 1989.

Hill, J. D. *Ritual and Rubbish in the Iron Age of Wessex*. Oxford: Tempus Reparatum (BAR International Series 602), 1995.

Hochseider, Peter, and Doris Knösel. *Les Taules de Menorca: Un Estudi Arqueoastronòmic*. Maó [Mahón], Menorca: Govern Balear, 1995. [An invaluable set of surveyed plans of Menorcan taulas accompanied by questionable interpretations. In Catalan.]

Hodder, Ian. *Symbols in Action: Ethnoarchaeological Studies of Material Culture*. Cambridge: Cambridge University Press, 1982. [Key ideas that strongly influenced the development of interpretative archaeology in the 1980s and beyond.]

Hodder, Ian, and Scott Hutson. *Reading the Past: Current Approaches to Interpretation in Archaeology* (3rd ed.). Cambridge: Cambridge University Press, 2003. [A thought-provoking exploration of different approaches to the interpretation of archaeological evidence. The first edition, by Hodder alone, appeared in 1986.]

Hodson, F. R., ed. *The Place of Astronomy in the Ancient World*. London: Royal Society, 1976. [A classic collection of papers on astronomy in both pre-literate and literate societies.]

Hoskin, Michael. *Tombs, Temples and Their Orientations*. Bognor Regis: Ocarina Books, England, 2001. [A synthesis of many years' fieldwork in southern Europe that presents and analyzes orientation data from some three thousand later prehistoric tombs and temples, mainly in the western Mediterranean.]

———, ed. *Cambridge Illustrated History of Astronomy*. Cambridge: Cambridge University Press, 1997. [Broad and richly illustrated introductory text for students in the history of science.]

———, ed. *The Cambridge Concise History of Astronomy*. Cambridge: Cambridge University Press, 1999. [Essentially the same as the *Cambridge Illustrated History of Astronomy* but produced in a more basic format without the high-quality illustrations.]

Hoskin, Michael, and William Waldren. *Taulas and Talayots*. Cambridge: Michael Hoskin, 1988. [A very useful little guidebook to these monuments on the island of Menorca.]

Hostnig, Rainer. *Arte Rupestre del Perú: Inventario Nacional*. Lima: Consejo Nacional de Ciencia y Tecnología, 2003. [Inventory of rock art in Peru. In Spanish.]

Hoyle, Fred. *From Stonehenge to Modern Cosmology*. San Francisco: Freeman, 1972. [The first half of the book describes the author's controversial and strongly criticized theories about Stonehenge.]

Hudson, Travis, and Ernest Underhay. *Crystals in the Sky: An Intellectual Odyssey Involving Chumash Astronomy, Cosmology, and Rock Art*. Socorro, NM: Ballena Press, 1978.

Hugh-Jones, Stephen. *The Palm and the Pleiades: Initiation and Cosmology in Northwest Amazonia*. Cambridge: Cambridge University Press, 1979. [This in-depth ethnographic study of the Barasana includes evidence for the timing of certain activities in relation to the movement of constellations.]

Hunger, Hermann, and David Pingree. *Astral Sciences in Mesopotamia*. Boston and Leiden: Brill, 1999.

Hunter, John, and Ian Ralston, eds. *The Archaeology of Britain*. London: Routledge, 1999. [Comprehensive introductory textbook for archaeology students.]

Hutton, Ron. *The Pagan Religions of the Ancient British Isles*. Oxford: Blackwell, 1991.

———. *The Stations of the Sun: A History of the Ritual Year in Britain*. Oxford: Oxford University Press, 1996.

Ingold, Tim. *The Perception of the Environment: Essays in Livelihood, Dwelling and Skill*. New York and London: Routledge, 2000. [A key work from a leading social anthropologist exploring the ways in which people perceive the world they inhabit.]

———. *Companion Encyclopedia of Anthropology* (rev. ed.). New York: Routledge, 2002. [A collection of substantial articles addressing key themes in social/cultural anthropology.]

Insoll, Timothy. *Archaeology, Ritual and Religion*. New York: Routledge, 2004. [Explores the question of what we can deduce about ancient religions from archaeological evidence.]

Iwaniszewski, Stanisław, ed. *Readings in Archaeoastronomy*. Warsaw: State Archaeological Museum and Warsaw University, 1992. [A collection of papers from a European archaeoastronomy meeting held in Warsaw in 1990. Contains articles in English and French.]

Iwaniszewski, Stanisław, Arnold Lebeuf, Andrzej Wierciński, and Mariusz Ziółkowski, eds. *Time and Astronomy at the Meeting of Two Worlds*. Warsaw: Centrum Studiów Latynoamerykańskich, 1994. [A collection of papers from a conference held in Poland in 1992, focusing on Mesoamerican and Andean archaeoastronomy and medieval astronomy in Europe. Contains articles in English and Spanish.]

James, Simon. *The Atlantic Celts: Ancient People or Modern Invention?* Madison: University of Wisconsin Press, 1999. [Anyone contemplating aspects of Celtic ritual or the Celtic calendar should first seriously consider this book, which challenges the whole notion of Celts.]

James, Van. *Ancient Sites of Oʻahu*. Honolulu: Bishop Museum Press, 1991. [A useful general guide to the archaeological sites on Oʻahu. For this island there also exists the older but more extensive gazetteer written by Sterling and Summers and published in 1978.]

———. *Ancient Sites of Hawaiʻi*. Honolulu: Mutual Publishing, 1995. [A useful general guide to the archaeological sites on the Big Island of Hawaiʻi. For this island there also exists the older gazetteer of temple sites written by Stokes in 1919 (although unpublished until 1991) and the description of petroglyph sites by Lee and Stasack published in 1999.]

Jaschek, Carlos. *Les Hommes regardent le Ciel*. Strasbourg: Observatoire Astronomique de Strasbourg, 1998. [Introduction to archaeoastronomy focusing on the different celestial bodies and methods for their naked-eye observation. In French.]

———, ed. *Reunion Européenne d'Astronomie et Sciences Humaines: European Meeting on Archaeoastronomy and Ethnoastronomy*. Strasbourg: Observatoire Astronomique, 1992. [A collection of papers from a European archaeoastronomy meeting. Contains articles in French and English.]

Jaschek, Carlos, and F. Atrio Barandela, eds. *Actes del IV Congreso de la SEAC «Astronomia en la Cultura.»* Salamanca, Spain: University of Salamanca, 1997. [A collection of papers from the SEAC (European Society for Astronomy in Culture) meeting held in Spain in 1996. Contains articles in Spanish, English, and French.]

Johnson, Dianne. *Night Skies of Aboriginal Australia: A Noctuary*. Sydney: Oceania Publications/University of Sydney, 1998.

Johnson, Matthew. *Archaeological Theory: An Introduction*. Oxford: Blackwell, 1999. [The best general introduction to different schools of thought regarding the interpretation of archaeological evidence. An introductory text for archaeology students but also essential reading for non-archaeologists entering a field such as archaeoastronomy.]

Johnson, Rubellite K. *Kumulipo: Hawaiian Hymn of Creation*. Vol. I. Honolulu: Topgallant Press, 1981. [An alternative translation and commentary to the classic version of Martha Beckwith.]

———. *The Kumulipo Mind: A Global Heritage*. Honolulu: Anoai Press, 2000. [Some challenging interpretations of the Kumulipo.]

Johnson, Rubellite K., and John K. Mahelona. *Nā Inoa Hōkū: A Catalogue of Hawaiian and Pacific Star Names*. Honolulu: Topgallant Press, 1975.

Jones, Donald W. *Peak Sanctuaries and Sacred Caves in Minoan Crete: A Comparison of Artifacts*. Jonsered, Sweden: Paul Åströms Förlag, 1999.

Kaeppler, Adrienne L., and H. Arlo Nimmo, eds. *Directions in Pacific Traditional Literature: Essays in Honor of Katharine Luomala*. Honolulu: Bishop Museum Press, 1976.

Kak, Subhash. *The Astronomical Code of the Ṛgveda*. New Delhi: Aditya Prakashan, 1994.

Kamakau, Samuel M. *The Works of the People of Old*. Honolulu: Bishop Museum Press, 1976. [One of a small number of vital accounts of native Hawaiian religious beliefs and practices as recalled and recorded in the nineteenth and early twentieth century.]

Kaurov, E. N., ed. "Archaeoastronomy: The Problems of Being." *Astronomical and Astrophysical Transactions* 17(6), 1999. [A special issue of the journal devoted to papers from a national conference on archaeoastronomy held in Moscow in 1997.]

Kelley, David H. *Astronomical Identities of Mesoamerican Gods*. Miami: Institute of Maya Studies (Contributions to Mesoamerican Anthropology), 1980.

Kelley, David H., and Eugene F. Milone. *Exploring Ancient Skies: An Encyclopedic Survey of Archaeoastronomy*. New York: Springer, 2004.

Kelley, Klara Bonsack, and Harris Francis. *Navajo Sacred Places*. Bloomington: Indiana University Press, 1994. [An ethnographic study seeking to identify culturally important places and to identify ways of protecting them.]

Kelly, Joyce. *An Archaeological Guide to Northern Central America: Belize, Guatemala, Honduras and El Salvador*. Norman: University of Oklahoma Press, 1996. [A very useful and relatively compact guidebook by a freelance writer.]

Keys, David. *Catastrophe*. London: Century Books, 1999. [An archaeological journalist explores climatic catastrophes that might be attributable to meteoric impacts.]

Kidger, Mark. *The Star of Bethlehem: An Astronomer's View*. Princeton, NJ: Princeton University Press, 1999. [Conventional discussion in terms of events in the sky that might have been conspicuous at the time.]

King, John. *The Celtic Druids' Year: Seasonal Cycles of the Ancient Celts*. London: Blandford, 1994. [A reconstruction of the Druidic seasonal calendar by a modern bard.]

Kirch, Patrick V. *Feathered Gods and Fishhooks: An Introduction to Hawaiian Archaeology and Prehistory*. Honolulu: University of Hawai'i Press, 1985.

———. *Legacy of the Landscape: An Illustrated Guide to Hawaiian Archaeological Sites*. Honolulu: University of Hawai'i Press, 1996. [An excellent guidebook.]

———. *On the Road of the Winds: An Archaeological History of the Polynesian Islands before European Contact*. Berkeley: University of California Press, 2000. [An authoritative and eminently readable broad overview of Pacific archaeology.]

Kirch, Patrick V., and Roger C. Green. *Hawaiki, Ancestral Polynesia: An Essay in Historical Anthropology*. Cambridge: Cambridge University Press, 2001. [Includes an analysis of sacred seasonal and calendrical rituals in an attempt to identify common ancestral practices.]

Knight, Chris. *Blood Relations*. New Haven, CT: Yale University Press, 1991. [A controversial interpretation of Palaeolithic rites and practices relating to the moon.]

Knight, Christopher, and Robert Lomas. *Uriel's Machine: The Prehistoric Technology that Survived the Flood*. London: Century, 1999. [Controversial!]

Kõiva, Mare, Harry Mürk, and Izold Pustõlnik, eds. *Cultural Context from Archaeoastronomical Data and the Echoes of Cosmic Catastrophic Events*. Tallinn, Estonia: Estonian Literary Museum, and Tartu, Estonia: Tartu Observatory, in press. [A collection of papers from the SEAC (European Society for Astronomy in Culture) meeting held in Estonia in 2002. May contain non-English articles.]

Koleva, Vesselina, and Dimiter Kolev, eds. *Astronomical Traditions in Past Cultures*. Sofia: Bulgarian Academy of Sciences, 1996. [A collection of papers from the SEAC (European Society for Astronomy in Culture) meeting held in Bulgaria in 1993. In English; contains abstracts in Bulgarian.]

Kowalski, Jeff Karl. *The House of the Governor: A Maya Palace of Uxmal, Yucatan, Mexico*. Norman and London: University of Oklahoma Press, 1987.

Krupp, Edwin C. *Echoes of the Ancient Skies*. Oxford: Oxford University Press, 1983. [Thematic introduction to archaeoastronomy aimed at a general audience. Somewhat dated in places.]

———. *Skywatchers, Shamans and Kings*. New York: Wiley, 1997. [Broad-ranging popular introduction to some themes in archaeoastronomy.]

———, ed. *Archaeoastronomy and the Roots of Science*. Washington, DC: Westview Press, 1984.

———, ed. *In Search of Ancient Astronomies*. New York: Doubleday, 1977. [This brings together chapters on British megaliths including Stonehenge, Medicine Wheels, Mesoamerica, and ancient Egypt. Much has now been superseded.]

Lancaster Brown, Peter. *Megaliths, Myths and Men*. Poole: Blandford Press, 1976. [Describes "megalithic astronomy," and in particular the theories of Hawkins and Thom, for a general audience. Of historical interest only.]

Larsson, Lars, and Berta Stjernquist, eds. *The World-View of Prehistoric Man*. Stockholm: Swedish Academy of Sciences, 1998.

Lattimore, Richmond, trans. *Hesiod*. Ann Arbor: University of Michigan Press, 1959.

Laval, P. Honoré. *Mangareva: L'Histoire Ancienne d'un Peuple Polynésien*. Braine-le-Comte, Belgium: Maison des Pères des Sacrés-Coeurs, 1938. [In French.]

Leandri, Franck. *Les Mégalithes de Corse*. Luçon, France: Editions Jean-Paul Gisserot, 2000. [Guide to the megalithic monuments of Corsica with photographs and, in a few cases, site plans. In French.]

Leandri, Franck, and Laurent Chabot. *Monuments de Corse*. Aix-en-Provence: Édisud, 2003. [Guide to the monuments of Corsica, covering all periods. In French.]

Le Cam, Gabriel. *Le Guide des Mégalithes du Morbihan*. Spézet, France: Coop Breizh, 1999. [Comprehensive pictorial guide to megaliths in the Carnac area. In French.]

Le Contel, Jean-Michel, and Paul Verdier. *Un Calendrier Celtique: Le Calendrier Gaulois de Coligny*. Paris: Éditions Errance, 1997. [A short description with photographs, accompanied by speculations that reach far beyond the evidence, e.g., about its use for calculating precession. In French.]

Lebeuf, Arnold. *Les Eclipses dans l'Ancien Mexique*. Kraków: Jagiellonian University Press, 2003. [In French.]

Lebeuf, Arnold, and Mariusz Ziółkowski, eds. *Actes de la Vème Conférence Annuelle de la SEAC*. Warsaw: University of Warsaw and Gdańsk: Central Maritime Museum, 1999. [Contains articles in English, German, and French.]

Lee, Georgia. *An Uncommon Guide to Easter Island*. Arroyo Grande, CA: International Resources, 1990. [A colorful and very useful little guidebook.]

Lee, Georgia, and Edward Stasack. *Spirit of Place: Petroglyphs of Hawai'i*. Los Osos, CA: Easter Island Foundation, 1999. [Documents several important petroglyph sites on the Big Island of Hawai'i.]

Legesse, Asmerom. *Gada: Three Approaches to the Study of African Society*. New York: Macmillan, 1973. [A social anthropological study containing the original account of the Borana calendar that led to a string of subsequent misunderstandings.]

Lekson, Stephen H. *Great Pueblo Architecture of Chaco Canyon, New Mexico*. Albuquerque: University of New Mexico Press, 1987. [A detailed survey and synthesis including descriptions and plans of several Great Houses.]

———. *The Chaco Meridian: Centers of Political Power in the Ancient Southwest*. Walnut Creek, CA: AltaMira Press, 1999. [A controversial but seriously challenging theory.]

Lekson, Stephen H., John R. Stein, and Simon J. Ortiz. *Chaco Canyon: A Center and Its World*. Santa Fe: Museum of New Mexico Press, 1994.

Lewis, David. *We the Navigators* (2nd ed). Honolulu: University of Hawai'i Press, 1994. [An account of Pacific navigation by a leading voyager who has not only researched and attempted to reconstruct indigenous Polynesian and Micronesian techniques of navigation but has also endeavored to put some of them into practice.]

Liller, William. *The Ancient Solar Observatories of Rapanui: The Archaeoastronomy of Easter Island.* Old Bridge, NJ: Cloud Mountain Press, 1993.

Littarru, Paolo, and Mauro Peppino Zedda. *Santu Antine: Guida Archeoastronomica al Nuraghe Santu Antine di Torralba.* Cagliari, Italy: Associazione Culturale Agora' Nuragica, 2003. [Archaeoastronomical ideas relating to one of the most impressive of the Sardinian nuraghi. In Italian.]

Littmann, Mark, Ken Willcox, and Fred Espenak. *Totality—Eclipses of the Sun* (2nd ed.), 1–53. New York: Oxford University Press, 1999. [The first part of the book explores cultural perceptions of, and reactions to, solar eclipses.]

Lockyer, Norman. *The Dawn of Astronomy.* London: Cassell, 1894. [Classic early work on astronomical alignments of Egyptian temples.]

———. *Stonehenge and Other British Stone Monuments Astronomically Considered* (2nd ed.). London: MacMillan, 1909. [A key work in the early historical development of British "megalithic astronomy." First published in 1906.]

Lucas, Gavin. *The Archaeology of Time.* New York: Routledge, 2005. [An exploration of what we can deduce about concepts of time from archaeological evidence.]

MacCana, Proinsias. *Celtic Mythology.* London: Hamlyn, 1970.

MacDonald, John. *The Arctic Sky: Inuit Astronomy, Star Lore and Legend.* Toronto: Royal Ontario Museum and Iqaluit: Nunavut Research Institute, 1998. [The only detailed study available of sky knowledge in this important culture area.]

MacDonald, William L. *The Pantheon: Design, Meaning and Progeny.* Cambridge, MA: Harvard University Press, 1976.

MacKie, Euan. *The Megalith Builders.* Oxford: Phaidon, 1977. [Broad background including reference to Alexander Thom's theories.]

———. *Science and Society in Prehistoric Britain.* London: Paul Elek, 1977. [Controversial interpretation of Alexander Thom's theories. MacKie was one of the very few archaeologists of his time who attempted to frame a social context for Thom's ideas, but he faced a barrage of criticism from his colleagues.]

Magini, Leonardo. *Astronomy and Calendar in Ancient Rome: The Eclipse Festivals.* Rome: «L'Erma» di Bretschneider, 2001. [Challenging new theories about the Roman calendar and its roots.]

Magli, Giulio. *Misteri e Scoperte dell'Archeoastronomia.* Rome, Italy: Newton & Compton Editori, 2005. [A survey of world archaeoastronomy by a physicist. In Italian.]

Mainfort, Robert, and Lynne Sullivan, eds. *Ancient Earthen Enclosures of the Eastern Woodlands.* Gainesville: University Press of Florida, 1998. [A collection of papers documenting recent research on the earthwork enclosures of the eastern United States.]

Makemson, Maud. *The Morning Star Rises: An Account of Polynesian Astronomy.* New Haven: Yale University Press, 1941. [Classic early work on Polynesian astronomy by an astronomer.]

Malmström, Vincent H. *Cycles of the Sun, Mysteries of the Moon: The Calendar in Mesoamerican Civilization.* Austin: University of Texas Press, 1997. [Argues that the Mayan tzolkin (260-day calendar round) derived from horizon calendars established as early as the second millennium B.C.E. Although many arguments remain about the detailed evidence, the general idea is not as highly controversial as it was when first presented in the 1970s.]

Malo, David. *Hawaiian Antiquities (Moʻolelo Hawaiʻi)* (2nd ed.). Honolulu: Bishop Museum Press, 1951. [One of a small number of vital accounts of native Hawaiian religious beliefs and practices as recalled and recorded in the nineteenth and early twentieth centuries.]

Malone, Caroline. *Avebury.* London: Batsford/English Heritage, 1994. [A good general introduction to the site, but now superseded by accounts drawing upon more recent excavations.]

Maltwood, Katherine. *A Guide to Glastonbury's Temple of the Stars.* London: James Clarke, 1964. Originally published in 1929. [A classic of "fringe" literature, presenting an idea that does not withstand scholarly critique yet achieved a great deal of popular attention.]

Malville, J. McKim, and Gary Matlock, eds. *The Chimney Rock Archaeological Symposium.* Fort Collins, CO: U.S. Department of Agriculture (Rocky Mountain Forest and Range Experiment Station, General Technical Report RM-227), 1993. [A collection of symposium papers including some archaeoastronomical theories.]

Malville, J. McKim, and Claudia Putnam. *Prehistoric Astronomy in the Southwest.* Boulder, CO: Johnson Books, 1989. [A useful overview of archaeoastronomical theories relating to the U.S. Southwest in the 1980s.]

Manzanilla, Linda, and Leonardo López Luján, eds. *Atlas Histórico de Mesoamérica.* Mexico City: Ediciones Larousse (2nd ed.), 2003. [A concise history of Mesoamerican cultures organized as a series of short articles. Good for reference, although it omits discoveries since first published in 1993. In Spanish.]

Maravelia, Amanda-Alice, ed. *Ad Astra per Aspera et per Ludum: European Archaeoastronomy and the Orientation of Monuments in the Mediterranean Basin.* Oxford: Archaeopress (BAR International Series, 1154), 2003. [A collection of papers presented at an archaeoastronomy session at the European Association of Archaeologists annual meeting at Thessaloniki, Greece, in 2002.]

Markey, T. L., and Greppin, A. C., eds. *When Worlds Collide: The Indo-Europeans and Pre-Indo-Europeans.* Ann Arbor: Karoma, 1990. [A mixed bag of papers that includes a discussion of European archaeoastronomy.]

Marshack, Alexander. *The Roots of Civilization.* New York: Weidenfeld and Nicolson, 1972. [This book first introduced to a wide public the author's analyses of markings on bone fragments tens of thousands of years old suggesting that Palaeolithic hunters used them to record lunar calendars.]

Martin, Simon, and Nikolai Grube. *Chronicle of the Maya Kings and Queens: Deciphering the Dynasties of the Ancient Maya.* London: Thames and Hudson, 2000. [A detailed history constructed with the help of recently deciphered Maya inscriptions.]

Masi, Fausto. *The Pantheon as an Astronomical Instrument.* Rome: Edizioni Internazionali di Letteratura e Scienze, 1996. [A short pamphlet containing sketches showing the position of the shaft of sunlight through the roof at the solstices and equinoxes. Versions also exist in Italian, French, etc.]

Matos Moctezuma, Eduardo. *The Great Temple of the Aztecs: Treasures of Tenochtitlan.* London: Thames and Hudson, 1988. [A highly readable and authoritative introduction to the Aztecs and Tenochtitlan in general and to the Templo Mayor in particular.]

Maude, Harry E. *The Gilbertese Maneaba.* Suva, Fiji: Institute of Pacific Studies and Tarawa: Kiribati Extension Centre, 1980. [An account of the protocols governing the construction of the focal building in the traditional Gilbertese village, relevant to astronomy since the design of house roofs influenced indigenous conceptions of the sky.]

McCluskey, Stephen C. *Astronomies and Cultures in Early Medieval Europe*. Cambridge: Cambridge University Press, 1998. [Covers the development of astronomies in Europe from Celtic Gaul through to late Medieval times. Combines an archaeoastronomical with a historical approach in tackling questions relating to solar rituals, calendrical development, and monastic computations.]

McCoy, Ron. *Archaeoastronomy: Skywatching in the Native American Southwest*. Flagstaff: Museum of Northern Arizona, 1992. [A good general introduction to archaeoastronomy in the U.S. Southwest.]

McCready, Stuart, ed. *The Discovery of Time*. Naperville, IL: Sourcebooks, 2001. [A broad introduction to calendars and timekeeping for a general audience.]

Mercer, Roger. *Causewayed Enclosures*. Princes Risborough, UK: Shire, 1990. [A short introduction for a general audience.]

Michell, John. *A Little History of Astro-Archaeology*. London: Thames and Hudson, 1989. [A survey of early developments in archaeoastronomy by an author best known for his involvement in "fringe" topics such as ley lines, but containing some informative material rarely found elsewhere, e.g., on Nazi archaeoastronomy.]

Milbrath, Susan. *Star Gods of the Maya: Astronomy in Art, Folklore and Calendars*. Austin: University of Texas Press, 1999.

Millon, René, ed. *The Teotihuacan Map* (in two parts). Austin: University of Texas Press, 1973.

Mithen, Steven. *The Prehistory of the Mind: The Cognitive Origins of Art, Religion and Science*. London: Thames and Hudson, 1999.

———, ed. *Creativity in Human Evolution and Prehistory*. London: Routledge, 1998. [A broad collection of papers from a session at the annual British Theoretical Archaeology Group conference on the theme of human creativity and its manifestation in the archaeological record.]

Mitton, Simon. *The Crab Nebula*. New York: Charles Scribner's Sons, 1978. [Astronomical background on the supernova of C.E. 1054.]

Molnar, Michael. *The Star of Bethlehem: The Legacy of the Magi*. New Brunswick, NJ: Rutgers University Press, 1999. [Innovative interpretation in terms of events in the sky and their astrological significance at the time.]

Monroe, Jean G., and Ray A. Williamson. *They Dance in the Sky: Native American Sky Myths*. Boston: Houghton Mifflin, 1987. [A collection of native North American sky myths aimed at readers aged 10–14.]

Montgomery, John. *Dictionary of Maya Hieroglyphs*. New York: Hypocrene Books, 2002. [A visual dictionary of over 1000 Maya glyphs.]

Morero Corral, Marco Arturo, ed. *Historia de la Arqueoastronomía en México*. Mexico City: Fondo de Cultura Económica, 1986. [In Spanish.]

Morieson, John. *The Night Sky of the Boorong*. Melbourne: Unpublished MA thesis, University of Melbourne, 1996.

Morris, Richard. *Churches in the Landscape*. London: Dent, 1989. [Some useful background for studies of English church orientations.]

Morrison, Tony. *The Mystery of the Nasca Lines*. Woodbridge, UK: Nonesuch Expeditions, 1987. [One of several popular books describing the Nasca geoglyphs and the life and theories of Maria Reiche, but preceding the work of Aveni and his team in the early 1980s. In many ways an update of *Pathways to the Gods: The Mystery of the Andes Lines*, first published in 1978.]

Mountford, Charles P. *Nomads of the Australian Desert*. Adelaide: Rigby, 1976. [This book, whose publication provoked some strong reactions from Aboriginal communities, contains information about indigenous sky myths that is hard to find elsewhere.]

Needham, Joseph. *Science and Civilisation in China, Vol. 3: The Sciences of the Heavens and Earth*. New York: Cambridge University Press, 1959. [Seminal work on ancient Chinese astronomy.]

Neugebauer, Otto. *The Exact Sciences in Antiquity*. Princeton: Princeton University Press, 1951. (Second edition published 1957 by Brown University Press, Providence; further corrected edition published 1969 by Dover, New York.)

Neugebauer, Otto, and Richard A. Parker. *Egyptian Astronomical Texts, I: The Early Decans*. Providence, RI: Brown University Press, 1960. [Seminal work on ancient Egyptian astronomy.]

———. *Egyptian Astronomical Texts, II: The Ramesside Star Clocks*. Providence, RI: Brown University Press, 1964. [Seminal work on ancient Egyptian astronomy.]

———. *Egyptian Astronomical Texts, III: Decans, Planets, Constellations and Zodiacs*. Providence, RI: Brown University Press, 1969. [Seminal work on ancient Egyptian astronomy.]

Nissen, Heinrich. *Orientation. Studien zur Geschichte der Religion* (3 vols.). Berlin: Weidmannsche Buchhandlung, 1906–1910. [A work of huge historical significance as part of the early development of orientation studies and archaeoastronomy. In German.]

Noble, David G., ed. *New Light on Chaco Canyon*. Santa Fe: School of American Research Press, 1984. [A collection of short introductory articles from different disciplinary perspectives, including a piece on archaeoastronomy. Although ideas have moved on in many areas, this still contains much useful background material.]

North, John D. *The Fontana History of Astronomy and Cosmology*. London: Fontana, 1994. [Authoritative overview of the history of astronomy.]

———. *Stonehenge: Neolithic Man and the Cosmos*. London: HarperCollins, 1996. [A novel but controversial interpretation of the astronomical significance of British megalithic monuments including Stonehenge that has drawn much criticism.]

Obeyesekere, Gananath. *The Apotheosis of Captain Cook: European Mythmaking in the Pacific* (repr.). Princeton, NJ: Princeton University Press, 1997. [A book that sparked a fierce debate by challenging Marshall Sahlins's view (see Sahlins 1985) concerning native perceptions of Captain Cook and the reasons for his apparent apotheosis and subsequent death in the Hawaiian Islands. Originally published in 1992, this edition contains a response to Sahlins' (1995) response.]

O'Brien, William. *Sacred Ground: Megalithic Tombs in Coastal South-West Ireland*. Department of Archaeology, Galway: National University of Ireland Galway, 1999. [A collection of papers on wedge tombs in this region, providing an archaeological background to discussions of their possible astronomical significance.]

O'Kelly, Claire. *Illustrated Guide to Newgrange and Other Boyne Monuments* (3rd ed.). Wexford: John English, 1978. [Popular account of the monument. The author assisted her husband in the excavations from 1962 to 1975.]

O'Kelly, Michael. *Newgrange: Archaeology, Art and Legend*. London: Thames and Hudson, 1982. [Definitive account of the excavations from 1962 to 1975. Essential background to discussions of the astronomical significance of the roof-box and other putative astronomical associations.]

———. *Early Ireland: an Introduction to Irish Prehistory*. Cambridge: Cambridge University Press, 1989. [A general introduction by the excavator of Newgrange.]

Oliveira, Jorge. *Sepulturas Megalíticas del Termino Municipal de Cedillo*. Cáceres, Spain: Edición Patrocinada/Ayuntamiento de Cedillo, 1994. [A gazetteer of megalithic tombs in the Cedillo area of western Spain. Includes photographs and plans. In Spanish.]

Olmsted, Garrett. *The Gaulish Calendar*. Bonn: Habelt, 1992.

Orefici, Giuseppe, and Andrea Drusini. *Nasca: Hipótesis y Evidencias de su Desarrollo Cultural*. Brescia, Italy: Centro Italiano Studi e Ricerche Archeologiche Precolombiane, 2003. [Account of recent archaeological investigations including excavations at the ceremonial center of Cahuachi. Also includes accompanying work in physical anthropology. In Spanish.]

Ortiz, Alfonso, ed. *Handbook of North American Indians, Volume 9: Southwest*. Washington, DC: Smithsonian Institution, 1979. [Contains several chapters on Hopi history and culture.]

Oswald, Alastair, Carolyn Dyer, and Martyn Barber. *The Creation of Monuments: Neolithic Causewayed Enclosures in the British Isles*. Swindon, UK: English Heritage, 2001.

Ott, Sandra. *The Circle of Mountains: A Basque Shepherding Community* (2nd ed.). Reno: University of Nevada Press, 1993. [An ethnographic investigation of a twentieth-century rural community in the mountainous Basque country.]

Parke, H. W. *The Delphic Oracle*. Oxford: Blackwell, 1956.

Parker Pearson, Michael. *Bronze Age Britain* (new ed.). London: Batsford/English Heritage 2005. [An excellent introduction to later prehistoric Britain, spanning (despite the title) the whole of the Neolithic as well as the Bronze Age.]

Parker Pearson, Michael, and Colin Richards, eds. *Architecture and Order: Approaches to Social Space*. London: Routledge, 1994. [A broad collection of conference papers.]

Parker, R. A. *The Calendars of Ancient Egypt*. Chicago: University of Chicago Press, 1950. [A classic.]

Pásztor, Emília, ed. *Archaeoastronomy from Scandinavia to Sardinia: Current Problems and Future of Archaeoastronomy 2*. Budapest: Roland Eötvös University, 1995. [A collection of papers from a European archaeoastronomy meeting held in Hungary in 1991.]

Patton, Mark. *Statements in Stone: Monuments and Society in Neolithic Brittany*. London: Routledge, 1993. [Not an introductory text.]

Peiser, Benny, Trevor Palmer, and Mark Bailey, eds. *Natural Catastrophes during Bronze Age Civilisations: Archaeological, Geological, Astronomical and Cultural Perspectives*. Oxford: Archaeopress (BAR International Series 728), 1998. [A motley collection of papers including much that is highly controversial.]

Piggott, Stuart. *The Druids*. London: Thames and Hudson, 1968. [A classic of its time.]

Plog, Stephen. *Ancient Peoples of the American Southwest*. London: Thames and Hudson, 1997. [Overview providing an essential archaeological background to archaeoastronomical work in this region.]

Politzer, Anie, and Michel Politzer. *Des Mégalithes et des Hommes*. Spézet, France: Coop Breizh, 2004. [An excellent story book, aimed both at children and adults, that reconstructs aspects of the lives of the megalith builders of Brittany and elsewhere around the archaeological facts, respecting and presenting those facts in a very well informed manner. In French.]

Pollard, Joshua, and Andrew Reynolds. *Avebury: Biography of a Landscape*. Mount Pleasant, SC: Tempus, 2002. [A description of Avebury and its landscape from prehistory into historical times.]

Potyomkina, Tamila M., and V. A. Jurevich. *Experiences of Archaeoastronomical Research at Archaeological Monuments (Methodological Aspects)*. Moscow: Institute of Archaeology, Russian Academy of Sciences, 1998. [An approach based on the identification of potentially astronomically significant alignments at excavated monuments. In Russian with a short summary in English.]

Potyomkina, Tamila M., and V. N. Obridko, eds. *Astronomy of Ancient Societies.* Moscow: Nauka, 2002. [A collection of papers from the SEAC (European Society for Astronomy in Culture) meeting held in Moscow in 2000. Contains articles in Russian and English, each with an abridged version in the other language.]

Power, Denis. *Archaeological Inventory of County Cork, Volume 1: West Cork.* Dublin: Stationery Office, 1992. [Inventory of monuments.]

———. *Archaeological Inventory of County Cork, Volume 2: East and South Cork.* Dublin: Stationery Office, 1994. [Inventory of monuments.]

———. *Archaeological Inventory of County Cork, Volume 3: Mid Cork.* Dublin: Stationery Office, 1997. [Inventory of monuments.]

Proverbio, Edoardo. *Archeoastronomia: Allo Ricerca delle Radici dell'Astronomia Preistorica.* Milan: Teti Editore, 1989. [Broad survey of archaeoastronomy, now dated. In Italian.]

Pryor, Francis. *Britain B.C: Life in Britain before the Romans.* London: HarperCollins, 2003. [A lively and eminently readable introduction to British (and Irish) prehistory by a well-respected British Iron Age specialist.]

Pukui, Mary K., E. W. Haertig, and Catherine A. Lee. *Nānā I Ke Kumu (Look to the Source), Volume I.* Honolulu: Hui Hānai, 1972. [An invaluable compendium of indigenous Hawaiian cultural concepts and practices.]

Quilter, Jeffrey, and Gary Urton, eds. *Narrative Threads: Accounting and Recounting in Andean Khipu.* Austin: University of Texas Press, 2002. [A collection of papers resulting from a round table on *quipu* (*khipu*) held in 1997.]

Rappenglück, Michael A. *Eine Himmelskarte aus der Eiszeit?: Ein Beitrag zur Urgeschichte der Himmelskunde und zur paläoastronomischen Methodik.* Frankfurt: Peter Lang, 1999. [Controversial interpretation of French Palaeolithic cave art. In German.]

Rebullida Conesa, Amador. *Astronomía y Religión en el Neolítico-Bronce.* Terrassa, Spain: Editorial Ègara, 1988. [One of numerous works claiming to have uncovered evidence of highly sophisticated ancient astronomy based upon fitting a detailed numerological interpretation to megalithic alignments and rock art. Such works typically pay little or no attention to testing the proposed explanation against alternatives that might fit the data just as well, or to broader evidence relating to the social context. In Spanish.]

Redman, Charles, ed. *Social Archaeology: Beyond Substance and Dating.* New York: Academic Press, 1978. [A collection of papers on what was a "hot topic" archaeologically in the 1970s.]

Reiche, Maria. *Mystery on the Desert* (4th ed.). Stuttgart: Heinrich Fink, 1982. [This short work, describing Maria Reiche's theories, was sold to tourists at Nasca for many years. In German, English, and Spanish.]

Reinhard, Johan. *The Nazca Lines: A New Perspective on their Origin and Meaning* (4th ed.). Lima: Editorial Los Pinos, 1988. [A classic work relating possible interpretations of the Nasca lines to historical and modern practices of mountain worship and to horizon astronomical observations.]

———. *Machu Picchu: The Sacred Center* (2nd ed.). Cusco: Instituto Machu Picchu, 2002. [A lusciously illustrated account including discussions of the significance of the Intihuatana stone in relation to the horizon and the sun, and Inca conceptions of the cardinal directions.]

Reinhard, Johan, and María Constanza Ceruti. *Investigaciones Arqueológicas en el Volcán Llullaillaco.* Salta, Argentina: Ediciones Universidad Catolica de Salta, 2000. [A report on the extraordinary excavation of an Incaic sanctuary on the summit of a 6,700-meter (22,000-foot) volcano, where the offerings included three infant/child sacrifices and hundreds of other objects. In Spanish. An account in English appeared in 1999 in the *National Geographical Magazine*.]

Renfrew, Colin. *Before Civilization*. Harmondsworth: Penguin Books, 1976. [This paperback, aimed at a general audience, included one of the very few efforts by any archaeologist in the early 1970s to seriously examine some of the implications of early ideas in "megalithic astronomy" and integrate them into a broader framework of interpretation.]

———, ed. *The Prehistory of Orkney* (repr.). Edinburgh: Edinburgh University Press, 1990. [A collection of papers providing an overview covering the first settlers through to the Viking period. First published in 1985.]

Renfrew, Colin, and Paul Bahn. *Archaeology: Theories, Methods and Practice* (4th ed.). London: Thames and Hudson, 2004. [A detailed overview widely used as a textbook for students taking their first courses in archaeology.]

———, eds. *Archaeology: The Key Concepts*. Abingdon: Routledge, 2005. [An extremely useful collection of fifty-five short articles on key themes in modern archaeology.]

Renfrew, Colin, and Ezra Zubrow, eds. *The Ancient Mind: Elements of Cognitive Archaeology*. Cambridge: Cambridge University Press, 1994. [A key work in the development of cognitive archaeology: a collection of papers on prehistoric beliefs and thought.]

Renshaw, Steven, and Saori Ihara. *Astronomy in Japan: Science, History, Culture*. http://www2.gol.com/users/stever/jastro.html.

Rice, Prudence M. *Maya Political Science: Time, Astronomy and the Cosmos*. Austin: University of Texas Press, 2004.

Richards, Julian C. *Stonehenge and its Environs*. Edinburgh: Edinburgh University Press, 1979. [An early study of the Stonehenge landscape including an inventory of monuments.]

———. *English Heritage Book of Stonehenge*. London: English Heritage/Batsford, 1991. [Authoritative general introduction to Stonehenge and its landscape.]

Ridpath, Ian, ed. *Norton's Star Atlas and Reference Handbook* (20th ed.). New York: Pi Press, 2004. [An essential sourcebook on background astronomy, including vital star maps.]

Ritchie, Anna. *Prehistoric Orkney*. London: Batsford/Historic Scotland, 1995. [A sound general introduction to the archaeology of the Scottish Orkney Islands.]

———, ed. *Neolithic Orkney in its European Context*. Cambridge: McDonald Institute for Archaeological Research, 2000.

Ritchie, Anna, and Graham Ritchie. *Scotland: An Oxford Archaeological Guide*. London: Thames and Hudson, 1998. [An excellent general introduction to Scottish prehistory and early history.]

Rochberg, Francesca. *The Heavenly Writing: Divination, Horoscopy, and Astronomy in Mesopotamian Culture*. Cambridge: Cambridge University Press, 2004.

Romain, William F. *Mysteries of the Hopewell: Astronomers, Geometers, and Magicians of the Eastern Woodlands*. Akron, Ohio: University of Akron Press, 2000. [A re-examination of the Hopewell earthworks, including some controversial astronomical and geometrical interpretations.]

Romano, Giuliano. *Archeoastronomia Italiana*. Padova, Italy: Cleup, 1992. [In Italian.]

Romano, Giuliano, and Gustavo Traversari, eds. *Colloquio Internazionale Archeologia e Astronomia*. Rome: Giorgio Bretschneider Editore, 1991. [A collection of papers from a meeting held in Venice in 1989. Several focus on Italy, but the collection as a whole spans many parts of the world. Contains articles in Italian, English, Spanish, and French.]

Roy, Archie E. "The Origin of the Constellations." *Vistas in Astronomy* 27 (1984), 171–197.

Royal Commission on the Ancient and Historic Monuments of Scotland (RCAHMS). *Argyll: An Inventory of the Ancient Monuments, Volume 1: Kintyre.* Edinburgh: Her Majesty's Stationery Office, 1971. [Inventory of monuments including Ballochroy.]

———. *Argyll: An Inventory of the Ancient Monuments, Volume 3: Mull, Tiree, Coll and Northern Argyll.* Edinburgh: Her Majesty's Stationery Office, 1980. [Inventory of monuments including the stone rows of northern Mull, which have been closely studied archaeoastronomically.]

———. *Argyll: An Inventory of the Ancient Monuments, Volume 6: Mid Argyll and Cowal, Prehistoric and Early Historic Monuments.* Edinburgh: Her Majesty's Stationery Office, 1988. [Inventory of monuments including Kintraw.]

Ruggles, Clive. *Megalithic Astronomy: A New Archaeological and Statistical Study of 300 Western Scottish Sites.* Oxford: British Archaeological Reports (British Series 123), 1984. [A detailed reassessment of Thom's theories that helped to reconcile archaeologists and astronomers in the 1980s.]

———. *Astronomy in Prehistoric Britain and Ireland.* New Haven: Yale University Press, 1999. [A comprehensive discussion and critical review of ideas concerning the possible astronomical significance of various prehistoric monuments in Britain and Ireland. It also tackles a number of broader issues. Written for a cross-disciplinary audience, the book contains a number of boxes that explain basic astronomical and statistical concepts as well as an appendix on fieldwork techniques.]

———, ed. *Archaeoastronomy in the 1990s.* Loughborough: Group D Publications, 1993. [A collection of papers from the third "Oxford" international symposium on archaeoastronomy held in Scotland in 1990.]

———, ed. *Records in Stone: Papers in Memory of Alexander Thom* (repr.). Cambridge: Cambridge University Press, 2002. [A diverse collection of papers forming a "posthumous festschrift." First published in 1988.]

Ruggles, Clive, Frank Prendergast, and Tom Ray, eds. *Astronomy, Cosmology and Landscape.* Bognor Regis: Ocarina Books, 2001. [A collection of papers from the SEAC (European Society for Astronomy in Culture) meeting held in Ireland in 1998.]

Ruggles, Clive, and Nicholas Saunders, eds. *Astronomies and Cultures.* Niwot: University Press of Colorado, 1993. [A collection of papers selected from the third "Oxford" international symposium on archaeoastronomy held in Scotland in 1990. They cover theory, method, and practice, providing a good insight into the scope of archaeoastronomy as well as some of the main issues of contention.]

Ruggles, Clive, and Gary Urton, eds. *Cultural Astronomy in New World Cosmologies.* Niwot: University of Colorado Press, in press. [A collection of papers from a symposium held in honor of Anthony Aveni in 2003.]

Ruggles, Clive, and Alasdair Whittle, eds. *Astronomy and Society in Britain during the Period 4000–1500 BC.* Oxford: British Archaeological Reports (British Series 88), 1981. [A collection of papers from both sides of the "megalithic astronomy" debate in the late 1970s.]

Sahlins, Marshall. *Islands of History.* Chicago: University of Chicago Press, 1985. [Discusses native perceptions of Captain Cook in relation to the calendrical festivals dedicated to the god Lono, in the course of arguing a deeper case about the interrelationship of anthropology and history.]

———. *How "Natives" Think: About Captain Cook, for Example.* Chicago: University of Chicago Press, 1995. [Response to Gananath Obeyesekere regarding the apotheosis of Captain Cook, thus continuing one of the great debates about the role of anthropology in allowing Western scholars to appreciate (and communicate) non-Western ways of understanding the world.]

Sampson, Ross, ed. *The Social Archaeology of Houses*. Edinburgh: Edinburgh University Press, 1990.

Saunders, Nicholas, ed. *Ancient America: Contributions to New World Archaeology*. Oxford: Oxbow Books, 1992.

Scarre, Chris, ed. *Monuments and Landscape in Atlantic Europe*. London: Routledge, 2002. [A collection of papers by European prehistorians on the megalithic monuments of the Iberian peninsula, France, Ireland, Britain, and Scandinavia and their relationship to the landscape and sky.]

Schaafsma, Polly. *Rock Art in New Mexico* (rev. ed.). Santa Fe: Museum of New Mexico Press, 1992.

Schiffman, Robert A., ed. *Visions of the Sky: Archaeological and Ethnological Studies of California Indian Astronomy*. Salinas, CA: Coyote Press, 1988.

Schlosser, Wolfhard, ed. *Proceedings of the Second SEAC Conference, Bochum, August 29th–31st, 1994*. Bochum, Germany: Astronomisches Institut der Ruhr-Universität, 1996. [Contains articles in English and German.]

Schlosser, Wolfhard, and Jan Cierny. *Sterne und Steine: Eine praktische Astronomie der Vorzeit*. Darmstadt: Wissenschaftliche Buchgesellschaft, 1996. [An introduction to archaeoastronomy focusing on European case studies and practical techniques. In German.]

Selin, Helaine, ed. *Astronomy across Cultures*. Dordrecht, Neth.: Kluwer, 2000. [A compendium of preliterate astronomical traditions among human societies in places as diverse as central and southern Africa, the Islamic world, southeast Asia, Aboriginal Australia, and ancient Polynesia.]

Serio, Salvatore, ed. *Proceedings of the INSAP III Conference*. Pisa: Società Astronomica Italiana (Memoire della Società Astronomica Italiana, special issue no. 1), 2002. [A collection of papers from the third INSAP (Inspiration of Astronomical Phenomena) conference held in Palermo in 2001. The remit of the INSAP conferences is very broad but includes aspects of cultural astronomy.]

Shank, Michael, ed. *The Scientific Enterprise in Antiquity and the Middle Ages*. Chicago: University of Chicago Press, 2000. [A compilation of papers by leading historians of science.]

Shaw, Ian, ed. *The Oxford History of Ancient Egypt*. Oxford: Oxford University Press, 2000. [An indispensable background to discussions of ancient Egyptian astronomy.]

Shee Twohig, Elizabeth, and Margaret Ronayne, eds. *Past Perceptions: The Prehistoric Archaeology of South-West Ireland*. Cork: Cork University Press, 1993. [Contains useful background material on the prehistoric monuments in this area that have possible astronomical significance.]

Siarkiewicz, Elżbieta. *El Tiempo en el Tonalamatl*. Warsaw: University of Warsaw (Cátedra de Estudios Ibéricos), 1995. [In Spanish.]

Silverman, Helaine, ed. *Andean Archaeology*. Malden, MA: Blackwell, 2004. [A collection of articles providing an excellent introduction to the archaeology of this region.]

Silverman, Helaine, and Donald A. Proulx. *The Nasca*. Oxford: Blackwell, 2002. [Essential archaeological introduction to the Nasca culture.]

Smith, A. T., and A. Brookes, eds. *Holy Ground: Theoretical Issues Relating to the Landscape and Material Culture of Ritual Space*. Oxford: Archaeopress (BAR International Series 956), 2001. [A collection of conference papers.]

Souden, David. *Stonehenge: Mysteries of the Stones and Landscape.* London: Collins and Brown/English Heritage, 1997. [A sound general introduction to the archaeology of Stonehenge and its landscape, including astronomy.]

Spalinger, Anthony, ed. *Revolutions in Time: Studies in Egyptian Calendrics (Varia Aegyptiaca).* San Antonio, TX: Van Siclen Press, 1994.

Šprajc, Ivan. *La Estrella de Quetzalcóatl: El Planeta Venus en Mesoamérica.* Mexico City: Editorial Diana, 1996. [In Spanish.]

———. *Orientaciones Astronómicas en la Arquitectura Prehispánica del Centro de México.* Mexico City: Instituto Nacional de Antropología e Historia (Colección Científica 427), 2001. [In Spanish.]

Squier, Ephraim, and Edwin Davis. *Ancient Monuments of the Mississippi Valley.* Washington, DC: Smithsonian Institution Press, 1998. (Originally published in 1848.) [A classic antiquarian text on the monumental earthworks in this part of native North America.]

Stanbury, Peter, and John Clegg. *A Field Guide to Aboriginal Rock Engravings.* Sydney: Sydney University Press, 1990. [An extremely useful field guide to Aboriginal rock engravings in the vicinity of Sydney. Also includes a list of other selected sites dispersed around Australia.]

Stănescu, Florin, ed. *Ancient Times, Modern Methods.* Sibiu, Romania: "Lucian Blaga" University, 1999. [A collection of papers from the SEAC (European Society for Astronomy in Culture) meeting held in Romania in 1995. Contains articles in English, German, and French.]

Steel, Duncan. *Rogue Asteroids and Doomsday Comets: The Search for the Million Megaton Menace that Threatens Life on Earth.* New York: Wiley, 1995. [Unsustainable speculations that Stonehenge was built to observe and predict cataclysmic impacts.]

Steele, John. *Observations and Predictions of Eclipse Times by Early Astronomers.* Dordrecht, Neth.: Kluwer, 2000.

Stephenson, Bruce, Marvin Bolt, and Anna Friedman. *The Universe Unveiled: Instruments and Images through History.* Chicago: Adler Planetarium and Astronomy Museum, and Cambridge: Cambridge University Press, 2000. [Richly illustrated account of early astronomical instruments.]

Stephenson, F. Richard. *Historical Eclipses and Earth's Rotation.* Cambridge: Cambridge University Press, 1997.

Stephenson, F. Richard, and David Clark. *Applications of Early Astronomical Records.* Bristol: Adam Hilger, 1978.

Stephenson, F. Richard, and David A. Green. *Historical Supernovae and their Remnants.* Oxford: Oxford University Press, 2002.

Sterling, Elspeth P. *Sites of Maui.* Honolulu: Bishop Museum Press, 1998. [Gazetteer of archaeological sites on the Hawaiian island of Maui.]

Sterling, Elspeth P., and Catherine C. Summers. *Sites of Oahu.* Honolulu: Bishop Museum Press, 1978. [Gazetteer of archaeological sites on the Hawaiian island of Oʻahu.]

Stokes, John. *Heiau of the Island of Hawaiʻi.* Honolulu: Bishop Museum Press, 1991. [A gazetteer of temple sites on the Big Island of Hawaiʻi. Written in c. 1919, but unpublished in the meantime.]

Stopford, Jennifer, ed. *Pilgrimage Explored.* Rochester, NY: Boydell and Brewer, 1999. [A set of conference papers, mostly about modern pilgrimage.]

Stout, Geraldine. *Newgrange and the Bend of the Boyne.* Cork: Cork University Press, 2002. [A richly illustrated, large-format book documenting the landscape of the Boyne valley from prehistoric into modern times.]

Summers, Catherine C. *Molokai: A Site Survey.* Honolulu: Department of Anthropology, Bernice P. Bishop Museum, 1971. [Gazetteer of archaeological sites on the Hawaiian island of Molokaʻi.]

Sutton, Douglas G., ed. *The Origins of the First New Zealanders.* Auckland: Auckland University Press, 1994. [A collection of conference papers addressing the issue of the colonization of New Zealand from various disciplinary perspectives.]

Swerdlow, Noel M., ed. *Ancient Astronomy and Celestial Divination.* Cambridge, MA: MIT Press, 1999.

Tava, Rerioteria, and Moses K. Keale. *Niʻihau: The Traditions of an Hawaiian Island.* Honolulu: Mutual Publishing, 1989. [A description of the island and its traditions by two native Hawaiians.]

Tedlock, Barbara. *Time and the Highland Maya* (rev. ed.). Albuquerque: University of New Mexico Press, 1982. [A classic work in ethnography, and indeed in ethnoastronomy.]

———. "Maya Astronomy: What We Know and Why We Know It." *Archaeoastronomy: The Journal of Astronomy in Culture* 14 (1) (1999), 39–58. [A very useful overview.]

Tena, Rafael. *El Calendario Mexica y la Cronografía.* Mexico City: INAH, 1987.

Thom, Alexander. *Megalithic Sites in Britain.* Oxford: Oxford University Press, 1967. [The first of Thom's three seminal works on "megalithic astronomy." Detailed and technical.]

———. *Megalithic Lunar Observatories.* Oxford: Oxford University Press, 1971. [The second of Thom's three seminal works on "megalithic astronomy." Detailed and technical.]

Thom, Alexander, and Archibald S. Thom. *Megalithic Remains in Britain and Brittany.* Oxford: Oxford University Press, 1978. [The third of Alexander Thom's three seminal works on "megalithic astronomy," coauthored with his son Archie. Detailed and technical.]

Thom, Alexander, Archibald S. Thom, and Aubrey Burl. *Megalithic Rings.* Oxford: British Archaeological Reports (BAR British Series 81), 1980. [Gazetteer, containing detailed data and site plans, of 229 British stone circles.]

———. *Stone Rows and Standing Stones.* Oxford: British Archaeological Reports (BAR International Series 560), 1990. [Gazetteer containing detailed data and site plans of several hundred British stone rows, stone pairs, and single standing stones. The gazetteer was superseded by Burl 1993.]

Thom, Archibald S. *Walking in All the Squares: A Biography of Alexander Thom.* Glendaruel, Scotland: Argyll Publishing, 1995. [A personal account of Alexander Thom's life by his son Archie.]

Thompson, J. Eric S. *A Commentary on the Dresden Codex: A Maya Hieroglyphic Book.* Philadelphia: American Philosophical Society, 1972. [A classic work on the Dresden Codex, preceding decades of more recent work on the astronomical tables and also preceding the deciphering of the Maya hieroglyphic script, but valuably including a facsimile of the Codex itself. A Spanish version published by the Fondo de Cultura Económica, Mexico City, in 1988 includes the facsimile as a separate book, folded to replicate the original concertina form.]

Thurston, Hugh. *Early Astronomy.* Berlin: Springer-Verlag, 1994. [Technical themes in the early history of astronomy, including British "megalithic astronomy."]

Tilley, Christopher. *A Phenomenology of Landscape: Places, Paths and Monuments.* Oxford: Berg, 1994. [A classic work applying a phenomenological approach to the interpretation of prehistoric landscapes, which strongly influenced the development of ideas in this field in the 1990s.]

———. *The Dolmens and Passage Graves of Sweden: An Introduction and Guide.* London: Institute of Archaeology, University College London, 1999. [Includes a gazetteer of 150 sites, selected from a total of about 500, with finder maps, detailed descriptions, and plans.]

Toal, Caroline. *North Kerry Archaeological Survey*. Dingle, Ireland: Brandon/Oifig na nOibreacha Poiblí [Office of Public Works], 1995. [Inventory of monuments.]

Tsenev, Gore. *Neboto Nad Makedonija*. Skopje, Macedonia: Mladinski Kulturen Tsentar, 2004. [A compilation of Macedonian folk beliefs relating to the sky, celestial objects, and meteorological phenomena. In Macedonian.]

Tuplin, C. J., and T. E. Rihll, eds. *Science and Mathematics in Ancient Greek Culture*. Oxford: Oxford University Press, 2002.

Ulansey, David. *The Origins of the Mithraic Mysteries: Cosmology and Salvation in the Ancient World*. New York and Oxford: Oxford University Press, 1989.

Urton, Gary. *At the Crossroads of the Earth and Sky: An Andean Cosmology*. Austin: University of Texas Press, 1981. [An in-depth ethnoastronomical study at Misminay, a remote village in the Peruvian Andes.]

———. *The Social Life of Numbers*. Austin: University of Texas Press, 1997. [A study of numerical knowledge and arithmetical practices among modern Quechua-speaking peoples.]

———. *Inca Myths*. Austin: University of Texas Press, and London: British Museum Press, 1999. [Excellent short introduction.]

———. *Signs of the Inka Khipu: Binary Coding in the Andean Knotted-String Records*. Austin: University of Texas Press, 2003. [Ground-breaking new ideas about the way in which information was stored on the Inca knotted-string devices known as *quipu* (*khipu*).]

Valeri, Valerio. *Kingship and Sacrifice: Ritual and Society in Ancient Hawaii*. Chicago: University of Chicago Press, 1985. [A detailed study of Hawaiian sacrificial rituals.]

Valev, Peter, ed. First National Symposium on Archaeoastronomy, Tolbukhin '88. *Interdisciplinary Studies,* vol. 15. Sofia: Archaeological Institute and Museum, 1988. [A motley collection of abstracts from the first meeting of its type in Europe, which led to subsequent annual conferences on archaeoastronomy in different countries and eventually to the formation of the European Society for Astronomy in Culture (SEAC). Contains abstracts in Bulgarian, Russian, and English.]

———. First National Symposium on Archaeoastronomy, Tolbukhin '88. *Interdisciplinary Studies,* vol. 17, 102–219. Sofia: Archaeological Institute and Museum, 1990. [Papers from the 1988 Tolbukhin meeting, mostly focusing on Bulgaria and central Europe. Contains articles in Bulgarian and English.]

———. First National Symposium on Archaeoastronomy, Tolbukhin '88. *Interdisciplinary Studies,* vol. 18. Sofia: Archaeological Institute and Museum, 1991. [Further papers from the 1988 Tolbukhin meeting, mostly focusing on Bulgaria and central Europe. Contains articles in Bulgarian, Russian, and English.]

Van der Waerden, Bartel L. *Geometry and Algebra in Ancient Civilizations*. Berlin: Springer-Verlag, 1983. [A pithy and technical synthesis by a leading Dutch mathematician and historian of science.]

Volkov, V. V., E. N. Kaurov, M. F. Kosarev, and T. M. Potyomkina, eds. *Archaeoastronomy: Emerging Problems*. Moscow: Institute of Archaeology, Russian Academy of Sciences, 1996. [A collection of papers from a conference held in Moscow in 1996. In Russian.]

Waddell, John. *The Prehistoric Archaeology of Ireland*. Galway: Galway University Press, 1998. [Detailed and authoritative.]

Walker, Christopher, ed. *Astronomy before the Telescope*. London: British Museum Press, 1996. [A broad range of case studies spanning archaeoastronomy, the history of astronomy, and indigenous astronomies. Provides an excellent overview.]

Watkins, Alfred. *The Old Straight Track*. London: Abacus, 1974. Originally published in 1925 by Methuen, London. [The original publication that gave rise to the ley line phenomenon in its various manifestations. Of historic interest only.]

Waugh, Albert E. *Sundials: Their Theory and Construction*. New York: Dover, 1973. [A technical overview of different types of sundials and how they function.]

Wheatley, Paul. *The Pivot of the Four Quarters: A Preliminary Enquiry into the Origins and Character of the Ancient Chinese City*. Edinburgh: Edinburgh University Press, 1971.

Whittington, E. Michael, ed. *The Sport of Life and Death: The Mesoamerican Ballgame*. London: Thames and Hudson, 2001. [Overview papers accompanying what started as an exhibition catalog, in large format with excellent color illustrations. An excellent introduction.]

Whittle, Alasdair. *Europe in the Neolithic: The Creation of New Worlds*. Cambridge: Cambridge University Press, 1996. [Detailed, wide-ranging, and authoritative synthesis with thought-provoking interpretations.]

———. *The Archaeology of People: Dimensions of Neolithic Life*. New York: Routledge, 2003. [An innovative interpretative framework by a leading British prehistorian offering fresh insights into daily life in early Neolithic Europe.]

Williamson, Ray A., ed. *Archaeoastronomy in the Americas*. Los Altos, CA: Ballena Press, and College Park, MD: Center for Archaeoastronomy, 1981.

———. *Living The Sky: The Cosmos of the American Indian*. Norman: University of Oklahoma Press, 1987.

Williamson, Ray A., and Claire R. Farrer, eds. *Earth and Sky: Visions of the Cosmos in Native American Folklore*. Albuquerque: University of New Mexico Press, 1992. [A collection of papers presenting ethnoastronomical studies in a range of native North American communities.]

Williamson, Tom, and Liz Bellamy. *Ley Lines in Question*. Tadworth, England: World's Work, 1983. [A critical but serious examination of ley lines and of "alternative archaeology" in general, by two archaeologists.]

Wood, John E. *Sun, Moon and Standing Stones*. Oxford: Oxford University Press, 1978. [One of a number of books describing archaeoastronomy for a general audience that followed in the wake of Alexander Thom's theories before their reassessment.]

Xu Zhentao, David Pankenier, and Jiang Yaotiao. *East Asian Archaeoastronomy: Historical Records of Astronomical Observations of China, Japan and Korea*. Amsterdam: Gordon and Breach, 2000. [An extremely useful sourcebook of references to astronomical phenomena among East Asian historical sources, including original texts, English translations, and commentaries.]

Zaldua, Luix Mari. *Saroeak Urnietan [Stone Octagons in Urnieta]*. Urnieta, Spain: Kulturnieta, 1996. [In Basque, Spanish, and English.]

Zedda, Mauro Peppino. *I Nuraghi: Il Sole La Luna*. Cagliari: Ettore Gasperini Editore, 1991. [Methodologically unconstrained identifications of numerous solar and lunar alignments between Sardinian nuraghi, written before the author embraced stricter procedures. In Italian.]

———. *I Nuraghi tra Archeologia e Astronomia*. Cagliari, Sardinia, Italy: Agorà Nuragica, 2004. [Detailed descriptions with the author's own interpretations. In Italian.]

Żerańska-Kominek, Sławomira, with Arnold Lebeuf. *The Tale of Crazy Harman*. Warsaw: DIALOG, 1997. [An epic narrative from Turkmenistan with commentary including interpretations of its astronomical/cosmological imagery. Originally published in Polish.]

Ziółkowski, Mariusz, and Robert Sadowski, eds. *Time and Calendars in the Inca*

*Empire*. Oxford: British Archaeological Reports (International Series 479), 1989. [A collection of papers presented at a conference in Bogotá in 1985. Contains articles in English and Spanish.]

Zuidema, R. Tom. *The Ceque System of Cuzco: The Social Organization of the Capital of the Inca*. Leiden: Brill, 1964. [Pioneering work on the Cusco *ceque* system, establishing their relation to concepts of space and time and social engineering. Reprinted in Spanish by Pontificia Universidad Católica del Perú, Lima, in 1995.]

# Glossary

*This glossary gives short definitions of certain terms that occur regularly within thematic entries and case studies. Some readers will find useful a fuller explanation of some of these concepts, and an asterisk indicates that the topic is elaborated further in the main body of the encyclopedia. Italicized words cross-refer to other glossary entries.*

*ACCURACY. The closeness of a measurement to the true value. Cf. *Precision*.

*ACRONICAL RISE. [Alternatively spelled "acronychal," for example in British usage.] Strictly, the rising of a star at sunset. This event is not directly visible and the term is sometimes loosely applied to the annual last visible rising of a star in or after the evening twilight.

*ACRONICAL SET. [Alternatively spelled "acronychal," for example in British usage.] Strictly, the setting of a star at sunrise. This event is not directly visible and the term is sometimes loosely applied to the annual first visible setting of a star in or before the morning twilight.

*ALTITUDE. The angle between the direction to an observed point and the horizontal plane through the observer, with a positive value indicating that the point is above the observer. Cf. *Elevation*.

ANTIZENITH. The point directly beneath the observer, also known as the *nadir*. Cf. *Zenith*.

ARC MINUTE. An angle equal to one sixtieth part of a degree, and hence 1 / (60 × 360) or one 21,600th part of the whole circle. This angle is roughly equivalent to the distance between the front and back of a sheep or similar-sized animal viewed side-on (one meter, or three feet) at a distance of about three kilometers (two miles). The apparent diameter of the sun and moon, by comparison, is about thirty arc minutes.

ARC SECOND. An angle of one sixtieth of an *arc minute*, and hence 1 / (60 × 60 × 360) or one 1,296,000th part of the whole circle. This angle is roughly equivalent to the distance between the eyes of a sheep or similar-sized animal viewed front-on (ten centimeters, or four inches) at a distance of about eighteen kilometers (eleven miles).

ASTERISM. In modern astronomy, a grouping or pattern of stars that does not form a (modern Western) constellation; in cultural astronomy the term may also be applied to a (historical or indigenous) constellation.

ASTROLABE. An instrument for measuring and determining the positions of the sun and bright stars as seen from given locations at given times; in many senses a precursor to the modern planisphere.

*AZIMUTH. Bearing clockwise from due north.

CELESTIAL NORTH POLE. Point on the *celestial sphere,* visible from any location in the northern hemisphere where the relevant part of the sky is unobscured, around which the celestial bodies appear to rotate daily (in a counterclockwise direction).

CELESTIAL SOUTH POLE. Point on the *celestial sphere,* visible from any location in the southern hemisphere where the relevant part of the sky is unobscured, around which the celestial bodies appear to rotate daily (in a clockwise direction).

*CELESTIAL SPHERE. Imaginary sphere surrounding the observer on which the celestial bodies can be considered to be positioned.

*CIRCUMPOLAR STAR. For a given location on earth, a star that is always above the horizon and so never sets or rises.

CODEX. In the Mesoamerican context, an indigenous "book" made from pages of painted tree bark folded together in concertina fashion.

CODICES. Plural of *Codex*.

COSMOGRAM. A symbolic representation of the cosmos, as found in some Mesoamerican *codices*.

CULMINATION. The moment when a celestial object reaches its highest *altitude*, which happens when it crosses the *meridian*.

*DECLINATION. "*Latitude*" on the spinning *celestial sphere*.

*DIURNAL MOTION. Daily rotation of the *celestial sphere*.

DOLMEN. Most usually, a *megalithic monument* comprising a large horizontal stone supported by a number of upright stones. Sometimes the term is applied to a wider range of megalithic tombs.

*ECLIPTIC. The annual path of the sun through the stars on the *celestrial sphere*.

*ELEVATION. Height above sea level. Cf. *Altitude*.

*EQUINOXES. Technically, the times when the sun's *declination* is zero, but often loosely taken to mean the halfway point in time between the *solstices*.

*EXTINCTION. The dimming of a star at low *altitude* due to the earth's atmosphere.

GNOMON. A rod, pole, or equivalent device whose function is to cast a shadow for the purposes of reckoning time, typically as part of a sundial.

*HELIACAL RISE. The annual first appearance of a star, or the first appearance of a planet, in the eastern sky before sunrise after a period of invisibility.

*HELIACAL SET. The annual last appearance of a star, or the last appearance of a planet, in the western sky after sunset before a period of invisibility.

HENGE. A prehistoric enclosure consisting of a circular ditch and adjacent external bank, of which many were constructed in Britain and Ireland during the Late Neolithic period. Associated rings of timber posts or stone circles were common but not universal, and were not always part of the original design. Although the term "henge" derives from Stonehenge (that is, like Stonehenge but without the stones), Stonehenge itself—ironically—is not now generally considered a henge, since the circular ditch and bank at the site are of early (Middle Neolithic) date and the bank is internal to the ditch.

HIEROPHANY. An appearance of the sacred, or the experience of this. The term is commonly used to describe a spectacular effect perceived as mystical, such as those produced by sunlight and shadow at ancient sites, but only on rare occasions.

INFERIOR PLANET. A planet whose orbit is closer to the sun than the Earth's. The two inferior planets, Venus and Mercury, are both visible to the naked eye, although Mercury, being closer to the sun, is more seldom seen. Cf. *Inner planet* and *Superior planet*.

INNER PLANET. One of the four planets closest to the sun, namely Mercury, Venus, Earth and Mars. Cf. *Inferior planet* and *Outer planet*.

KIVA. A subterranean building in the southwestern United States, roofed and used as a meeting place or for sacred ceremonies.

LATITUDE. Angular distance from the equator of a location on the earth. Because of a confusion of terminology it is better to speak of *declination* on the spinning *celestial sphere*.

LUNATION. The time that it takes the moon to complete its cycle of phases. Equivalent to *synodic month*.

MAJOR STANDSTILL LIMITS. The most northerly or southerly possible rising or setting positions of the moon, which can only be reached around the time of a "major standstill," which only occurs every 18.6 years. See the encyclopedia entry Moon, Motions of.

MEGALITHIC MONUMENT. Any monument built of large stones, applied particularly to prehistoric tombs and temples built in Europe during the Neolithic and Early Bronze Age.

*MERIDIAN. Imaginary line across the sky, joining the north point on the horizon to the south point and passing through the *zenith*.

MIDSUMMER SUNRISE. Popular term for sunrise at the *summer solstice*.

MIDSUMMER SUNSET. Popular term for sunset at the *summer solstice*.

MIDWINTER SUNRISE. Popular term for sunrise at the *winter solstice*.

MIDWINTER SUNSET. Popular term for sunset at the *winter solstice*.

NADIR. The point directly beneath the observer, also known as the *antizenith*. Cf. *Zenith*.

NEOLITHIC. Broadly speaking, a prehistoric society whose subsistence is based on animal herding or agriculture but lacks any knowledge of metallurgy. Meaning "New Stone Age," the term is a remnant of the "three-age" classification (Stone Age, Bronze Age, Iron Age) that has long been superseded as a primary indicator of social development but is often still retained for convenience. The term is seldom used in reference to the Americas.

*OBLIQUITY OF THE ECLIPTIC. The amount by which the earth's axis is "tilted" with respect to the plane of its orbit about the sun. More precisely, the angle between the earth's axis and the direction perpendicular to the plane of its orbit.

OUTER PLANET. One of the five planets furthest from the sun, namely Jupiter, Saturn, Uranus, Neptune and Pluto. Only Jupiter and Saturn are visible to the naked eye. Cf. *Inner planet* and *Superior planet*.

*PARALLAX. The difference between the position of an object in the sky as viewed (conceptually) from the center of the earth and (actually) from somewhere on the earth's surface. It is only significant in the case of the moon.

*PRECESSION. Short for "precession of the equinoxes." A gradual change in the orientation of the earth's axis with respect to distant space, which results in the positions where the solstices and equinoxes occur slowly shifting along the earth's orbit. This significantly alters the position of any given star in the sky over a timescale of centuries.

*PRECISION. The degree of consistency of several measurements of the same thing by a particular method. Cf. *Accuracy*.

QUADRANT. An instrument for measuring the *altitude* of a celestial body.

QUADRIPARTITE COSMOLOGY. A perception of the world, or cosmos, as viewed from a central place as being divided horizontally into four quarters, each with distinctive properties and qualities.

*REFRACTION. The bending of light, in particular as it passes through the atmosphere, which causes astronomical objects near to the horizon to appear higher than they actually are.

SARSEN. A type of large sandstone block found extensively in southern England and used in the construction of Stonehenge.

*SOLSTICES. The times of year when the sun reaches its highest or lowest *declination* and moves on its most extreme daily path across the sky (which, outside the tropics, is its highest or lowest). One solstice occurs on or close to June 21 and the other on or close to December 22.

SUMMER SOLSTICE. June *solstice* (northern hemisphere) or December *solstice* (southern hemisphere).

SUPERIOR PLANET. A planet whose orbit is farther out from the sun than the earth's. The superior planets visible to the naked eye are Mars, Jupiter, and Saturn. Cf. *Inferior planet* and *Outer planet*.

SYNODIC MONTH. The time that it takes the moon to complete its cycle of phases. Equivalent to *lunation*.

TRILITHON. A formation of three stones, with two vertical uprights holding up a horizontal lintel, as at Stonehenge and Ha'amonga-a-Maui.

TROPICAL YEAR. The seasonal year, equal to the time taken by the sun to complete one circuit around the *celestial sphere*. It is not quite the same as the time taken for the earth to complete an orbit of the sun (the "sidereal year") because of *precession*.

WINTER SOLSTICE. December *solstice* (northern hemisphere) or June *solstice* (southern hemisphere).

ZENITH. The point in the sky directly above the observer. Cf. *Antizenith* or *nadir*.

ZODIAC. A set of *asterisms* spread around the *ecliptic* and hence useful as reference points for the sun's annual motion through the stars. The term is commonly used to refer to the particular set of twelve zodiacal constellations that has come to be recognized in the modern Western world.

# Topic Index

*This list indicates the different types of entries to be found in this encyclopedia.* Concepts *includes broad definitions and explanations of* key concepts, *together with nontechnical explanations of more* basic concepts, *mostly astronomical, often elaborating on short definitions to be found in the glossary. Some entries elaborate upon key* themes *and* issues *relating to ancient astronomy.* Case studies, *which illustrate these and other key issues and themes, are organized in this list by continent: the Americas are split into North, South, and Mesoamerica, and Europe is divided into Britain and Ireland and Continental Europe. The entries also include a very small sample of* people *important in the development of our knowledge of ancient astronomy, and an equally small sample of* sky objects and events and their cultural significance. *Items such as Sirius and the Pleiades that do not receive entries of their own may nonetheless appear in others, and these can be located using the general index. Finally, there are brief descriptions of a number of* procedures and techniques *that are used in the modern practice of archaeoastronomy.*

**Case Studies**

*Africa*
  Ancient Egyptian Calendars
  Borana Calendar
  Coffin Lids
  Egyptian Temples and Tombs
  Mursi Calendar
  Nabta Playa
  Namoratung'a
  Pyramids of Giza

*America (North)*
  Cahokia
  Casa Rinconada
  Chaco Canyon
  Chaco Meridian
  Chaco Supernova Pictograph
  Fajada Butte Sun Dagger
  Hopewell Mounds
  Hopi Calendar and Worldview
  Inuit Cosmology
  Lakota Sacred Geography
  Medicine Wheels
  Navajo Cosmology
  Navajo Hogan
  Navajo Star Ceilings

  Pawnee Cosmology
  Pawnee Earth Lodge
  Pawnee Star Chart
  Presa de la Mula
  Sky Bears

*America (South)*
  Andean Mountain Shrines
  Barasana "Caterpillar Jaguar" Constellation
  *Ceque* System
  Cusco Sun Pillars
  Island of the Sun
  Misminay
  Nasca Lines and Figures
  Quipu
  Yekuana Roundhouses

*Asia*
  Angkor
  Babylonian Astronomy and Astrology
  Carahunge
  Chinese Astronomy
  Crucifixion of Christ
  Islamic Astronomy
  Javanese Calendar

Land of the Rising Sun
Rujm el-Hiri
Star of Bethlehem

*Australasia and Oceania*
Aboriginal Astronomy
Easter Island
Emu in the Sky
Haʻamonga-a-Maui
Hawaiian Calendar
Kumukahi
Kumulipo
Mangareva
Nā Pali Chant
Navigation in Ancient Oceania
Necker Island
Polynesian and Micronesian Astronomy
Polynesian Temple Platforms and Enclosures
Star Compasses of the Pacific
Zenith Stars in Polynesia

*Britain and Ireland*
Avebury
Axial Stone Circles
Ballochroy
Beltany
Boyne Valley Tombs
Brainport Bay
Brodgar, Ring of
Bush Barrow Gold Lozenge
Callanish
Church Orientations
Circles of Earth, Timber, and Stone
Clava Cairns
Cumbrian Stone Circles
Cursus Monuments
Drombeg
Fiskerton
Iron Age Roundhouses
Kintraw
Maes Howe
Megalithic Monuments of Britain and Ireland
Newgrange
Recumbent Stone Circles
Short Stone Rows
Stone Circles
Stonehenge
Thornborough
Tri-Radial Cairns
Wedge Tombs

*Continental Europe*
Abri Blanchard Bone
Antas
Church Orientations
Crucuno
Delphic Oracle
Grand Menhir Brisé
Gregorian Calendar
Is Paras
Julian Calendar
Minoan Temples and Tombs
Mithraism
Nebra Disc
Nuraghi
Pantheon
Prehistoric Tombs and Temples in Europe
Roman Astronomy and Astrology
Sarmizegetusa Regia
Saroeak
Sky Bears
Son Mas
Swedish Rock Art
Taulas
Temple Alignments in Ancient Greece

*Mesoamerica*
Aztec Sacred Geography
Cacaxtla
Caracol at Chichen Itza
Dresden Codex
Governor's Palace at Uxmal
Group E Structures
Horizon Calendars of Central Mexico

Kukulcan
Maya Long Count
Mesoamerican Calendar Round
Mesoamerican Cross-Circle Designs
Teotihuacan Street Grid
Venus in Mesoamerica
Zenith Tubes

## Concepts

*Basic Concepts*
Altitude
Azimuth
Celestial Sphere
Circumpolar Stars
Declination
Diurnal Motion
Ecliptic
Extinction
Heliacal Rise
How the Sky Has Changed over the Centuries
Inferior Planets, Motions of
Lunar Parallax
Lunar Phase Cycle
Meridian
Moon, Motions of
Obliquity of the Ecliptic
Precession
Refraction
Solstices
Star Names
Star Rising and Setting Positions
Sun, Motions of
Superior Planets, Motions of
Years B.C.E. and Years before 0

*Key Concepts*
Alignment Studies
Archaeoastronomy
Archaeotopography
Astro-Archaeology
"Brown" Archaeoastronomy
Cosmology
Ethnoastronomy
"Green" Archaeoastronomy
"Megalithic Astronomy"

## Issues
Astrology
Astronomical Dating
Celtic Calendar
Ethnocentrism
Ley Lines
"Megalithic" Calendar
Megalithic "Observatories"
Methodology
Nationalism
Science or Symbolism?

## People
Cobo, Bernabé (1582–1657)
Hesiod (Eighth Century B.C.E.)
Lockyer, Sir Norman (1836–1920)
Nissen, Heinrich (1839–1912)
Somerville, Boyle (1864–1936)
Thom, Alexander (1894–1985)

## Procedures and Techniques
Compass and Clinometer Surveys
Field Survey
GPS Surveys
Precision and Accuracy
Statistical Analysis
Theodolite Surveys

## Sky Objects and Events and Their Cultural Significance
Antizenith Passage of the Sun
Catastrophic Events
Comets, Novae, and Meteors
Equinoxes
Lunar Eclipses
Magellanic Clouds
Mid-Quarter Days
Orion
Solar Eclipses
Solstitial Directions
Zenith Passage of the Sun

**Themes**
- Calendars
- Cardinal Directions
- Christianization of "Pagan" Festivals
- Cognitive Archaeology
- Constellation Maps on the Ground
- Eclipse Records and the Earth's Rotation
- Landscape
- Lunar and Luni-Solar Calendars
- Monuments and Cosmology
- Navigation
- Orientation
- Palaeoscience
- Pilgrimage
- Power
- Sacred Geographies
- Solstitial Alignments
- Space and Time, Ancient Perceptions of
- Star and Crescent Symbol
- Symbols

# Geographical Index

*This index lists the various case studies according to the modern country within which the culture area in question falls. Where they span more than one, only the main one (or ones) is listed. Entries not confined to a single country or even a few countries are listed separately at the end.*

**Argentina**
Andean Mountain Shrines

**Armenia**
Carahunge

**Australia**
Aboriginal Astronomy
Emu in the Sky

**Bolivia**
Andean Mountain Shrines
Island of the Sun

**Cambodia**
Angkor

**Canada**
Inuit Cosmology
Medicine Wheels

**Chile**
Andean Mountain Shrines
Easter Island

**China**
Chinese Astronomy

**Colombia**
Barasana "Caterpillar Jaguar" Constellation

**Egypt**
Ancient Egyptian Calendars
Coffin Lids
Egyptian Temples and Tombs
Nabta Playa
Pyramids of Giza

**Ethiopia**
Borana Calendar
Mursi Calendar

**France**
Abri Blanchard Bone
Crucuno
Grand Menhir Brisé

**French Polynesia**
Mangareva

**Germany**
Nebra Disc

**Greece**
*Mainland*
Delphic Oracle
Hesiod (Eighth Century B.C.E.)
Temple Alignments in Ancient Greece

*Crete*
Minoan Temples and Tombs

**Guatemala**
Group E Structures
Maya Long Count
Mesoamerican Calendar Round
Venus in Mesoamerica

**Indonesia**
Javanese Calendar

**Iraq**
Babylonian Astronomy and Astrology

**Ireland**
  Axial Stone Circles
  Beltany
  Boyne Valley Tombs
  Drombeg
  Megalithic Monuments of Britain and Ireland
  Newgrange
  Short Stone Rows
  Stone Circles
  Wedge Tombs

**Israel and Palestinian Territories**
  Rujm el-Hiri
  Crucifixion of Christ
  Star of Bethlehem

**Italy**
*Mainland*
  Julian Calendar
  Pantheon

*Sardinia*
  Is Paras
  Nuraghi

**Japan**
  Land of the Rising Sun

**Kenya**
  Borana Calendar
  Namoratung'a

**Mexico**
  Aztec Sacred Geography
  Cacaxtla
  Caracol at Chichen Itza
  Dresden Codex
  Governor's Palace at Uxmal
  Horizon Calendars of Central Mexico
  Kukulcan
  Maya Long Count
  Mesoamerican Calendar Round
  Mesoamerican Cross-Circle Designs
  Presa de la Mula

  Teotihuacan Street Grid
  Venus in Mesoamerica
  Zenith Tubes

**Peru**
  Andean Mountain Shrines
  *Ceque* System
  Cusco Sun Pillars
  Misminay
  Nasca Lines and Figures
  Quipu

**Portugal**
  Antas

**Romania**
  Sarmizegetusa Regia

**Spain**
*Mainland*
  Antas
  Saroeak

*Balearic Islands*
  Son Mas
  Taulas

**Sweden**
  Swedish Rock Art

**Tonga**
  Ha'amonga-a-Maui

**United Kingdom**
*England and Wales*
  Avebury
  Bush Barrow Gold Lozenge
  Circles of Earth, Timber, and Stone
  Cumbrian Stone Circles
  Cursus Monuments
  Fiskerton
  Iron-Age Roundhouses
  Megalithic Monuments of Britain and Ireland
  Stone Circles

Stonehenge
  Thornborough
  Tri-Radial Cairns

*Scotland*
  Ballochroy
  Brainport Bay
  Brodgar, Ring of
  Callanish
  Circles of Earth, Timber, and Stone
  Clava Cairns
  Iron-Age Roundhouses
  Kintraw
  Maes Howe
  Megalithic Monuments of Britain and Ireland
  Recumbent Stone Circles
  Short Stone Rows
  Stone Circles

**United States of America**
*Mainland*
  Cahokia
  Casa Rinconada
  Chaco Canyon
  Chaco Meridian
  Chaco Supernova Pictograph
  Fajada Butte Sun Dagger
  Hopewell Mounds
  Hopi Calendar and Worldview
  Lakota Sacred Geography
  Medicine Wheels
  Navajo Cosmology
  Navajo Hogan
  Navajo Star Ceilings

  Pawnee Cosmology
  Pawnee Earth Lodge
  Pawnee Star Chart

*Hawaiian Islands*
  Hawaiian Calendar
  Kumukahi
  Kumulipo
  Nā Pali Chant
  Necker Island

**Venezuela**
  Yekuana Roundhouses

**Wide Geographical Extent**
*The following broad case studies are not confined to a single country or even a few countries, and in some cases not even to a single continent:*
  Church Orientations
  Gregorian Calendar
  Islamic Astronomy
  Mithraism
  Navigation in Ancient Oceania
  Polynesian and Micronesian Astronomy
  Polynesian Temple Platforms and Enclosures
  Prehistoric Tombs and Temples in Europe
  Roman Astronomy and Astrology
  Sky Bears
  Star Compasses of the Pacific
  Zenith Stars in Polynesia

# Chronological Index

This list provides a broad indication of the date range, or the principal data range, within which particular case studies fall. The ranges chosen are unavoidably arbitrary, and in such a very wide variety of examples the accuracy to which a date can be quoted varies considerably, both because of the inherent nature of the example concerned and the different types of dating evidence available in any given case. As a general rule we can only present a best estimate on the basis of current evidence. Some items span more than one date range, and the "Modern indigenous" category includes several examples where cultural traditions demonstrably extend back well into the past. Entries not confined to a single date range or even a few such ranges are listed separately at the end.

Labels such as Palaeolithic, Neolithic, Copper/Bronze/Iron Age, etc. have been avoided here on the grounds that they reflect developments occurring at (sometimes very) different dates in different parts of the world and place undue emphasis upon purely technological innovation. Readers interested in the correlation between astronomical practice and the nature of the human society concerned are referred to the Cultural Index that follows.

**c. 30,000 B.C.E.**
  Abri Blanchard Bone

**Fourth millennium B.C.E.**
  Antas
  Boyne Valley Tombs
  Cursus Monuments
  Megalithic Monuments of Britain and Ireland
  Nabta Playa
  Newgrange
  Prehistoric Tombs and Temples in Europe
  Thornborough

**Third millennium B.C.E.**
  Ancient Egyptian Calendars
  Antas
  Avebury
  Babylonian Astronomy and Astrology
  Beltany
  Brodgar, Ring of
  Callanish
  Carahunge
  Circles of Earth, Timber, and Stone
  Coffin Lids
  Crucuno
  Cumbrian Stone Circles
  Egyptian Temples and Tombs
  Grand Menhir Brisé
  Maes Howe
  Megalithic Monuments of Britain and Ireland
  Prehistoric Tombs and Temples in Europe
  Pyramids of Giza
  Recumbent Stone Circles
  Rujm el-Hiri
  Stone Circles
  Stonehenge
  Thornborough
  Wedge Tombs

**Second millennium B.C.E.**
  Ancient Egyptian Calendars
  Axial Stone Circles
  Babylonian Astronomy and Astrology
  Ballochroy

Chronological Index 495

Brainport Bay
Bush Barrow Gold Lozenge
Circles of Earth, Timber, and Stone
Clava Cairns
Coffin Lids
Drombeg
Egyptian Temples and Tombs
Is Paras
Kintraw
Megalithic Monuments of Britain and Ireland
Minoan Temples and Tombs
Nebra Disc
Nuraghi
Prehistoric Tombs and Temples in Europe
Short Stone Rows
Son Mas
Stone Circles
Swedish Rock Art
Taulas
Tri-Radial Cairns

**First millennium** B.C.E.
*General*
Ancient Egyptian Calendars
Babylonian Astronomy and Astrology
Chinese Astronomy
Delphic Oracle
Fiskerton
Hesiod (Eighth Century B.C.E.)
Hopewell Mounds
Horizon Calendars of Central Mexico
Iron Age Roundhouses
Mesoamerican Calendar Round
Prehistoric Tombs and Temples in Europe
Taulas
Temple Alignments in Ancient Greece
Venus in Mesoamerica

*Second and first centuries* B.C.E.
Julian Calendar
Roman Astronomy and Astrology
Sarmizegetusa Regia
Star of Bethlehem

**First millennium** C.E.
*General*
Chinese Astronomy
Easter Island
Group E Structures
Hawaiian Calendar
Hopewell Mounds
Javanese Calendar
Land of the Rising Sun
Maya Long Count
Mesoamerican Calendar Round
Mesoamerican Cross-Circle Designs
Nasca Lines and Figures
Navigation in Ancient Oceania
Necker Island
Polynesian and Micronesian Astronomy
Polynesian Temple Platforms and Enclosures
Saroeak
Teotihuacan Street Grid
Venus in Mesoamerica
Zenith Stars in Polynesia
Zenith Tubes

*First to fourth centuries* C.E.
Crucifixion of Christ
Mithraism
Pantheon
Roman Astronomy and Astrology

*Seventh to tenth centuries* C.E.
Angkor
Cacaxtla
Caracol at Chichen Itza
Governor's Palace at Uxmal
Kukulcan

**Second millennium** C.E.
*General*
Chinese Astronomy
Church Orientations

Islamic Astronomy
Island of the Sun
Kumukahi
Land of the Rising Sun
Medicine Wheels
Nā Pali Chant
Navigation in Ancient Oceania
Polynesian and Micronesian Astronomy
Polynesian Temple Platforms and Enclosures
Saroeak
Star Compasses of the Pacific
Zenith Stars in Polynesia

*Eleventh to fourteenth centuries C.E.*
Angkor
Cahokia
Caracol at Chichen Itza
Casa Rinconada
Chaco Canyon
Chaco Meridian
Chaco Supernova Pictograph
Dresden Codex
Fajada Butte Sun Dagger
Ha'amonga-a-Maui
Kukulcan
Maya Long Count
Mesoamerican Calendar Round
Venus in Mesoamerica

*Fifteenth and sixteenth centuries C.E.*
Aztec Sacred Geography
*Ceque* System
Cusco Sun Pillars
Gregorian Calendar
Mesoamerican Calendar Round
Quipu
Venus in Mesoamerica

*Seventeenth century onward*
Gregorian Calendar
Kumulipo
Navajo Star Ceilings
Pawnee Star Chart

*Modern indigenous*
Aboriginal Astronomy
Andean Mountain Shrines
Barasana "Caterpillar Jaguar" Constellation
Borana Calendar
Emu in the Sky
Hopi Calendar and Worldview
Inuit Cosmology
Lakota Sacred Geography
Mangareva
Misminay
Mursi Calendar
Navajo Cosmology
Navajo Hogan
Pawnee Cosmology
Pawnee Earth Lodge
Yekuana Roundhouses

**Unknown or Wide Chronological Span**
*The following items are undated within wide bounds or cover an extremely wide chronological range:*
Namoratung'a
Presa de la Mula
Sky Bears

# Cultural Index

*In this list, case study entries are classified according to the type of society concerned. Such categorizations inevitably oversimplify, in that they ignore and obscure many important variations between one human society and another. In the 1960s and 1970s, many archaeologists believed that one could classify human societies in straightforward terms—the main ones being band, tribe, chiefdom, and state—arguing a strong correlation between the size of the political unit, social and economic organization, and means of subsistence. Such naïve ideas have largely been superseded. One problem is that they carry the implication that we can measure every human society's progress along a single inevitable path of social development. Added to this, in the case of past societies, the very categorization is a matter of interpretation from the available archaeological or historical evidence and (particularly in prehistory) can in itself be highly questionable. The extent to which (for example) communities in Neolithic Britain developed from segmentary societies (i.e., with no social group or village dominating over the others) into chiefdoms has been a matter of intense debate, and even to focus on this question is arguably to be distracted from more interesting and more complex issues of settlement organization and the development of social differentiation.*

*For these reasons, the set of categories used in this list is intended merely to give the broadest indication of the nature of the human society responsible for a particular set of practices relating to the sky. Thus the* pastoral/agricultural *category covers, on the one hand, communities that are primarily mobile pastoralists and herders, and on the other, societies whose principal means of subsistence is agriculture and who live in fixed homesteads, whether scattered through the landscape or clustered together into villages; the common feature is that they are not organized into chiefdoms or larger political units. It is a broad category but not easy to subdivide, since there is no clear demarcation between mobile pastoralism and sedentary agriculturalism. Likewise the "state" category spans a range from small, competing city-states (as, for example, in pre-Hispanic Mesoamerica) to large empires.*

**Hunter-Gatherer**
- Aboriginal Astronomy
- Abri Blanchard Bone
- Barasana "Caterpillar Jaguar" Constellation
- Emu in the Sky
- Inuit Cosmology
- Presa de la Mula

**Pastoral/Agricultural**
*Prehistoric*
- Antas
- Avebury
- Axial Stone Circles
- Ballochroy
- Beltany
- Boyne Valley Tombs
- Brainport Bay
- Brodgar, Ring of
- Callanish
- Carahunge
- Circles of Earth, Timber, and Stone
- Clava Cairns
- Crucuno
- Cumbrian Stone Circles
- Cursus Monuments
- Drombeg
- Grand Menhir Brisé

Is Paras
Kintraw
Maes Howe
Megalithic Monuments of Britain and Ireland
Nabta Playa
Namoratung'a
Newgrange
Nuraghi
Prehistoric Tombs and Temples in Europe
Recumbent Stone Circles
Short Stone Rows
Son Mas
Stone Circles
Stonehenge
Swedish Rock Art
Taulas
Thornborough
Tri-Radial Cairns
Wedge Tombs

*Native American*
Andean Mountain Shrines
Hopewell Mounds
Hopi Calendar and Worldview
Horizon Calendars of Central Mexico
Lakota Sacred Geography
Medicine Wheels
Misminay
Navajo Cosmology
Navajo Hogan
Navajo Star Ceilings
Pawnee Cosmology
Pawnee Earth Lodge
Pawnee Star Chart
Yekuana Roundhouses

*Oceanic*
Easter Island
Mangareva
Navigation in Ancient Oceania
Necker Island
Polynesian and Micronesian Astronomy
Polynesian Temple Platforms and Enclosures
Star Compasses of the Pacific
Zenith Stars in Polynesia

*Other*
Borana Calendar
Javanese Calendar
Mursi Calendar
Saroeak

**Chiefdom**
*Prehistoric*
Brodgar, Ring of*
Bush Barrow Gold Lozenge
Circles of Earth, Timber, and Stone*
Fiskerton
Iron Age Roundhouses
Maes Howe*
Megalithic Monuments of Britain and Ireland*
Nebra Disc
Prehistoric Tombs and Temples in Europe*
Rujm el-Hiri
Sarmizegetusa Regia
Stone Circles*
Stonehenge*

*Native American*
Cahokia
Casa Rinconada
Chaco Canyon
Chaco Meridian
Chaco Supernova Pictograph
Fajada Butte Sun Dagger
Hopewell Mounds
Nasca Lines and Figures

*Polynesian*
Ha'amonga-a-Maui
Hawaiian Calendar
Kumukahi
Kumulipo
Nā Pali Chant

Polynesian and Micronesian
  Astronomy*
Polynesian Temple Platforms and
  Enclosures*

*Indicates entries that span societies of which only some are generally supposed to be chiefdoms, or where the chiefdom categorization is particularly debatable. They are also included under pastoral/agricultural.

## State
*Near and Far Eastern Civilizations*
  Ancient Egyptian Calendars
  Angkor
  Babylonian Astronomy and
    Astrology
  Chinese Astronomy
  Coffin Lids
  Egyptian Temples and Tombs
  Land of the Rising Sun
  Pyramids of Giza

*Classical Civilizations (Greece and Rome)*
  Crucifixion of Christ
  Delphic Oracle
  Hesiod (Eighth Century B.C.E.)
  Julian Calendar
  Minoan Temples and Tombs
  Mithraism
  Pantheon
  Roman Astronomy and Astrology
  Star of Bethlehem
  Temple Alignments in Ancient
    Greece

*Mesomerican City-States*
  Aztec Sacred Geography
  Cacaxtla
  Caracol at Chichen Itza
  Dresden Codex
  Governor's Palace at Uxmal
  Group E Structures
  Kukulcan
  Maya Long Count
  Mesoamerican Calendar Round
  Mesoamerican Cross-Circle Designs
  Teotihuacan Street Grid
  Venus in Mesoamerica
  Zenith Tubes

*Inca State*
  *Ceque* System
  Cusco Sun Pillars
  Island of the Sun
  Quipu

## Wide Cultural Extent
*The following items cover a wide range of human societies:*
  Church Orientations
  Gregorian Calendar
  Islamic Astronomy
  Sky Bears

# Index

Aboriginal astronomy. *See* Australian Aborigines
Abri Blanchard bone, **5–7**, 228
Accuracy and precision, **347–348**
Acronical rise and set, 7, 180. *See also* Heliacal rise and set
Acrux, 129
Adena, 183
Africa:
   Borana calendar, **45–46**, 60, **284–285**
   burial orientations, 320
   Namoratung'a, 60, **284–285**
   *See also specific people or locations*
Aldebaran, 199, 246–247, 285
Alignment studies, **8**, 20
   archaeological verification at Kintraw, 211–213
   archaeotopography, **23**
   astro-archaeology, **24**
   astronomical dating, **27–29**
   field survey, **158–160**
   GPS surveys, 165
   green archaeoastronomy, 20–21, 169, **169**, 189, 247, 399
   methodological issues, 159, 262–263
   Neolithic "megalithic calendars" hypothesis, 248–250
   orientation, **319–321**
   parallax corrections, 234–235
   precession and, 346–347
   science versus symbolism debate, 375–376
   statistical analysis, **399–401**
   *See also* British megalithic structures; Equinoctial alignments; Lunar alignments; Solar alignments; Solstitial alignments; Stone circles; Thom, Alexander; *specific sites, structures*

Alignment studies, Alexander Thom and. *See* Thom, Alexander
Alnilam, 199
Alnitak, 199
Altair, 199
Altar tomb, 435–436
Altitude, **8–9**, 129, 159, 424
   compass and clinometer surveys, 112
   extinction angle, 154
Amun-Ra, temple of, 144
Anasazi (Pueblo culture), Chaco Canyon. *See* Chaco Canyon
Andean cultures. *See* Inca
Andean mountain shrines, **13**
Andesite Sun, 372
Andorra, 380
Angkor, **14–16**
   "constellation map," 16, 113–114, 262, 355
Angkor Wat, 14–15
*Anglo-Saxon Chronicle*, 232
Antas, **17–18**, 350
Antizenith passage of the sun, **19**, 127
Aotearoa, 301
Applied historical astronomy, 142
Arab/Islamic astronomy, **199–202**
Archaeoastronomy, **19–23**
   applied historical astronomy, 142
   astro-archaeology, **24**
   brown, 20–21, **52**
   definition, 19
   ethnoastronomy, **152**
   ethnocentrism, 21
   field survey, **158–160**
   green, 20–21, 169, 189, 247, 399
   journals, 20
   statistical analysis, **399–401**
Archaeoastronomy, alignment studies. *See* Alignment studies

Archaeology:
   cognitive, **108–109**
   environmental, 108
   post-processual, 21–22
   processual, 21
   social, 108
Archaeotopography, **23**
Arcturus, 4, 396
Armenia, 65–67
Armenoi, 266
Ashanti, 320
Astro-archaeology, **24**
Astrology, **24–27**
   Babylonian, 39–40
   Chinese traditions, 24, 90, 92
   ephemerides, 40
   horoscopes, 26, 39–40
   Roman, 366
Astronomical artifacts:
   Abri Blanchard bone, **5–7**
   Bush Barrow gold lozenge, **52–54**
   Inca *quipu,* 78, 291, **357–359**
   Nebra disc, **304–307**
   Pawnee star chart, **332–333**
   Täi plaque, 7
Astronomical dating, **27–29**
Astronomical instruments:
   ancient China, 91, 94
   navigation gourds, 395
   precision and accuracy, **347–348**
   zenith tubes, 188, 344, 444, **445–448**
Astronomy, applied historical, 142
Atkinson, Richard, 225
Atmospheric extinction, 153. *See also* Extinction
Atmospheric refraction, **364–365**
Aubrey Holes, 382–383, 406–407
Auchmallidie, 363
Australian Aborigines, **1–5**
   celestial emu, 4, 22, **147–148**
   dreaming paths, 3, 223
   Dreamtime, 2
   Magellanic Clouds and, 241
   rock art, 3–4
Autumnal equinox, 148–149. *See also* Equinoctial alignments
Avebury, **29–30**, 84, 100, 101, 123, 250
Aveni, Anthony, 20, 171, 262
Axial stone circles, **30–31**, 134–136, 361, 435
Azimuth, 33, 129, 159, 200, 268–269, 424

   compass and clinometer surveys, 112
   GPS surveys, 165
Aztecs, 255
   Calendar Round, 256
   calendar system, 34
   eclipse records, 382
   lunar eclipses and, 232
   meteors and, 344
   rain and fertility beliefs, 56
   sacred geography, **33–35**, 223
   space and time perceptions, 391
   Venus and, 433

Babylon, **37–41**, 233
   astrology, 39–40
   calendar system, 38
   writing and number systems, 37–38
Ballochroy, **41–42**, 252–253, 347, 377
Balnuaran, 103–105, 117
Baltic cultures, 26
Barasana celestial caterpillar, 27, **43**, 116
Basque stone octagons (*saroeak*), 372–374
Bassi, Marco, 285
Bauer, Brian, 79
Bauval, Robert, 355
Bayesian paradigm, 401
B.C.E., 439
Bear, celestial, **378–380**
Bellatrix, 285
Belmonte, Juan Antonio, 11
Beltaine, 75, 265
Beltany, Ireland, **44**
*Bencet,* 208
Berrybrae, 363
Betelgeuse, 396
Bethlehem, Star of, **396–398**
Big Horn, 246, 387
*Bintang Weluku,* 207–208
Bog causeway (Fiskerton), 160–162, 439
Bohonagh, 32
Boorong, 4
Borana calendar, **45–46**, 60, 284–285
Boyne Valley tombs, **46–47**, 150. *See also* Newgrange, Ireland
Bradley, Richard, 101, 104, 363
Brainport Bay, Scotland, **48–50**, 150, 204, 249–250, 386
British bog causeway, 160–162
British burial mounds (Bush Barrow), 52–54
British causewayed enclosures, 99–100

British cursus monuments, 123–126, 334, 427
British henge monuments, 125, 323, 335
    Avebury, 29–30
    Ring of Brodgar, **50–52**
    Thornborough, 125, 323, 335, **427–429**
British Iron-Age roundhouses, **195–197**, 320
British megalithic structures, 8, **250–252**, 271–272, 401–404, 405–409
    Ballochroy, **41–42**, 252–253, 347, 377
    Callanish, **61–63**, 250, 389
    Cumbrian stone circles, **122**
    Kintraw, **211–213**, 253, 386
    linear monuments, 84–85, 123
    "megalithic calendar" hypothesis, **248–250**
    Ring of Brodgar, **50–52**
    stone transport, 402, 407
    *See also* Ireland; Scotland; Stone circles; Stonehenge; *specific sites or structures*
British megalithic structures, Thom's alignment studies of. *See* Thom, Alexander
British Neolithic calendar, 237
British tri-radial cairns, **429–430**
Brittany:
    Carnac, 119, 166, 348
    Crucuno megalithic enclosure, **119–121**
    Grand Menhir Brisé, **166–168**
    tombs and cromlechs, 166
Broda, Johanna, 189
Brodgar, Ring of, **50–52**, 253
Brown archaeoastronomy, 20–21, 52
Bruck, Peter, 242
Burial mounds:
    Bush Barrow, 52–54
    Hopewell assemblage, **183–185**
    *See also* Cairns; Tombs
Burl, Aubrey, 30, 61, 104, 122, 361
Bush Barrow gold lozenge, **52–54**

Cacaxtla, Mexico, **55–57**
Cahokia, **57–59**, 183
Cahuachi, 335
Cairns:
    Clava, Scotland, **103–105**, 117, 435
    tri-radial, **429–430**
Calendar systems, **59–61**, 230
    ancient Egyptian, **9–12**, 107, 229
    ancient Egyptian, coffin lids, **106–108**

ancient Greece, 130, 230
ancient perceptions of space and time, 390
Aztec, 34
Babylon, 38
Borana, Africa, **45–46**, 60, 284–285
"Celtic," 44, 47, 60, **75–76**, 95, 249, 265
Christianization of "pagan" festivals, 95, 270
dating Christ's crucifixion, 118
diagonal star clocks, 107
functions, 60
Gregorian, 118, **169–170**, 209
Hawaiian, 27, **176–179**, 338, 343
Hopi, **186–187**, 387
horizon calendars of central Mexico, **188–189**, 258
Inca *quipu,* 358
Islamic, 201
Javanese, **207–208**, 323
Jewish, 118
Julian, 118, 169–170, 208–209, 365
latitude and, 60
leap year, 208
Mesoamerican Calendar Round, **255–258**
Mursi of Ethiopia, **274–277**
names of days, 229
Neolithic "megalithic" structures, 248–250
Oceanic peoples, 337–338
peg-hole star calendars, 130, 230
sidereal month, 45, 79
Callanish, **61–63**, 250, 389
Cambodia. *See* Angkor
Canoe, Polynesian, 301
Canopus, 202
Caracol at Chichen Itza, **64–65**
Carahunge, **65–67**
Cardinal directions and alignments, **67–69**, 254, 326
    Chaco Canyon, 71, 82, 83–86
    Chinese traditions, 222
    church orientations, 68, 96–99, 151
    Egyptian temple and tomb alignments, 145, 353
    intercardinal directions and Skidi Pawnee cosmology, 330
    Navajo hogans, 295–296
    Orkney house and tomb orientations, 239–240

Cardinal directions *(cont.)*
    quadripartite cosmology, 68–69, 82–83, 122, 187, 240
    solstitial directions, 388–389
Carlson, John, 55
Carnac, 119, 166, 348
Caroline Islands, 302, 323, 394–395
Casa Rinconada, **69–72**, 84
Cassiopeia, 294
Castlerigg, 122
Catastrophic events, **72–74**, 190
    end of world prediction, 245
    *See also* Comets; Meteors
Caterpillar Jaguar constellation, 27, 43, 116
Causeway, Iron Age site at Fiskerton, 160–162, 439
Causewayed enclosures, 99–100
Cehtzuc, 163
Celestial divination. *See* Astrology
Celestial equator, 129. *See also* Obliquity
Celestial sphere, **74–75**, 409
    changes over time, 190
    declination, 129
    diurnal motion, 131
    ecliptic, 142
    heliacal rise, 180
    meridian, **253–254**
Celtic calendar, 44, 47, 60, **75–76**, 249, 265
    Christianization of "pagan" festivals, 95
    equinoxes, 148
Centaurus, 418
*Ceque* system, 77–80, 106, 226, 291
Chaco Canyon, **80–83**
    cardinal orientations, 71
    Casa Rinconada, 69–72, 84
    computer simulation, 157
    Fajada Butte sun dagger, 81, 89, 155–157, 262, 328
    Great North Road, 71, 84, 85, 254
    roads and trackways, 335
    simulation, 159
    supernova pictograph, **86–89**, 111–112, 394
Chaco Meridian, 71, **83–86**
Chamberlain, Von Del, 328
Chauncy, Henry, 96
Chichen Itza, 243
    Caracol at, **64–65**
    Kulkulkan pyramid, 213–214, 262, 335

Child sacrifice, 34, 56
China, ancient, 26, **90–94**
    astronomical instruments, 91, 94
    astronomical versus divinatory (astrological) traditions, 92
    cardinal alignments, 222
    eclipse records and predictions, 92, 233, 382
    Imperial Palace (Forbidden City), 93–94, 115, 343
    lunar lodge system, 92–93
    oracle bones, 24, 90, 92, 233
    solstice ceremonies, 93–94
Christ, date of crucifixion, **117–119**
Christian church orientation, 68, **96–99**, 151
Christian traditions, Star of Bethlehem, 396–398
Christianization of "pagan" festivals, **95**, 270
Chumash, 385
Church orientation, 68, **96–99**, 151
Circles of earth, timber, and stone, **99–102**. *See also* British megalithic structures; Stone circles; *specific sites, structures*
Circumpolar stars, **102–103**, 145
Clava cairns, Scotland, 103–105, 117, 364, 435
Clinometer surveys, 113
Clocks, night, 11
Coal sack, 147–148
Cobo, Bernabé, 77–78, **105–106**, 126, 203
Coffin lids, **106–108**
Cognitive archaeology, **108–109**
Coligny calendar, 75
Comet Encke, 413
Comets, 72–73, **110–112**, 190, 344
Compass and clinometer surveys, **112–113**, 424
Compass, star, 298, 302, 337, **394–395**
Computer simulation, 157
Coneybury, 102
Constellation maps on the ground, **113–115**, 262, 323
    Angkor (Draco), 16, 113–114, 262, 355
    Pyramids of Giza (Orion's Belt), 16, 114, 323, 355
Constellations, 321–322
    Aboriginal mythology, 3

Barasana caterpillar, 27, 43, 116
bears, **378–380**
changes over time, 190
dark cloud emu, 4, 22, 147–148
Inuit cosmology, 194
Javanese stellar calendar, 207–208
Lakota, 142, 219, 223–224
Navajo star ceilings, 296–297
star naming conventions, 396
*See also specific constellations*
Cook, James, 1, 178–179, 217, 300, 302, 395
Corona Borealis, 332, 333
Corvus, 294
Cosmograms, 257, **259–261**
Cosmology, **115–117**
  Inuit, **193–194**
  monuments and, **271–272**
  Navajo, 293–295
  orientation and, 319
  Pawnee, **329–331**
  quadripartite, 68–69, 82–83, 122, 187, 240, 388
  Yekuana roundhouse, 439–441
Coyote, 294
Crab Nebula, 87
Crete, 266, 419–420
Cromlechs, 119, 166
Cross-circle designs, 257, **259–261**
Crucifixion of Christ, **117–119**, 232
Crucuno, Brittany, **119–121**
Crump, Thomas, 326
Cuicuilco, 189
Cultural astronomy, 152
Cultural relativism, 153
Cumbrian stone circles, **122**
Cup-marked stones, 44, 50, 114, 363
Cursus monuments, **123–126**, 334, 427
Curvigrams, 400–401
Cusco (or Cuzco), 267
  antizenith solar alignments, 19, 127
  *ceques*, **77–80**, 106, 226, 291
  sun pillars, **126–127**

Dating conventions, B.C.E. and years before 0, **439**
Dating techniques:
  astronomical dating, **27–29**
  crucifixion of Jesus, 117–119, 232
  dendrochronology, 73–74, 160
Days, names of, 229
Dead reckoning, 299, 303, 337

December solstice (winter solstice), 317, 385, 388, 409, 443. *See also* Solstices; Solstitial alignments
Declination, **129–130**, 159, 317, 409–410, 424
  compass and clinometer surveys, 112
  equinoxes and, 148
  lunar parallax, 234–235
  precession and, 346
  refraction and, 365
Delphic oracle, **130–131**, 335
Delphinus, 130–131, 335
Dendrochronology, 73–74, 160
Diagonal star clocks, 107
Diodorus, 61
Diurnal motion, **131**
Divination, celestial. *See* Astrology
Dolmen, Portuguese antas, 17–18, 350
Dolmenic hypogea, 350
Donald, Merlin, 325
Dorado constellation, 240
Dorset cursus, 124, 125
Dowth, Ireland, 46–47, 265
Doyle, Laurence, 285
Draco, 16, 262, 353
Dreaming paths, 3, 223
Dreamtime, 2
Dresden Codex, 20, 25, 59, **132–134**, 257, 382
Dromberg, Ireland, 8, 32, **134–136**, 263–264, 389

Earth lodges, Pawnee, 115, 196, 319, 328, **331–332**
Earth's rotation, eclipse records and, **141–142**
East-west alignments, 68–69, 149, 151. *See also* Cardinal directions and alignments; Equinoctial alignments; Orientation; Solar alignments
Easter, 151, 170
Easter Island (Rapa Nui), **137–140**, 241, 301, 340
Eclipses. *See* Lunar eclipses; Solar eclipses
Ecliptic, **142**, 317
  obliquity of the, 190, 263, 273, 308, **317–319**, 410
Effigy mounds, 183, 184
Egypt, ancient, 323
  calendar system, **9–12**, 107, 229
  coffin lids, **106–108**
  Menorca taulas and, 418

Egypt, ancient *(cont.)*
    Pyramids of Giza, 16, 28, 114, 143, 145, 254, 323, **353–356**
    Saharan Nabta Playa megaliths, **282–284**
    temples and tombs, **143–146**, 228, 313
Electronic distance measurement (EDM), 424–425
Elevation, 9
Emory, Kenneth, **307–309**
Emu, 4, 22, **147–148**
Encke Comet, 413
End of the world prediction, 245
Environmental archaeology, 108
Epagomenal days, 11, 107
Equinoctial alignments, 150–151, 392
    Angkor, 15
    cursus monuments, 125
    Fajada Butte sun dagger, 155, 157, 328
    Irish axial stone circles, 32–33
    Kulkulkan pyramid at Chichen Itza, 213–214
    Thom equinox, 149–150
Equinox hierophany, 213–214, 262
Equinoxes, **148–151**
    mid-quarter days, 265, 436
    ritual significance, 150–151
    *See also* Equinoctial alignments
Er Grah, 166
Eskimos. *See* Inuit
Ethnoastronomy, 21, **152**. *See also* Archaeoastronomy
Ethnocentrism, 21, **152–153**
Etruscans, 366
Extinction, **153–154**, 298, 398
Extinction angle, 154

Fajada Butte sun dagger, 81, 89, **155–157**, 262, 328
Féjérvary-Mayer Codex, 257
Field surveys, **158–160**
    compass and clinometer surveys, 112–113, 424
    electronic distance measurement, 424–425
    GPS surveys, **165–166**
    precision and accuracy, 347–348
    theodolite, 112, 113, 165, 423–425, 426
    Thom's methodology, 426
Fiskerton, England, **160–162**, 439
Flooding, 27

Frank, Roslyn, 380
Fritz, John, 71
Gambier Islands, 241–242
Gemini, 219
Geocentric lunar declination, 235
Geoglyphs, Nasca plains of Peru, 286–292, 335
Geographical Information Systems (GIS), 159
Georgia, 95
Giant's tombs, Sardinia, 349
Gilbert Islands, 339, 394
Giza. *See* Pyramids of Giza
Glastonbury zodiac, 114
Global Positioning System (GPS) surveys, **165–166**
Goal Year Texts, 40
Godmanchester, 265
Golan Heights, 366–368
Goodenough, Ward H., 394
Governor's palace at Uxmal, 8, 28, **163–164**
GPS surveys, **165–166**
Grand Menhir Brisé, **166–168**
Great North Road, 71, 82, 85, 254
Greece, ancient:
    calendar system, 230
    Delphic oracle, **130–131**, 335
    equinoxes and, 149
    Hesiod, 130, **181–182**, 207, 230
    Islamic astronomy and, 199
    temple alignments, **419–421**
Green archaeoastronomy, 20–21, **169**, 189, 247, 399
Gregorian calendar, 118, **169–170**, 209
Group E structures, **170–173**
Guo Shou Jing, 94

Ha'amonga-a-Maui, **175–176**
Hale-Bopp comet, 111
Halley's Comet, 111, 344
Hancock, Graham, 15–16
Hartung, Horst, 171
Hawai'i, 301
    calendar system, 27, **176–179**, 229, 343
    celestial tropics names, 337
    eclipse records, 381
    hula chants (Na Pali), 279–282, 338–339
    Kahikinui, 342–343
    Kumukahi, **215–216**, 341

*Kumulipo* chant, **217–218**
    Magellanic Clouds and, 241
    Mangareva, 341
    natural alignments, 264
    navigation gourds, 395
    Necker Island, 216, **307–309**, 318, 340, 444
    solar zenith passage and, 443–444
    solstitial alignments, 264, 341
    temple platforms and enclosures, 340–343
Hawkins, Gerald, 20, 61, 289, 382, 406–407
Heelstone, 405
Heilige Linien, 225
Heliacal rise and set, **180–181**, 182, 398
    Delphinus and the Delphic oracle, 130–131
    Egyptian calendar and, 9, 11, 229
    Japanese agricultural calendar and, 222
    Javanese stellar calendar, 207
    medicine wheel alignments, 247
    Misminay astronomy and, 267
    Mursi calendar, 275–276
    planetary, 192, 411, 433
Henge monuments:
    Avebury, 29–30, 100
    cursus monuments, 123–126, 334, 427
    Ring of Brodgar, 50–52
    Thornborough, 125, 323, 335, **427–429**
Henriksson, Göran, 413
Hesiod, 130, **181–182**, 207, 230
Heyerdahl, Thor, 301
Hipparchus, 149
Hogans, Navajo, 115, 196, **295–296**
Hopewell mounds, **183–185**
Hopi, 115, **186–187**, 223, 387, 388, 391
Horizon calendars of Mexico, **188–189**, 258
Horoscopes, 26, 39–40
Hoskin, Michael, 17, 23, 321, 349, 418
House orientations, 239, 319
    British Iron-Age roundhouses, 195–197, 320
    Navajo hogans, 295–296
    Pawnee earth lodges, 331–332
How the sky has changed over time, **190**
Hoyle, Fred, 382, 406–407
*Huacas*, 77–78
Hula, 279–282, 338–339
Human sacrifice, 34, 56

Hyades, 220

Imbolc, 75, 265
Imhotep, 418
Imperial Palace, Chinese Forbidden City, 343
Inca:
    Andean mountain shrines, 13
    Bernabé Cobo and, 77–78, **105–106**, 126, 203
    *ceques*, **77–80**, 106, 226, 291
    Cusco sun pillars, **126–127**
    *huaca* shrines, 77–78
    Island of the Sun, **203–204**
    *quipu*, 78, 291, **357–359**
Indonesia, Javanese calendar, **207–208**
Inferior conjunction, 191
Inferior planets, motions of, **191–192**
Intercalary months, 9, 38, 75, 130, 201, 229, 243
Inuit, **193–194**
    stellar navigation, 300
Ireland:
    axial stone circles, 30–31, 134–136, 361, 435
    Beltany, 44
    Boyne Valley tombs, **46–47**
    Dowth, 265
    Knowth passage tombs, 46–47, 150
    Newgrange passage tombs, 158, 228, 238, 250, 262, **309–312**, 386, 415
    short stone rows, 8, 263–264, 376–378, 389
    wedge tombs, **435–436**
Iron-Age roundhouses, **195–197**, 320
Iroquois, 380
Is Paras, **197–199**, 314
Islamic astronomy, **199–202**
Islamic calendar, 201, 229
Islamic star and crescent, 201, 393
Island of the Sun, **203–204**
Isle of Lewis, 61–63

Japan, **221–223**
Javanese calendar, **207–208**, 323
Jewish calendar, 118
Johnson, Rubellite Kawena, 218
Julian calendar, 118, 169–170, 208–209, 365
June solstice, 317, 385, 388, 409, 443. *See also* Solstices; Solstitial alignments

Jupiter, 397, 410–412

Kaaba, 202
Kahikinui, 342–343
Kapingamarangi, 323
Karnak, 144, 145
Kennet Avenue at Avebury, 84–85, 123
Kenya, Namoratung'a, 60, **284–285**
Kepler's Supernova, 111
Khafre, 353
Khmer civilization. *See* Angkor
Khoisan, 241
Khufu, pyramid of, 145, 254, 353, 354. *See also* Pyramids of Giza
Kintraw, Scotland, **211–213**, 253, 386
Kirch, Patrick, 242
Kiribati, 394
Kitora Kofun, 222
Kivas, 69–72, 81
Knowth, Ireland, 46–47, 150
*Kon Tiki,* 301
Kosok, Paul, 289
Kukulcan, **213–214**, 262, 335
Kumukahi, **215–216**, 341
Kumulipo, **217–218**
Kyoto, Japan, 222

La Malinche, 56
Lakota, 116, 323
  sacred geography, **219–221**, 223–224
  space and time perceptions, 391
  Sun Dance, 220
  zodiacal constellations, 142
Landscape, **223–224**. *See also* Sacred geographies
Latitude, 129
  calendar correlation, 60
  declination, 129
  navigation, 299, 303
  zenith stars, 444–445
Latvia, 26
Laval, Honoré, 241–242
Leap year, 208
Legesse, Asmerom, 285
Levantine megaliths, Rujm el-Hiri, 366–368
Lewis, David, 394, 444
Ley lines, 189, **224–226**
Liller, William, 138
Linear monuments, cursus, **123–126**, 427
Lithuania, 26
Llama, celestial, 267–269

Lockyer, Norman, 20, 146, 224, **227–228**, 248, 250, 313, 389, 425
Long Meg and Her Daughters, 122
Longitude, determining, 303
Lughnasa, 75, 265
Lunar alignments, 426
  Callanish, off Scotland, 61–63, 389
  Casa Rinconada, Chaco Canyon, 69–72
  Chaco Canyon structures, 156–157
  Clava cairns, Scotland, 364
  dating issues, 28–29
  Grand Menhir Brisé, 166
  obliquity of the ecliptic, 317
  Ring of Brodgar, 51
  Scottish recumbent stone circles, 361, 363–364
Lunar calendars, **228–230**
  Abri Blanchard bone, 6
  ancient Greece, 130
  appearance of new moon, 228–229
  Babylonian system, 38
  Borana system, 45–46
  Coligny calendar, 75
  Egyptian system, 9–12, 229
  Hawaiian system, 177–178, 229
  intercalary months, 9, 38, 75, 130, 229, 243
  Islamic system, 201
  Jewish system, 118
  Mayan Long Count, **243–246**
  Metonic cycle, 230
  Mursi system, 274–277
  Polynesian, 242–243
  Presa de Mula petroglyph, Mexico, 352
  Roman, 208
  sidereal month, 45–46
Lunar eclipses, **230–231**
  Chinese records, 92
  dating Christ's crucifixion, 118, 232
  predicting, 232–233
  records and predictions, **141–142**
  Saros cycle, 39, 161, 233
  social impact of, 231–232, 344
Lunar lodges, 92–93
Lunar mansions, 200, 222
Lunar node cycle, 61, 155–156, 273
Lunar parallax, **234–235**
Lunar phase cycle, 59, 228, **235–236**, 272
  Abri Blanchard bone, 6
  correlations with personal

characteristics, 26
new moon, 6
Lunar standstill, 157, 167, 273
Lunar stations (*anwâ*), 200
Lunation, 235
Luni-solar calendar, 229. *See also* Lunar calendars

Machu Picchu, 267
MacKie, Euan, 49, 212, 237
Maes Howe, 103, **237–240**, 250
Magellan, Ferdinand, 300
Magellanic Clouds, 3, **240–241**
Magi, 397
Magnetic declination, 112
Magnetic north, 112, 129
Major standstill, 273
Majorville Cairn, 246
Mallorca, 28, 347, 390
Malville, Kim, 282
Mana, 443
Mangareva, **241–242**, 338, 341
Maori:
    lunar eclipses and, 232
    Pleiades and, 337
Mars, 410–412
    synodic period, 257, 411
Marshack, Alexander, 5
Maya, 243
    ball game, 172–173
    building and city orientations, 258
    Calendar Round, 256
    Caracol, **64–65**
    celestial divination, 25–26
    Chichen Itza, 64–65, 213–214, 243, 262
    Dresden Codex, 20, 25, 59, **132–134**, 257, 382
    eclipse records, 382
    end of world prediction, 245
    Governor's palace at Uxmal, 8, 28, **163–164**
    Group E structures, **170–173**
    Kulkulkan pyramid, **213–214**, 262, 335
    language and script, 133
    Long Count, **243–246**
    lunar calendar (the Long Count), 244–246
    lunar eclipses and, 232, 233
    Venus and, 433
McCluskey, Stephen, 98, 187

Measurement concepts, 326
Mecca, 202
Medicine wheels, **246–248**, 387
"Megalithic astronomy," 248. *See also* Alignment studies; British megalithic structures; Thom, Alexander; *specific sites, structures*
"Megalithic" calendar, **248–250**
Megalithic monuments of Britain and Ireland, **250–252**. *See also* British megalithic structures; *specific sites or structures*
Megalithic "observatories," **252–253**. *See also* British megalithic structures; *specific sites or structures*
Megalithic yard, 121, 326, 403, 425, 426
Menkaure, 353
Menorca, 351, **417–418**
Mercury, 191, 192
Meridian, **253–254**
Meridian, Chaco, 71, **83–86**
Mesoamerican astronomy and culture:
    Cacaxtla, **55–57**
    Calendar Round, 188, **255–258**
    Caracol "observatory," 64–65
    cross-circle designs, 257, **259–261**
    eclipse records, 382
    horizon calendars of central Mexico, **188–189**, 258
    Teotihuacan, 260, 421–423
    Venus alignments and calendars, 28, 55, 134, 163–164, 257, **433–434**
    Xochicalco, 188, 344, 444, 445–448
    zenith tubes, 188, 344, 444, 445–448
    *See also* Aztecs; Maya
Meteorites, 110
Meteors, 72–74, 110, 190, 344
Methodology, **261–264**
    field surveys, 159
    simulation, 157, 159
    statistical analysis, **399–401**
Metonic cycle, 230
Mexico:
    Cacaxtla, **55–57**
    horizon calendars of, **188–189**
    pecked cross-circle designs, 257, 259–261
    Presa de Mula petroglyph, **351–352**
    Teotihuacan, 260, 421–423
    Xochicalco, 188, 344, 444, 445–448
Mexico, Aztec culture and astronomy. *See* Aztecs

Micronesia. *See* Polynesia and Micronesia
Mid-quarter days, **265**, 436
Middle Eastern megaliths, Rujm el-Hiri, 366–368
Midsummer's Day, 385–386
Milky Way:
   dark cloud emu, 147–148
   Egyptian lunar calendar and, 10
   Magellanic Clouds, **240–241**
   Misminay astronomy and, 267–269
Mi'kmaq, 380
Minard, 48–50
Minoan temples and tombs, **266**, 419–420
Minor standstill, 273
Misminay, 21, **267–269**, 388
Mississippian culture, 57–59, 183
Mithraic cult, 366
Mithraism, **269–270**
*Moai*, 137
Mohenjodaro, 326
Molnar, Michael, 397
Monk's Mound, 58
Monte Alban, 444, 446
Monte-Carlo methods, 400
Monuments and cosmology, **271–272**
Moon, motions of, **272–273**. *See also* Lunar calendars; Lunar phase cycle
Moon Spirit, 194
Moose Mountain, 246
Morieson, John, 4
Mosque orientations, 202
Mountain shrines, Andean, 13
Mull, Isle of, 378
Mulloy, William, 138
Murray, Breen, 351–352
Mursi, 27, 60, 116, 228, 229, **274–277**

Na Pali chant, **279–282**, 338–339
Nabta Playa, **282–284**
Nadir, 200
Names of days, 229
Namoratung'a, 60, **284–285**, 347
Nasca lines and figures, **286–292**, 335
National socialists (Nazis), 292
Nationalism, **292–293**
Native Americans:
   Hopewell mounds, **183–185**
   medicine wheels, **246–248**, 387
   northern "bear" constellations, 380
   sacred geography, 219–221, 223–224
   space and time perceptions, 391

   *See also* Hopi; Navajo; Pawnee; *specific peoples*
Native Americans, Puebloan culture. *See* Chaco Canyon
Navajo:
   cosmology, **293–295**
   hogan design, 115, 196, **295–296**
   star ceilings, **296–297**
Navigation, **298–300**
   ancient Oceanic peoples and, 299, 300–303, 337, 394–395
   Australian Aborigines and, 4
   Oceanic peoples and, 398, 445
   Roman, 366
   zenith stars, 445
Navigation gourds, 395
Nebra disc, **304–307**
Necker Island, Hawai'i, 216, **307–309**, 318, 340, 444
New moon, 6, 228–229
New Zealand, 301
Newgrange, Ireland, 46–47, 103, 158, 228, 238, 250, 262, **309–312**, 386, 415
Newton, Isaac, 117, 118
Night "clocks," 11, 107
Nissen, Heinrich, 146, **312–313**, 419
North celestial pole, 129, 254, 299, 343, 346, 378–379
North, magnetic versus true, 112, 129
North Road, Chaco Canyon, 71, 82, 85, 254
North-south alignments, 67
   Chaco Canyon structures, 82, 83–86
   *See also* Cardinal directions and alignments
North Strone, 363
North, true, 112
Northern "bear" constellations, 378–380
Novae, 110, 111
Number concepts, 326
Nuraghi, 197–199, **313–314**
Nut, 9–10, 143

Ó Nualláin, Seán, 361
Obeyesekere, Gananath, 179
Obliquity of the ecliptic, 190, 263, 273, 308, **317–319**, 410
Occam's razor, 430
Oceania. *See* Hawai'i; Polynesia and Micronesia
Oceania, ancient navigation in, 299, 300–303, 337, 394–395, 398, 445

O'Kelly, Michael, 310
Olmeca-Xicalanca, 55, 255
Oracle bones, 24, 90, 92, 233
Orientation, **319–321**
  archaeotopography, **23**
  British Iron-Age roundhouses, 195–197, 320
  Christian churches, 68, **96–99**, 151
  cosmology and, 115
  house and tomb similarities, 239
  Mayan structures, 258
  mosques, 202
  political power and, 343–344
  prehistoric European tombs and temples, 320–321, 348–351
  *See also* Alignment studies; Cardinal directions and alignments; *specific alignments, places or structures*
Orientation signature, 321, 350
Orion, **321–324**
  Aboriginal mythology, 3
  Javanese calendar, 207–208, 323
  Lakota and, 323
  Navajo cosmology, 294
Orion's Belt, 322
  alignments and ceremonies or pilgrimages, 335–336
  Borana calendar and, 285
  cursus alignments, 125–126, 428
  Japanese agricultural calendar and, 222
  Pyramids of Giza and, 16, 114, 145, 323, 355
Orkney Islands, 237–240, 237–240
  Ring of Brodgar, 50–52

Pagan festivals, Christianization of, **95**
Palaeoscience, **325–328**
  science versus symbolism debate, 375–376
  space and time perceptions, **390–392**
  symbols, **414–415**
Pantheon, **328–329**
Paquime, 85
Parallax, lunar, **234–235**
Parapegmata, 130, 230
Parker, R. A., 11
Parpalló, Spain, 318, 386
Parthenon, 419
Passage tombs, 250
  Brittany, 166
  Clava cairns, Scotland, 103–105, 117
  Dowth, Ireland, 265
  Knowth, Ireland, 46–47, 150
  Maes Howe, Orkney Islands, **237–240**
  Newgrange, Ireland, 46–47, 103, 158, 238, 250, 262, **309–312**, 386, 415
  wedge tombs of Ireland, **435–436**
Pawnee:
  cosmology, **329–331**
  earth lodges, 115, 196, 319, 328, **331–332**
  star chart, **332–333**
Pecked cross-circle designs, 257, **259–261**
Peg-hole star calendars, 130, 230
Penumbra, 231
Persia, Mithraic cult, **269–270**
Peru, Inca cultures. *See* Inca
Peru, Nasca lines and figures, **286–292**
Petroglyphs. *See* Rock art
Pilgrimages, **333–336**
  Andean Island of the Sun, 203–204
  Andean mountain shrines, 13
  Thornborough, 427
Planetary alignment, Venus. *See* Venus
Planetary motions, inferior planets, **191–192**. *See also* Venus
Planetary motions, superior planets, 410–412
Plate bearing, 424
Pleiades, 182
  Aboriginal mythology, 3
  Borana calendar and, 285
  declination, 129
  Hawaiian calendar and, 177–178, 229, 343
  Japanese agricultural calendar and, 222
  Lakota sacred geography, 222
  Misminay astronomy and, 267, 268–269
  names for in Oceania, 337
  Navajo cosmology, 294, 296
  Nebra disc and, 304
  Orion and, 322
  Pawnee star chart artifact, 333
  Teotihuacan street orientation and, 422
Polaris:
  declination, 129
  meridian and, 254
  Navajo cosmology, 294
  navigation and, 299, 303
Political power, 343–344
Polynesia and Micronesia, **336–339**
  calendar system, 337–338
  canoes, 301

Polynesia and Micronesia *(cont.)*
   lunar calendar, 242–243
   Mangareva, **241–242**
   navigation, 299, **300–303**, 337, 394–395, 398, 445
   Orion and, 323
   Pleiades and, 337
   solstitial alignments, 242, 338
   temple platforms and enclosures, 340–343
   Tongan *Haʻamonga-a-Maui,* 175–176
   zenith stars, **444–445**
   *See also* Hawaiʻi
Portuguese antas, 17–18, 350
Post-processual archaeology, 21–22
Poverty Point, 183
Power, **343–344**
*Pranotomongso,* 208
Prayer directions, Islamic, 201–202
Precession, 190, 263, 318, 319, **345–346**, 379, 390, 398–399
Precision and accuracy, **347–348**
Presa de Mula, **351–352**
Processual archaeology, 21
Pueblo culture, Chaco Canyon. *See* Chaco Canyon
Pyramid of the Flowers, 56
Pyramid of the Sun, Teotihuacan 421
Pyramid texts, 106, 145, 355
Pyramids, Egyptian, 143
Pyramids of Giza, 16, 28, 114, 143, 145, 254, 323, **353–356**
Pythagorean proportions, 119–121

Quadripartite cosmology, 68–69, 82–83, 122, 187, 240, 388
Quetzalcoatl, 214
*Quipu,* 78, 291, **357–359**

Ra, 9, 143, 145
Ramadan, 201
Rapa Nui (Easter Island), **137–140**, 145, 241, 301, 340
Rectangular megalithic enclosure, Crucuno, 119
Recumbent stone circles, 30–31, 272, 351, **361–364**, 402, 435
Refraction, **364–365**
Reiche, Maria, 287, 289–290
Reinhard, Johan, 291
Retrograde motion, 411–412
Rigel, 247, 396

Ring of Brodgar, **50–52**, 253
Rock art:
   Australian Aborigines, 3–4
   Chaco supernova pictograph, **86–89**, 394
   Presa de Mula, Mexico, **351–352**
   supernova pictograph, 111–112
   Swedish, **412–414**
   symbolism, 415
Roman calendar (Julian). *See* Julian calendar
Roman Pantheon, **328–329**
"Romanian Stonehenge" (Sarmizegetusa Regia), 370–372, 386
Rome, ancient:
   astronomy and astrology, **365–366**
   Mithraic cult, 269–270, 366
Roundhouses, Iron-Age (Britain), **195–197**, 320
Roundhouses, Yekuana, 115, **439–441**
Rujm el-Hiri, **366–368**

Sacred calabash, 395
Sacred geographies, 223, **369–370**
   Aztec, **33–35**, 223
   Cacaxtla, Mexico, 56
   Hopi calendar and worldview, 186–187
   Inca *ceques,* 77–80, 106, 226, 291
   Inca *huacas,* 77–78
   Lakota, **219–221**, 223–224
   Navajo cosmology, 293–295
   *See also* Chaco Canyon; *specific cultures or sites*
Sacsahuaman, 126
Sahara Desert, Egyptian Nabta Playa megaliths, 282
Sahlins, Marshall, 179
Saiph, 285
Samhain, 75, 95, 265, 436
Santu Antine, 314
Sardinia, 197–199, **313–314**, 349
Sarmizegetusa Regia, 370–372, 386
Saroeak, **372–374**
Saros cycle, 39, 161, 233
Saturn, 410–412
Science and technology of prehistoric man, 325–328
Science versus symbolism debate, **375–376**
Scorpius, 267, 268, 294, 323
Scotland:
   Ballochroy, 41–42, 253–254, 347, 377

514   Index

Brainport Bay, 48–50, 150, 249–250, 386
Callanish, **61–63**, 250, 389
Clava cairns, Scotland, 117, 364, 435
Kintraw, **211–213**, 253, 386
Maes Howe, **237–240**, 250
recumbent stone circles, 30–31, 272, 351, 402, 435
Temple Wood, 101
*See also* British megalithic structures
Scottish quarter days, 265
Serpent Mound, 183, 184, 386
Shinto, 221
Shooting stars, 110
Short stone rows, 8, **376–378**
Sidereal month, 45–46, 79
Simulations, 157
Sirius, 396
  alignments and ceremonies or pilgrimages, 335–336
  Borana calendar and, 285
  cursus alignments, 428
  declination, 129
  Egyptian calendar and, 9, 229
  Inuits and, 194
  medicine wheel alignments, 246
  Orion and, 322
Skara Brae, 239
Skidi Pawnee. *See* Pawnee
Sky bears, **378–380**
Social archaeology, 108
Society Islands, 340
Solar alignments:
  ancient Greek temple alignments, 419–421
  Andean Island of the Sun, 203–204
  antizenith passage, 19, 127
  Australian structures, 4
  Ballochroy, 41–42
  Beltany, Ireland, 44
  Boyne Valley tombs, 47
  Brainport Bay, Scotland, 48–50
  Bush Barrow lozenge and, 53
  Casa Rinconada, Chaco Canyon, 69–72
  church orientations, 68, 96–99, 151
  Clava cairns, Scotland, 104–105
  Cumbrian stone circles, 122
  dating issues, 28–29
  Dromberg axial stone circle, 135–136
  Easter Island structures, 137–140
  Fajada Butte sun dagger, 155–156
  Hopi calendar and, 186–187
  horizon calendars of central Mexico, 188–189, 258
  Maya Group E structures, 171–173
  medicine wheels, 246
  obliquity of the ecliptic, 317–319
  Portuguese antas, 17–18
  records and predictions, **141–142**
  *See also* Equinoctial alignments; Solstitial alignments
Solar eclipses, **380–384**
  Chinese records, 382
  Mesoamerican records, 382
  predicting, 382–383, 406
  social impact of, 231
  Swedish rock art, 413
Solar zenith passage. *See* Zenith passage of the sun
Solstice ceremonies:
  ancient China, 93–94
  Christianization of "pagan" festivals, 95
  Hopi, 186
  Inca, 126
  Inuit, 194
  Tongan, 175
Solstices, 317, **384–385**, 409
  equinoxes and, 149–150
  Javanese calendar, 207
  mid-quarter days, 265, 436
  Misminay astronomy and, 268
  solar antizenith passage, 19, 127, 281
  solar zenith passage, 281, 308, 443
  *See also* Solstitial alignments
Solstitial alignments, **385–387**, 392
  Angkor, 15
  Ballochroy, 41–42
  Boyne Valley tombs, 47
  Brainport Bay, Scotland, 48–50, 150, 204
  Clava cairns, Scotland, 104–105
  cross-circle designs, 260
  Crucuno, Brittany, 119–121
  cursus monuments, 125
  Cusco sun pillars, 126–127
  Dromberg axial stone circle, 135–136
  Easter Island structures, 137–140
  Egyptian temples and tombs, 145–146
  Fajada Butte sun dagger, 155, 157
  field survey, 158
  Hawaiian occurrences, 264, 341
  Hopewell mounds, 184, 185

Index 515

Solstitial alignments *(cont.)*
  Hopi calendar and, 186–187, 387
  Irish axial stone circles, 32
  Kintraw, 211–213
  Maes Howe, 237–238
  Mangareva, Polynesia, 242, 338
  Maya structures, 171, 258
  "megalithic observatories" hypothesis, 252–253
  methodological issues, 262–264
  Na Pali hula chant, 281
  Necker Island, Hawai'i, 308
  Neolithic "megalithic calendars," 249
  Newgrange, Ireland, 158, 238, 309–312
  obliquity of the ecliptic, 318
  Parpalló, Spain, 318, 386
  Rujm el-Hiri, 367
  Saharan Nabta Playa megaliths, 284
  Sardinian nuraghi, 197–199, 314
  Stonehenge, 386, 408–409
  Tongan *Ha'amonga-a-Maui*, 175–176
  tri-radial cairns, 429–430
  Yekuana roundhouse, 440
Solstitial directions, **388–389**
Somerville, Boyle, 135, **389**, 436
Son Mas, 28, 347, **390**
Song lines, 3
South celestial pole, 129, 254, 299
Southern Cross, 129, 299
  Aboriginal "constellation maps," 114
  Aboriginal mythology, 3
  Magellanic Clouds and, 240
  Menorca taulas and, 417–418
  Mursi calendar and, 275
  Son Mas, Mallorca alignment, 390
South-west aligned stone circle, 134–136
Space and time, ancient perceptions of, **390–392**
Spain, *saroeak* (Basque stone octagons), 372–374
Sprajc, Ivan, 189
Spring (vernal) equinox, 148
Star and crescent, **392–394**
  Chaco supernova pictograph, 86–89, 394
  Islamic, 201, 393
Star calendars (parapegmata), 130, 230
Star ceilings, Navajo, **296–297**
Star chart, Pawnee, **332–333**
Star compass, 298, 302, 337, **394–395**
Star names, **396**

Arab/Islamic origins, 199–200
Star of Bethlehem, **396–398**
Star rising and setting positions, **398–399**
  changes over time, 190
  extinction, **153–154**, 398
  Japanese agricultural calendar and, 222
  medicine wheel alignments, 246–247
  Mursi calendar, 275–276
  precession and, 319, 346–347, 398–399
  *See also* Heliacal rise and set; Navigation
Statistical analysis, **399–401**
Stellar navigation. *See* Navigation
Stenness, Stones of, 50
Stephen, Alexander, 186
Stone canoes, 337, 338, 394
Stone circles, 101, 271–272, **401–404**
  Beltany, Ireland, **44**
  Callanish, **61–63**, 250, 389
  Clava cairns, Scotland, 103
  *cromlechs* of Brittany, 166
  Cumbria, **122**
  Dromberg, Ireland, 8, 32, 263–264, 389
  Irish axial stone circles, 30–31, 134–136, 361, 435
  recumbent stone circles, 30–31, 272, 351, **361–364**, 402, 435
  Ring of Brodgar, **50–52**
  Saharan Nabta Playa megaliths, **282–284**
Stone rectangles, Brittany, 119–121
Stone rows, short, 8, **376–378**
Stone transport, 402, 407
Stonehenge, 100, 101, 250, 405–409
  Aubrey Holes and eclipse prediction, 382–383, 406–407
  avenue leading to, 101, 335
  cognitive archaeology, 109
  ley lines, 225
  midsummer sunrise alignments, 386
  popular literature, 20
  Pythagorean ratios, 121
  ritual evidence, 407
  solstitial alignments, 386, 408–409
  station stones, 408
  stone transport, 402, 407
Stonehenge cursus, 124
Street orientation, Teotihuacan, **421–423**
Stukeley, William, 124
Summer solstice (June solstice), 317, 385, 409, 443. *See also* Solstices

Sun, Island of the, **203–204**
Sun cults, Mithraism, **269–270**
Sun dagger at Chaco Canyon, 81, 89, **155–157**, 159, 262, 328
Sun Dance (Lakota), 220
Sun gods, 9, 143, 145, 203, 221, 270
Sun, motions of, **409–410**. *See also* Equinoxes; Solar alignments; Solar eclipses; Zenith passage of the sun
Sun pillars, Cusco, **126–127**
Sun Spirit, 194
Sun-watching station, 89
Sundials, 366, 372
Sunrise alignments. *See* Solar alignments; Solstitial alignments
Superior conjunction, 191
Superior planets, motions of, **410–412**
Supernovae, 111
   Chaco pictograph, **86–89**, 111–112, 394
   Star of Bethlehem, 397
Swedish rock art, **412–414**
Symbolism, 375–376
Symbols, **414–415**
Synodic cycles of Venus, 257
Synodic month, 59, 228, 235, 272, 352
Synodic period of Jupiter, 412
Synodic period of Mars, 257, 411

Table des Marchand, 166
Täi plaque, 7
Takamatsu Zuka Kofun, 222
Talayotic culture, 417
Taulas of Menorca, 351, **417–418**
Temple alignments, ancient Greece, **419–421**
Temple platforms and enclosures, Polynesian, **340–343**
Temple Wood, 101
Temples and tombs:
   ancient Egypt, **143–146**, 228
   Minoan, **266**, 419–420
   prehistoric Europe, **348–351**
   *See also* Tombs
Tenochtitlan, 33–34, 344
Teotihuacan, 260, **421–423**
Theodolite surveys, 112, 113, **423–425**
   GPS surveys versus, 165
   precision and accuracy, 347–348
   Thom's methodology, 426
Thom, Alexander, **425–426**
   archaeoastronomy development and, 8, 20
   Avebury and, 30
   Ballochroy and, 41
   Castlerigg and, 122
   Clava cairns and, 104
   equinoctial alignments and, 149
   ethnocentrism, 21
   Grand Menhir Brisé and, 166–168
   Kintraw and, 211
   Lockyer and, 228, 425
   lunar alignments and, 61
   lunar standstill and, 167, 273
   megalithic measurement and astronomy hypotheses, 29, 47, 53, 120–121, 169, 237, 249, 250, 252–253, 310, 326, 403, 425–426, 436
   methodological issues, 262
   Ring of Brodgar and, 51
   science versus symbolism debate, 375
   short stone rows and, 377
   Somerville and, 389
   statistical analytical method, 399–401
   theodolite survey methods, 426
Thom, Archie, 53
Thom equinox, 149–150
Thornborough, 125, 323, 335, **427–429**
Thuban, 145, 353, 355
Timber circles, 101
Tombs:
   Boyne Valley, Ireland, 46–47
   Egyptian temples and tombs, **143–146**, 353–356
   orientations, 319, 320
   prehistoric European tombs and temples, **348–351**
   Sardinian *tombi di giganti*, 349
   similar house and tomb design and orientation, 239
   wedge tombs of Ireland, **435–436**
   *See also* Burial mounds; Cairns; Passage tombs; Temples and tombs
Tomnaverie, 363
Tonga, 175–176, 302, 394
Tongatapu, 175–176
Total solar eclipse. *See* Solar eclipses
Transylvania, 370–372
Tri-radial cairns, **429–430**
Triangulum, 45, 201
Tropic of Cancer, 19, 308, 318, 340, 409, 443–444
Tropic of Capricorn, 308, 309, 318, 409, 443
True north, 112, 129

Tswana, 323
Tupaia, 302
Tutankhamun's tomb, 144

Uaxactún, 170–173, 259
Umbra, 231
Ursa Major, 194, 294, 333, 378–380
Ursa Minor, 378, 380
Urton, Gary, 267
Uxmal, 8, 28, **163–164**

Venus, 191–192
  equinox hierophany and, 214
  Mesoamerican astronomy and culture, 28, 55, 134, 163–164, 257, **433–434**
  star and crescent, 201, **392–394**
  synodic cycles, 59
Venus skirts, 55
Vernal equinox, 148. *See also* Equinoctial alignments
Vijayanagara, 343
Vitruvius, 419
Vogt, David, 247
Volcano (La Malinche), 56

Walpi, 115, 186, 223
Watkins, Alfred, 224–225

Wedge tombs, Ireland, **435–436**
Williamson, Ray, 71
Winter solstice (December solstice), 317, 385, 409, 443. *See also* Solstices
Wisconsin effigy mounds, 183, 184
Wooden causeway, Fiskerton, 160–162, 439
Woodhenge, 58, 101, 102
Wordsworth, William, 96

Xesspe, Toribio Mejía, 289
Xochicalco, 188, 344, 444, 445–448
Xochitecatl, 55–56

Year dating conventions, B.C.E. and years before 0, **439**
Yekuana roundhouse, 115, **439–441**

Zenith, 200
Zenith passage of the sun, 281, **443–444**
  antizenith passage, **19**, 127
  Hawaiian alignments, 281, 308
Zenith stars in Polynesia, **444–445**
Zenith tubes, 188, 344, 444, **445–448**
Zodiac, 142
Zuidema, Tom, 79
Zuni sun-watching stations, 89